Antimicrobials in Livestock 1: Regulation, Science, Practice

Lucie Pokludová
Editor

Antimicrobials in Livestock 1: Regulation, Science, Practice

A European Perspective

Springer

Editor
Lucie Pokludová
Institute for State Control of Veterinary
Biologicals and Medicines
Brno, Czech Republic

ISBN 978-3-030-46720-3 ISBN 978-3-030-46721-0 (eBook)
https://doi.org/10.1007/978-3-030-46721-0

This Springer imprint is published by the registered company Springer Nature Switzerland AG.
The registered company address is: Gewerbestrasse 11, 6330 Cham, Switzerland

To the memory of my grandparents, Božena and Ladislav, to whom I am grateful for directing my journey through life.

Preface

Thinking about antimicrobials, some questions necessarily should come that touch the essence of this term and related issues. For some of these questions, scientists have the answer or at least a science-based hypothesis.

What does antibiosis mean? What are antibiotics (or the currently used broader term "antimicrobials")? Where and when they came from? Why some microbes synthesize them? How they interact and keep the balance in microorganisms' ecosystems? What is resistance to antimicrobials? Where it comes from? Why we started to use antimicrobials in animals not only for treatment of diseases? And what is the impact of that broad use both in human and veterinary medicine? What is current knowledge on the use of antimicrobials in veterinary medicine and how can we minimize the need for the use of antimicrobials in animals and pressure on selection of resistance? What happened with antimicrobials after administration to animals? Are their residues harmful to the health of consumers? And are the residues risky from the perspective of resistance spread? Which concentrations of residues we should consider of concern? What is the load on the environment caused by naturally, semi-synthetically and synthetically produced antimicrobials used in human and veterinary medicine and plant protection? And what is the extent of influence of both residues of antimicrobials and resistance on the environment?

Even some of those questions sound very simple and easy to answer; they are asked with increased frequency during the last decade not only by veterinarians and farmers but also by the public, as the awareness of antimicrobial resistance as well as the need for responsible use of antimicrobials is growing. Therefore, it seems that there is a growing need to come to different stakeholders involved in different areas linked in some extent to solving an issue of AMR as well as to the students and broader public with examples, explanations and answers that can also be used in everyday practice.

The authors of this book believe that a lot of questions, including those mentioned as examples above, are fully valid and need to be addressed using the current optics of knowledge. Therefore, the following pages are provided to the readers not only as suggested answers but also with the intention to recognize cross-links and complexity of use of antimicrobials as well as matter for further thinking that is intended to be provoked by the text of this book, which is divided into two volumes.

The first volume intends to start with more general parts, but continues with the exact topics covering the use, safety and efficacy of antimicrobials, as well as aspects of laboratory testing that is described considering especially need of practice. The first volume is divided into nine chapters.

After the first introductory chapter comes the second one, which is intended to cover *status quo* within the international context, covering briefly the very recent update on the existing global activities. Despite the fact that the issue of antimicrobial resistance is tackled from different perspectives by different international bodies and institutions and can be seen by someone as over discussed, the perspective of the global threat especially in human medicine is of paramount importance.

The third chapter is related to the second one, but brings more targeted insights to the European Union and policies, activities as well as regulatory surroundings considering especially new legal provisions on veterinary medicinal products and animal health, but also highlights the importance of the "soft law" as guidelines both at the European and national levels.

The fourth chapter is targeted on the use of antimicrobials in animals as well as on the description of the projects and methodologies of measuring of the extent of the use of antimicrobials in food-producing animals.

The fifth chapter reflects possible risks from the use of antimicrobials and concepts of approaching these risks. It involves especially parts focused on the impact on food chain safety, including considerations of the possible residues of antimicrobials (i.e. perspective of "chemical safety") as well as considering food-borne resistance linked with extensive farming of animals producing food for human consumption (i.e. perspective of "microbiological safety").

The sixth chapter focuses on prevention and alternative tools, which seem to be essential to avoid the broad use of antimicrobials. Among the other factors the chapter predominantly involves information on biosecurity and hygiene considerations and examples, vaccination, welfare and other tools to keep animals healthy. As an important factor influencing the scale of prescription in different countries, socio-economic aspects are considered.

The seventh chapter describes prophylaxis, metaphylaxis and off-label use, as practices that were or still are extensively used in some countries/sectors, but highlights the dangers that are accompanied with them. On the other hand, the chapter also comes with the examples of the future treatment strategies counting with antimicrobial stewardship approaches.

Within the eight chapter authors give an overview of laboratory investigation and results of antimicrobial susceptibility testing interpretation. A critically important step for the whole laboratory testing is the decision when to collect sample and proper sampling.

Chapter nine focuses on antimicrobial resistance, providing summary both from the general perspective and from the perspective of pathogen resistance of four major livestock species on which the second volume is targeted.

The second volume of the book is intended to be more pragmatic and targeted on specificities of individual sectors—pigs, poultry, cattle and horses, but starts with an introductory chapter that pays specific attention to pharmacological characteristics, especially pharmacokinetics and pharmacodynamics.

Brno, Czech Republic
2020

Lucie Pokludová
On behalf of the author's team

List of Abbreviations (Vol I Chapters I–X: All Chapters Together)

AACTING	Antimicrobial usage at herd level and analysis, communication and benchmarking
ADD	Animal daily dose
ADI	Acceptable daily intake
ADME	Absorption, distribution, metabolism, elimination
AGISAR	Advisory Group on Integrated Surveillance of Antimicrobial Resistance (WHO)
AGP	Antimicrobial growth promoter
AHAW	Animal Health and Welfare EFSA Panel
AHL	Animal Health Law
AMC	Antimicrobial consumption
AMCRA	Antimicrobial Consumption and Resistance in Animals (Belgium)
AMDUCA	Animal Medicinal Drug Use Clarification Act
AMEG	Antimicrobial Advice Ad Hoc Expert Group
AMR	Antimicrobial resistance
APBP	Aminophenyl boric acid
APP	*Actinobacillus pleuropneumoniae*
ARfD	Acute reference dose
ARGs	Antibiotic resistance genes
ARR	Absolute risk reduction
ASF	African swine fever
AST	Antimicrobial susceptibility testing
ATI	Animal Treatment Index
ATU	Area of technical uncertainty
AUC	Area under curve
BIOHAZ	Biological Hazards EFSA Panel
BMRGs	Biocide/metal resistance genes
BRD	Bovine respiratory disease
BRSV	Bovine respiratory syncytial virus
BSAC	British Society for Antimicrobial Chemotherapy
BVDV	Bovine viral diarrhoea virus
CAC	Codex Alimentarius Commission

CAMHA	Cation-adjusted Mueller-Hinton agar
CA – SFM	Comité de l'Antibiogramme de la Société Française de Microbiologie
CA TFAMR	Codex Alimentarius *Ad Hoc* Intergovernmental Task Force on Antimicrobial resistance
CBPP	Contagious bovine pleuropneumonia
CBPs	Clinical breakpoints
CCRVDF	Codex Committee on Residues of Veterinary Drugs in Food
CIAs	Critically important antimicrobials
CEESA	Centre Europeen d'Etudes pour la Sante Animale
CLSI	Clinical Laboratory Standards Institute
CoNS	Coagulase negative Staphylococci
CRG	Commissie Richtlijnen Gevoeligheidsbepalingen
CVMP	Committee for Veterinary Medicinal Products
DAEC	Diffusely adherent *E. coli*
DCD	Defined course dose
DCP	Decentralized procedure
DDD	Defined daily dose
DDM	Disc diffusion method
DDST	Double disc synergy test
DIN	Deutsches Institut für Normung
DNA	Deoxyribonucleic acid
EAEC	Enteroaggregative *E. coli*
EARS-net	European Antimicrobial Resistance Surveillance Network
EASSA	European Antimicrobial Susceptibility Surveillance in Animals
ECDC	European Centre for Disease Control
ECHA	European Chemicals Agency
ECOFF	Epidemiological cut-off
EDTA	Ethylenediaminetetraacetic acid
EEA	European Economic Area
EFSA	European Food Safety Authority
EHEC	Enterohaemorrhagic *E. coli*
EIEC	Enteroinvasive *E. coli*
ELDU	Extra-label drug use
EMA	European Medicine Agency
EP	European Parliament
EPEC	Enteropathogenic *E. coli*
ERA	Environmental Risk Assessment
ESAC-net	European Surveillance of Antimicrobial Consumption Network
ESBL	Extended spectrum beta-lactamase
ESC(K)APE	*Enterococcus faecium, Staphylococcus aureus, Clostridium difficile (Klebsiella pneumoniae), Acinetobacter baumannii, Pseudomonas aeruginosa* and *Enterobacteriaceae*

ESVAC	European Surveillance of Veterinary Antimicrobial Consumption
ETEC	Enterotoxigenic *E. coli*
EUCAST	European Committee on Antimicrobial Susceptibility Testing
EU – JAMRAI	EU Joint Action on Antimicrobial Resistance and Healthcare-Associated Infections
EUROSTAT	Statistical Office of the European Union
FAO	Food and Agriculture Organization
FARAD	Food Animal Residue Avoidance Databank
FIC	Fractional inhibitory concentration
FDA	Food and Drug Administration
FEEDAP	Additives and Products or Substances used in Animal Feed EFSA Panel
GF -TADs	Global Framework for the Control of Transboundary Animal Diseases
GHSA	Global Health Security Agenda
GLASS	Global Antimicrobial Resistance Surveillance System
GLEWS	Global Early Warning and Response System
GMP	Good manufacturing practice
HGT	Horizontal gene transfer
HLAR	High-Level aminoglycoside resistance
HPLC	High-performance liquid chromatography
IACG	Interagency Consultation Group
IBDV	Infectious bursal disease virus
IBR	Infectious bovine rhinotracheitis
ICEs	Integrative conjugative elements
IMI	Innovative Medicines Initiative
IMI	Imipenemase (carbapenemase type)
IPCC	Intergovernmental Panel on Climate Change
IPFSAPH	International Portal on Food Safety, Animal and Plant Health
IS	Insertion sequences
JECFA	Joint Expert Committee on Food Additives (Joint = FAO/WHO)
JIACRA	Joint Inter-Agency Antimicrobial Consumption and Resistance Analysis
JPIAMR	Joint Programming Initiative on Antimicrobial Resistance
JSC	Japanese Society for Chemotherapy
KPC	*Klebsiella pneumoniae* carbapenemase
LA-MRSA	Livestock-associated methicillin-resistant *Staphylococcus aureus*
LC	Liquid chromatography
LPS	Lipopolysaccharide
MALDI TOF	Matrix-assisted laser desorption ionization time of flight
MALDI TOF MS	Matrix-assisted laser desorption ionization time of flight mass spectrometry

MBLs	Metallo-beta-lactamases
MDR	Multidrug resistance
MF	Medicated feed
MFP	Membrane fusion protein
MGEs	Mobile genetic elements
MI	Medically important
MIC	Minimum inhibitory concentration
MICEs	Mycoplasma ICEs
MLST	Multilocus sequence typing
MLVA	Multiple-locus variable number tandem repeat analysis
MNEC	Meningitis-associated *E. coli*
MRL	Maximum residue limit
MRP	Mutual recognition procedure
MRSA	Methicillin-resistant *Staphylococcus aureus*
MS	Mass spectrometry
MSCRAMMCs	Microbial surface components of *S. aureus* recognizing the adhesive matrix molecular components
NDM	New Delhi metallo-beta-lactamase
NGS	New (Next) generation genome sequencing
NMI	Not medically important (MI)
NNT	Number needed to treat
NOAEC	No observable adverse effect concentration
NOAEL	No observable adverse effect level
OIE	World Organisation for Animal Health
OMP	Outer membrane protein
OTC	Ove-the-counter
OXA	Carbapenem-hydrolysing oxacillinase
PCR	Polymerase chain reaction
PCU	Population correction unit
PCV2	Porcine circovirus type 2
PD	Pharmacodynamic
PE	Proliferative enteropathy
PFGE	Pulsed-field gel electrophoresis
PI3	Parainfluenza 3 virus
PK	Pharmacokinetic
PMQR	Plasmid-mediated quinolone resistance
PPDS	Postpartum dysgalactia syndrome
PRDC	Porcine respiratory disease complex
PRRS	Porcine reproductory and respiratory syndrome
RAPD	Random amplified polymorphic DNA
RNA	Ribonucleic acid
RND	Resistance nodulation–cell division
QSIs	Quorum sensing inhibitors
RCT	Randomized clinical trial
RfD	Reference dose
RPA	Reference points for action
RONAFA	Reduction of the Need for Antimicrobials in Food-Producing Animals and Alternatives

SCCmec	Staphylococcal cassette chromosome mec
SCENHIR	Scientific Committee on Emerging and Newly Identified Health Risks
SIV	Swine influenza virus
SNP	Single-nucleotide polymorphism (typing)
SPC	Summary of Product Characteristics
SRGA	Swedish Reference Group for Antibiotics
TATFAR	Transatlantic Taskforce on AMR
TF	Treatment frequency
TFEU	Treaty on the Functioning of the European Union
TMDI	Theoretical maximum daily intake
TRACES	Trade Control and Expert System
UNGA	United Nations General Assembly
UPEC	Uropathogenic *E. coli*
VAV	Veterinary autogenous vaccines
VetCAST	Veterinary Committee on Antimicrobial Susceptibility Testing
VFM	Veterinary fastidious medium
VICH	International Cooperation on Harmonisation of Technical Requirements for Registration of Veterinary Medicinal Products
VIM	Verona integron-encoded metallo-beta-lactamase
VMP	Veterinary medicinal products
WHO	World Health Organization
WHO GAP	World Health Organization Global Action Plan
WAHID	World Animal Health Information Database
WAHIS	World Animal Health Information System
WGS	Whole genome sequencing
WMS	Whole metagenome sequencing
WHA	World Health Assembly
WP	Withdrawal period
WTO	World Trade Organization

List of Abbreviations (Vol I Chapters I–X: By Chapters)

Introduction

AMR	Antimicrobial resistance
MRL	Maximum residue limit

Status Quo in International Context

AGISAR	Advisory Group on Integrated Surveillance of Antimicrobial Resistance (WHO)
CAC	Codex Alimentarius Commission
CA TFAMR	Codex Alimentarius Ad Hoc Intergovernmental Task Force on Antimicrobial resistance
CCRVDF	Codex Committee on Residues of Veterinary Drugs in Foods
CIAs	Critically important antimicrobials
FAO	Food and Agriculture Organization
GLASS	Global Antimicrobial Resistance Surveillance System
GLEWS	Global Early Warning and Response System
GF-TADs	Global Framework for Progressive Control of Transboundary Animal Diseases
GHSA	Global Health Security Agenda
IACG	Interagency Consultation Group
IPCC	Intergovernmental Panel on Climate Change
IPFSAPH	International Portal on Food Safety, Animal and Plant Health
NGS	New generation genome sequencing
OIE	World Organisation for Animal Health
TATFAR	Transatlantic Taskforce on AMR
UNGA	United Nations General Assembly
WHO	World Health Organization
WHO GAP	World Health Organization Global Action Plan
WAHID	World Animal Health Information Database
WAHIS	World Animal Health Information System
WHA	World Health Assembly
WTO	World Trade Organization

EU Policies and Regulatory Surroundings

AGP	Antimicrobial growth promoter
AHAW	Animal Health and Welfare EFSA Panel
AMCRA	Antimicrobial Consumption and Resistance in Animals (Belgium)
AMEG	Antimicrobial Advice Ad Hoc Expert Group
BIOHAZ	Biological Hazards EFSA Panel
CEESA	Centre Europeen d'Etudes pour laSant_e Animale
CVMP	Committee for Veterinary Medicinal Products
EARS-net	European Antimicrobial Resistance Surveillance Network
EASSA	European Antimicrobial Susceptibility Surveillance in Animals
ECDC	European Centre for Disease Control
EEA	European Economic Area
EFSA	European Food Safety Authority
EMA	European Medicine Agency
EP	European Parliament
ERA	Environmental Risk Assessment
ESAC-net	European Surveillance of Antimicrobial Consumption Network
ESVAC	European Surveillance of Veterinary Antimicrobial Consumption
EU-JAMRAI	EU Joint Action on Antimicrobial Resistance and Healthcare-Associated Infections
FEEDAP	Additives and Products or Substances used in Animal Feed EFSA Panel
GMP	Good manufacturing practice
IMI	Innovative Medicines Initiative
JIACRA	Joint Inter-Agency Antimicrobial Consumption and Resistance Analysis
MF	Medicated feed
MRSA	Methicillin-resistant *Staphylococcus aureus*
RONAFA	Reduction of the Need for Antimicrobials in Food-Producing Animals and Alternatives
SCENHIR	Scientific Committee on Emerging and Newly Identified Health Risks
SPC	Summary of Product Characteristics
TFEU	Treaty on the Functioning of the European Union
VICH	International Cooperation on Harmonisation of Technical Requirements for Registration of Veterinary Medicinal Products
VMP	Veterinary medicinal products

Use of Antimicrobials in Practice (Targeted on Cattle, Pigs, Poultry, Horses)

AACTING	Antimicrobial usage at herd level and analysis, communication and benchmarking
ADD	Animal daily dose
AMC	Antimicrobial consumption
APP	*Actinobacillus pleuropneumoniae*
ATI	Animal Treatment Index
BRD	Bovine respiratory disease
DCD	Defined course dose
DCP	Decentralised procedure
DDD	Defined daily dose
ESBL	Extended spectrum beta-lactamase
FDA	Food and Drug Administration
EUROSTAT	Statistical Office of the European Union
JPIAMR	Joint Programming Initiative on Antimicrobial Resistance
LA-MRSA	Livestock-associated methicillin-resistant *Staphylococcus aureus*
MI	Medically important
MRP	Mutual recognition procedure
NMI	Not medically important (MI)
PCU	Population correction unit
PE	Proliferative enteropathy
PPDS	Postpartum dysgalactia syndrome
PRDC	Porcine respiratory disease complex
TRACES	Trade Control and Expert System
TF	Treatment frequency

Considerations Reflecting Possible Risks from Use of Antimicrobials

ADI	Acceptable daily intake
ADME	Absorption, distribution, metabolism, elimination
ARfD	Acute reference dose
CA TFAMR	Codex Alimentarius *Ad Hoc* Intergovernmental Task Force on Antimicrobial resistance
ECDC	European Centre for Disease Control
ECHA	European Chemicals Agency
EEA	European Economic Area
EFSA	European Food Safety Authority
ESBL	Extended spectrum beta-lactamase
ESC(K)APE	*Enterococcus faecium, Staphylococcus aureus, Clostridium difficile (Klebsiella pneumoniae), Acinetobacter baumannii, Pseudomonas aeruginosa* and *Enterobacteriaceae*
FAO	Food and Agriculture Organization
HPLC	High-performance liquid chromatography
IMI	Imipenemase (carbapenemase type)

JECFA	Joint Expert Committee on Food Additives (Joint = FAO/WHO)
KPC	*Klebsiella pneumoniae* carbapenemase
LA-MRSA	Livestock-associated methicillin-resistant *Staphylococcus aureus*
LC	Liquid chromatography
MBLs	Metallo-beta-lactamases
MLVA	Multiple-locus variable number tandem repeat analysis
MRL	Maximum residue limit
MS	Mass spectrometry
MIC	Minimum inhibitory concentration
NDM	New Delhi metallo-beta-lactamase
NOAEC	No observable adverse effect concentration
NOAEL	No observable adverse effect level
OXA	Carbapenem-hydrolysing oxacillinase
PFGE	Pulsed-field gel electrophoresis
RfD	Reference dose
RPA	Reference points for action
SCCmec	Staphylococcal cassette chromosome mec
SNP	Single-nucleotide polymorphism (typing)
TGE	Transmissible gastroenteritis
TMDI	Theoretical maximum daily intake
VICH	International Cooperation on Harmonisation of Technical Requirements for Registration of Veterinary Medicinal Products
VIM	Verona integron-encoded metallo-beta-lactamase
WGS	Whole genome sequencing
WHO	World Health Organization
WP	Withdrawal period

Prevention Is Better Than Cure

APP	*Actinobacillus pleuropneumoniae*
ASF	African swine fever
BRD	Bovine respiratory disease
BRSV	Bovine respiratory syncytial virus
BVDV	Bovine viral diarrhoea virus
CBPP	Contagious bovine pleuropneumonia
IBDV	Infectious bursal disease virus
IBR	Infectious bovine rhinotracheitis
PCV2	Porcine circovirus type 2
PI3	Parainfluenza 3 virus
PRRS	Porcine reproductory and respiratory syndrome

QSIs	Quorum sensing inhibitors
SIV	Swine influenza virus
TGE	Transmissible gastroenteritis
VAV	Veterinary autogenous vaccines

Mass Medications: Prophylaxis and Metaphylaxis Cascade and Off-label Use Treatment Guidelines and Antimicrobial Stewardship

AMDUCA	Animal Medicinal Drug Use Clarification Act
ARR	Absolute risk reduction
AUC	Area under curve
BRD	Bovine respiratory disease
CIAs	Critically important antimicrobials
ELDU	Extra-label drug use
FARAD	Food Animal Residue Avoidance Databank
MIC	Minimum inhibitory concentration
NNT	Number needed to treat
OTC	Ove-the-counter
PCR	Polymerase chain reaction
PD	Pharmacodynamic
PK	Pharmacokinetic
RCT	Randomized clinical trial
SPC	Summary of product characteristics
WHO	World Health Organization

Laboratory Investigations and Result Interpretation

APBP	Aminophenyl boric acid
AST	Antimicrobial susceptibility testing
ATU	Area of technical uncertainty
BSAC	British Society for Antimicrobial Chemotherapy
CAMHA	Cation-adjusted Mueller-Hinton agar
CA – SFM	Comité de l'Antibiogramme de la Société Française de Microbiologie
CBPs	Clinical breakpoints
CLSI	Clinical Laboratory Standards Institute
CoNS	Coagulase negative Staphylococci
CRG	Commissie Richtlijnen Gevoeligheidsbepalingen
DDST	Double disc synergy test
DIN	Deutsches Institut für Normung
ECOFF	Epidemiological cut-off
EDTA	Ethylenediaminetetraacetic acid
ESBL	Extended spectrum beta-lactamase
EUCAST	European Committee on Antimicrobial Susceptibility Testing
FIC	Fractional inhibitory concentration
HLAR	High-level aminoglycoside resistance

IEF	Isoelectric focusing
JSC	Japanese Society for Chemotherapy
MALDI TOF MS	Matrix-assisted laser desorption ionization time of flight mass spectrometry
MDR	Multidrug resistance
MHA	Mueller-Hinton agar
MIC	Minimum inhibitory concentration
MLST	Multilocus sequence typing
MRSA	Methicillin-resistant *Staphylococcus aureus*
PCR	Polymerase chain reaction
PFGE	Pulsed-field gel electrophoresis
SRGA	Swedish Reference Group for Antibiotics
VetCAST	Veterinary Committee on Antimicrobial Susceptibility Testing
VFM	Veterinary fastidious medium

Antimicrobial Resistance from the Molecular Biology Perspective and Implications for Clinical Practice

AMR	Antimicrobial resistance
ARGs	Antibiotic resistance genes
BMRGs	Biocide/metal resistance genes
CBPs	Clinical breakpoints
CoNS	Coagulase negative Staphylococci
DNA	Deoxyribonucleic acid
HGT	Horizontal gene transfer
ICEs	Integrative conjugative elements
IS	Insertion sequences
MALDI TOF	Matrix-assisted laser desorption ionization time of flight
MALDI TOF MS	Matrix-assisted laser desorption ionization time of flight mass spectrometry
MDR	Multidrug resistance
MFP	Membrane fusion protein
MGEs	Mobile genetic elements
MLST	Multilocus sequence typing
MLVA	Multiple-locus variable number tandem repeat analysis
NGS	New/next generation sequencing
PCR	Polymerase chain reaction
PFGE	Pulsed-field gel electrophoresis
PMQR	Plasmid-mediated quinolone resistance
RAPD	Random amplified polymorphic DNA
RNA	Ribonucleic Acid
RND	Resistance nodulation–cell division
SCCmec	Staphylococcal cassette chromosome mec
SNP	Single-nucleotide polymorphism (typing)
WGS	Whole genome sequencing

WMS	Whole metagenome sequencing

Molecular Biology Perspective of Susceptibility and Resistance in Main Target Pathogens in the Respective Species and Antimicrobials of Concern

AST	Antimicrobial susceptibility testing
DAEC	Diffusely adherent *E. coli*
DDM	Disc diffusion method
EAEC	Enteroaggregative *E. coli*
EHEC	Enterohaemorrhagic *E. coli*
EIEC	Enteroinvasive *E. coli*
EPEC	Enteropathogenic *E. coli*
ETEC	Enterotoxigenic *E. coli*
HGT	Horizontal gene transfer
LPS	Lipopolysaccharide
MDR	Multidrug resistance
MGEs	Mobile genetic elements
MFP	Membrane fusion protein
MICEs	Mycoplasma ICEs
MNEC	Meningitis-associated *E.coli*
MSCRAMMCs	Microbial surface components of *S. aureus* recognizing the adhesive matrix molecular components
MSSA	Methicillin-sensitive *Staphylococcus aureus*
OMP	Outer membrane protein
PCR	Polymerase chain reaction
PMQR	Plasmid-mediated quinolone resistance
RND	Resistance nodulation–cell division
UPEC	Uropathogenic *E. coli*
XDR	Extensively drug resistant

Acknowledgments

Many thanks to all the members of my family, especially Tom, Barbora and Jakub for their great support.

Big thanks also belong to the contributing authors who shared their experience, expertise, effort and time.

My gratitude also goes to the colleagues I have met in the institute, at the university and in everyday practice, from whom I have learned a lot.

Contents

About the Editor

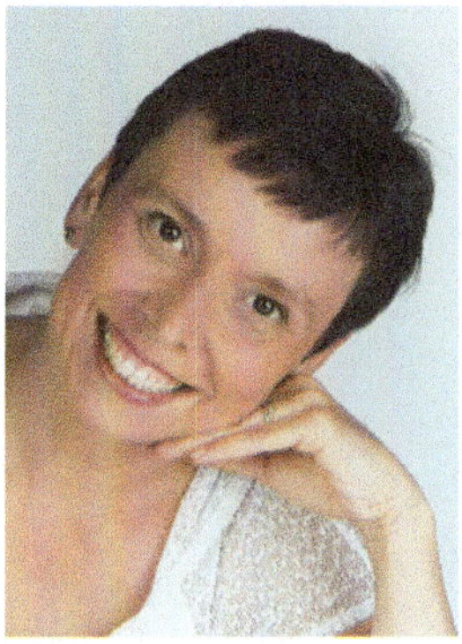

Lucie Pokludová had graduated as microbiologist at the Faculty of Science at Masaryk University in Brno, Czech Republic, and finished PhD studies in analytical microbiology and computer image analysis. She worked as microbiologist and then residue assessor, and her current experience is as an assessor of efficacy of antimicrobials and AMR at the Institute for State Control of Veterinary Biologicals and Medicines. She represents the Czech Republic on different global and European fora, as e.g. EMA—ESVAC, EU JAMRAI project EARS-vet, AACTING and VETCAST, and for more than 10 years she has participated in the One Health Czech National Action Plan on AMR. She teaches part-time at Masaryk University in Brno, but also performs regular lectures in European AMR BTSF programmes as well as invited lectures on different occasions, enjoying teaching as well as communication and sharing the views with students, veterinarians and further audience bringing her different experiences from practice. She loves her jobs but also her family, being married and having two children.

Introduction

Lucie Pokludová

Bacteria are more flexible and clever, than we, as human beings, are. Despite decades of scientific investigations, we sometimes can feel as beginners. We cannot win the battle over bacteria; we can only try to find a way of coexistence with them as well with the whole nature considering ourselves to be an integral part and respecting the rules of nature. But it is a long way in front of us to learn these rules.

Abstract

The introductory part of the book starts with the known facts from the deep history, when people started to use some product of plant origin (e.g. extracts from the barks of *Cinchona officinalis*) to treat certain diseases. Later on, mainly secondary metabolic products of both bacteria and fungi of soil origin were recognised having an antimicrobial effect and after several decades from their discoveries, they started to be used therapeutically. New century, especially thanks to the advance in the genome sequencing, also brings publications providing evidence of the natural and cosmopolitan presence of resistance, which has been on the Earth prior to the use of antimicrobials by humans. Despite this knowledge, still should be kept in mind the imperative of the responsible use of any antimicrobials and substances with the potential to select or co-select resistance in any sector—human and veterinary medicine as well as plant protection. Differences in the use of antimicrobials in veterinary sector from history to current days are described. Also, food safety from not only "chemical" but also "biological" perspective is discussed, especially due to the fact that the book is targeted on livestock. Questions remaining to be answered are defined as the basement for further considerations.

In the deep history people started to use some products of plant origin to treat certain diseases. The use of extracts from the barks of *Cinchona officinalis* can be given as an example. Those extracts (containing quinine together with the other alkaloids) were used for treatment of malaria. In the nineteenth century several observations confirmed antagonistic interactions among microbes described as antibiosis (terms defined by Vuillemin 1889), among microorganisms (Pasteur and Joubert 1877). As one of the exact examples that can be mentioned is pyocyanase produced by *Pseudomonas aeruginosa* inhibiting *Bacillus anthracis*, described by Charrin and Guignard in 1889 (Lochmann 1999).

L. Pokludová (✉)
Institute for State Control of Veterinary Biologicals and Medicines, Brno, Czech Republic
e-mail: pokludova@uskvbl.cz

L. Pokludová (ed.), *Antimicrobials in Livestock 1: Regulation, Science, Practice*,
https://doi.org/10.1007/978-3-030-46721-0_1

Originally the antimicrobials were recognised as products of mainly secondary metabolism of different soil microorganisms, and most common were both bacteria (e.g. *Bacillus* and *Streptomyces*) and fungi (e.g. *Penicillium*, *Cephalosporium*, *Fusidium* and *Pleurotus*). Most recently, with the need of new antimicrobials, there were identified also other microbes, some of them from bacterial species hardly to be cultured and might be therefore waiting such a long time for discovery (e.g. *Eleftheria terrae*, Ling et al. (2015), discovered as producer of antituberculotic teixobactin). Some others on the other hand are produced by well-known commensal species as e.g. *Staphylococcus lugdunensis*, which produces lugdunin, thiazolidine—containing cyclic peptide antibiotic blocking colonisation by *S. aureus* (including strains resistant to methicillin) as published by Zipperer et al. (2016).

Summarising above mentioned, it seems proven that antibiosis as well as substances responsible for this antagonistic interaction—antibiotics (more generally antimicrobials) are natural part of most of ecosystems in the Earth having a great influence on the microbes and the vital balance among themselves. One microbe produces antimicrobial, another one to survive, tries to find mechanisms, how to protect itself. And this is in fact the essence of resistance—survival of individual cell, of the cells community, of the microbiome, of the ecological niche. And here should be highlighted, that a lot of mechanisms of resistance are natural characteristic developed before start of any clinical use of antimicrobials. The exact proof of evidence for above hypothesis was given by the authors of the several studies as this confirming resistance in bacteria from permafrost (D'Costa et al. 2011) as well as another one confirming resistance of bacteria in cave microbiome isolated from any influence for over 4 million years (Bhullar et al. 2012). Despite all these facts creating together mosaic of knowledge of "natural and cosmopolitan" occurrence of resistance, all of us should keep in mind that any use of antimicrobials either in human, veterinary or plant protection area can cause increase of selection pressure, can promote the spread of resistance and occurrence of new, emerging combinations of resistance mechanisms.

Scientists were not satisfied with just recognising the antimicrobials as such, but with discoveries of the causative pathogens associated with different diseases of human and animals, started to perform experiments on use of antimicrobials as effective agents for treatment of the diseases, which cause severe health damages or even death. The most famous are probably first cases of infections effectively treated by penicillin saving the human lives. This "success story" continues and with the discovery of many new antimicrobials was seemed for several decades that the issue of infectious diseases can be covered fully with this specific group of active substances involved in both human and veterinary medicinal products. But being more and more broadly used, the issue of antimicrobial resistance has started to be urgent.

Despite the fact that the resistance and some of its mechanisms had been recognised and proved earlier, the issue became being serious in clinical practice, relevant measures to be implemented to decrease broad use of antimicrobials and slow down the spread of resistance a little bit, have not yet become as a common practice by many clinicians and vets. In 1969, Swann Report, in which one of the main thoughts was to start to use antimicrobials more responsibly and as a certain practical outcome to limit the use of therapeutically important antimicrobials as prescription only, has been published. It can be considered as a milestone starting dozens of documents that have been issued, coming with imperative on rational, responsible, or prudent use of antimicrobials since the 1960s. Some people, who are in close touch with antimicrobials, therefore may feel that the issue of resistance threat is "overdiscussed" in the current days. But considering the data on consumption of human and veterinary (and in some scale also plant protective) antimicrobials not only in some parts of Europe, but worldwide, it seems that the period of raising awareness is not at the end, reaching sufficient level of both knowledge and willingness to change behaviour frameworks to use antimicrobials only in necessary cases and as properly as current scientific knowledge allows.

Ways how antimicrobials are used in animals differ in some scale from their use in human medicine. The original purpose of these substances as specific antiinfectives—effective drugs started to be changed early after their introduction into veterinary practice, in the 1950s. In this period, characterised by the increase of the demand of food and therefore starting of extensive farming systems, it has been discovered that except direct treatment effect, antimicrobials can also have specific growth and production promoting action. The additional effects were recognised when used fermentation waste of fungal production of antimicrobials contained not only vitamins, but also residual amount of antimicrobials. As proved later on, better performance was caused by, e.g. chlortetracycline, that started to be broadly used as a growth promoter in broiler chickens as well as in fattening pigs. With intensive farming, the era of the big farms developed (with significantly higher infection pressure). Increased rate of transports of animals and merging the animals of different origin (with significant mixing not only animals but also causative agents of infections within the herds like feedlot cattle, day-old chicks or weaned piglets) led to disease outbreaks and economical lost. The farmers and vets needed to prevent such situations and the era of routine preventive (prophylactic) use began. As herds and flocks of animals became bigger and bigger, mass medication (long course with lower doses, or short term with therapeutic doses) started to be a routine practice. The practice of using antimicrobials as growth promoters was stopped by a ban coming into force at the beginning of 2006 in the EU. Significant reduction of prophylaxis and metaphylaxis is foreseen to come once new veterinary medicinal product regulation will come into force in EU (i.e. since 2022). Also in other parts of the world activities, including those legally based, targeted on ban or phasing out of antimicrobial growth promoters and minimising of routine prophylaxis started to be more frequent during the last years.

Arising pressure on use of alternatives have become real. This approach will be beneficial being aware that also alternatives could have harmful effects. One of the examples can be zinc oxide and its use in pig production after the ban of growth promoters, what has been later recognised as a risky practice (co-selection of resistance and risks for the environment).

The use of antimicrobials in a broad scale necessarily led to afraid that antimicrobials used in food producing animals do not disappear from their tissues (meat and offals) and products (milk, eggs, honey) and that residues of them and their persistence should be investigated. Therefore, the concept of setting of the maximum residue limit (MRL as a value that can be considered as safe from the perspective of exposure of the potential human consumer) was established. In the case of such specific substances, like antimicrobials, with potential to have not only toxic, but also microbiological adverse effects, the specific concept of the so-called microbiological acceptable daily intake (mADI) was agreed internationally. Investigations targeted on influence on human gut microflora were involved in the assessment of mADI and MRL for individual substances in different species and tissues. Moreover, once the substance became a part of veterinary medicinal product, specific studies are carried out to establish withdrawal period, as the minimum time between the last administration of this product to an animal and the production of foodstuffs from that animal, which ensure that such foodstuffs do not contain residues exceeding MRLs, i.e. in quantities harmful to potential consumer. In those cases, when residues in certain food commodity exceed MRLs, the product should not be used for human consumption and producers are usually subjects of inspections or even penalised. This is the concept of "chemical safety" that is necessary to avoid residues—but does not protect against the transfer of the antimicrobial resistance (either via resistant commensal or zoonotic bacteria, or via genes of resistance). Issues linked to resistance associated with the food production should be part of the concept of "biological safety", which is quite complicated as for the establishment of the checking mechanism together with the measures to be put in place internationally by global consensus.

Summarising the above information can be for this moment concluded that despite the fact that it has been recognised and agreed by worldwide consensus on the need for rational and responsible use of antimicrobials and that level of awareness increased dramatically during the last years, there is still a lot of work to be done. The change of behaviour of each of us is considered essential.

It seems apparent that there is still a lack of information and important questions are waiting to be answered. Among these questions belong information on how to effectively disconnect links of transfer of resistance (especially as not all links are already sufficiently precisely described); what is the level of contribution of veterinary use of antimicrobials on the issue of antimicrobial resistance in human medicine and contribution to resistome pool in the environment; and what alternatives to veterinary antimicrobials can be used safely, effectively and sustainably (also from economical perspective). Also some more "practical" issues have been recognised and on some of them the work already has been started—e.g. clinical breakpoints for pathogens of concern in animals, discovery and introduction in routine practice of rapid antimicrobial susceptibility testing, rapid and reliable diagnostic tests as well as updating the doses of therapeutically used "old, narrow spectrum" antimicrobials to ensure efficacious treatment with concurrent minimising selection pressure for antimicrobial resistance.

Think About 1: Searching New Antimicrobials, but Do Not Forget to Revisit Those We Already Have

One of the cornerstones of each strategy of the last years is calling for the search of new antimicrobials. It is generally agreed as vital to search for the new substances with completely new mechanisms of actions against bacteria (especially Gram-negatives). But we should not also forget revisiting the "old molecules", especially on veterinary side. Such revisiting can be considered also from the perspective of the performance of the new studies on pharmacokinetic and pharmacodynamic and, including into those studies also newly invented pharmaceutical forms/carriers improving the properties of the medicinal product and e.g. moving the active moiety in more targeted way to the tissues affected by infection. Incentives to perform such studies should be provided in balanced manner to those new antimicrobials.

References

Bhullar K, Waglechner N, Pawlowski A, Koteva K, Banks ED, Johnston MD, Barton HA, Wright GD (2012) Antibiotic resistance is prevalent in an isolated cave microbiome. PLoS One 7:e34953

Charrin A, Guignard L (1889) Action du bacille pyocyanique sur la bactéridie charbonneuse. CR Hebd Seances Acad Sci. Paris 108:764–766

D'Costa VM, King CE, Kalan L, Morar M, Sung WWL, Schwarz C, Froese D, Zazula G, Calmels F, Debruyne R, Golding GB, Poinar HN, Wright GD (2011) Antibiotic resistance is ancient. Nature 477:457–461

Ling LL, Schneider T, Peoples AJ, Spoering A, Engels I, Conlon BP, Mueller A, Schäberle TF, Hughes DE, Epstein S, Jones M, Lazarides L, Steadman VA, Cohen DR, Fetterman KA, Millett WP, Nitti AG, Zullo AM, Chen C, Lewis K (2015) A new antibiotic kills pathogens without detectable resistance. Nature 517:455–459

Lochmann O (1999) Základy antimikrobní terapie, 2nd edn. Triton, Praha

Pasteur L, Joubert J (1877) Charbonne et septicémie. CR Hebd Seances Acad Sci. Paris 85:101–115

Vuillemin P (1889) Antibioses et symbiose. CR de l'Assoc. Française pour l'Avancem. des Sci. Paris Pt (2) 18:525–543

Zipperer A, Konnerth MC, Laux C, Berscheid A, Janek D, Weidenmaier C, Burian M, Schilling NA, Slavetinsky C, Marschal M, Willmann M, Kalbacher H, Schittek B, Brötz-Oesterhelt H, Grond S, Peschel A, Krismer B (2016) Human commensals producing a novel antibiotic impair pathogen colonization. Nature 535:511–516

Status Quo in International Context

Gérard Moulin and Lucie Pokludová

Abstract

International organisations such as WHO, FAO, OIE have starting working on antimicrobial resistance more than 25 years ago and have published many guidelines and standards on surveillance of antimicrobial resistance and usage, prudent use of antimicrobials, risk assessment. Cooperation between the international organisations improved following a tripartite meeting (WHO/FAO/OIE). In 2011, WHO, FAO and OIE adopted a tripartite agreement in order to extend their collaboration in sense of "One Health" concept. This increased collaboration led to the publication in May 2015 of the WHO Global Action Plan. In September 2016, during the United Nation General Assembly, Member States adopted a political declaration to commit to fight antimicrobial resistance together with the aim to ensure sustained effective global action to address antimicrobial resistance. On the animal side, minimising the use of antimicrobials and using them prudently is the cornerstone of the fight against antimicrobial resistance. International recommendations on these subjects need to be implemented in the field.

Keywords

Prudent use · WHO Global Action Plan · Critically important antimicrobials · OIE guidelines · International cooperation · FAO Codex Alimentarius · Antimicrobials

1 Status Quo in International Context

Despite the "50 years anniversary" of the milestone in recognising the resistance issue as a potential threat in 2019 (The Swann Report, UK To Parliament 1969), we are still on the way to find the most appropriate international platform that will help the "One Health World" to find an effective set of tools and a coordinated approach to keep antimicrobials working, as one of the most powerful medicines used for treatment of infectious diseases. Due the time course, starting with phenotype microbiological methods and going through the successful history of new methods development to new generation of genome sequencing (NGS) and further more sophisticated molecular biology, genetic and even physico-chemical methods, we have started to recognise how rich and complex the world of microbes is, and which areas need to be covered

G. Moulin (✉)
French Agency for Food, Environmental and Occupational Safety, National agency for veterinary Medicinal Products, Fougères, France
e-mail: Gerard.Moulin@anses.fr

L. Pokludová
Institute for State Control of Veterinary Biologicals and Medicines, Brno, Czech Republic
e-mail: pokludova@uskvbl.cz

L. Pokludová (ed.), *Antimicrobials in Livestock 1: Regulation, Science, Practice*,
https://doi.org/10.1007/978-3-030-46721-0_2

to contain antimicrobial resistance. As different areas need specific approaches, but also some interfaces to work together, broad spectrum of international platforms, bodies and activities tackling the antimicrobial resistance appeared on international scene targeting the AMR. Due the time course it was also recognised that AMR is like a tip of the iceberg, and there is a need to start solving this issue from a broad basement, within the concept covering human, animal and environmental aspects.

Due to concerns that have arisen, especially in connection with hospital-acquired infections, first activities started mainly in human area. But during the mid-1950s, following long-standing complaints by dairy industry about antibiotics' disruption of cheese production, consumers were shocked to learn that up to 10% of US milk samples were contaminated with penicillin (Kirchhelle 2018). Under intense pressure, the FDA introduced the first national monitoring programme for penicillin residues in milk in 1960 (Smith-Howard 2017). Six years later, similar public concerns and new residue detections resulted in the first national monitoring programme for antibiotics in meat and license withdrawals for antibiotic preservatives in the United States (Kirchhelle 2018). After spot tests revealed considerable residues in German meat, the governmental monitoring programme was launched in the mid-1970s also in Germany (Kirchhelle 2016; Thoms 2017). In this period British inhabitants and food producing companies started to think about antimicrobials, but due to the strong position of the farmers and vets, first Netherthorpe Committee Report was not too strict, endorsed existing antibiotic use, but recommended restrictions of future antibiotics. This initial compromise came under scrutiny in 1964 when Ruth Harrison's bestseller "Animal Machines" attacked alleged welfare abuses, drug overuse and AMR selection on "factory farms" (Harrison 1964). In the 1950s Andersen's report brought revolutionary thoughts and evidence of fragment DNA, that can be spread and promote AMR. According to Anderson, medically important antibiotics had to be restricted before uncontrolled agricultural use allowed more dangerous pathogens like *Salmonella enterica serovar Typhi* (typhoid) to acquire multiple resistance (Kirchhelle 2018). Finally Swann report came and recommended a series of reforms of which the restriction of medically important antibiotics to veterinary prescription was the most significant.

How years passed, it was not only scientists, who began to point out, how food with residues of antimicrobials can be dangerous and what other risks, including antimicrobial resistance, can be associated with use of antimicrobials in animals (see also Chap. 5). Due to increasing public pressure politicians also identify the risks associated with use of antimicrobials in animals and possible impacts on the public as well as animal health. In some references (Sample 2013; Kirchhelle 2018) it is considered as one of the impulses of Britain's Chief Medical Officer Dame Sally Davies lobbing to include AMR in the United Kingdom's National Risk Register of Civil Emergencies as a threat comparable to major coastal flooding or a catastrophic terrorist attack. Davies' strong warnings were followed by a row of expert reports and finally also national action plans, and pledges to reduce antibiotic use by members of WHO in 2015, FAO in 2016 and G20 in 2017. All these reports and strategies show that a more broad concept of the policies targeted on combating against AMR including not only human, but also animal sector led to One Health concept to be set. Finally, the last years are characterised by raising attention to the questions linked to the environment that creates global living space for microbes, human, animals and plants. Despite the fact that resistance is considered to be as old as the bacteria in the world, the extensive use of antimicrobials across the sectors make a big selection pressure and can finally "change the Earth". Therefore, it seems that except the One Health Concept, thinking about even broader concept of "One Earth" might be an option how all anthropogenic activities as well as systems of nature to be involved and approached in complexity. It might be surprising that the idea of "One Health" or "One Earth" or at least interlinks among human and animal health is not as new as it could seem.

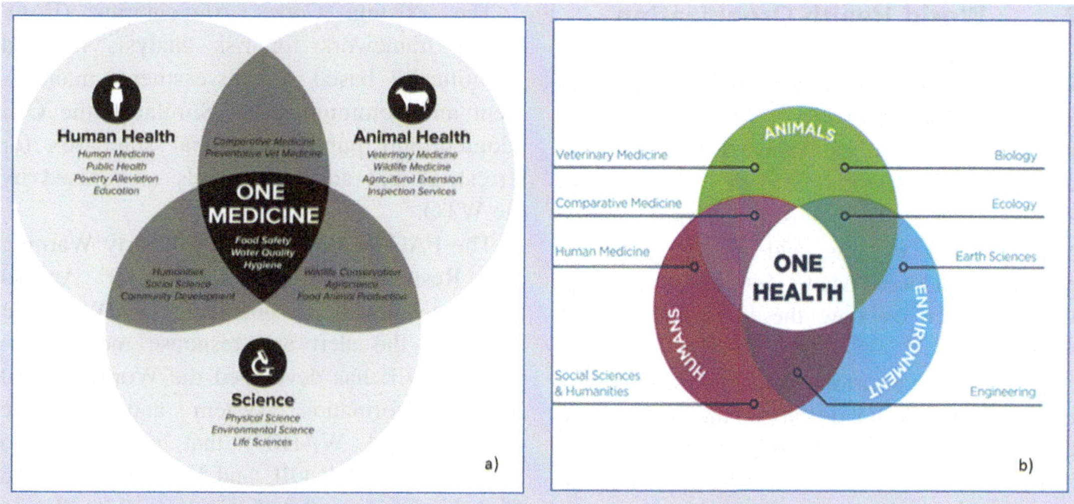

Fig. 1 Comparison of the "One Medicine concept" (**a**) by Schwabe, 1964, reflecting input of expertise from different areas of science and "One Health concept" (**b**) as published recently (https://www.ucdavis.edu/one-health/collaborations) and accepted also by different international bodies worldwide

Probably the first scientist who noticed it was Rudolf Virchow (by noticing relation among human and animal diseases and introducing the term "zoonose"). The concept of "One Medicine" was introduced by Charles Schwabe's vision in 1964 (in the textbook Veterinary Medicine and Human Health), where Human Health, Animal Health and Science (including environmental science) were involved and complexity and overlapping and cross links were clearly identified and described (please compare schemes in Fig. 1).

A big debate at the leading worldwide acting international bodies under the roof of the United Nations and the tripartite as World Health Organization (WHO), Food and Agriculture Organization (FAO) and World Organisation for Animal Health (OIE) started approximately 5 years ago accompanied by an avalanche of plans, strategies, guidances and manuals tackling issues related to AMR. Under the umbrella of FAO and WHO, Codex Alimentarius has started to work at first on the issue of residues of veterinary medicinal products—within the Codex Committee on Residues of Veterinary Drugs in Food (CCRVDF) and since 2010 the work on food-borne antimicrobial resistance has started, including the publication of the Guidelines (Codex Alimentarius Ad Hoc Intergovernmental Task Force on Antimicrobial resistance). Later on were also established other multilateral initiatives as Global Health Security Agenda (GHSA), Transatlantic Task Force on AMR (TATFAR), G7 and G20 and OECD activities. Further parts of this chapter will in stepwise manner approach to either individual bodies, or merged activities (Tripartite FAO/OIE/WHO) as well as key documents from the European regulatory surrounding will be introduced on next pages (Fig. 1).

In 2015, in a United Nation summit, countries adopted the 2030 Agenda for Sustainable Development and its 17 Sustainable Development Goals. A number of these goals are relevant for health and antimicrobial resistance issues.

The Interagency Coordination Group produced a report adopted in 2019 that gives recommendations to provide practical guidance for approaches needed to ensure sustained effective global action to address AMR. This report makes Global governance recommendations and links AMR activities to the Sustainable Development Goals.

2 World Health Organization

WHO works in close cooperation with the Food and Agriculture Organization (FAO) and the World Organisation for Animal Health (OIE) to promote cross-sectoral holistic approach to address risks from zoonoses and other public health threats existing and emerging at the human–animal–ecosystems interface, and provide guidance on how these risks should be reduced.

In 2004, a consortium of agencies, including FAO, OIE and WHO, developed the International Portal on Food Safety, Animal and Plant Health (IPFSAPH), an online source to facilitate international trade in food and agricultural products.

In 2011, WHO, FAO and OIE adopted a Tripartite Agreement in order to extend their collaboration in sense of "One Health" concept. The Agreement document sets a strategic direction for FAO–OIE–WHO to take together and proposes a long-term basis for international collaboration aimed at coordinating global approach for a complementary agenda, bringing new synergies among FAO, OIE and WHO that will include normative work, public communication, pathogen detection, risk assessment and management, technical capacity building and research/development activities and their support. Based on the Agreement those tripartite organisations work to achieve alignment and coherence of related global standard setting activities (Codex Alimentarius, OIE and Intergovernmental Panel on Climate Change) referred to in the World Trade Organization (WTO) Agreement on the Application of Sanitary and Phytosanitary Measures.

In May 2018, WHO, FAO and OIE signed a Memorandum of Understanding (MoU) to strengthen their long-standing partnership, with a strong focus on tackling antimicrobial resistance (AMR). Following the MoU, the tripartite have also engaged closely with the United Nations Environment Programme (UNEP) to strengthen the integration of environment in their collective work. In this context, the WHO, FAO and OIE have collaborated to develop a Tripartite Workplan on antimicrobial resistance in 2019–2020, with the involvement of UNEP to ensure that all relevant dimensions are considered.

The existing Codex Alimentarius (FAO/WHO) framework for risk analysis founded scientifically based risk assessment, management and communication. Similarly, the OIE adopted and published global standards for terrestrial and aquatic animals recognised by the WTO.

The FAO–OIE–WHO Global Early Warning and Response System for Major Animal Diseases, including zoonoses, (GLEWS) that combines the alert and response mechanisms was set. OIE has developed the World Animal Health Information System and Database (WAHIS and WAHID) that contributes to GLEWS. In 2004, OIE and FAO launched the Global Framework for the Control of Transboundary Animal Diseases (GF-TADs), which provides a clear vision and framework to address endemic and emerging infectious diseases, including zoonoses. WHO is associated with this mechanism through GLEWS, in the case of zoonoses, where information exchange occurs daily.

One of the cornerstones is WHO Global Action Plan (2015) with five strategic objectives:

- To improve awareness and understanding of antimicrobial resistance
- To strengthen knowledge through surveillance and research
- To reduce the incidence of infection
- To optimise the use of antimicrobial agents
- Develop the economic case for sustainable investment that takes account of the needs of all countries, and increase investment in new medicines, diagnostic tools, vaccines and other interventions

Principles applied are whole-of-society engagement, promote prevention as the first priority, make antimicrobials available, but not excessively used, keep the antimicrobials working—sustainability of the system.

As listed in Table 1, certain projects were launched and under the leadership of WHO as *GLASS* (surveillance on AMR in humans) and *AGISAR* (List of Critically Important Antimicrobials in human medicine; Integrated Surveillance on AMR).

Table 1 Summary of WHO (and Tripartite related) key strategic activities and key working documents of importance for the last 5 years

Year	Activity	Key outcomes
2014	67th World Health Assembly WHO (2014a)	Resolution on AMR (WHA67.25) STAG and FAO/OIE to develop an Action Plan on AMR
2014	Antimicrobial resistance: global report on surveillance WHO (2014b)	The report makes a clear case that AMR to common bacteria has reached alarming levels in many parts of the world and that in some settings, few, if any, of the available treatment options remain effective Another important finding: surveillance of AMR was neither coordinated nor harmonised; there are many gaps in information on bacteria of major public health importance. Need for establishment of GLASS indicated
2015	68th World Health Assembly WHO (2015a)	Adopt Global Action Plan on AMR (GAP)—supported by FAO and OIE Call all Member States of the WHO to put in place national action plans against AMR by mid-2017 Library on National action plans on AMR is available at: https://www.who.int/antimicrobial-resistance/national-action-plans/library/en/
2015	Global Antimicrobial Resistance Surveillance System Manual for Early Implementation WHO (2015b)	The Global Antimicrobial Resistance Surveillance System (*GLASS*) to facilitate and encourage a standardised approach to resistance surveillance in common human bacterial pathogens globally and in turn support the implementation of the GAP on AMR. Targeted on public health professionals and health authorities responsible for national AMR surveillance (standards, road map for implementation, for the period 2015–2019)
2015	About Antimicrobial resistance for Policy Makers WHO (2015c)	Banners and infographic to draw attention of politicians and public to set activities to combat AMR
2016	69th World Health Assembly WHO (2016)	Addressed drug-resistant pathogens Call for setting a global framework to Combat AMR
2017	WHO guideline on use of medically important antimicrobials in food producing animals WHO (2017a)	These guidelines present evidence-based recommendations and best practice statements on use of medically important antimicrobials in food-producing animals, based on the WHO CIA List
2017	Global Framework for Development and Stewardship to Combat AMR WHO (2017b)	Document describing current state of play and the way forward with respect to the establishment of a global framework for development and stewardship to combat AMR. Developed in collaboration with OIE and FAO. It builds upon the options for the development of such a framework presented to the 69th World Health Assembly
2018	Tackling AMR Together Working paper 1.0: Multisectoral coordination WHO (2018a)	Practical tips and suggestions on how to establish and sustain the multisectoral coordination needed to develop and implement National Action Plans on AMR (NAPs). Published literature + the operational experience of four "focal countries" (Ethiopia, Kenya, Philippines and Thailand) lead to summary on lessons learned and the latest thinking on multisectoral working to achieve effective AMR action
2018	Assessing entry points and options for increasing investments in AMR in low- and middle-income	Three country case studies were commissioned by the WHO AMR Secretariat—in Ghana, Nepal and Nigeria—to assist teams working on AMR to explore

(continued)

Table 1 (continued)

Year	Activity	Key outcomes
	countries WHO (2018b)	the scope to scale up delivery of AMR activities through existing programmes and projects and those that are under development (human and animal) Next steps: the development of a Guidance Note "Getting more done to combat AMR: resource mobilisation in low- and middle-income countries"
2018	Monitoring Global Progress on Addressing AMR WHO/FAO/OIE (2018)	Analysis report of the second round of results of AMR country self-assessment survey: 100 countries recognised to have developed National action plans (NAPs), 51 under development While 9 of the top 10 chicken-, pork- and cattle-producing countries developed a national action plan, the survey response shows that in almost all domains—surveillance, education, monitoring and regulating consumption and use—more activities can be seen in the human sector. An urgent need for resource prioritisation and more action in the animal and food sectors highlighted. Only 41.6% (64 countries) have limited the use of CIAs (human and animal) for growth promotion in agriculture. Substantial data is also missing from the environment and plant sectors
2018	Tackling AMR together Working Paper 5.0: Enhancing the focus on gender and equity WHO (2018c)	Aims to assist countries to take the first step towards better considering gender and equity issues in their efforts to tackle AMR, to inform the implementation of strategies in national action plans and contribute to improved reach and effectiveness of AMR efforts in the longer term. (Considering e.g. Women's risk of exposure to AMR during pregnancy, abortion and childbirth. Urinary Tract Infection (UTI) higher prevalence in females than males overall and especially at younger ages, in older vice versa. Risk for woman as frontline health workers. Risk (MRSA)—for man farmers/vets in agriculture in pigs/cattle and woman rather in poultry sector)
2018	WHO list of critically important antimicrobials for human medicine (6th revision) WHO (2018d)	In the latest version of the CIA list (6th revision, 2018), the "Highest Priority Critically Important Antimicrobials" are: quinolones, third and higher generation cephalosporins, macrolides and ketolides, glycopeptides and polymyxins
2019	Turning plans into action for AMR Working Paper 2.0: Implementation and coordination WHO (2019)	This paper was developed to support AMR coordination committees and others tasked with addressing AMR at country level Six key strategies are pointed in the document for success and offers a series of practical tips and suggestions on how to implement each one: Establish AMR coordination committee roles and responsibilities; prioritise AMR activities; get AMR into plans; make the case for investment; engage stakeholders; tailor the message
2019	Tripartite Monitoring and Evaluation (M&E) framework for the Global Action Plan on Antimicrobial Resistance WHO/FAO/OIE (2019a, b)	Aims: to generate data to assess the delivery of GAP objectives, and inform operational and strategic decision-making on AMR for the next 5 years: Track 1 focuses on the inputs, activities and outputs of the GAP: monitoring the progress of different stakeholders in implementing the GAP, and to evaluate how to improve the collective response

(continued)

Table 1 (continued)

Year	Activity	Key outcomes
		Track 2 focuses on GAP outcomes and impact goals: assessing the effectiveness of GAP implementation efforts, including monitoring their results, and evaluating their impact on, for example, AMR, appropriate use of antimicrobials, and burden of diseases
2019	Reports of the Executive Board of WHO WHO (2019a)	Report of the Executive Board of WHO at its 144 Session describes: country-level progress combating AMR (112 countries have developed NAP, 65 under development), progress in implementing GAP (antibiotic Handle with Care campaign with 131 countries participating), multisectoral Tripartite collaboration (FAO/OIE/WHO), ongoing challenges and emerging threats
2019	72nd World Health Assembly of a resolution on antimicrobial resistance WHO (2019a, b)	The access to medicines, vaccines and other health products as highlight (Draft road map). AMR issue addressed in extent equal to above-mentioned Report of the Executive Board of WHO 144 and its recommendations for actions: stewardship framework development/improvement, further NAP implementation, strengthening links between plans for combating AMR and plans for universal health coverage, health security and multisectoral action

WHO also set *Global framework for development and stewardship to combat AMR* targeted on Implementation of GAP, in collaboration with two other tripartite bodies FAO and OIE that has three key objectives:

- *Stewardship*: preserve antimicrobial medicines by taking measures to promote control, appropriate distribution and appropriate use
- *Research and Development*: develop new antimicrobial medicines, diagnostic tools, vaccines and other interventions for detecting, preventing and controlling antimicrobial resistance
- *Access*: promote affordable access to existing and new antimicrobial medicines, vaccines and diagnostic tools

3 Food and Agriculture Organization

Food and Agriculture Organization (FAO) declares supportive role towards governments, producers, traders and other stakeholders to enforce the responsible use of antimicrobials in agriculture, thus helping reduce antimicrobial resistance in agricultural systems. FAO also hosts the Secretariat of the Codex Alimentarius and of the International Plant Protection Commission.

FAO's 39th Conference (in June 2015) adopted Resolution 4/2015 on AMR that flags as an urgent concern on growing levels of AMR in disease- and infection-causing microorganisms, as they become less responsive to treatment, making infections or diseases more difficult or impossible to cure. It highlights the need for the prudent and responsible use of antimicrobials in agriculture. To support the implementation of Resolution 4/2015, the *FAO Action Plan on AMR* addresses four major Focus Areas:

- Improve awareness on AMR and related threats
- Develop capacity for surveillance and monitoring of AMR and AMU (antimicrobial use) in food and agriculture

- Strengthen governance related to AMU and AMR in food and agriculture
- Promote good practices in food and agricultural systems and the prudent use of antimicrobials

Just very recently on 19 June 2019, the Tripartite—a joint effort by the FAO/OIE/WHO launched the new project—*AMR Multi-Partner Trust Fund* (AMR MPTF 2019) with a 5-year scope, through 2024, and aims to scale up efforts to support countries to counter the immediate threat of AMR. Fund budget will be used to support countries and the implementation of the Tripartite's AMR Workplan 2019–2020, particularly in providing technical support to countries designing National Action Plans on AMR and to scale up local action. The Fund will take into account the recommendations highlighted in the recently released *Interagency Coordination Group (IACG) report* on AMR. This report highlights the need for coordinated and intensive efforts, acknowledging AMR as a major barrier to the achievement of many of the Sustainable Development Goals, including universal health coverage, secure and safe food, sustainable farming systems and clean water and sanitation.

Codex Alimentarius started to work under umbrella of FAO. Food Standards Programme established by FAO and WHO with objectives to protect consumer health and promote fair practices in food trade was set and standards, guidelines and codes of practice were adopted by the Codex Alimentarius Commission (CAC). Global Action Plan on AMR called Codex to update its standards/codes on AMR and therefore work on Code of Practice to Minimise and Contain AMR (CAC/RCP 61-2005) and Guidelines on Risk Analysis of Foodborne AMR (CAC/GL 77/2011) started in 2016. Work in the time frame of maximum 4 years of Codex Alimentarius Ad Hoc Intergovernmental task Force on Antimicrobial Resistance is foreseen to update the Code of Practice to Minimise and Contain AMR and to draft a new Guideline on Integrated surveillance on AMR.

4 World Organisation for Animal Health

One of the key players from the global perspective considering animal health and welfare is the World Organisation for Animal Health (OIE), which has been working on the AMR issue for a long time. Activities in fight against AMR are built on logical concept of the standard setting—a wide range of international standards on antimicrobial agents, in particular on responsible and prudent use was developed by OIE (Please refer to Table 2 for key OIE documents and respective chapters of them related to AMR). These standards are regularly reviewed and updated through the transparent and inclusive process of expert advice and member consultation before presentation for adoption to the World Assembly of Delegates from our 180 Member Countries each year. The OIE also works with its Member countries in a comprehensive and continuous capacity building process for their Veterinary Services.

Resolution combating AMR and promoting prudent use of antimicrobial agents in animals was released by OIE (May 2015, 83rd General Session World Assembly of Delegates). OIE Strategy on AMR (May 2016, 84th General Session World Assembly of Delegates) was unanimously adopted as Resolution 36, which mandates that OIE compiles AMR activities into a strategy. The OIE Strategy on Antimicrobial Resistance is aligned with the WHO Global Action Plan and recognises the importance of a "One Health" approach—involving human and animal health, agricultural and environmental needs. OIE strategy has four main objectives:

- Improve awareness and understanding
- Strengthen knowledge through surveillance and research
- Support good governance and capacity building
- Encourage implementation of international standards

Table 2 OIE key documents related to AMR issue

OIE document	Key chapters
Terrestrial Animal Health Code (2018)	6.7. Introduction to the recommendations for controlling AMR 6.8. Harmonisation of national AMR surveillance and monitoring programmes 6.9. Monitoring of the quantities and usage patterns of antimicrobial agents used in food-producing animals 6.10. Responsible and prudent use of antimicrobial agents in veterinary medicine 6.11. Risk analysis for AMR arising from the use of antimicrobial agents in animals
Aquatic Animal Health Code (2018)	6.1. Introduction to the recommendations for controlling antimicrobial resistance 6.2. Principles for responsible and prudent use of antimicrobial agents in aquatic animals 6.3. Monitoring of the quantities and usage patterns of antimicrobial agents used in aquatic animals 6.4. Development and harmonisation of national antimicrobial resistance surveillance and monitoring programmes for aquatic animals 6.5. Risk analysis for antimicrobial resistance arising from the use of antimicrobial agents in aquatic animals
Manual of Diagnostic Tests and Vaccines for Terrestrial Animals (2019)	2.1.1. Laboratory methodologies for bacterial antimicrobial susceptibility testing Diagnostic tests (section and selected chapters) Vaccines (section and selected chapters)
OIE List of Antimicrobials of Veterinary Importance (2019a)	All antimicrobial agents used in food-producing animals in a comprehensive list, divided into critically important, highly important and important antimicrobial agents. Fluoroquinolones, third and fourth generation of Cephalosporins, Colistin are in line with WHO sixth updated list (WHO 2018d) critically important both for human and animal health and for veterinary use following recommendations are released by OIE: • Not to be used as preventive treatment applied by feed or water in the absence of clinical signs in the animal(s) to be treated • Not to be used as a first-line treatment unless justified, and when used as a second-line treatment, it should ideally be based on the results of bacteriological tests • Extra-label/off label use should be limited and reserved for instances where no alternatives are available. Such use should be in agreement with the national legislation in force • Urgently prohibit their use as growth promotors
Prioritisation of diseases for which vaccines could reduce antimicrobial use in animals (2015/2018)	Infections where new or improved vaccines would significantly reduce the need for antibiotic use in: • Chickens, Swine, Fish (OIE 2015) • Cattle, Sheep, Goats (OIE 2018a)
OIE reports on use of antimicrobials in animals (2016, 2017, 2018b, 2019b)	First, Second, Third and Fourth OIE Annual report on the use of antimicrobial agents in animals: Big achievement: 1st report 130 Member Countries submitted completed reports—increase in 4th report 152 Member countries. Most of the countries cannot indicate the quantities of antimicrobial agents used by animal groups as defined by OIE, or the routes of administration, and cannot distinguish therapeutic use from use in growth promotion

5 United Nations General Assembly

In September 2016 political Declaration on AMR (Resolution A/RES/71/3) at the General Assembly of the UN committed on the high political level to follow One Health approach. It also reaffirms Blueprint AMR is GAP and call WHO, FAO, OIE to work together on finalisation of a global development and stewardship framework.

It also established Interagency Coordination Group (IACG) with the objectives to follow implementation of GAP and political Resolution of United Nations General Assembly (UNGA), and link to the Sustainable Development Goals (SDG). The United Nations, international agencies and experts released in April 2019 IACG report demanding immediate, coordinated and ambitious action to avert a potentially disastrous drug-resistance crisis. One of the purposes is the report consideration by 73rd session of the General Assembly in 2019 on the implementation of the political declaration and on further developments and recommendations emanating from the IACG, including on options to improve coordination, considering the 2015 Global Action Plan on Antimicrobial Resistance.

6 Global Health Security Agenda

It was launched in February 2014 as a 5-year initiative, growing partnership due the time course: 50 nations and international organisations as well as non-governmental stakeholders are involved. Global Health Security Agenda (GHSA) pursues a multilateral and multi-sectorial approach to strengthen both the global capacity and nations' capacity to prevent, detect and respond to human and animal infectious diseases threats whether naturally occurring or accidentally or deliberately spread. Strenghtening of sustained country-level health security capacity belong to one of the important target to be achieved.

Action Package AMR GHSA

- Target for 5 years period: Develop an integrated and global package of activities to combat AMR spanning human, animal, agricultural, food and environmental aspects.
- Evaluation parameter: Number of comprehensive plans to combat antimicrobial resistance agreed and implemented at a national level, and yearly reporting against progress towards implementation at the international level.
- Number of countries actively participating in a twinning framework, with countries agreeing to assist other countries:
  - *Leading countries*: Canada, Germany, Japan, the Netherlands, Sweden, the United Kingdom
  - *Contributing countries*: Australia, Bangladesh, Colombia, Cote d'Ivoire, India, Indonesia, Italy, Norway, Portugal, Saudi Arabia, South Africa, Switzerland, Thailand, the United States, Zimbabwe
  - *Contributing international organisations*: FAO, OIE, WHO

7 Transatlantic Task Force on AMR

It was established in 2009 in the United States and European Union Summit. TATFAR's technical experts from Canada, the European Union, Norway, and the United States collaborate and share best practices to strengthen domestic and global efforts. For example, members share technical guidance, alerts of emerging trends and scientific recommendations that can translate globally, increasing the impact of efforts to combat AMR exponentially. Collaborations have resulted in increased information exchange, understanding best approaches and development of peer relationships.

Three key areas (18 recommendations) were established and are available on the TATFAR webpage together with indication of implementer organisations:

- Improve appropriate therapeutic use of antimicrobial drugs in medical and veterinary communities
- Prevent healthcare- and community-associated drug-resistant infections
- Develop strategies for improving the pipeline of new antimicrobials

8 G7 and G20 Engagement on AMR and Declarations Released

The G7 has consistently committed to tackling global health challenges, including the fight against infectious diseases, and positioned itself as a leading partner in reaching health-related Millennium Development Goals. Antimicrobial resistance has been defined as one of the emerging areas.

Declaration by the G7 Health Ministers on AMR (Berlin, October 2015), officially recognises the various threats associated with AMR, both to health and to the economy, as well as the range of actions required to address the issue. The G7 Health Ministers called for a High Level Meeting on AMR in 2016 at the United Nations General Assembly to promote increased political awareness, engagement and leadership on antimicrobial resistance among Heads of States, Ministers and global leaders.

Declaration of G7 Agriculture Ministers (Niigata, 2016) within the commitment on improving of sustainable agricultural production/productivity and food supply capacity, combating AMR was also addressed. As Japan (chairing G7 in 2016) also prioritised AMR as a national and international health concern, G7 Ise-Shima Vision for Global Health was released, where the G7 leaders commit to take the following concrete actions for advancing global health:

- Reinforcing the Global Health Architecture to strengthen responses to public health emergencies
- Attaining United Health Care (UHC) with strong health systems and better preparedness for public health emergencies
- Antimicrobial resistance (AMR)
- Research and development and innovation

In 2017, a common approach of G7 CVOs on the definition of therapeutic, responsible and prudent use of antimicrobials in animals was released and after that Communiqué of G7 Health Ministers was released in November 2017 (Milano) with the key message of "United towards Global Health".

Declaration by the G20 Health Ministers and Agriculture Ministers (Berlin, 2017), the Action Plan of G20 Agriculture Ministers (2017) was approved including plan for new R&D Collaboration Hub. Further meeting in Buenos Aires (July 2018) discussed how to achieve a sustainable food future and in this context adopted a declaration, which supports the OIE in its activities, including tackling AMR. Mar del Plata (October 2018) meeting recalled to previous G20 activities and looked also to the future, supporting activities related to combating AMR.

9 Organisation for Economic Co-operation and Development

It has collaboration with G7, G20, EC. In 2018 it released the report—"Stemming the superbug Tide, *Just a few dollars more*" (OECD 2018), where based on new forecasts of the growth in

Table 3 Other organisations and partners of importance with regard to *AMR* as indicated by the WHO (2019) and organisations and partners of veterinary importance *non-exhaustive list* https://www.who.int/antimicrobial-resistance/Other_Organizations_and_Partners/en/

Antimicrobial Resistance Advocacy and Education, Standards
ESCMID (EUCAST/VETCAST): European Society of Clinical Microbiology and Infectious Diseases *CLSI*: Clinical Laboratory Standards Institute *APUA*: Alliance for the Prudent Use of Antibiotics *AWARE*: Alliance Working for Antibiotic Resistance Education *CDC*: US Centers for Disease Control and Prevention (Antibiotic Page) *EAAD*: European Antibiotic Awareness Day *ReACT*: Action on Antibiotic Resistance
Antimicrobial Resistance Surveillance
ANSORP: Asian Network for Surveillance of Resistant Pathogens *AURA*: Antibiotic Use and Resistance in Australia *ECDC*: European Centre for Disease Control *CIPARS*: Canadian Integrated Program for Antimicrobial Resistance Surveillance *EARS-Net*: European Antimicrobial Resistance Surveillance Network *EFSA*: Harmonised monitoring of AMR in Zoonotic, Indicator and Commensal Bacteria *ICARE*: Intensive Care Antimicrobial Resistance *JAMRAI-WP 4.7.2*: Planned EARS-Vet (target pathogens) *NARMS*: National Antimicrobial Resistance Monitoring System (USA)
Antimicrobial use
ESAC: European Surveillance of Antimicrobial Consumptions *INRUD*: International Network for the Rational Use of Drugs *ESVAC*: European Surveillance of Veterinary Antimicrobial Consumptions *OIE*: Annual reports on the use of antimicrobial agents in animals *AACTING*: Network on quantification of veterinary *A*ntimicrobial usage at herd level and *A*nalysis, *C*ommunica*T*ion and benchmark*ING* to improve responsible usage
Other veterinary organisations/activities
Partners in Tripartite: FAO/OIE *EEFORT*: Ecology from Farm to Fork of microbial drug Resistance and Transmission *EPRUMA*: The European Platform for the Responsible Using of Medicines in Animals *FVE*: Federation of Veterinarian of Europe *VICH*: International Cooperation on Harmonisation of technical requirement for Registration of Veterinary Medicinal Products *Animal Health*: Association representing manufacturers of animal medicines, vaccines and other animal health products

resistance to 2050, outlines how countries can tackle AMR and significantly reduce the personal and economic costs. The report considers data from EARS-Net (human medicine). As the issue of AMR is an issue of great importance and complexity it can be hardly achieved exhaustive list and description of all activities worldwide. Despite that fact Table 3 brings a set of examples of further international bodies, organisations and partners as indicated by WHO (2019).

10 Conclusion

During last years, a considerable amount of documents have been published on antimicrobial resistance by global organisations, national governments, consortia and private sector bodies.

Long time ago it has been recognised that the benefits of antimicrobials use was connected also with significant risks and despite this knowledge, progress worldwide appears to be slow.

In agricultural sector, throughout the time, two main drivers appeared to slow down responsible use policies and approaches. Those two most important drivers seem to be that many countries have historically favoured reliable access to cheap meat over broader agricultural and antibiotic reform and of great importance is, and historically was, ask for proteins ad satisfaction of the world market with products of animal origin (Kirchhelle 2018). Both of the above-mentioned drivers lead to pressure for more and more intensive farming and cause economical pressure. Also this made responsible use of antimicrobials in individual countries more difficult to be performed, as their agricultural sectors, are faced with cheaper imports of product from countries still using antimicrobials as growth promotors.

Considering the price of antimicrobials and reactive approach in comparison to other "alternative" proactive approaches, which all seem to be more costly and in many cases still with more uncertain effect comparing to antimicrobials, it is difficult to change the world.

To keep a little bit more optimism, it should be noted that despite no "revolution" in use of antimicrobials after the release of all the above-mentioned documents, there are steps that show improvement [e.g. ban of the use of medically important antimicrobials as growth promoters, or at least ban of such use in critically important antimicrobials that have been currently announced by countries with biggest food production (e.g. ban of colistin as feed additive in China, medically important antimicrobials banned as growth promoters in the United States)].

The adoption of the WHO Global Action plan prepared in collaboration with FAO and OIE was an important step in the fight against antimicrobial resistance and has contributed to maintain antimicrobial resistance on the political agenda.

The recent report of the IACG "No time to wait—securing the future from Drug resistant infections" makes proposals to tackle the problem globally and to link antimicrobial resistance to the sustainable development goals including in health, food security, clean water and sanitation, responsible consumption and production, and poverty and inequality.

Dealing with antimicrobial resistance and taking into account at the same time the real background issues such as production of cheap products is an important point to make real progress.

References

AMR MPTF (2019) AMR multi-partner trust fund: new multi-partner trust fund launched to combat antimicrobial resistance globally. http://www.oie.int/en/for-the-media/press-releases/detail/article/new-multi-partner-trust-fund-launched-to-combat-antimicrobial-resistance-globally/ and http://mptf.undp.org/. Accessed 26 June 2019

Harrison R (1964) Animal machines: the new factory farming industry. Vincent Stuart, London

Kirchhelle C (2016) Toxic confusion: the dilemma of antibiotic regulation in West German food production (1951–1990). Endeavour 40(2):114–127

Kirchhelle C (2018) Pharming animals: a global history of antibiotics in food production (1935–2017). Palgrave Communications, 4 Paper 96. https://doi.org/10.1057/s41599-018-0152-2

OECD (2018) Stemming the superbug tide: just a few dollars more. http://www.oecd.org/health/health-systems/Stemming-the-Superbug-Tide-Policy-Brief-2018.pdf. Accessed 26 June 2019

OIE (2015) Report of the meeting of OIE *ad hoc* group on prioritisation of diseases for which vaccines could reduce antimicrobial use in animals. http://www.oie.int//fileadmin/SST/adhocreports/Diseases%20for%20which%20Vaccines%20could%20reduce%20Antimicrobial%20Use/AN/AHG_AMUR_Vaccines_Apr2015.pdf. Accessed 26 June 2019

OIE (2016) World Organisation for animal health: OIE annual report on the use of antimicrobial agents in animals. First Report. http://www.oie.int/fileadmin/Home/eng/Our_scientific_expertise/docs/pdf/AMR/Survey_on_monitoring_antimicrobial_agents_Dec2016.pdf. Accessed 26 June 2019

OIE (2017) World Organisation for animal health: OIE annual report on the use of antimicrobial agents in animals. Second Report. http://www.oie.int/fileadmin/Home/eng/Our_scientific_expertise/docs/pdf/AMR/Annual_Report_AMR_2.pdf. Accessed 4 June 2019

OIE (2018a) Report of the meeting of OIE *ad hoc* group on prioritisation of diseases for which vaccines could reduce antimicrobial use in cattle, sheep and goats. http://www.oie.int//fileadmin/SST/adhocreports/Diseases%20for%20which%20Vaccines%20could%20reduce%20Antimicrobial%20Use/AN/AHG_AMUR_Vaccines_ruminants_May2018.pdf. Accessed 26 June 2019

OIE (2018b) World Organisation for animal health: OIE annual report on the use of antimicrobial agents in animals. Third Report. http://www.oie.int/fileadmin/Home/eng/Our_scientific_expertise/docs/pdf/AMR/Annual_Report_AMR_3.pdf. Accessed 4 June 2019

OIE (2019a) World Organisation for animal health: OIE list of antimicrobial agents of veterinary importance. http://www.oie.int/fileadmin/Home/eng/Our_scientific_expertise/docs/pdf/AMR/A_OIE_List_antimicrobials_June2019.pdf. Accessed 26 June 2019

OIE (2019b) World Organisation for Animal Health: OIE Annual report on the use of antimicrobial agents in animals. Fourth Report. https://www.oie.int/fileadmin/Home/eng/Our_scientific_expertise/docs/pdf/A_Fourth_Annual_Report_AMU.pdf Accessed 30 May 2020

Sample IAN (2013) In Kirchhelle 2018, antibiotic-resistant diseases pose 'apocalyptic threat', top expert says. Guardian [Online], 23 Jan 2013

Smith-Howard K (2017) Healing animals in an antibiotic age: veterinary drugs and the professionalism crisis, 1945–1970. Technol Cult 58:722–748

Swann (1969) Report of the joint committee on the use of antibiotics in animal husbandry and veterinary medicine, 1969–1970. HMSO, London

Thoms U (2017) Antibiotika, agrarwirtschaft und politik in Deutschland im 20. und 21. Jahrhundert. Z für Agrargesch und Agrarsoziol 65:35–52

WHO (2014a) World Health Organization: 67th World Health Assembly

WHO (2014b) World Health Organization: antimicrobial resistance: global report on surveillance. ISBN: 978-92-4-156474-8. https://apps.who.int/iris/bitstream/handle/10665/112642/9789241564748_eng.pdf?sequence=1 Accessed 26 June 2019

WHO (2015a) World Health Organization: 68th World Health Assembly

WHO (2015b) World Health Organization: global antimicrobial resistance surveillance system manual for early implementation

WHO (2015c) World Health Organization: about antimicrobial resistance for policy makers. https://www.who.int/antimicrobial-resistance/policy-package-july2016.pdf?ua=1. Accessed 26 June 2019

WHO (2016) World Health Organization: 69th World Health Assembly

WHO (2017a) WHO guideline on use of medically important antimicrobials in food producing animals. https://apps.who.int/iris/bitstream/handle/10665/258970/9789241550130-eng.pdf;jsessionid=F5EB043E153C22C0190139C171FFF8A3?sequence=1. Accessed 28 June 2019

WHO (2017b) World Health Organization: global framework for development & stewardship to combat AMR. https://www.who.int/phi/implementation/research/WHA_BackgroundPaper-AGlobalFrameworkDevelopmentStewardship.pdf?ua=1. Accessed 26 June 2019

WHO (2018a) World Health Organization: tackling antimicrobial resistance (AMR) together. Working paper 1.0: Multisectoral coordination. (WHO/HWSI/AMR/2018.2), Geneva

WHO (2018b) World Health Organization: assessing entry points and options for increasing investments in AMR in low- and middle-income countries. https://www.who.int/antimicrobial-resistance/national-action-plans/countrycasestudiesinvestmentsinAMRincountries/en/. Accessed 26 June 2019

WHO (2018c) World Health Organization: tackling antimicrobial resistance (AMR) together. Working paper 5.0: enhancing the focus on gender and equity. (WHO/HWSI/AMR/2018.3), Geneva

WHO (2018d) World Health Organisation: WHO list of critically important antimicrobials 6th revision. http://who.int/foodsafety/cia/en. Accessed 26 June 2019

WHO (2019) World Health Organization: turning plans into action for antimicrobial resistance (AMR). Working paper 2.0: implementation and coordination. (WHO/WSI/AMR/2019.2), Geneva. Licence: CC BY-NC-SA 3.0 IGO

WHO/FAO/OIE (2018) World Health Organization (WHO), Food and Agriculture Organization of the United Nations (FAO) and World Organisation for Animal Health (OIE): Monitoring Global Progress on Addressing AMR. ISBN: 978-92-4-151442-2

WHO (2019a) Executive Board 144 session: Follow-up to the high-level meetings of the United Nations general assembly on health-related issues – antimicrobial resistance - report by the director-general. https://apps.who.int/gb/ebwha/pdf_files/EB144/B144_19-en.pdf. Accessed 30 May 2020

WHO (2019b) 72nd WHA session: access to medicines and vaccines - report by the director-general. https://apps.who.int/gb/ebwha/pdf_files/WHA72/A72_17-en.pdf?ua=1. Accessed 30 May 2020

WHO/FAO/OIE (2019a) World Health Organization (WHO), Food and Agriculture Organization of the United Nations (FAO) and World Organisation for Animal Health (OIE): Tripartite Monitoring and Evaluation (M&E) framework for the Global Action Plan on Antimicrobial Resistance

WHO/FAO/OIE (2019b) World Health Organization (WHO), Food and Agriculture Organization of the United Nations (FAO) and World Organisation for Animal Health (OIE): Memorandum of understanding between FAO, WHO and OIE regarding cooperation to combat health risks at the Animal – Human – Ecosystems interface in the context of the "One Health" approach and including antimicrobial resistance. https://www.who.int/zoonoses/MoU-Tripartite-May-2018.pdf?ua=1. Accessed 28 June 2019

EU Policies and Regulatory Surroundings

Lucie Pokludová and Jiří Bureš

Abstract

European Union plays an active role in the fight against antimicrobial resistance both in human and veterinary medicine. Intensity of activities supported by the legal provisions, soft law, political declarations and expressed commitments, regulatory requirements, growing body of scientific evidence as well as business policies and models has been increasing. European Commission and its decentralised agencies: European Medicine Agency, European Food Safety Authority and European Centre for Disease Prevention and Control cooperate closely in sake of "One Health concept". Active roles play not only these EU public bodies and public bodies in the Member States, but also academia and actors in various areas of the private sector. New regulation on Veterinary Medicinal Products, together with previously approved "Animal Health Law" and regulation on Medicated Feed will to the large extent determine how veterinary medicinal products will be authorised, prescribed, used and controlled in the EU. The newly adopted rules include a number of positive elements with the potential to contribute to containment of antimicrobial resistance in the EU, but also contain a number of pitfalls, which can significantly curtail the positive potential of the new legal settings. The final outcome will depend on a number of additional measures and policies implemented both in the EU and at the global scene. From the perspective of policy documents three documents of high importance have to be mentioned. First one, most political—Council conclusions on the next steps under a One Health approach to combat antimicrobial resistance has been adopted in the mid of 2016 and, followed up by another one from Council conclusions from June 2019. Following the evaluation of the European Action Plan 2011–2016, new European Action Plan has been released in June 2017. Third document, Prudent Use Guideline has been adopted in 2015 and creates the platform highlighting the role of different stakeholders, what seems key for any effective action. Reports targeted on issues related to antimicrobial resistance and use of antimicrobials were released (JIACRA II, RONAFA). It should be noted that a load of everyday practical work rests on the shoulders of veterinarians and farmers that will be driven by the national strategies respecting the local conditions.

Keywords

Prudent use guideline · European action plan · National policies · JIACRA reports · RONAFA report · European Commission ·

L. Pokludová (✉) · J. Bureš
Institute for State Control of Veterinary Biologicals and Medicines, Brno, Czech Republic
e-mail: pokludova@uskvbl.cz; bures@uskvbl.cz

L. Pokludová (ed.), *Antimicrobials in Livestock 1: Regulation, Science, Practice*,
https://doi.org/10.1007/978-3-030-46721-0_3

European Medicine Agency · European Food Safety Authority

1 Introduction

The European Union, especially during the last two decades, has started to play more and more active role in the fight against antimicrobial resistance both in human and veterinary medicine. Within the European region, Nordic countries were pioneers with respect to the responsible use of antimicrobials. They considered antimicrobial resistance as risks and gradually set national antibiotic policies with exact measures and legal provisions coming into practice. Those countries can be considered as the area of the world with the lowest antimicrobials use. Successful measures like switch from use of antimicrobials to vaccination since 1994 that fish farmers across Norway had made as well as the Swedish approach, where antimicrobials growth promoters were banned in 1986 (in Finland in 1996) gave the positive examples on how to tackle the use of antimicrobials in the most responsible way. In 1998 The Microbial Threat conference in Copenhagen started the period of actions.

During the time course more countries in the EEA set antibiotic programmes and policies and adopted national (legal) rules to use antimicrobials more responsibly (e.g. Denmark: robust system of collection and analysing the data on the use of antimicrobials and benchmarking with "yellow cards" for those exceeding the threshold, pioneering in ban of the antimicrobial growth promoters; the Czech Republic: National Antibiotic Programme (join human and veterinary medicine) since 1995 and due to the public health concerns establishment of "prudent use regimen" (responsible use warning, including antimicrobial susceptibility testing) introduced into summary of product characteristics of (fluoro)quinolones, cephalosporins third and fourth generations, rifaximin and aminoglycosides of high generations (amikacin, gentamicin kanamycin) and enforcement of prudent use by National Decree in 2008).

2 Political Commitments to Recognise the Issue and Fight Against AMR

Common agreement among Member States at the Council level as well as the effort of the European Commission led to political commitments (Table 1), that were initially mainly targeted on human sector, and at later stages the veterinary sector was also covered. The European Commission, together with regulatory bodies such as European Medicine Agency, European Food Safety Authority and European Centre for Disease Prevention and Control were active in establishment of several important activities as listed in Table 2.

It should be noted that, in the EU, actions tackling AMR are more harmonised and also more regulated in veterinary area compared to human medicine. They are, in many cases, based on legal provisions making them obligatory. This is not the case in human medicine, mainly due to Article 168 TFEU. Under this Treaty of Lisbon, human health is a policy area where the Union supports, complements or supplements the actions of the Member States (Article 6 TFEU). However, common safety concerns in public health matters are an area where competence is shared between the Union and the Member States (Article 4 TFEU). The dual nature of the competences in the area of public health is reflected in the different types of measures that the EU can take under Article 168 TFEU.

Within the second EU Action Plan ambitions increased to willingness of making the European Union the best practice region with respect to responsible use of antimicrobials. The strong message was also released to the individual Member States to set their national action plans that should be tailored for the local conditions, considering not only types of animal husbandries, but also national socio-economic surrounding. Positive information is that some of the Member States in the past issued either national action plans and/or worked on their own strategies, guidelines and campaigns to promote responsible use of antimicrobials leading to drop down of the

Table 1 Political commitments targeted on combating AMR in the European Union (Council conclusions, Council Recommendations, European Parliament Resolutions adopted since 2009)

Date	Document type	Document
2009	Council Recommendation	Council Recommendation of 9 June 2009 on patient safety, including the prevention and control of healthcare associated infections *OJ C 151, 3.7.2009, p. 1–6*
2011	EP Resolution	Antibiotic resistance European Parliament resolution of 12 May 2011 on antibiotic resistance *OJ C 377E, 7.12.2012, p. 131–135*
2011	EP Resolution	European Parliament resolution of 27 October 2011 on the public health threat of antimicrobial resistance
2012	EP Resolution	European Parliament resolution of 11 December 2012 on the Microbial Challenge—Rising threats from Antimicrobial Resistance *OJ C 434, 23.12.2015, p. 49–58*
2012	Council conclusion	Council conclusions of 22 June 2012 on the impact of antimicrobial resistance in the human health sector and in the veterinary sector—a "One Health" perspective *OJ C 211, 18.7.2012, p. 2–5*
2015	EP Resolution	European Parliament resolution of 19 May 2015 on safer healthcare in Europe: improving patient safety and fighting antimicrobial resistance
2016	Council conclusion	Council conclusions on the next steps under a One Health approach to combat antimicrobial resistance *OJ C 269, 23.7.2016, p. 26–30*
2018	EP Resolution	European Parliament resolution of 13 September 2018 on a European One Health Action Plan against Antimicrobial Resistance (AMR)
2019	Council conclusion	Council conclusions on the next steps towards making the EU a best practice region in combatting antimicrobial resistance *OJ C 214.1, 25.6. 2019, p. 1–7*

antimicrobial consumption (well documented, e.g. in the ESVAC reports for individual Member States during the time course). Thanks to activities of the individual Member States and involved stakeholders, Europe can be considered as an area with advanced, responsible approach to antimicrobials.

3 Key Policy and Scientific Bodies to Tackle AMR at EU Level

Except the policy/decision maker bodies (European Commission, European Council and European Parliament), that create provisions within the European legal framework, the importance of scientific and regulatory bodies (European Medicine Agency—EMA, European Food Safety Authority—EFSA, European Centre for Disease Prevention and Control—ECDC) that can analyse scientific, surveillance and other evidence-based data both from human and veterinary area, became the key elements of the last decade. Performing such analyses arising due to mandate or request issued by the Commission revealed many gaps still waiting for satisfactory response by scientists—but in close cooperation with people from everyday practice to make the measures recommended more pragmatic. Currently the European Economic Area and their Member States have been able, thanks to the different surveillance/monitoring systems, to achieve huge amount of the more or less harmonised data. Sets of data related to antimicrobial consumption, both in human (ESAC-net) and animals (ESVAC) as well as data on antimicrobial resistance in human (EARSS-net data on AMR from bacterial isolates coming from serious infections—hospitals) and on veterinary side (data on AMR in zoonotic, indicator and commensal bacteria isolated from animals and food—lead by EFSA) are collected and analysed. On the other hand, data on AMR of target veterinary pathogens are not collected in harmonised manner and only a few European countries have, with certain level of complexity, such data available (Schrijver et al. 2017). The most difficult issue is how, from such available data, sufficiently robust

Table 2 EU agencies and their selected projects related to antimicrobials (mainly from vet perspective)

EU agencies	Activity	Summary
EMA–EFSA–ECDC–SCENHIR	Joint opinion AMR zoonotic 2009 ECDC/EFSA/EMA/SCENHIR (2009)	Joint Opinion on antimicrobial resistance (AMR) focused on zoonotic infections Request from the Commission for: Opinion on antimicrobial resistance in foodborne zoonotic agents (*Campylobacter*, *Salmonella*, *E. coli*, MRSA), on extent of use of antimicrobials and biocides, on emerging risks, multidrug resistance, on the data gaps and needs for further research, innovation and EU surveillance(s)
EMA–EFSA–ECDC	JIACRA JIACRA I 2011–12 JIACRA II 2013–15 (ECDC/EFSA/EMA (2015) ECDC/EFSA BIOHAZ/CVMP (2017)	Joint Interagency Antimicrobial Consumption and Resistance Analysis: Request from the Commission for: • Integrated analysis of the consumption of antimicrobials and occurrence of antimicrobial resistance in bacteria of human and food-producing animals origin • Analysis of antimicrobial use and resistance from human and food-producing animals
	RONAFA (EMA and EFSA 2017)	Reduction of the Need for Antimicrobials in Food animals and Alternatives: containing the reviews on: • Measures that have been taken by MSs to reduce the use of antimicrobials in food-producing animals • Alternatives to the use of antimicrobials • Impact assessment of the measures and alternatives on the antimicrobial resistance occurrence • Conclusion on recommended options to reduce use and for responsible use of antimicrobials
	Outcome indicators (ECDC/EFSA/EMA 2017)	List of harmonised outcome indicators to assess progress in reduction of antimicrobial use and resistance: • Indicators for human and animal sectors • Indicators for AM consumption and resistance • Indicators for community, in hospitals and in food-producing animals
EMA	CVMP strategy on antimicrobials for 2016–2020 (EMA 2016)	As a key part of the strategy CVMP aims to ensure the availability of effective antimicrobial medicines for the treatment of important infectious diseases of animals, while minimising the risks to animals or humans arising from their use
	Authorisation of VMPs	One of the key elements of the above commented Strategy is the revised CVMP guideline for the demonstration of efficacy for VMPs containing antimicrobials Also further scientific guidelines (e.g. revision of guideline on Summary of Product Characteristics of antimicrobial VMPs; safety, residue and efficacy guidelines) are of importance
	Referrals	One of the powerful legal tools for revisiting of (not only) antimicrobial VMPs, when risk/s for human, animal health or environment are identified
	ESVAC	European Surveillance of Veterinary Antimicrobials Consumption based on data on sales of antimicrobials in EEA states in relation to population/biomass of animals
	AMEG	Antimicrobial Advice ad hoc Expert Group: • Scientific advice on the impact on public health and animal health of the use of antibiotics in animals • Categorisation of antimicrobials reflecting: – The need for the antimicrobials in human medicine in the EU—sole therapy or few alternatives – Probability of the transfer of resistance from bacteria in animals to human (mechanisms of resistance, mobile genetic elements transfer, possibility of food-borne transmission)

(continued)

Table 2 (continued)

EU agencies	Activity	Summary
EFSA	BIOHAZ on AMR	Panel on Biological Hazards (e.g. harmonised monitoring of AMR in animals and food)
	AHAW	Animal Health and Welfare Panel (help with elaboration of RONAFA report with examples of farm health programmes or disease eradication programmes)
	FEEDAP	Additives and Products or Substances used in Animal Feed (help with elaboration of RONAFA report: examples of alternatives—probiotics, prebiotics, minerals and other possible feed additives)
	Other panels	Contaminants in the food chain (including residues of VMPs (antimicrobials) GMO Plant Protection product and their residues

proof of evidence on the possible correlation among the consumption and resistance data and on possible cross links among the human and veterinary medicine could be gained (please refer to Think About Box). Despite the difficulties certain analyses have already been performed as, e.g. the JIACRA I and II reports (ECDC/EFSA/EMA 2015, 2017).

Key regulatory/scientific bodies in the EU that are involved in tackling AMR with a main focus on veterinary side:

Many tasks of EMA covering both the human and veterinary medicines sectors are linked to antimicrobial resistance and are legally based on the Directives 2001/83/EC (human) and 2001/82/EC (veterinary, currently replaced by the regulation on VMPs) and also on the Regulation 726/2004 (currently updated by Regulation (EU) 2019/5). Some of the activities that are of importance are listed in Table 2. In veterinary medicine, EMA, through its scientific committee for veterinary medicinal products (CVMP), has been setting the regulatory guidelines for authorisation of antimicrobial veterinary medicinal products and their post-authorisation management, authorises veterinary medicinal products by means of centralised marketing authorisation procedure, provides scientific advice and/or opinions on "benefit:risk balance" issues. Especially of importance is that at the time of granting of the marketing authorisation, the veterinary medicinal product has positive benefit:risk. It in fact means that any veterinary medicinal product for which serious risks for human and animal health or the environment are identified must not be authorised. Issues related to antimicrobial resistance due to use of veterinary medicinal products as well as consumer safety considering residues of veterinary medicinal product is thoroughly assessed. EMA has also been putting together data collected by the Member States on the sales of veterinary antimicrobials in the European Economic Area. One of the main projects with the great achievement—European Surveillance of Veterinary Antimicrobial Consumption—was founded in 2009 (meeting in Marienbad under the Czech Republic Presidency in the Council of the EU). Due to the big effort of both the Member States and coordination role of EMA practically full coverage of the EEA (30 Member States/2018 with data 2016) on consumption of veterinary medicinal products containing antimicrobials has been attained. By means of this system, continuous drop of sales of antimicrobial veterinary medicines in the EEA could be evidenced.

As for EFSA, the body targeted on food and feed safety, several scientific panels are working and having their regulatory as well as scientific role (the most linked with the AMR is the panel on Biological hazards (BIOHAZ)), which lead, amongst the others, project of AMR detection in zoonotic, indicator and commensal bacteria from animals and food. In the broader sense, the activities of the other panels are also directly or indirectly linked with antimicrobial resistance. As BIOHAZ panel works on the issue of zoonosis as well as antimicrobial resistance; it has also close cooperation, in line with One Health concept, with another important European body—European Centre for Disease Prevention and Control. The examples of this cooperation are reports JIACRA I and II (see Table 2).

Except BIOHAZ two further panels also cooperated on elaboration of the RONAFA report. Keeping animals healthy and ensuring their well-being can minimise the need for use of antimicrobials, therefore several examples were provided by Animal Health and Welfare panel (AHAW). Under the scope of the panel "Additives and Products or Substances used in Animal Feed" (FEEDAP) belong, e.g. evaluation and approval of the feed additive microorganisms with the function of beneficial probiotic microbes. One of such assessments of FEEDAP, except the other tasks, is targeted on the proof that those viable microorganisms used as the active agent (s) in feed additives not introduce any additional resistance genes to the pool of antimicrobial resistance genes already present in the gut bacterial population or otherwise increase the risk of transfer of drug resistance.

Also other panels either play or can play the role related to antimicrobial resistance:

Panel on Contaminants in the Food Chain (tackling the issue of residues including residues of veterinary medicinal product containing antimicrobials in food of animal origin)

Panel on Genetically Modified Organisms (dealing with a controversial issue of genes of antimicrobial resistance—neomycin, kanamycin, streptomycin as a marker of GMO; statement on the risk posed to humans by vitamin B2 produced by a genetically modified strain of *Bacillus subtilis* and assessment on AMR gene transfer to human-associated bacteria) (EFSA 2019a)

Panel on Plant Protection Products and their Residues (new assessment of glyphosate, a substance originally invented as antimicrobial with potential for resistance co-selection, is foreseen to start in December 2019. The assessment of the application dossier and preparation a draft renewal assessment report to be reviewed by EFSA in 2021) (EFSA 2018b)

4 Recent Legislative Tackling the Issues of Antimicrobial Resistance

Legal restrictions were established in the EU to avoid misuse and overuse of antimicrobials in animals. One of the first was Regulation (EC) No 1831/2003 (Article 11), in which it is stipulated that since 1 January 2006, the use of antimicrobial growth promoters as feed additives has been banned.

As having the biggest impact for at least nearest future decade can be considered Regulation EU (2019/6) on veterinary medicinal products (VMPs), which lays down rules for the placing on the market, manufacturing, import, export, supply, distribution, pharmacovigilance, control and use of veterinary medicinal products, including those containing antimicrobials. For those antimicrobial veterinary medicinal products many parts of preamble as well as body of the regulation stipulates specific rules (please refer Tables 3 and 4). There should be highlighted some parts of this new regulation that can be considered as the breakpoints with respect to use of antimicrobials containing VMPs in practice. Therefore, special attention is to be paid to Article 57 and new obligation of collection and providing by the Member States of the data on *sales AND use* of antimicrobials in animals to create European statistics related to more precise antimicrobial consumption per animal species that are aimed to be compared especially with the antimicrobial resistance data in those animal species. Principal importance of Article 105 is on veterinary prescription including rules for antimicrobials as well as Article 107 with specific, newly much more restrictive, rules for use of antimicrobial VMPs, especially in relation to minimising prophylactic and metaphylactic use (please refer below to highlighted parts of Table 4). There is also emphasised and kept the ban of antimicrobials for promoting growth or to increase yield. Despite the fact that many of the Articles of this Regulation tackling directly an issue of antimicrobials, their use, authorisation, prescription, consumption rates etc., there are also some articles tackling different areas of VMPs

Table 3 Regulation (EU) 2019/6 on VMPs—Preamble points related directly to antimicrobials (modified according Regulation (EU)2019/6)

Preamble	
(14)	Avoiding cross-contamination—reduce selection of antimicrobial resistance (when used orally veterinary medicinal products (VMPs) administered via drinking water or feed)
(25)	Special care to be taken, when prescribed antimicrobials under the "cascade" principles
(33)	Innovative industry data to be protected for certain period: • Especially in products for minor use and antimicrobials, with aim to stimulate research and innovation within these areas and help to ensure availability • Similar protection of investments should be applied to studies supporting a new pharmaceutical form, administration route or dosage that reduces the antimicrobial or antiparasitic resistance or improves the benefit–risk balance
(41)	Declaring antimicrobial resistance as Union and worldwide problem: • Complexity of the problem, its cross-border dimension and the high economic burden: its impact goes beyond its severe consequences for human and animal health and has become a global public health concern that affects the whole of society and requires urgent and coordinated inter-sectoral action in accordance with the *"One Health" approach* • Actions as strengthening of the prudent use of antimicrobials, avoiding their routine prophylactic and metaphylactic use • Actions to restrict the use in animals of antimicrobials that are of critical importance for preventing or treating life-threatening infections in humans • Encouraging and incentivising the development of new antimicrobials • Include appropriate warnings and guidance on the labels of veterinary antimicrobials • Restriction of use that is not covered by the terms of the marketing authorisation of certain new or critically important antimicrobials for humans • Advertising rules for veterinary antimicrobials should be tightened • Authorisation requirements should sufficiently address the risks and benefits of antimicrobial veterinary medicinal products
(42)	Risk mitigation for HMPs/VMPs: • Application for an antimicrobial VMP should contain information about the potential risks that use of that medicinal product may lead to the development of antimicrobial resistance in humans or animals or in organisms associated with them • Antimicrobial VMPs only to be authorised with careful scientific benefit–risk assessment • Restrictive conditions on the use of the VMP, if necessary (restrictions on the use of the VMP outside of the terms of authorisation, especially not in line with the summary of product characteristics)
(43)	The combined use of several antimicrobial active substances may represent a particular risk with respect to the development of antimicrobial resistance: • Need for specific assessing whether to authorise a VMP with such combination
(44)	The development of new antimicrobials has not kept pace with the increase of resistance to existing antimicrobials • Essentiality of maintenance of existing antimicrobials due to limited innovation in developing new antimicrobials • Use of antimicrobials in animals may accelerate the emergence and spread of resistant microorganisms and may compromise the effective use of the already limited number of existing antimicrobials to treat human infections • Misuse of antimicrobials should not be allowed • *Antimicrobials—prophylaxis* only in well-defined cases for the administration to an individual animal or restricted number of animals when the risk for infection is very high or its consequences are likely to be severe • *Antibiotics—prophylaxis* only in exceptional cases only for the administration to an individual animal • *Antimicrobials—metaphylaxis* only when the risk of spread of an infection or of an infectious disease in a group of animals is high and no appropriate alternatives available • Such restrictions should allow the decrease of prophylactic and metaphylactic use in animals towards representing a smaller proportion of the total use of antimicrobials
(45)	Restriction/prohibition of the use of antimicrobials essential for treatment of infections in humans • Member States should be allowed, therefore, following scientific recommendations, to define restrictive conditions for their use—e.g. conditioning their prescription to test antimicrobial susceptibility to ensure that there are no other antimicrobials available that are sufficiently effective or appropriate to treat diagnosed disease

(continued)

Table 3 (continued)

Preamble	
(46)	In order to preserve as long as possible the efficacy of certain antimicrobials in the treatment of infections in humans, it may be necessary to reserve those antimicrobials for humans only: • Decision that certain antimicrobials should not be available on the market in the veterinary sector • Such decision following the scientific recommendations of the EMA, EFSA and other relevant EU agencies and also take into account any relevant recommendations from international organisations, such as WHO, OIE and the Codex Alimentarius
(47)	If an antimicrobial is administered or used incorrectly, this presents a risk to public or animal health. Antimicrobial VMPs should only be available on veterinary prescription. Veterinarians have a key role in ensuring prudent use of antimicrobials and consequently they should prescribe the antimicrobial medicinal products based on: • Knowledge of antimicrobial resistance • Epidemiological and clinical knowledge • Understanding of the risk factors for the individual animal or group of animals • Respecting their professional code of conduct • Avoiding conflict of interest when prescribing medicinal products, while recognising their legitimate activity of retail in accordance with national law • Independency on economic incentives (direct/indirect) when prescribing those medicinal products • VMPs supply restrictions to the amount required for treatment of the animals under the care of the respective veterinarian
(48)	The prudent use of antimicrobials as key in addressing antimicrobial resistance: • All the stakeholders to promote prudent use of antimicrobials • Guidance on the prudent use of antimicrobials in veterinary medicine be taken into account and further elaborated • Identification of risk factors and the development of criteria for the initiation of administration of antimicrobials, as well as the identification of alternative measures, could help in avoiding the unnecessary use of antimicrobial medicinal products, including through metaphylaxis Member States should be allowed to take further restrictive measures to implement national policy on the prudent use of antimicrobials, provided that those measures do not unduly restrict the functioning of the internal market
(49)	Importance of international dimension of the development of antimicrobial resistance when assessing the benefit–risk balance of certain veterinary antimicrobials in the EU • Measures restricting the use of antimicrobials based on scientific advice and considered in the context of cooperation with third countries and international organisations • Non-discriminatory and proportionate manner respecting certain basic conditions relating to antimicrobial resistance for animals and products of animal origin exported to the Union by the operators from third countries • Relevant international agreements should be reflected • Contribution of above mentioned to the international fight against antimicrobial resistance, in particular in line with the WHO Global Action Plan and the OIE Strategy on Antimicrobial Resistance and the Prudent Use of Antimicrobials
(50)	There is still a lack of sufficiently detailed and comparable data at the EU level to: • Determine the trends • Identify possible risk factors that could lead to the development of measures to limit the risk from antimicrobial resistance and to monitor the effect of measures already introduced It is therefore important to: • Continue the collection of *data on sales of antimicrobials* • Further develop it in line with a stepwise approach collection of *antimicrobial use data*, under conditions of agreed technical rules, with responsibility of Member States for collecting and Agency responsible for coordination. Reliability and validity of the data should be ensured That data, when available, should be analysed with data on the use of antimicrobials in humans and data on antimicrobial resistant organisms found in animals, humans and food

Table 4 Regulation (EU) 2019/6 on VMPs—Articles directly mentioning[a] rules for antimicrobials (modified according Regulation (EU)2019/6)

Articles	
(4)	Definition of "antimicrobial resistance" as the ability of microorganisms to survive or to grow in the presence of a concentration of an antimicrobial agent which is usually sufficient to inhibit or kill microorganisms of the same species Definition of "antimicrobial" as any substance with a direct action on microorganisms used for treatment or prevention of infections or infectious diseases, including antibiotics, antivirals, antifungals and antiprotozoals
(8)	Data to be submitted with the application for marketing authorisation in the case of antimicrobial VMP, in addition to the information, technical documentation and summary of pharmacovigilance master file: (a) Documentation on the direct or indirect risks to public or animal health or to the environment of use of the antimicrobial VMP in animals (b) Information about risk mitigation measures to limit antimicrobial resistance development related to the use of the VMP
(34)	Antimicrobial VMP classified as subject to veterinary prescription
(35)	Specific parts related to AMR to be mentioned in the Summary of product characteristic of the antimicrobial VMPs: • Special conditions for use, including restrictions on the use of antimicrobial and antiparasitic VMP in order to limit the risk of development of resistance
(36)	Where the application concerns an antimicrobial veterinary medicinal product, the competent authority or the Commission, as applicable, may require the marketing authorisation holder to conduct post-authorisation studies in order to ensure that the benefit–risk balance remains positive given the potential development of antimicrobial resistance
(37)	Marketing authorisation can be refused where: • The application concerns an antimicrobial VMP presented for use as performance enhancer in order to promote the growth of treated animals or to increase yields from treated animals • The risk for public health in case of development of antimicrobial resistance or antiparasitic resistance outweighs the benefits of the veterinary medicinal product to animal health • If the antimicrobial contained on the VMP of concern is reserved for treatment of certain infections in humans as will be stipulated by implementing acts that designate antimicrobials or groups of antimicrobials reserved for treatment of certain infections in humans, to preserve their efficacy
(39)	Fourteen year period of protection of the technical documentation for antimicrobial VMPs for cattle, sheep for meat production, pigs, chickens, dogs and cats containing an antimicrobial active substance which has not been an active substance in a VMP authorised within the Union on the date of the submission of the application
(40)	Prolongation of the period of protection of technical documentation for the cases of variation/s that involve a change/s to the pharmaceutical form, administration route or dosage provided that is assessed as demonstrated reduction of antimicrobial or antiparasitic resistance
(57)	*Collection of data on antimicrobial medicinal products used in animals:* • Relevant and comparable data on the volume of sales and on the use of antimicrobial medicinal products used in animals to be collected allowing direct and indirect evaluation of on farm use • Collated data on the volume of sales and the use per animal species and per types of antimicrobial medicinal products used in animals should be sent to the Agency by Member States The European Medicines Agency shall: • Cooperate with Member States/other Union Agencies to analyse data/publish an annual report • Take into account those data when adopting any relevant guidelines and recommendations The Commission shall adopt delegated acts as regards: • The types of antimicrobial medicinal products used in animals for which data shall be collected • The quality assurance to ensure quality and comparability of data • The rules on the methods of gathering data and on the method of transfer of data to the Agency The Commission shall adopt implementing acts as regards to the format of the data collected Member States shall be allowed to apply a progressive stepwise approach: • Within 2 years from 28 January 2022 (i.e. **2024**) data shall be collected at least for the species and categories included in Commission Implementing Decision 2013/652/EU (wording 2018) • Within 5 years from 28 January 2022 (i.e. **2027**), data shall be collected for all food-producing animal species • Within 8 years from 28 January 2022 (i.e. **2030**) data shall be collected for other animals which are bred or kept

(continued)

Table 4 (continued)

Articles	
(105)	*Veterinary prescription for an antimicrobial VMP shall*: • Only be issued after a diagnosis of the infectious disease by a veterinarian who shall be able to provide justification for those veterinary prescription, in particular for metaphylaxis and for prophylaxis, for which only for a limited duration to cover the period of risk can vet prescribe • Be issued only after a clinical examination or any other proper assessment of the health status of the animal or group of animals by a veterinarian • Contain except other primary information also warnings to ensure prudent use of antimicrobials • Be valid for 5 days from the date of its issue Note: Member State may allow a veterinary prescription to be issued by a professional, other than a veterinarian, who is qualified to do so in accordance with applicable national law, such prescription shall be valid only in this respective Member State
(107)	*Use of antimicrobial medicinal products* Antimicrobial medicinal products: • Shall not be applied routinely nor used to compensate for poor hygiene, inadequate animal husbandry or lack of care or to compensate for poor farm management • Shall not be used in animals for the purpose of *promoting growth* nor to *increase yield* • Shall not be used for *prophylaxis* other than in exceptional cases, for the administration to an individual animal or a restricted number of animals when the risk of an infection or of an infectious disease is very high and the consequences are likely to be severe—for *antibiotics* (=antibacterials) such use to be limited for individual animals only • Shall be used for *metaphylaxis* only when the risk of spread of an infection or of an infectious disease in the group of animals is high and where no other appropriate alternatives are available Member States should provide guiding on appropriate alternatives and shall actively support the development and applications of guidelines which promote the understanding of risk factors associated with metaphylaxis and include criteria for its initiation For antimicrobials with critical importance/reserved for human medicine only is not possible to use those antimicrobials/medicinal products *under the cascade rules* (see in Articles 112, 113 and 114) The Commission may by means of implementing act set the rules for antimicrobials: • Shall not be used under the cascade/or only under some conditionsAnd should consider in setting those acts: • Risks to animal or public health if the antimicrobial is used under the cascade • Risk for animal or public health in case of development of antimicrobial resistance • Availability of other treatments for animals • Availability of other antimicrobial treatments for humans • Impact on aquaculture and farming if the animal affected by the condition receives no treatment A Member State may further restrict or prohibit the use of certain antimicrobials (based on justification) in animals on its territory if the administration of such antimicrobials to animals is contrary to the implementation of a national policy on prudent use of antimicrobials. The Commission should be informed about the above measures
(118)	*Animals or products of animal origin imported into the Union* Article 107 (2—growth promoters) shall apply, mutatis mutandis, to operators in third countries and those operators shall not use the designated antimicrobials referred to in Article 37(5-antimicrobials of critical importance/reserved for human medicine only), insofar as relevant in respect of animals or products of animal origin exported from such third countries to the Union. The Commission shall adopt delegated acts related to above rules
(119)	*Advertising of antimicrobial VMPs*: shall not be distributed for promotional purposes as samples or in any other presentation
(141)	Tasks for the EMA Committee for Veterinary Medicinal Products (CVMP): provide scientific advice on the use of antimicrobials and antiparasitics in animals in order to minimise the occurrence of resistance in the Union, and update that advice when needed;

[a]Prophylaxis, Metaphylaxis, see chapter "Mass Medications: Prophylaxis and Metaphylaxis Cascade and Off-label Use Treatment Guidelines and Antimicrobial Stewardship" for further detail information in relation to regulation (EU)2019/6

regulation, but having influence both on the use of antimicrobials and antimicrobial resistance. Among those articles, indirectly targeting AMR areas belong those containing, e.g. rules for vaccines (either those authorised and manufactured as VMPs or veterinary autogenous vaccines) and other immunologicals, and also those covering issue of availability of certain VMPs on the Member States internal markets that can be affected by new rules for centralised authorisations as well as by articles on harmonisation of Summary of Product Characteristics (SPC). The data protection rules for the innovative molecules of antimicrobials, or new/innovative routes of administration or pharmaceutical forms for certain antimicrobial or combinations of antimicrobials with specific carriers, can have influence on the authorisation procedures and finally also impacting the use of antimicrobials in practice.

As listed above in comments towards Regulation (EU) 2019/6 on veterinary medicinal products also in Regulation (EU) 2019/4 on medicated feed are set obligations that shall be interpreted as *de facto* ban of the prophylactic use of veterinary *antibiotics* containing medicinal products via medicated feed—as this commodity is mostly intended for the mass medication of the herds or flocks of animals (Article 17, Regulation on medicated feed). The prophylactic use of veterinary *antimicrobials* (including e.g. antiparasitics) should be due to the cross link to Article 107 of Regulation on VMPs very restricted for specific cases only. New rules for metaphylaxis are stipulated (Please refer for the comments with regard to prophylaxis, metaphylaxis and prescription for those cases and exact rules in the chapter "Mass Medications: Prophylaxis and Metaphylaxis Cascade and Off-label Use Treatment Guidelines and Antimicrobial Stewardship"). Provisions that allow for the possibility to reserve certain antimicrobials for human use only, in order to better preserve their efficiency is listed in Regulation on veterinary medicinal products with link to the medicated feed Regulation. Press release 333/18 of the Council of the EU also draws the attention to the parts of the Regulation targeted on improvement of protection of the European consumers against the risk of the spread of antimicrobial resistance through imports of products of animal origin. It also creates a level playing field between the EU and third country operators insofar as the latter will have to respect the ban on antibiotics for growth promotion, as well as the restriction on antimicrobials reserved for use in humans.

In consistency with regulation on veterinary medicinal product, further Regulation (EU) 2019/4 on medicated feed set also the provisions for use, prescription and avoiding of cross-contamination for feed medicated by premixes (as specific pharmaceutical form of veterinary medicinal products intended to be used by incorporation to feed in licenced feed mills or—according to the new legal rules, also in mobile mixers). As the premixes and finally medicated feed can contain antimicrobials, the rules for them should be clearly stipulated and linked to the Regulation on VMPs (please refer to Tables 5 and 6).

The European Parliament and the Council adopted the Regulation (EU) 2016/429 on transmissible animal diseases ("Animal Health Law") in March 2016, with the date to be applicable 2021. This Regulation lays down rules for the prevention and control of animal diseases which are transmissible to animals or to humans. These above rules are stipulated to take into account also the public health and antimicrobial resistance, also tackling the issue, e.g. via surveillance of animal pathogens resistant to antimicrobial agents (if those pathogens are considered eligible for listing in accordance with the criteria laid down in this Regulation). As indicated in the Answer on behalf of the European Commission E-004664/2018, European Food Safety Authority (EFSA) will perform the assessment to identify the pathogens of concern, for which should be conducted and where necessary harmonised rules on their surveillance may be developed. That will complement the monitoring of AMR in zoonotic and commensal bacteria in necessary scale. Also several other parts both in preamble and the body of the Regulation (see Table 7) touching issues of antimicrobial resistance. Once

Table 5 Regulation on MF (EU) 2019/4—Preamble points related directly to antimicrobials (modified according Regulation (EU)2019/4)

Preamble	
(4)	Prevention of disease is better than cure. Medicinal treatments, especially with antimicrobials, should never replace good husbandry, bio-security and management practices
(22)	Importance of consideration the international dimension of the development of AMR Antimicrobial resistant organisms can spread to humans and animals in the Union and third countries through consumption of products of animal origin, from direct contact with animals or humans or by other means This has been recognised in Article 118 of Regulation on VMPs (EU) 2019/6 which provides that operators in third countries are to respect certain conditions relating to AMR for animals and products of animal origin exported from such third countries to the Union. This is to be taken into consideration also in respect of the use of antimicrobial medicinal products concerned if they are administered via medicated feed Furthermore, in the context of international cooperation and in line with the activities and policies of international organisations such as WHO Global Action Plan and the Strategy on Antimicrobial Resistance and Prudent use of Antimicrobials of the OIE, steps restricting the use of medicated feed containing antimicrobials in order to prevent a disease should be considered worldwide for animals and products of animal origin exported from third countries to the Union
(25)	A prescription for medicated feed issued by a professional person, other than a veterinarian, should be valid only in the Member State of "non-veterinarian professional" and should exclude the prescription of medicated feed containing antimicrobial VMPs and of any other VMPs where a diagnosis by a veterinarian is necessary
(27)	Taking into account the serious public health risk posed by antimicrobial resistance, it is appropriate to limit the use of medicated feed containing antimicrobials for animals Prophylaxis or use of medicated feed to enhance the performance of animals should not be allowed, except, in certain cases, as regards medicated feed containing antiparasitics and immunological veterinary medicinal products Metaphylactic use of medicated feed should only be allowed when the risk of spread of an infection or of an infectious disease is high, in accordance with Regulation 2019/6
(29)	In accordance with Regulation (EC) No 1831/2003, the ban on the use of antibiotics as growth promoting agents as from 1 January 2006 should be strictly adhered to and properly enforced
(30)	Ensuring the prudent use of antimicrobials in food-producing animals considering the "One Health" concept, endorsed by the WHO and the World Organisation for Animal Health (OIE), recognises that human health, animal health and ecosystems are interconnected
(31)	Recall to Council conclusion (2016) and Resolution of EP (2018) on European One Health Action Plan against Antimicrobial Resistance
(33)	Commission is obliged via delegated act establish specific maximum levels of cross-contamination for active substances in non-target feed and methods of analysis for active substances in feed and of the amendment to the Annexes to this Regulation (including among the others, the list of antimicrobial active substances which are most commonly used in medicated feed)

of the importance is the inclusion to the list of the diseases the disease agent that has developed resistance to treatments which poses a significant danger to public and/or animal health.

In connection with previously mentioned Animal Health Law and the EFSA activities and JIACRA reports, since 2013 has become valid Commission Implementing Decision 2013/652/EC on the monitoring and reporting of AMR in zoonotic and commensal bacteria that lays down rules for the harmonised monitoring and reporting carried out by Member States. Defined zoonotic and indicator bacteria obtained from samples from certain food-producing animal populations and food are monitored based by rules set by this implementing decision, it means that representative isolates of *Salmonella* spp., *Campylobacter jejuni*, indicator commensal *E. coli* and ESBL-, AmpC- or carbapenemase-producing *E. coli* shall be collected by MSs, which can also voluntarily collect isolates of *Campylobacter coli* and indicator commensal *Enterococcus faecalis* and *Enterococcus faecium*. Isolates should be collected from caecal samples and carcasses, depending on the animal species, which include *Gallus gallus*—laying hens, broilers; fattening turkeys; fattening pigs and bovines under 1 year of age.

Table 6 Regulation on MF (EU) 2019/4—Articles/Annexes directly mentioning rules for antimicrobials in medicated feed (modified according Regulation (EU)2019/4)

Articles	
(3)	Definitions of "antimicrobial", "antiparasitic", "antibiotic", "metaphylaxis", "prophylaxis" are cross-linked to the Regulation on VMPs (EU) No 2019/6
(7)	The Commission shall (by 2023), adopt delegated acts in accordance with Article 20 in order to supplement this Regulation by establishing, as regards the antimicrobial active substances listed in Annex II, specific maximum levels of cross-contamination for active substances in non-target feed and methods of analysis for active substances in feed
(11)	Advertisement: Medicated feed containing antimicrobial VMPs shall not be distributed for promotional purposes as samples or in any other presentation
(16)	*Prescription*: The duration of a treatment shall comply with the summary of product characteristics of the VMP incorporated in the feed and, where not specified, shall not exceed 1 month, or 2 weeks in case of a medicated feed containing antibiotic VMP Validity of prescription: medicated feed containing antimicrobial VMPs—from the date of its issuance for a maximum period of 5 days The veterinarian shall not prescribe medicated feed with more than one veterinary medicinal product containing antimicrobials
(17)	Use of antimicrobial/s containing medicated feed: Medicated feed containing antimicrobial VMPs shall be used in accordance with Article 107 of Regulation (EU) 2019/6, except as regards Paragraph 3 thereof, and shall not be used for prophylaxis
Annex	
II	List of antimicrobial active substances as referred to in Article 7
III	Specific labelling requirements referred to in Article 9: Except the other requirements should be included information that inappropriate disposal of medicated feed poses serious threats to the environment and may, where relevant, contribute to antimicrobial resistance
IV	Permitted tolerances for the compositional labelling of medicated feed or intermediate products as referred to in Article 9: Where the composition of a medicated feed or an intermediate product is found to deviate from the amount of an antimicrobial active substance indicated on the label, a tolerance of 10% shall apply
V	Information to be included in the veterinary prescription for medicated feed as referred to in Article 16: Specific veterinary prescription for medicated feed is established and include among the others rules tackling the antimicrobials/antiparasitics: • Specification of diagnosed disease to be treated. In the case of immunological VMPs or antiparasitics without antimicrobial effects, disease to be prevented • Any warnings necessary to ensure the proper use (if relevant prudent use of antimicrobials)

Under the new European One Health action plan against antimicrobial resistance, adopted by the Commission, and in order to strengthen One Health surveillance and reporting of antimicrobial resistance and antimicrobial use, the Commission committed to review Decision 2013/652/EC by 2021, to take into account new scientific developments and data collection needs. In March 2019 therefore EFSA provided for consultation draft of above asked revision based on status of scientific and technical knowledge, and considering also new legal surrounding.

During different negotiations both on political level and the expert level the issue of the surveillance of the target veterinary pathogens become an important question waiting to be answered via starting the EU project that can merge existing national systems of monitoring of antimicrobial resistance in target veterinary pathogens and help to establish new systems in those countries that have not started yet. It is organised since 2018 under the umbrella of the EU-JAMRAI platform WP 7.4.2 (Grant Agreement No 761296, 2014), where is planned to set a system methodologically considering the experience of the human medicine within the EARS project. Until now, the most comprehensive overview of the existing surveillance systems of monitoring of resistance in pathogens of veterinary importance in diseased animals in Europe is given in the article from

Table 7 Regulation on transmissible animal diseases preamble and articles tackling AMR

Preamble	
(32)	Antimicrobial resistance: • Definition (the ability of microorganisms to survive or to grow in the presence of a concentration of an antimicrobial agent which is usually sufficient to inhibit or kill microorganisms of the same species) • Increase of AMR • Action No 5 of EU Action Plan emphasises the preventive role to be played by this Regulation and the consequent expected reduction of the use of antibiotics in animals • AMR complicates the treatment, threat for animal/human health—microorganisms with should be treated as if they were transmissible diseases, and thus covered by the scope of this Regulation—this will enable action to be taken against antimicrobial-resistant organisms where appropriate and necessary
(59)	A key purpose of disease notification and reporting is to generate reliable, transparent and accessible epidemiological data: • Effective collection and management of surveillance data should be established at Union level also for antimicrobial-resistant pathogens • Process operated through the database of the OIE • Consistency in exchanges of information in accordance with Directive 2003/99/EC
(81)	Veterinary medicinal products such as vaccines, hyper-immune sera and antimicrobials important for the prevention and control of transmissible animal diseases • Importance of vaccines as a tool in the prevention, control and eradication of animal diseases
(83)	This Regulation should therefore provide for rules on the use of veterinary medicinal products for the prevention and control of certain listed diseases and for harmonised criteria to be taken into consideration when determining whether or not to use, and how to use, vaccines, hyper-immune sera and antimicrobials
Articles	
(1)	The scope of the Regulation lays down rules for the prevention and control of animal diseases which are transmissible to animals or to humans: (b) Take into account the relationship between animal health and among the other factors also AMR
(5)	disease shall be included on the list as set out in the list in Annex II of this Regulation in the case that meet the condition that the disease agent has developed resistance to treatments which poses a significant danger to public and/or animal health in the Union
(7)	There are defined assessment parameters for the listing of diseases, among these parameters also the resistance to treatments, including antimicrobial resistance is involved
(11)	Operators and animal professionals shall have adequate knowledge of resistance to treatments, including antimicrobial resistance, and its implications
(12)	Veterinarians shall play an active role raising awareness of resistance to treatments, including antimicrobial resistance, and its implications

Schrijver et al. (2017). The authors summarised that at the national level, AMR surveillance systems in livestock apply heterogeneous sampling, testing and reporting approaches, leading in results that cannot be compared. Most reports are not easily available, also due to the reason that they are written in a local language. From the information available to date of this monography with the EU region in following countries monitoring/surveillance systems collecting data on antimicrobial resistance gained from isolates from diseased animals run: BE, CZ, DE, DK, FI, FR, NL and SE. In other EU countries individual studies/projects or just EU harmonised monitoring of zoonotic and indicator bacteria are in place.

There is also existing industry-funded monitoring system undertaken by the Centre European d'Etudes pour laSant_e Animale (CEESA) that conducts four AMR resistance surveillance and monitoring programmes across Europe: *European Antimicrobial Susceptibility Surveillance in Animals* (EASSA): examines the antimicrobial susceptibility of zoonotic and commensal bacteria in healthy food-producing animals. *VetPath* (15 years programme) performing antimicrobial susceptibility of bacterial pathogens causing disease in food-producing animals. *ComPath* project, which is targeted on diseased companion animals and *MycoPath* project targeted on mycoplasma coming from major diseases in food-producing animals and their susceptibility (Schrijver et al. 2017).

5 European Action Plans Tackling Antimicrobial Resistance

Action Plan against the rising threats from antimicrobial resistance (COM (2011) 748 was the first one, designed for 5 years period, containing seven key pillars and 12 exact actions (Fig. 1), it highlighted the need for holistic approach and the plan symbolised EU political commitment and strategy with exact actions to be done to contain antimicrobial resistance.

The first Action plan was evaluated and several recommendations were established to further continue the actions already started. It was decided that EU should continue progress and play an active role globally, but also be targeted on the support to Member States in development and implementing National Action Plans fighting against antimicrobial resistance. International cooperation both within the Union and its regulatory and scientific bodies as well as worldwide should be promoted. There was also imperative on creation of the new or updated legislative tools that enhance surveillances on antimicrobial use and antimicrobial resistance. It was agreed that "One health" approach should be continued, but environmental actions should be involved. Research and innovations targeted on new treatments, vaccines, alternatives, diagnostics, but also social factors were proposed as the future plan cornerstones. As a reaction to the first Action Plan, the world's biggest public–private partnership New Drugs for Bad Bugs (ND4BB)

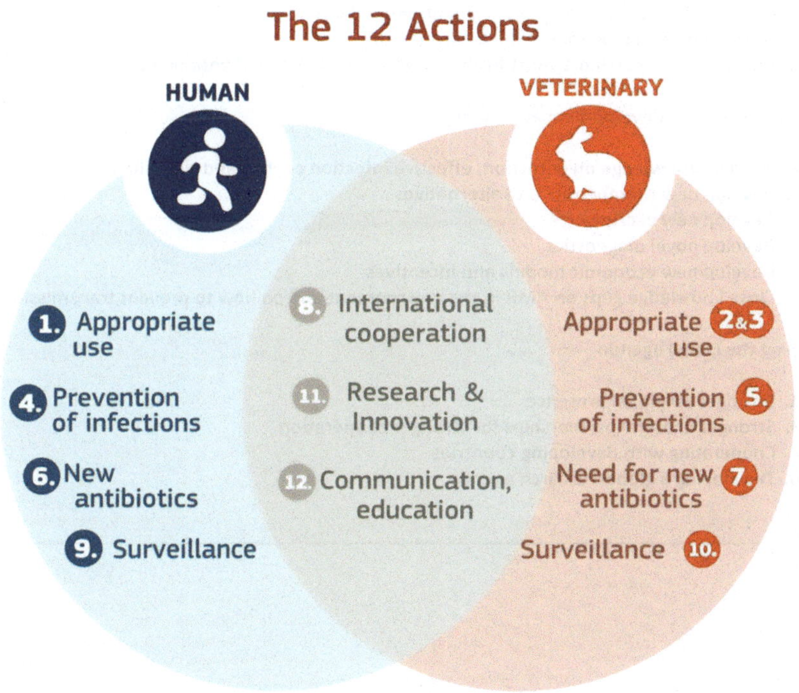

Fig. 1 EU Action plan against rising threats AMR (2011–2016), 12 key actions (Van Dyck 2017)

programme (based on Council Regulation (EC) No 73/2008) was launched in May 2012 within the Innovative Medicines Initiative (IMI). With this programme, academic and other public partners and pharmaceutical companies team up to advance the development of new antibiotics with a total budget of about 700 million euros. The Joint Programming Initiative on AMR (JPIAMR 2019) was also established to integrate research efforts across national borders via alignment and research funding, and to create a common research agenda. The JPI AMR involves 27 countries in 2019.

Within the new, second EU Action plan, improving of communication with stakeholders and the wider public about current and future EU activities with the area of antimicrobial resistance was recommended. Based on this last mentioned recommendation information gained in the Public Consultation and feedback from Roadmap was used to define *(second) European One Health Action Plan against Antimicrobial Resistance*. This plan is built on three pillars, each including proposal for improvement/development of certain areas:

Three pillars of the Second European One Health Action Plan against Antimicrobial Resistance

Making the EU a best practice region

A. Better evidence and awareness of the challenges of AMR
- Strengthen One Health surveillance and reporting of AMR and antimicrobial use
- Benefit from the best evidence-based analysis and data
- Increase awareness and understanding

B. Better coordination and implementation of EU rules to tackle AMR
- Improve the coordination of Member States' One Health responses to AMR
- Better implementation of EU rules

C. Better prevention and control of AMR
- Strengthen infection prevention and control measures
- Promote the prudent use of antimicrobials

D. Better addressing the role of the environment

E. A stronger partnership against AMR and better availability of antimicrobials

Boosting research, development and innovation

A. Improve knowledge on detection, effective infection control and surveillance
B. Develop new therapeutics and alternatives
C. Develop new vaccines
D. Develop novel diagnostics
E. Develop new economic models and incentives
F. Close knowledge gaps on AMR in the environment and on how to prevent transmission

Shaping the global agenda

A. Stronger EU global presence
B. Stronger bilateral partnerships for stronger cooperation
C. Cooperating with developing countries
D. Developing a global research agenda

6 Role of the Soft Law and Guidelines

Despite the fact that in veterinary side, especially in agriculture and veterinary medicine of food-producing animals were (within the last several years) approved important legal provisions, these mostly restrictive legal tools seem to be not only possibility how to fight against antimicrobial resistance. Moreover, people from the real practice search for more targeted and practical advices, what to exactly do and what are the really working tools to reduce the need of antimicrobials as well which antimicrobial, which route of administration, dose and duration of treatment are the most proper one in situation, when use of antimicrobials is necessary. Evidence-based and scientifically justified treatment guidelines can therefore assist veterinarians to prescribe and choose the most relevant antimicrobial considering except the other factors are also results of national or regional surveillance of antimicrobial resistance of target pathogens which is up to date. Moreover not in all clinical circumstances performance of antimicrobial susceptibility testing is possible, mostly because the causative pathogen cannot be routinely and easily cultured (e.g. in pathogens like *Lawsonia intracellularis*, *Mycoplasma* spp., anaerobes as *Dichelobacter* spp., *Fusobacterium* spp., *Ornithobacterium rhinotracheale* etc.) and in such cases examples from these guidance documents can help veterinarians in proper decision significantly. Several examples of the soft law, especially at national level, are of importance.

Guidelines covering diseases of certain animal species (including exact recommendations for "combination" disease/microorganism/recommended first, second and last choice of antimicrobials) have been issued mostly at national level. Those guidelines differ and in some countries are available for food-producing animals as listed, e.g. in the examples in RONAFA report for BE, DK, FI, NL, SE (EMA and EFSA 2017) as well as for companion animals (Danish Small Animal Veterinary Association 2013; Swedish Veterinary Association 2009). An example can be given of Belgium AMCRA that produced guidelines (dal Pozzo 2018), in which antimicrobials are colour-coded according to their importance for human health.

Also farmers' associations' commitments and food-chain industry requirements are of growing importance. Pharmaceutical industry also identified that broader spectrum of vaccines can help to minimise the use of antimicrobials. Examples can be listed of differently scoped guidelines on good husbandry practices (ASEAN 2015), code of practices for care and handling of animals—here pigs (NFAAC 2014), animal welfare (Buller et al. 2018; Down et al. 2016), vaccination practices (Small et al. 2017) or good practice of administration of antimicrobials (Landbrug Fødevarer 2011). Positive impact and success of such soft law document is dependent, among the other factors, on specificity/tailoring for exact husbandry/sector conditions, knowledge of national habits and behavioural formula and socio-economic aspects, considering what can be practically achieved. The authors of such guidances should base their recommendations on both scientific evidence, consideration of risks of AMR from "One health perspective", but also considering what is necessary for treatment and welfare of animals.

7 Guidelines for the Prudent Use of Antimicrobials in Veterinary Medicine

The Commission Notice No. 2015/C 299/04 is not a binding legal obligation, but despite that, it provides advices and examples for different stakeholders as for practical guidance/development of national strategies, promotion of the prudent use of antimicrobials in veterinary sector and contribution/complementing of control of antimicrobial resistance in humans. There are highlighted principles of the prudent (appropriate/responsible/rational) use of antimicrobials that should be targeted on justified cases and finally should lead to the overall reduction of the total use, considering that any use of antimicrobials can increase selection pressure and lead to the spread of resistance. The ultimate objective of this guideline is to reduce the need for antimicrobials by preventing diseases. For cases, when it is decided by the responsible veterinarian, those antimicrobials should be used following principles are recommended to be followed according to this guideline: Diagnosis and prescription made by veterinarian and delivery should be based on independent professional judgement and taking into account legal provisions and treatment guidelines. Where appropriate, diagnosis should be based on clinical examination accompanied with laboratory tests including antimicrobial susceptibility testing. If metaphylaxis—clinical findings of disease in flock/herd level should be confirmed prior start of treatment. Routine prophylaxis should be avoided, consideration of prophylactic use only for exceptional cases. Herd/flock medication should be avoided. For treatment preferably narrow spectrum, single-substance antimicrobial VMPs should be used according to the prescription (route of administration, dose, interval, duration). Special care/restrictions should be in any use of antimicrobials in food-producting animals, when considered use of critically important antimicrobials and in exceptional cases, when used "off-label"/under cascade. There should be considered and promoted alternative disease control strategies.

Special attention is paid to the advice that critically important antimicrobials should only be used when and where the veterinarian not find any effective non-critically important alternative available based on antimicrobial susceptibility testing and relevant epidemiological data. The guideline recommends prohibition as for the use of third- and fourth-generation cephalosporins in poultry (including eggs), but surprisingly the text recommends avoidance of injections of antimicrobials into eggs and 1-day-old chicks and at the same time describes acceptance of such practices in cases justified by regional/national guidelines. Fluoroquinolones are advised only to be used in poultry when treatment response to other antimicrobials is poor or expected to be poor, and only after susceptibility testing has been performed.

The prevention of infection and keeping of the good health status of the animals as well as following welfare principles is recommended as the best way to reach reduction and minimise the need to use antimicrobials. Key activities and principles are identified to be followed: hygiene and biosecurity measures implementation at farm level and in all related activities; infectious disease prevention protocols, infection control and hygiene; husbandry systems improvement; integrated production systems: mixing animals to be avoided and transport minimalised; stress to be avoided/minimised; high quality of feed and water; animal health control programmes—cooperation of vets and farmers; alternative tools to antimicrobials (including other VMPs—e.g. vaccines, or other as probiotics, prebiotics, minerals, vitamins, phytoadditives etc.).

The guidelines addressing responsibilities to different stakeholders, including authorities to clearly show the need for multidisciplinary and complex approach and roles (examples in brackets) of each subjects: main role of "prescriber" (in most countries it should be attending licensed veterinarian as the independent decision maker for diagnosis and where relevant antimicrobial prescription, considering the local policies/risks of AMR, advising the administrator proper dosing scheme), administrator of the antimicrobial (technically advanced, key for following veterinarian's advices on proper dosing/administration, receiving medicines from authorised sources, watching intake of medicated feed/water by the animals), farmer cooperating with veterinarian and laboratories (performing diagnostic tests including antimicrobial susceptibility testing, having quality assurance systems and providing proper interpretation of laboratory results). Active participation also from pharmaceutical industry, pharmacists, retailers, wholesalers (no improper advertisement, providing medicines/antimicrobials by approved channels only based on prescription); feed business operators (comply with legal requirements: prescription, hygiene, GMP, formulation, ingredients, production of medicated feed: from authorised VMP only in line with prescription and with appropriate labelling); food business operators (ensure maximum hygiene, no misleading advertisement); veterinary faculties and agriculture schools (set appropriate curricula containing lessons on antimicrobial resistance, prudent use, alternative tools, preventive programmes); veterinary professional associations (specific guidelines tailored for the species/sectors/husbandry types); industry stakeholder associations (promote quality schemes); farmers' associations (education, guidelines for good practices to avoid need for antimicrobials) and last, but very important competent authorities (setting national strategies, promoting of surveillance/monitoring; enforcement, if needed penalties/sanctions). This guideline also emphasises the need for awareness raising by targeted campaigns as well as strengthen education and training and promoting use of the national/sector/species specific guidelines.

8 Guidelines of EMA and EFSA

Despite the fact that those guidelines are not directly linked to the practice of the use of antimicrobials, they belong to the regulatory area that has a great influence, e.g. on portfolio of authorised veterinary medicinal products as well as on the essential parts of the product texts, i.e. information for veterinarians and farmers on how the VMPs should be used (indication, route of administration, dose, interval and total treatment duration, advice on the proper administration, withdrawal period in the case of food-producing animals as well as warnings and additional information essential for the proper use of antimicrobials containing VMPs). These guidelines are so called scientific guidelines and are produced either directly by *EMA*-European Medicines Agency in cooperation with national experts nominated for Committee for Veterinary Medicinal Products—Working Parties, or more internationally by *VICH* (International Cooperation on Harmonisation of Technical Requirement for Registration of Veterinary Medicinal Products). They serve as a guiding document for both pharmaceutical industry and assessors and they reflect a harmonised approach of the EU Member States and the Agency on how to interpret and apply the requirements for the demonstration of quality, safety and efficacy set out in the Community legal provisions (EMA web portal 2019). Main areas they are covering (considering the pharmaceuticals-antimicrobials) are:

- *Quality* (for example: assessing specifically the products administered in drinking water, or assessing stability of the product etc.)
- *Safety and residues*
  - *Toxicology* (e.g. mutagenicity, genotoxicity, carcinogenicity, chronic and acute toxicity with respect to use of VMPs impact on residues and food safety)
  - *User safety* (generally for pharmaceuticals, specific—topically administered VMPs)
  - *Environmental risk assessment* (assessing effects of persistence, bioaccumulation, toxicity, different considerations/guidelines as for groundwater, soil, manure etc.)
  - *Consumer safety* (guidelines related to residues, withdrawal periods, maximum residue levels establishment)
  - *Target animal safety* (included in the assessment of the clinical part of the dossiers, specific guideline, e.g. for local tolerance of intramammary VMPs)
- *Efficacy*:
  - *Preclinical* (pharmacokinetic, pharmacodynamics, mechanism of action, resistance development, dose determination and confirmation)
  - *Clinical* (efficacy—specific guidelines released, e.g. for intramammary products or fixed combinations)
- *Texts of SPC*: specific guidance for the VMPs containing antimicrobials

As also certain feed additives have antimicrobial properties, therefore should not be forgotten to list the *EFSA guidance documents* detailing for the feed additives licencing applicants how to compile dossiers for submission and the information and studies required for the evaluation of the feed additives. Those guidances cover identity, characterisation and conditions of use of feed additives, dealing with assessment of safety for target species, for users as well as consumer safety. Also risks for the environment are covered by the respective guidance document. Efficacy guidances are elaborated considering aquatic and terrestrial animals (EFSA 2019b).

In the case that as a feed additive (e.g. with probiotic effect) is microorganism or also for the cases of feed additives obtained by fermentation of a production strain and covers the safety aspects directly linked to the production strain, there exists the guidance (EFSA 2018a) on the characterisation of such microorganism/fermentation products including parts where attention is paid to evaluation of:

- *Identification/taxonomical classification*, including whole genome sequencing (WGS) analysis required for characterisation of bacteria, taxonomical classification (using, e.g. 16S rRNA techniques).
- *Antimicrobial susceptibility*: In the cases of viable microorganism as feed additives, considering antimicrobials with importance for human and veterinary medicine sectors, intrinsic resistance mainly not considered as an issue, but acquired resistance evaluated; phenotypic testing of AMR via determination of MICs required (and analysis if they are exceeding cut-off values proposed by FEEDAP) as well as WGS for the presence of known AMR genes.
- *Antimicrobial production*—inhibitory substances identified as produced by the respective microbes to be further investigated; for ionophoric coccidiostats produced from species known to produce other antimicrobials of clinical relevance, the presence of antimicrobial activity not related to the ionophore in the fermentation/final product should be investigated, e.g. by comparing the inhibitory spectrum of the pure ionophore with that of the additive).
- *Toxigenicity and pathogenicity*—Information relating to toxigenicity and virulence for humans and target species should be provided for active agents and production strains, including history of use of the strain or any close relative. This should be based on updated literature searches. Impact on gut microbiota is also assessed.

For fermentation products, test for absence of the product strains and also possible presence of DNA from the production strain or information on *genetic modification*.

Think About

Data on AMR in food-producing animals come mostly from phenotypical testing of isolates from animals—samples are not collected at farm level, but come from caecal samples of healthy animals at slaughterhouses. It is assumed that those data can reflect antimicrobial resistance considering as one of the most selective factor is the use of antimicrobials. The questions that can be linked to this are:

- What the samples reflect—situation on farms in the respective country or partly also situation reflecting transport and especially conditions at slaughterhouses, where animals from different sources (imported animals) are slaughtered?
- Is not the only reasonable way to cross link the data on antimicrobial resistance at certain farm (especially once have the farm closed turnover) with the data on antimicrobial use at the same farm (but considering also co-selection of resistance by other means, e.g. disinfectants; considering also "input" data, i.e. bacterial microbiome at arrival of 1-day-old chicks, e.g. in broilers for fattening the data from parent flock is of importance)?
- What is the level of reasonability to compare the aggregated data on AMR from zoonotic bacteria in the system of sampling (see comment above) to aggregated data from human monitoring of AMR in bacterial isolates coming from invasive isolates from blood and cerebrospinal fluid sampled in patients in hospitals?
- Is not it the reasonable way to be concentrated on casuistics where is possible real traceability of food from "farm to fork" and infection of human and where phenotyping as well as genotyping testing of causative bacterial agent can confirm the direct link?

Table 8 Examples of measures implemented by individual Member States based on national legal provisions/national guidelines/other documents/commitments in relation to CIAs (non-exhaustive list)

Measures
National/regional Treatment Guidelines/formularies that propose which antimicrobial to use for which disease, taking into account risk from use of CIAs amongst other factors: BE, DK, NL, SE, NO
National targets to reduce the use of CIAs: BE, FR, NL
Classification of antimicrobials as first, second or third choice (or colour-coding) According to the associated risk to public health BE: Colour-coding system according of importance (red substances of highest importance to human medicine) DK: Treatment guidelines for pigs: Colour-coding of antimicrobials that takes account of risk to public health NL: "Traffic light" (green, orange, red) visual scoring and benchmarking
Antimicrobial Susceptibility Testing prior to use of CIAs DK: Mandatory prior to use Fluoroquinolones FR: 3/4 generation cephalosporins and fluoroquinolones NL: 3/4 generation cephalosporins and fluoroquinolones SE: 3/4 generation cephalosporins and fluoroquinolones CZ: 3/4 generation cephalosporins and fluoroquinolones, ansamycins, aminoglycosides of high generations
Notification of use of CIAs to authorities DK—Mandatory for Fluoroquinolones
Voluntary species sector bans/limitations on the use of CIAs DK: Pig industry ban—use of 3/4 generation cephalosporins 2010, Dairy industry ban in 2014 FR: Pig industry ban—use of 3/4 generation cephalosporins 2010 IT: Voluntary withdraw—use of 3/4 generation cephalosporins in rabbit and poultry production NL: Pig sector—ban 3/4 generation cephalosporins and fluoroquinolones) UK: Poultry sector—ban 3/4 generation cephalosporins CZ: Poultry sector Quality programme, ban 3/4 generation cephalosporins, ban of preventive use, limitation for fluoroquinolones ES: Pig sector—limitation of use of colistin
Ban on Cascade ("off label") use of CIAs: FI, NL Ban on preventive use of CIAs: FR
Higher taxes on CIAs: BE, DK

9 Role of Individual Member States

Despite the fact that the chapter is targeted on the EU policies and regulations, the essential and key role of the individual Member States should be highlighted. National legal provisions, National action plans against AMR as well as activities of the different bodies (Chambers of veterinarians, Associations of veterinarians/specialists for different animals sectors and farmers associations) and its importance cannot be forgotten. Despite the fact that Nordic countries, with the lowest antimicrobial consumption, long history of antimicrobial policies are considered as positive examples of the success, in the current period many of the other European countries follow their success stories. Except National Action Plans, some of the countries also have the treatment guidelines, formularies or recommendations or also legal provisions defining national rules for use of antimicrobials—usually with special attention to limitation of the use of critically important antimicrobials (CIAs). Examples can be given as for approaching the antimicrobials considered as critically important in different Member States (see Table 8).

10 Conclusion

The EU is, according to the second Action Plan, continuing to aspire to be a leading world region in the fight against AMR. The tricky issue with regard to AMR and considering human as well as animal health is that AMR is cross-border threat and no one region in the world can stop AMR alone. Within the international transport and travelling of people, as well with the trade of animals and food of animal origin, also bacteria and mobile genetic elements travel as the "black" passengers. Moreover, for the EU to become "a best practice region" for real, the key is to reduce the wide and pronounced disparities among and within the EU countries concerning antibiotic consumption within the veterinary sector as well as human sector. Even more challenging task is to reduce significantly AMR in all parts of Europe. One Health policy responses to AMR, and providing targeted and sustained support and resources to those countries, which need it the most, should be considered; in other words, some balance as for sustainable financing of the effective tools and programmes should be found to tackle AMR equally according to identified issues in individual Member States. One of the biggest tasks will be to proceed from achieving isolated success and best practices in individual countries to good/best practices becoming a standard in all countries. This will need not only stricter European legal rules, but also active participation in the negotiations on global fora to ensure that the standards for health and welfare conditions of animals producing food for human consumption will be pretty similar worldwide. This should not only enforce competiveness, but at the first place maximise health protection from the perspective of consumer chemical (residues) and biological (pathogens/AMR) safety. Europe should use all the effort and knowledge capacity available for building the sustainable system of animal husbandry sector under the umbrella of One Health concept.

References

ASEAN (2015) ASEAN good animal husbandry practices for layers and broilers food safety module. https://www.asean.org/wp-content/uploads/images/Community/AEC/AMAF/OtherDocuments/ASEAN%20Food%20Safety%20Module%20GAHP%20For%20Layers%20and%20Broilers.pdf. Accessed 4 June 2019

Buller H, Blokhuis H, Jensen P, Keeling L (2018) Towards farm animal welfare and sustainability. Animals (Basel) 8(6):81 https://www.ncbi.nlm.nih.gov/pmc/articles/PMC6025272/. Accessed 15 June 2019

Council Regulation (EC) No 73/2008 of 20 December 2007 setting up the Joint Undertaking for the implementation of the Joint Technology Initiative on Innovative Medicines. https://eur-lex.europa.eu/legal-content/EN/TXT/PDF/?uri=OJ:L:2008:030:FULL&from=EN. Accessed 29 June 2019

Dal Pozzo F (2018) Antibiotic consumption and resistance in veterinary medicine in Belgium. http://www.resistenciaantibioticos.es/es/system/files/content_images/eucha_madrid_20181029-30_fabiana_dal_pozzo.pdf. Accessed 4 June 2019

Danish Small Animal Veterinary Association (SvHKS) (2013) Antimicrobial Use Guidelines for Companion Animal Practice. http://www.fecava.org/sites/default/files/files/DSAVA_AntibioticGuidelines%20-%20v1-1_3(1).pdf. Accessed 4 June 2019

Down PM, Bradley AJ, Breen JE, Hudson CD, Green MJ (2016) Current management practices and interventions prioritised as part of a nationwide mastitis control plan. Vet Rec 178:449

ECDC/EFSA BIOHAZ/CVMP (2017) European Centre for Disease Prevention and Control/ European Food Safety Authority Panel on Biological Hazards and/EMA Committee for

ECDC/EFSA/EMA (2015) European Centre for Disease Prevention and Control/European Food Safety Authority/European Medicine Agency: ECDC/EFSA/EMA first joint report on the integrated analysis of the consumption of antimicrobial agents and occurrence of antimicrobial resistance in bacteria from humans and food-producing animals. EFSA J 13:4006

ECDC/EFSA/EMA, European Centre for Disease Prevention and Control/European Food Safety Authority/European Medicines Agency (2017) ECDC/EFSA/EMA second joint report on the integrated analysis of the consumption of antimicrobial agents and occurrence of antimicrobial resistance in bacteria from humans and food-producing animals – Joint Interagency Antimicrobial Consumption And Resistance Analysis (JIACRA) report. EFSA J 15(7):4872

ECDC/EFSA/EMA (2017) Joint Scientific Opinion on a list of outcome indicators as regards surveillance of antimicrobial resistance and antimicrobial consumption in humans and food-producing animals. EFSA J 15(10):5017

ECDC/EFSA/EMA/SCENHIR (2009) Joint Opinion on antimicrobial resistance (AMR) focused on zoonotic infections – Scientific Opinion of the European Centre for Disease Prevention and Control; Scientific Opinion of the Panel on Biological Hazards; Opinion of the Committee for Medicinal Products for Veterinary Use; Scientific Opinion of the Scientific Committee on Emerging and Newly Identified Health Risks. EFSA J 7(11):1372 Question No. EFSA-Q-2008-781 (European Medicines Agency Reference EMEA/CVMP/447259/2009)

EFSA (2018a) European Food Safety Authority: guidance on the characterisation of microorganisms used as feed additives or as production organisms. EFSA J 16 (3):5206

EFSA (2018b) European Food Safety Authority: scientific report on evaluation of the impact of glyphosate and its residues in feed on animal health. EFSA J 16(5):5283

EFSA (2019a) European Food Safety Authority: EFSA statement on the risk posed to humans by a vitamin B2 produced by a genetically modified strain of *Bacillus subtilis* used as a feed additive. EFSA J 17(1):5615

EFSA (2019b) European Food Safety Authority: feed additive applications: regulations and guidance. https://www.efsa.europa.eu/en/applications/feedadditives/regulationsandguidance. Accessed 4 June 2019

EMA, European Medicine Agency (2016) CVMP strategy on antimicrobials 2016-2020. https://www.ema.europa.eu/en/documents/scientific-guideline/cvmp-strategy-antimicrobials-2016-2020_en.pdf. Accessed 4 June 2019

EMA, European Medicine Authority and EFSA, European Food Safety Authority (2017) Joint scientific opinion on measures to reduce the need to use antimicrobial agents in animal husbandry in the European Union, and the resulting impacts on food safety (RONAFA). EFSA J 15(1):4666

Grant Agreement No 761296 (2014) Joint Action on Antimicrobial Resistance and healthcare-Associated Infections (EU-JAMRAI) EU co-funded framework of the Health Program (2014–2020): EU-JAMRAI Platform 7.4.2 Surveillance of AMR in clinical pathogens of animals

JPIAMR (2019) Joint programming initiative on antimicrobial resistance. https://www.jpiamr.eu/activities/. Accessed 29 June 2019

Landbrug Fødevarer (2011) Guidelines on good antibiotic practice. Vindecenter for Svineproduktion, Danemark. https://svineproduktion.dk/. Accessed 4 June 2019

NFAAC (2014) Code of practice for the care and handling of pigs. National Farm Animal Care Council and Canadian Pork Council. ISBN 978-0-9936189-3-2. https://www.nfacc.ca/pdfs/codes/pig_code_of_practice.pdf. Accessed 15 June 2019

Regulation (EU) 2019/4 of the European Parliament and of the Council of 11 December 2018 on the manufacture, placing on the market and use of medicated feed, amending Regulation (EC) No 183/2005 of the European Parliament and of the Council and repealing Council Directive 90/167/EEC. Off J Eur Union 4:1–23

Regulation (EU) 2019/6 of the European Parliament and of the Council of 11 December 2018 on veterinary medicinal products and repealing Directive 2001/82/EC. Off J Eur Union 4:43–167

Schrijver R, Stijntjes M, Rodríguez-Baño J, Tacconelli E, Babu Rajendran N, Voss A (2017) Review of antimicrobial resistance surveillance programmes in livestock and their meat in Europe, with a focus on antimicrobial resistance patterns in humans. Clin Microbiol Infect 24(6):1–14

Small S, Williams P, Neto R, Butler N, FitzGerald J, Berret DC, Lovatt F, Carroll I (2017) Looking beyond antibiotics – time to vaccinate. MSD Animal Health. https://www.farmantibiotics.org/wp-content/uploads/2018/08/MSD-Animal-Health-Looking-beyond-antibiotics-80_216960.pdf. Accessed 15 June 2019

Swedish Veterinary Association (2009) Guidelines for the clinical use of antimicrobials in the treatment of dogs and cats. www.svf.se. Accessed May 2019

Van Dyck K (2017) TATFAR policy dialogue –video conference: the new EC action plan against antimicrobial resistance; 2011–2016 EU Action Plan against the threats of AMR – 12 Actions. https://ec.europa.eu/health/amr/sites/amr/files/ev_20170927_pres_02.pdf. Accessed 15 June 2019

Use of Antimicrobials in Practice (Targeted on Cattle, Pigs, Poultry, Horses)

Nancy De Briyne, Despoina Iatridou, Wannes Vanderhaeghen, and Kristine Ignate

Abstract

Worldwide, antibiotics are used to treat and prevent bacterial infections, both in humans and animals. In Europe, 2014 data (2017 JIACRA) estimate that the average antimicrobial consumption (AMC) in animals was higher in animals (152 mg/kg) than in humans (124 mg/kg), but the opposite applied to the median AMC (67 and 118 mg/kg, respectively). In 18 of 28 European countries, AMC was lower in animals than in humans. Most European countries have taken extensive actions to promote responsible and prudent use of antibiotics in animals.

Since 2011, the overall sales of veterinary antibiotics in EU/EEA countries are decreasing [EMA European Medicines Agency: answer to the request from the European Commission for updating the scientific advice on the impact on public health and animal health of the use of antibiotics in animals - categorisation of antimicrobials (EMA/CVMP/CHMP/682198/2017), 2019], and this is mainly accounted for reduction in antibiotic use in food-producing animals. Nevertheless, there seems to be still space for improvement, especially in certain pharmaceutical forms used for mass medication, consumptions of which create big part of the total figures of overall sales. The use in animals like dogs and cats in tonnes is relatively low. ESVAC sales data indicated that the majority of consumption of veterinary medicinal products used in 2017 (excluding topical formulation) can be allocated to food-producing animals in most of the EEA countries. The sales, in mg/PCU (Population Correction Unit), of antimicrobial veterinary medicinal products differ extensively between EU/EEA countries. This can partly be explained, among other factors, by the differences in the animal demographics, production systems and dosing of the various antimicrobials.

Keywords

Antimicrobial use · Livestock · Cattle · Pigs · Poultry · Horses · ESVAC · JIACRA

N. De Briyne (✉) · D. Iatridou
Federation of Veterinarians of Europe (FVE), Brussels, Belgium
e-mail: nancy@fve.org

W. Vanderhaeghen
Data Analysis Unit, Centre of Expertise on Antimicrobial Consumption and Reduction in Animals (AMCRA), Brussels, Belgium

K. Ignate
European Medicines Agency, Amsterdam, The Netherlands

L. Pokludová (ed.), *Antimicrobials in Livestock 1: Regulation, Science, Practice*,
https://doi.org/10.1007/978-3-030-46721-0_4

1 Background on Use of Antimicrobials in Animals

1.1 History of Use of Antimicrobials

In the first part of the twentieth century, around the world wars, the world suffered from disastrous food shortages. Farmers could not keep pace with the needs of the urban, non-food-producing majority. Demand routinely outstripped supplies and exorbitant food prices confounded policy makers and enraged consumers. Much pressure was put on farmers and the animal health sector to produce more food for lower prices. The result was industrialisation of farming with larger number of animals becoming raised faster in confined facilities. This was only possible due to advances in technology and science. As John Ikerd, Professor Emeritus from the University of Missouri quoted "*We bent nature to serve our needs. We achieved the economies of large-scale, specialized production as we applied the principles, strategies, and technologies of industrialization to farming*" (Ikerd 2008; MacKenzie 2015).

The use of antimicrobial agents in animals played an important role in this achievement. The first use of antibiotics in livestock dates back to the late 1940s when scientists found that adding antibiotics to livestock feed accelerated animals' growth and cost less than conventional feed supplements. Farmers embraced this finding and started administering antibiotics to healthy animals. Intensification of agriculture created also the need for better business management. "*Prevention is better than cure*" also applied at that time; however, the concept behind was different and rather equivalent to "*treat before get sick*".

In 1951, the United States Food and Drug Administration (FDA) approved, for the first time, the use of the antibiotics penicillin and chlortetracycline—Aureomycin (Picture 1) as animal feed "supplements", what in fact means, as antimicrobial growth promoters (AGPs). Also in Europe, penicillin and tetracycline started to be broadly used. At the same time, antimicrobials were recognised as substances allowing the treatment of animal diseases that were incurable before. Antimicrobial use in animals combined with progress made in genetics and technologies led to a substantial increase in global livestock production since the 1960s, improving food security and safety as well as economic growth, comparing to the first years coming after World War II (Kirchhelle 2019).

Global livestock production has increased substantially between 1960s and the early twenty-first century. Between 1960 and 2010, beef production more than doubled, while over the same time chicken meat production increased by a factor of nearly 10, made up of increases in both number of animals and productivity. From the early 1960s to the mid-2000s carcass weights increased by approximately 30% for both chicken and beef cattle and by approximately 20% for pigs (FAO 2010). Increases in milk production per dairy animal have amounted to about 30% for cows' milk, about the same as for increases in egg production per chicken over the same time period (FAO 2010).

Antimicrobials have been widely used worldwide since introduced, both in people and in animals and have contributed to better human and animal health as well as animal welfare. Productivity also greatly improved due to their use, improving food security, nutrition as well as economic growth. Unfortunately, these improvements came with a price, as the widespread use of antimicrobials in the past decades, in animals as well as in humans, have led to a dramatic emergence of antimicrobial resistance. In 1969, the UK Swann Committee reported that there was a significant problem with regard to antimicrobial (mis)use in both human and veterinary medicine and recommended that the UK Government establish a committee that should have overall responsibility for the whole field of antimicrobial use (Swann et al. 1969). Worst case scenarios estimate that antimicrobial resistance can contribute to 10 million human deaths per year and 10% production loss in the livestock sector in low income countries by 2050 if the emergence is not mitigated (O'Neil 2016).

Picture 1 Aureomycin, farm feed advertisement, 1954 (Fine Chemicals (1951)

1.2 Ways Antimicrobials Are Used in Animals

> ***Ways of antimicrobial use:***
>
> - Preventive ('prophylaxis')
> - Control ('metaphylaxis')
> - Curative ('therapeutic')
> - Growth promotion (low dose, long duration)

Globally, antimicrobials are used for various purposes such as preventive use (prophylaxis), control use (metaphylaxis), curative (therapeutic) treatment and growth promotion use. The European Platform for Responsible Use of Medicines (EPRUMA), being a multi-stakeholder platform amongst all animal health sector actors, adopted the following definitions (EPRUMA 2008).

Preventive treatment is also called "Prophylaxis"; derived from the Ancient Greek πρό (pró, "before") + φύλαξις (phúlaxis, "a watching, guarding"). Regulation *6/2019/EC* uses the term "prophylaxis" and defines it as administration of a medicinal product to an animal or group of animals before clinical signs of a disease are observed, in order to prevent the occurrence of disease or infection. Routine prophylaxis is considered as the most controversial. Such routine either can replace other more costly preventive measures (e.g. vaccination, good husbandry and nutrition, biosecurity, hygienic and animal welfare measures), or is used when there is a high risk for disease, for example before mixing of animals, when transporting a combination of 1-day-old chicks or veal calves from different sources or around the time of weaning when the animals are not yet ill but the farmer knows a risky period is coming. Currently, there is an increasing trend of phasing out such use. It is therefore vital to link the reduction of antibiotic usage with the improvement of animal welfare and other best practices in farming. Per species, some key animal welfare criteria should be monitored (e.g. cows: mastitis and lameness, pigs: mortality, tail/ear/flank biting) to ensure welfare improvements go hand in hand with antimicrobials use reduction (FVE 2016, 2019). These should be combined with good hygiene, biosecurity measures, good nutrition and other best Practices (EPRUMA 2008). Dry cow therapy is a specific case whereprophylactic use may be necessary. Nevertheless, the current approach is to move towards selective dry cow treatment after monitoring of the health status of the cows, evaluating incidence of mastitis and whenever possible using teat sealants without antimicrobials (Rabiee and Lean 2013).

Specific *perioperative prophylaxis* is considered usually less risky when current approaches of minimal duration are followed (For more information please refer to Chap. 7).

Control treatment is also called "Metaphylaxis"; derived from the Ancient Greek μετά- (meta, "after"/"beyond") + φύλαξις (phúlaxis, "a watching, guarding"). Regulation *6/2019/EC* uses the term "metaphylaxis" and defines it as the administration of a medicinal product to a group of animals after a diagnosis of the establishment of clinical disease in part of the group, with the aim of treating the clinically sick animals and controlling the spread of the disease to animals in close contact and at risk, which may already be infected, but are without clinical signs. Animals in close contact are the prerequisite of spreading of the diseases. Control treatment is regularly done in holdings, where a large number of animals are kept in confined places, as happens with broilers and pigs.

Curative treatment is also called therapeutic treatment, derived from the Ancient Greek θεραπευτικός (therapeutikós, "attentive, helpful, obliging, curative"). Curative treatment is the treatment of an ill animal or group of animals, when the diagnosis of disease or infection has been made.

Growth promoter use is the use of antimicrobials in the absence of disease, for a long duration at low doses in animal feed to improve the growth or yield performance. While any use of antimicrobial agents selects for resistance, use of low doses for a long time is considered particularly risky. A study showed that the use of antimicrobial agents for growth promotion in Denmark, Finland, and Norway have selected for resistance to most of these drugs among Enterococcus faecium in pigs and broilers (Aarestrup et al. 2000). Shorter courses can prevent selection for resistance to concentration-dependent antimicrobials if sufficiently high dose is chosen—the case of fluoroquinolones (Rees et al. 2015), while low doses and long treatment may be of particular concern from the perspective of resistance selection (Lin et al. 2014).

The EU Feed Additives Regulation (Regulation 1831/2003/EC) banned the use of antibiotics as growth promoters in the EU from 1 January 2006. In the United States those called medically important (MI) have been banned since 2017, the latter as a result of new FDA Veterinary Feed Directive (AVMA 2017). According to a survey done by the World Animal Health Organisation (OIE) in 2017, a total of 45 out of 155 responding OIE Member Countries (29%) reported the use of antimicrobial agents for growth promotion in animals in their countries. This demonstrates an important decrease compared to 2012 and to 2015 data, when 49% and 74% of the countries, respectively, declared usage of antimicrobial agents as growth promoters. The OIE also asked its

Member Countries what antimicrobial agents were authorised as growth promoters. The most frequently quoted antimicrobial substances for this purpose were tylosin (17 countries) and bacitracin (18 countries). Colistin was mentioned by 12 countries (OIE 2017, 2018a, b). For more details related to antimicrobial growth promoters and their use and authorisation please refer to Chap. 7.

Although no precise data exist on the ratio of prophylactic, metaphylactic or curative use of antimicrobials in Europe, some assumptions can be made, based e.g. on the reports of the European Commission on country fact finding missions performed in 2016 and 2017 (European Union, Health and Food Safety Directorate-General 2018). One assumption is that the vast majority of medicated premixes was or is used as prophylaxis or metaphylaxis. This is medically supported at least by the fact that clinically ill animals have decreased intake of the feed; therefore, they are rather treated via medicated drinking water or in case of individual/small groups of animals parenterally. The report brings also the message that most of the visited Member States have notably shifted away from medicated feed to using of powders or concentrates for medication of drinking water and therefore still keep in high percentage oral administration of antimicrobials. Five countries noted in their responses to the questionnaire that prophylactic use is either not permitted in general or it is restricted, either through legally prohibiting the preventive use of medicated feed or through voluntary bans implemented in the poultry and rabbit sectors. However, it is not known if this approach has simply led to an increase in the administration of antimicrobials via drinking water which, as one country noted, in the absence of disease in all or the majority of animals, would indeed constitute prophylactic or metaphylactic use. In two of the visited Member States which report the highest sales of antimicrobials, prophylactic use via medicated feed was widespread.

Since 2011, the European Surveillance of Veterinary Antimicrobial Consumption (ESVAC) from the European Medicines Agency (EMA) annually reports national sales figures of veterinary antimicrobial agents in food-producing animals (overall sales data). As stated in the new regulation EU 2019/6 on veterinary medicinal products, collection of data on antimicrobial medicinal products used in animals should not only be based on volume of sales but also on the use of antimicrobial medicinal products. New EU legislation on veterinary medicines foresees that Member States shall be collecting "use" data by animal species. Availability of "use" data would allow for integrated analysis of the consumption of antimicrobial agents and occurrence of antimicrobial resistance. It would also enable to monitor use patterns over time and the effect of implemented measures regarding, e.g. prudent use of antimicrobials (EMA 2018a, b, c). Some European Countries have already set and used detailed monitoring systems, such as Denmark (DANMAP). However, even if there are available detailed data on use of antimicrobials in animals gained, e.g. from pharmacies or from end users, distinguishing between prophylactic, metaphylactic or targeted treatment use will be difficult.

Governmental and industry bodies in different countries have started to take greater care on food quality in connection with responsible and reduced used of antimicrobials. Such systems, e.g. Q-CZ system in producers of poultry (EAGRI-CZ 2018) define exact conditions of use, monitoring and handling of antibiotics, e.g. antimicrobials cannot be used in flocks preventively, no use of cephalosporins of third and fourth generation is allowed at any stage of production (i.e. neither (grand) parents flocks, nor broilers for fattening), critically important antimicrobials as fluoroquinolones and macrolides can be used only prudently - when narrow spectrum first choice antimicrobials do not work and after performing susceptibility testing (rule enforced also by the Czech Republic Decree No 344/2008).

Increased pressure from consumers on food retailers such as McDonalds and KFC have promoted a shift to antibiotic-free animal food products (Baertlein 2015) and major food producers such as Tyson Foods and Perdue Farms have declared they will stop using antibiotics in broiler chicken production (FAO 2018).

In conclusion, it can be said that many stakeholders, worldwide, at the European level (enforced by the new legislation on VMPs and medicated feed(European Commission 2019a, b)) or at the national level (governmental bodies, farmers associations) have started to take the use of antimicrobials and antimicrobial resistance seriously. Even private sector, especially those targeted on food production and processing endorse and implement the requirements aiming to decreasing the use of antimicrobials in practice.

1.3 Quantification of Antimicrobials Used

> ***Terms to quantify antimicrobials:***
>
> **Consumption data:**
> general term used for sales, prescription or use data
>
> **Sales data:**
> antimicrobials sold
>
> **Use data:**
> End-user data, e.g. prescription data or farm records

Data on the *consumption*, e.g. sales data and use data, of antimicrobial agents as well as data on *resistance* are essential components to inform policies and strategies for the containment of antimicrobial resistance (see below). The data on either sale or use, including their trends, allow the evaluation of success of measures that were taken to minimise the use of antimicrobials as well as of the efficacy of prudent use campaigns.

To get representative and reliable data on consumption of antimicrobials is complex. It needs the development of robust systems, the corporation of many stakeholders and good statistics not only as for the amounts of antimicrobials consumed (numerator), but also on the number of animals being at risk (denominator).

2 Estimating Antimicrobial Consumption

Within the European Union, ESVAC (European Surveillance of Veterinary Antimicrobials Consumption) system was established in 2009. During ESVAC 10- year history starting from the originally participating 9 countries, it developed and achieved the data on sales from 30 countries of the European Economic Area. Surveillance data of antimicrobial consumption, such as the ESVAC data, are estimated on the basis of collected data on the number of packages of veterinary antimicrobial medicinal products sold. These data are provided by wholesaler-distributors, feed mills, manufacturers, pharmacies and/or marketing authorisation holders (EMA 2019a).

The raw data provided to the ESVAC at the level of individual packages of veterinary medicinal products are validated and analysed and finally the exact amount of active ingredients is calculated to express *numerator*—i.e. the amount of antimicrobial agents sold. The amounts are not expressed for individual presentations of veterinary medicinal products but are calculated and expressed aggregated per pharmacological groups of antimicrobials. The portfolio of both pharmaceutical forms, presentations and pharmacological groups of veterinary antimicrobial agents sold, are influenced by "market depending" factors and the other factors, linked to the animal demographics in each country, which differ across Europe. However, sales data of veterinary medicinal products do not allow to specify accurate amounts of antimicrobials used per species of animals (as most of the products are authorised for several species and finally it is hard to trace the indication for which each product was sold and whether it was really administered to an animal). Therefore, interpretation of the sales data should be done with caution. Due to the complexity of the overall process of data collection, it normally takes a couple of years before data collection systems in individual countries are validated and therefore gaining of the stabile European system was a matter of several years (EMA 2019a).

As noted above, to calculate correctly the use of antimicrobials, not only the consumption data, is very complex and requires specifically the consideration of the number of animals at risk, as well as of the population and/or biomass of animals. Therefore, within the ESVAC project, the Population Correction Unit (PCU) has been established as the technical proxy—expressing the denominator. The input data for calculations coming from Statistical Office of the European Union (EUROSTAT) whenever available, or can be supplemented by the national statistics for certain species (e.g. fish, rabbits). The PCU for each animal category is calculated by multiplying numbers of livestock animals (dairy cows, sheep, sows and horses) and slaughtered animals (cattle, goat, pigs, sheep, poultry, rabbits and turkeys) by the theoretical weight at the most likely time for treatment (refer for more details in Table 1). For fish biomass live-weight slaughtered (EUROSTAT data whenever available) is used for calculation of the total PCU. To consider also the export/import patterns data from TRACES (TRAde Control and Expert System, run by the European Commission's DG SANTE) are used.

It is generally considered that more detailed data reflecting the use in individual farming sectors, specifically in major food-producing animals as pigs, poultry (chickens and turkeys), cattle, but also in generally less represented species (with different importance in different EU Member States) as horses, sheep, goat, rabbits, fish, are needed. Monitoring of exact end-use data requires high initial investments. Setting up the whole data collection system, including, e.g. training of farmers/veterinarians in delivery of the end-use data, can be quite challenging. Some Member States may find it more difficult to create such new systems of data collection on use of antimicrobials at farm level. There is also a need for sufficient validation of already existing national systems on sales of antimicrobials. Therefore, in 2018 the European Medicines Agency started the pilot project on stratification of the sales data which provides an estimation of the antimicrobial consumption per species based on an approximate allocation of the proportion of total sales to each of the species for which a veterinary medicinal product (VMP) containing antimicrobials is authorised. This project is an interim approach to estimate consumption of antimicrobials per animal species for those countries that have not started yet to collect use data and provides complementary information for those countries that have the data for major species to estimate the use in minor species (EMA/284404/2018).

As stipulated by the veterinary medicinal product legislation (Regulation EU 2019/6), collection of use of antimicrobials by animal species at farm

Table 1 List of species/categories of animals including counting of living and slaughtered animals as well as exports/imports serving for calculation of PCU (modified according to EMA 2019a)

Species	Slaughtered/fattening/living	Categories
Cattle (heads/number of animals)	Slaughtered	Cows, heifers, bullocks, bulls, calves, young cattle
	Slaughtered/fattening bovine	Export/import
	Living	Dairy cows
Pigs (heads/number of animals)	Slaughtered	Pigs
	Slaughtered/fattening pigs	Export/import
	Living	Sows
Poultry (heads/number of animals)	Slaughtered	Broilers and turkeys
	Slaughtered poultry	Export/import
Caprinae (heads/number of animals)	Slaughtered	Sheep and goat
	Slaughtered/fattening sheep	Export/import
	Slaughtered/fattening goat	Export/import
	Living	Sheep
Equidae (heads/number of animals)	Living	Horses
Rabbits (heads/number of animals)	Slaughtered	Rabbits
Fish (tonnes)	Biomass fish live weight	

level should be implemented in all Member States by 2030. It is of importance that "use" data allow more detailed analysis, such as to identify animal species, production type, which regions, and, in cases where indications are recorded, also for which diseases the antimicrobials are used. If performed at farm level, analysis allows benchmarking and evaluation of the effect of measures taken. "Use" data may also record "cascade" or "off label" use of antimicrobials (using antimicrobials for animals differently from approved claims, i.e. outside of the terms of marketing authorisation). To collect the relevant "use" data variables, data sources other than those providing details on sales are needed, such as farm level data, e.g. prescription data or using the veterinary or farm records. The new Regulation on VMPs (6/2019/EC) brings the framework in Article 57 on data to be collected. More technical details will be given via a delegated act. It is expected that in the first phase (by 2024) data from major food-producing species (cattle, pigs, poultry) and the respective production categories will be obligatory collected for all antibacterials used for treatment of those animals. Next steps will cover the addition of other food-producing animals—expected till 2027. A remaining part, mostly targeted on antimicrobials' use data in companion animals as dogs and cats, but also other animal species (fur animals), should follow by 2030 as stipulated by the Regulation. Data can be collected via a system which covers most of the animal production system in a country.

The growing pressure from consumers have led some private food processing companies to start monitorin the use of antimicrobials at the level of primary producers e.g. QS system Germany, which declared in 2019, that since 2014 the amount of antibiotics used in the QS scheme (calculated in tonnes, 2014–2018) has been reduced by 35.7% (Nienhoff 2019) or AB Register Belgium (AB Register vzw 2019).

A successful example is the Dutch public–private partnership set up between government and stakeholders from the major livestock sectors (pigs, broilers, veal calves and dairy cattle) and the Royal Dutch Veterinary Association (KNMvD). This system was established in 2010 to collect data on antimicrobial consumption on farms, establish benchmark indicators for individual major livestock sectors and analyse trends in consumption. They set up the Dutch independent Veterinary Medicines Authority (SDa) with three objectives: firstly to collect and report antimicrobial use data from farms and veterinarians; secondly to set annual targets for antimicrobial use in each livestock sector; and thirdly to set species-specific benchmarks that differentiate between moderate, high and very high users (farmers) and prescribers (veterinarians) (Bos et al. 2013).

Speaking about certain systems and utilisation of the data on use, there should be also mentioned the example of the behaviour change campaign in Denmark via the introduction of a 'Yellow Card' system. The system helps raising awareness about antimicrobial overuse by giving veterinarians a 'Yellow Card' if they use antimicrobials in a quantity two times higher than the national average. To be able to recognise the patterns of use as well as to perform benchmarking among veterinarians, the system of collection of the data on use should be sufficiently robust. Danish VETSTAT system was built many years ago and allows precise analysis. The system has been associated with an overall reduction of 22% in antibiotics use in pigs for the period 2009–2015 (Ministry of Environment and Food of Denmark 2016). The use data are in Denmark were utilised also within the Danish Integrated Antimicrobial Resistance Monitoring and Research Programme (DANMAP), as this programme reports on usage and on the occurrence of antimicrobial resistance in zoonotic, indicator and pathogenic bacteria from animals, food and humans.

In 2017, a project was launched with funding from the Joint Programming Initiative on AntiMicrobial Resistance (JPIAMR), called AACTING (Network on quantification of veterinary **A**ntimicrobial usage at herd level and **A**nalysis, **C**ommunica**T**ion and benchmark**ING** to improve responsible usage). Among the aims of this project were to provide an overview of existing farm-level monitoring systems for antimicrobial use, as well as to provide practical guidelines for setting up such systems. An overview of the systems is available online, as a searchable database (http://www.aacting.org/monitoring-systems/) as well as a pdf (AACTING 2019a). Currently (March 2020)

38 systems from 16 countries are included; it is meant to provide an overview of all worldwide existing systems and is not limited to Europe. A system from Canada is also included in this network. Table 2 provides a summary of the characteristics of the systems that are of main relevance.

The AACTING guidelines address data collection and analysis, i.e. the calculation of indicators to quantify use of antimicrobials, along with reporting of the results and benchmarking as an important part of the antimicrobial stewardship in veterinary medicine. The guidelines were written based on the experiences of the countries participating in AACTING, and meant to provide tips and tricks as well as to suggest "best practices" to deal with the set-up or revision of monitoring systems: http://www.aacting.org/guidelines/ (AACTING 2019b).

In several European countries, farm level antimicrobial use monitoring systems have been implemented to a certain degree. Data can be collected via a system which covers all or most of the animal production sector in a country or via a sample survey by collecting data from a well-designed random selection of farms. The data collection systems are managed and controlled by the government or private stakeholders, e.g. quality schemes. In some countries, these data are used to benchmark farms and veterinarians, so that they can see how they perform compared to their colleagues (Table 2).

Think About 1

Antimicrobials use patterns:

How far we can go in reduction of use of antimicrobials?
What is necessary use and what is overuse or misuse?
How to define thresholds and what to prevent promoting wrong practices?

Despite the fact that there is a significant decrease in antimicrobial consumption in certain EU countries and a broad consensus that the use of antimicrobials can still be lowered in EU, the question is to what extent antimicrobial use is needed and how far can we go in minimising it. Costs and achievements should be thoroughly considered and balanced.

Different indicators are measured to benchmark the use of antimicrobials at farm level and scoring prescriptions of vets. Some of those systems are linked to legislation and include fines for users higher than country/sector threshold. Sometimes it is very tricky for the auditors to assess what is necessary use and what is overuse or misuse. There were also signals from practice that very strict thresholds can lead to wrong practices as shortening of the duration of treatment and/or minimising the dose used to fulfil the thresholds. This can be finally even worse than the correct duration and dose of antimicrobial, especially considering selection of resistance. **The reduction of antibiotics should be achieved by reducing the need for antimicrobials, rather than by reducing the amount or duration of necessary antimicrobial treatments of properly diagnosed bacterial infections** (Federation of Veterinarians of Europe 2016a, b)

Therefore it should be considered not only to set the thresholds and to benchmark, but also to perform audits on how reduction of the use was achieved, if it was accompanied by use of "alternative tools" and further measures that helped to improve animal health status and welfare and finally to also investigate the health status of animals on the farm and at the slaugherhouse (e.g. meat inspections, lung lesions score and scoring of signs of chronic enteritis/peritonitis (Federation of Veterinarians of Europe (2018)). More evidence proved studies are needed to show positive examples on how to reduce the use of antimicrobials at farm level without compromising animal health and welfare food safety as well as keeping the system sustainable from an economic perspective.

Table 2 Summary of the characteristics of the farm-level monitoring systems for antimicrobial use that are of main relevance and are listed under the AACTING project

	N		
	Countries	Systems	Importance and remarks
Countries (total)	16		As has been the case for sales-data collection, Europe has also been front-running with respect to farm-level use data collection.
Inside Europe	15		Most systems are in Western and northern European countries, while central, southern and especially Eastern Europe have fewer countries with systems.
Outside Europe	1		In Canada; it is known that some small private systems exist in the United States but no data has been provided to AACTING so far.
Systems (total)		38	
Systems per country		1–5	Some countries have separate systems for different species/sectors, systems with different funding/ownership, "test" systems and "real" systems
Main funding/ ownership			Not all systems are fully financed by either private or governmental funds; main funding generally means leading and managing the system.
Government	14	19	
Private	10	19	Private systems can be quality assurance schemes, industry organisations, farmers associations
Coverage			Coverage is the % of farm(er)s that (should) have their antimicrobial use data in the system.
Full coverage	11	12	This refers to systems that cover, in theory, 100% of a sector.
Partial coverage	8	16	This refers to systems that cover a non-random, large part of a sector, mostly on a compulsory base.
Sample/survey	7	10	This refers to systems that cover a small, random, representative part of sector, mostly on a voluntary base.
Species/sectors			
Pigs	16	24	Different categories can be distinguished: Sucklers, weaners, fatteners, sows/boars, gilts.
Poultry	14	22	Different categories can be distinguished: *Gallus gallus*, Turkeys: Broilers, laying hens, breeding animals
Calves	11	16	Mostly industrial veal calves.
Dairy cows	11	13	
Fish	7	7	
Others	11	15	This can be beef, horses, sheep, goats, deer, rabbits, minks, cats and dogs
Analysis			Analysis refers to the calculation of an indicator from the raw usage data, mostly comprising a number of units of measurements related to (an amount of) antimicrobials used (nominator) standardised by an animal population (denominator). The denominator, i.e. animal population/ biomass or another consensual technical unit can be calculated from the raw data on the produced kg animal biomass (e.g. at slaughter) and/or the average number of animals present at the farm
Mass-based	9	17	Typically mg/kg or mg/PCU (population correction unit)
Dose-based	10	18	The unit of measurement is dose-based, e.g. used daily dose, defined daily dose, defined course dose
Count-based	7	13	Antimicrobial use is calculated through the number of treated animals or treatment days.
Benchmarking			Benchmarking is the comparison of antimicrobial use with a reference group, with the antimicrobial use of the whole group calculated in a similar way.
Farmers	10	16	
Vets	4	4	

2.1 Challenges to Measure Antimicrobial Use in Animals on a Global Level

Compared to the European Union, surveillance of antibiotic consumption is much less evolved in many other parts of the world. According to a survey done by the World Animal Health Organisation (OIE) in 2015, it is very challenging for low- and middle-income countries to provide accurate quantitative data on the use of antimicrobial agents in animals. Despite this, the number of reporting countries is increasing. Within the third report, data from completed reports submitted by 155 countries was analysed; in the first report 130 Member Countries submitted completed reports. While the majority (55.6%) of the 181 OIE member countries provided quantitative data on the consumption of antimicrobial agents in animals, most of these countries cannot indicate the quantities of antimicrobial agents used by animal groups as defined by OIE, or the routes of administration, and cannot distinguish therapeutic use from use for growth promotion. In order to enable accurate comparisons among countries, it is important to take into account the animal population of the country that can be potentially exposed to treatment. Development of a suitable denominator (animal biomass) is vital (OIE 2017). The good news in this respect is that according to the third OIE report (2018a, b) the situation is under the stepwise development and is assumed to be improved in near future. FAO aims to stimulate reporting to OIE by supporting the countries to provide data on import, national sales and distribution of antimicrobials (OIE 2017). FAO contributesa to the development of the methodology for data collection at the farm level for those countries building the systems of surveillance of the antimicrobials use.

The amount of the antimicrobial agents intended for use in animals in kilograms are reported based on the source data as number of packages of a given pharmaceutical preparation sold. OIE provides guidance by giving detailed instructions on mathematical calculations to obtain quantities of the active ingredients from VMPs containing the antimicrobial agents sold. All antimicrobial agents destined for use in animals and contained in the OIE List of Antimicrobial Agents of Veterinary Importance (OIE 2018a, b), in addition to certain antimicrobial agents used only for growth promotion, were reportable.

Animal biomass is currently employed by OIE as a *denominator* in analysis of quantitative antimicrobial use data by other national and regional antimicrobial use surveillance groups, such as the European Surveillance of Veterinary Antimicrobial Consumption (ESVAC), the U.S. Food and Drug Administration (FDA), the Canadian Integrated Program for Antimicrobial Resistance Surveillance (CIPARS), and the Japanese Veterinary Antimicrobial Resistance Monitoring System (JVARM). Animal biomass is calculated as the total weight of the live domestic animals in a given population and year. Calculated number is considered a proxy representing those likely exposed to the quantities of antimicrobial agents reported. Animal biomass for the purpose of the OIE reports is the total weight of that country's production animals, but not incorporating companion animals (Table 3). The formulas for calculating biomass by species were developed using the two globally available datasets, WAHIS and FAOSTAT, and the results were compared to references from countries where more detailed animal population data by production class were available (e.g. from the EU countries EUROSTAT). All weights and biomass figures are measured in kilograms within the OIE methodology and report.

One of the specificities of the OIE template is the inclusion of a question for countries to report any antimicrobial agent *authorised or used* in animals as *growth promoters (AGPs)*. Ionophores are excluded from reporting as they are mostly used for parasite control and have different regulatory classifications in different countries. Despite that, some countries mentioned ionophores as AGPs as they are authorised as such.

Among the 116 countries providing quantitative data, tetracyclines (34.5%) were the most

Table 3 Species/categories involved in the OIE template as were grouped for further analysing (modified according to OIE 2019)

Species	Categories
Poultry	Layers (eggs commercial production)
	Broilers (meat commercial production)
	Other production poultry
	Backyard poultry
Bovines	Cattle
	Buffalos (excluding *Syncerus cafer*)
Sheep and goats	Sheep and goats (separately or together (mixed herds))
Pigs	Commercial pigs
	Backyard pigs
Aquaculture production	Fish
	Crustaceans
	Mollusc
	Amphibians
Equidae	
Rabbits/hares	
Bees	Bees (honey production)
Cervidae	Animals kept on farms
Camelidae	
Reptiles	Food producing (e.g. crocodiles)

commonly *reported* antimicrobial class, followed by penicillins (15.2%), polymyxins (10.0%) and macrolides (9.8%). As for aquatic species, where just nine countries reported quantitative data (four of them EU), florfenicol was the first one.

Expressing validly global consumption of antimicrobials in food animal production is difficult, as data on antimicrobial use in food-producing animals is scarce in many parts of the world. Correct expression of the animal population at risk is also challenging. Evidence-based predictions are hard to be done considering not only the above listed difficulties, but as well the assumption that use practices do not always follow regulations. Despite the fact that it is difficult to foresee the future, the following data were published and used by many references: estimates that consumption might rise by 67% to 105,596 (±3605) tonnes, by 2030, with the largest users then being China (30%), the United States (10%), Brazil (8%), India (4%), and Mexico (2%) (Van Boeckel et al. 2015). Looking per species, the global average annual consumption of antimicrobials per kilogram of animal produced is estimated as 45 mg/kg, 148 mg/kg, and 172 mg/kg for cattle, chicken, and pigs, respectively. The study used data from 32 high-income countries, interpolated these data among other high-income countries and subsequently extrapolated the data to estimate antimicrobial consumption in intensive production systems of low- and middle-income countries. It is recognised that the results are based on assumptions and have limitations. To decrease global use of antimicrobials in animals, several solutions have been suggested, such as to promote responsible use through regulation (as done currently in Europe and the United States), to promote low-animal-protein diets (currently done in China) or to charge a user fee, paid by veterinary drug users, on sales of antimicrobials for nonhuman use (Van Boeckel et al. 2017).

As noted above and will also be explained further in this chapter, numerous actions have been or are being taken worldwide to decrease the levels of antimicrobial use in animals. As the issue of minimising the use of antimicrobials is a global one some very recent activities have been launched as the project Healthy Livestock (http://healthylivestock.net/the-project) to tackle antimicrobial resistance (AMR), in which experts and scientists from the European Union and China work together in the EU—Chinese research subprojects, where more than 20 partners (from academia, research institutes, private partners and companies) join forces to improve the health and welfare of pigs and poultry (Wageningen University and Research 2018). Therefore, we should probably rather wait for the evidence-based statistics evaluating the complexity of the situation and aiming for changes not only in Europe, but worldwide.

2.2 Consumption of Antimicrobials in Animals Within European Economic Area

The development of an approach for collection of harmonised data on sales of antimicrobials in animals at European level started in 2009, when the European Medicines Agency launched the European Surveillance of Veterinary Antimicrobial Consumption (ESVAC) project. The denominator, Population Correction Unit (PCU) is established as a proxy for the animal population at risk to be treated (please refer also for more details above).

The sales data are expressed by the indicator mg/PCU, i.e. the amount of veterinary antimicrobials quantified in mg of active ingredient is normalised by the PCU, or in tonnes.

Detailed sales data at veterinary product presentation level is yearly gathered from individual EU and EEA countries and subsequently published in the ESVAC report, publically available on the official ESVAC webpage (https://www.ema.europa.eu/en/veterinary-regulatory/overview/antimicrobial-resistance/european-surveillance-veterinary-antimicrobial-consumption-esvac).

As shown in the ESVAC report published in 2019 (EMA 2019a), the average sales for all 31 countries which delivered data compiled from 2017 was 107.0 mg/PCU, while the median was 61.9 mg/PCU. Large differences were seen between the most- and least-selling countries (range 3.1–423.1 mg/PCU). These large differences can only partly be explained by, among others, differences in the animal demographics, production systems, size of the herds and in dosing of the various antimicrobials between countries.

The trends of sales of antimicrobials in individual countries can be followed within Fig. 1 that also indicates for waste majority of the countries the decrease in sales of antimicrobials in last years.

The most sold antimicrobials in Europe in 2017, expressed as a proportion of overall mg/PCU, were tetracyclines (30.4%), penicillins (26.9%) and sulfonamides (9.2%). Overall, these three classes accounted for 66.5% of the total sales in the 31 countries that submitted data. The proportion of sales of the highest priority critically important antimicrobials (CIAs) for human medicine included in the World Health Organization (WHO) list (6th revision) was low. The sales for food-producing animals of third- and fourth-generation cephalosporins, quinolones, fluoroquinolones, polymyxins and macrolides accounted for 0.2%, 0.4%, 2.2%, 3.4%, and 7.4%, respectively, of the total sales in the 31 European countries. Again, large variations in sales of the above noted antimicrobial classes were observed between countries (EMA 2019a).

In 2017, the major proportion of antimicrobials sold (mg/PCU) were products for group treatment (89.4%), premixes accounted for 28.8%, oral powders for 9.9% and oral solutions for 50.7%. The proportion accounted for by pharmaceutical forms for group treatment varied substantially between countries, ranging from 3% to 96% (EMA 2019a).

Antimicrobial use in animals has decreased in Europe in the last years, including sales of CIAs. For 25 countries reporting sales data to ESVAC for the years 2011–2017, an overall **decline in antimicrobial consumption (mg/PCU) of 32.5% was observed**. Total sales fell from 162.0 mg/PCU in 2011 to 109.3 mg/PCU in 2017 (EMA 2019a) (Fig. 2).

Important hereby to note is that the overall livestock population stayed stable in the EU/EEA countries covered by ESVAC during the period

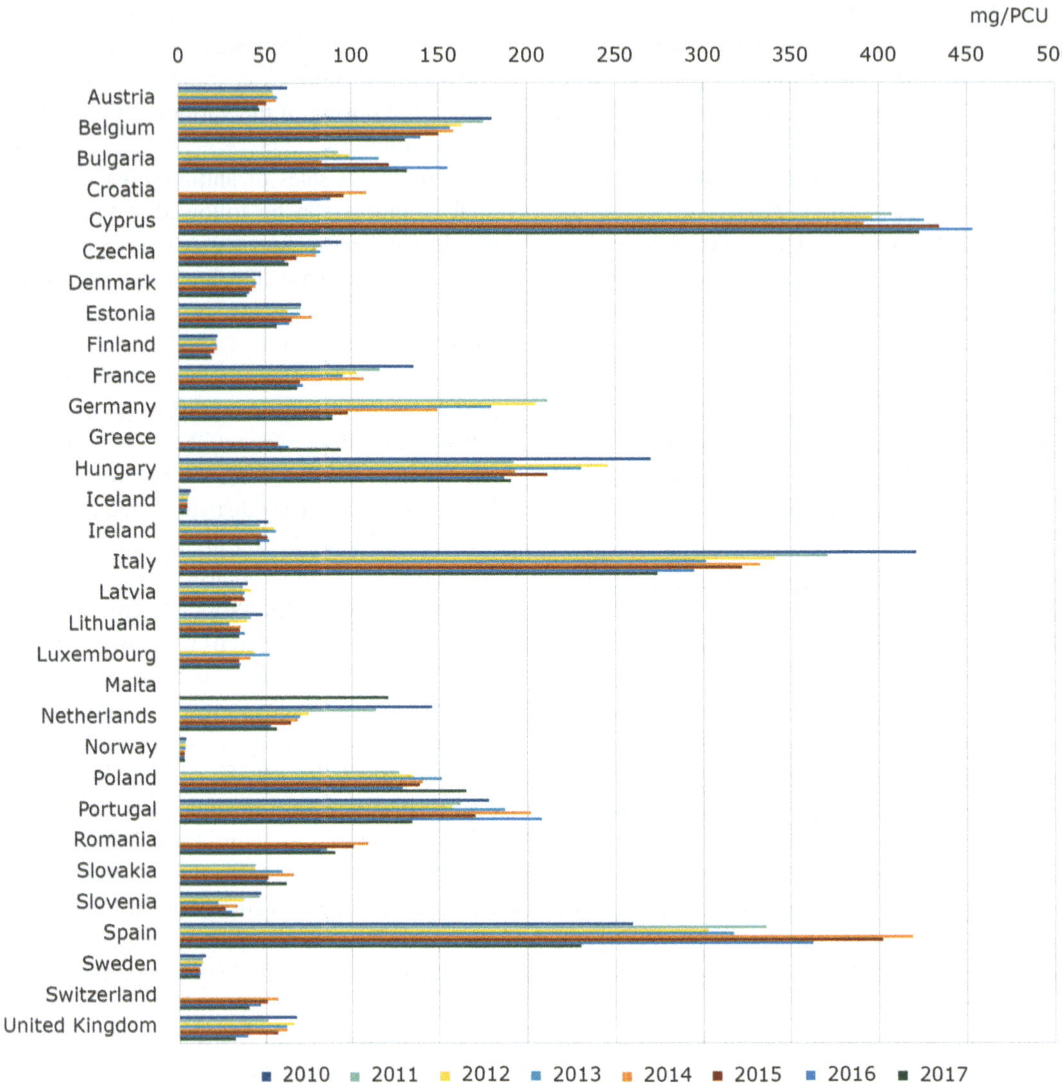

Fig. 1 Trends in overall sales of antimicrobials in EU/EEA countries 2010–2017 (modified according to the on-line ESVAC data available). Figure should be interpreted with caution due to factors impacting the sales (e.g. animal species demography, husbandry management/technologies and intensity of farming, portfolio of available VMPs as well as socio-economic factors). (1) Corrections to sales data and/or PCU data published in the ESVAC 2016 report are described in Chapter 1.5. of the ESVAC (2019) report. (2) Under-reported for Bulgaria for 2011, 2012 and 2014 as several wholesalers failed to report data. (3) Strength reported as the base for most VMPs for 2011–2012 for Czechia; for 2013–2017, strength reported as in the label of the VMPs. (4) Strength reported as the base for some VMPs for 2011–2012 for the Netherlands; for 2013–2017, strength reported as in the VMPs' label. (5) For Portugal, under-reporting has been identified for 2010–2014 and 2017. (6) For Romania, 2014 data were updated, as wholesalers initially failed to deliver all sales data. (7) For Slovakia, for 2011 and 2012, the data represent antimicrobial VMPs imported by wholesalers; from 2013, data represent all sales from wholesalers to end-users (veterinarians, pharmacies, producers of medicated feeding stuffs and farmers, obtained by import and from national manufacturers). (8) For Spain, under-reporting for the years 2010 to 2013 has been identified (underestimated) and the data provider for 2017 data was changed from MAHs to retailers. (9) For the United Kingdom, high sales of certain tetracycline-containing products late in 2010 were probably used in 2011 and thus their use has been underestimated for 2011

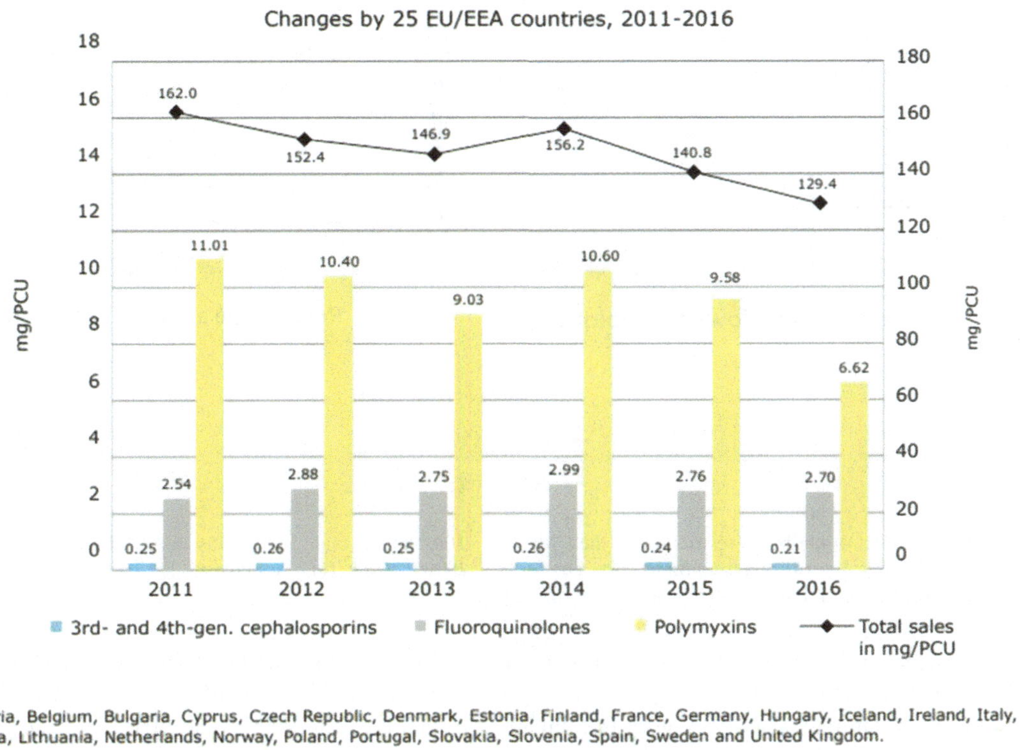

Fig. 2 Changes in aggregated overall sales and sales of third- and fourth-generation cephalosporins, other quinolones, fluoroquinolones and polymyxins for 25 EU/EEA countries, from 2011 to 2017 (EMA 2019b)

2011–2016. The reduction of antimicrobials sold was possible seemingly without detrimental effects on the livestock production.

It is very promising that in some EU countries a positive correlation can be seen between the reduced use of antibiotics and the levels of antimicrobial resistance (Chantziaras et al. 2014;-European Union 2017; EFSA/ECDC 2020). While considering big benefit of minimising the selection pressure for antimicrobial resistance spread and development by dropping down the use of antimicrobials, we should also be aware of the "costs of the success". All measures should be considered very thoroughly and in complexity and the role of well-educated veterinarian as well as farmers or generally people taking care of animals should be highlighted. Among "Cost of success" should be considered both impact on the health status (either clinical or subclinical diseases incidence rate (Maas 2014)), increased rate of lesions at slaughter (Alban et al. 2013) and economical costs (what can be illustrated by Fig. 3 (Maas 2014)).

3 Human Versus Animal Antimicrobial Consumption in Light of One Health Concept

Europe has a unique collaboration of three European Agencies collaborating together to compare the use of antimicrobials and the resistance levels in both animals and humans. These agencies are the European Centre for Disease Prevention and Control (ECDC), the European Food Safety Authority (EFSA) and European Medicines Agency (EMA). In 2017, the second Joint Interagency Antimicrobial Consumption and Resistance Analysis (JIACRA) report was published (JIACRA 2017). This report estimated that the average antimicrobial consumption in animals was 151.5 mg/kg and in humans

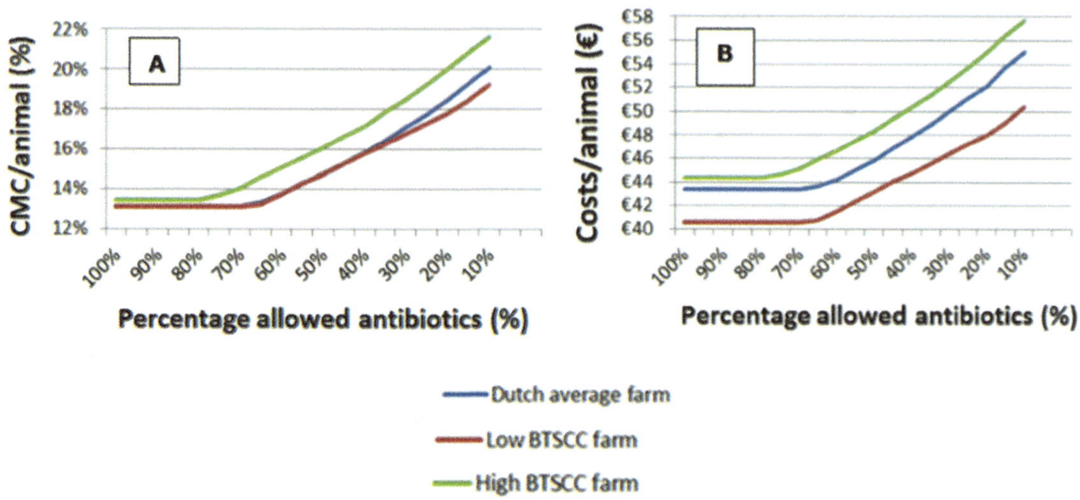

Fig. 3 The effect of antibiotic restriction on the CMC—clinical mastitis cases (**a**) and economic costs (**b**) for a Dutch average-, high BTSCC- and low BTSCC farm with 10% more clinical mastitis risk (Maas 2014). Note: *BTSCC* Bulk Tank Somatic Cells Count.

123.7 mg/kg (see Fig. 4 underneath). Looking at the median antimicrobial consumption, this was 67 mg/kg in animals and 118 mg/kg in humans. In 18 of the 28 countries, antimicrobial consumption was lower in animals than in humans. However, due to a number of countries in Europe with very high antimicrobial usage levels in animals, the average animal consumption is much higher than the median human consumption rate.

It should be recognised that comparison between human and animal antimicrobial use should be done with great care. Next to ESVAC, the European Surveillance on Antimicrobial Consumption (ESAC-net) collects data on human antimicrobial use. However, some countries do not provide full data coverage but e.g. hospital care data only or data on antimicrobials from reimbursement (insurance) systems. Therefore, validity and accuracy of comparison of use of antibiotics in human and veterinary side remain one of the big challenges.

In the last decade, the European Union, more than any other part of the world, has successfully taken many initiatives and measures to reduce the use of antibiotics both in food-producing animals and human sector. Especially in the veterinary field, major improvements (European Court of Auditors report) have been made in recent years, with a 32.5% reduction of sales between 2011 and 2017, including a significant decrease of use of critically important antimicrobials in some countries (European Court of Auditors 2019, EMA 2019a).

Within the One Health concept, with human and veterinary sectors involved, the outcome indicators including both quantitative and qualitative aspects of antimicrobial consumption and antimicrobial resistance seems to be very useful for evaluation of the progress achieved by the measures implemented at different levels. Therefore EMA, EFSA and ECDC network developed this list of below-mentioned harmonised outcome indicators (Table 4) for community (primary care), hospitals and food-producing animals.

3.1 What Data and for Which Purpose to Be Collected and Analysed

There can be different purposes and outcomes, why to collect and analyse data on antimicrobial consumption. *Sales data* are usually suitable for analysis of the trends within the time frame for individual countries (for more comments refer to subchapter above). *Use data* can be however used

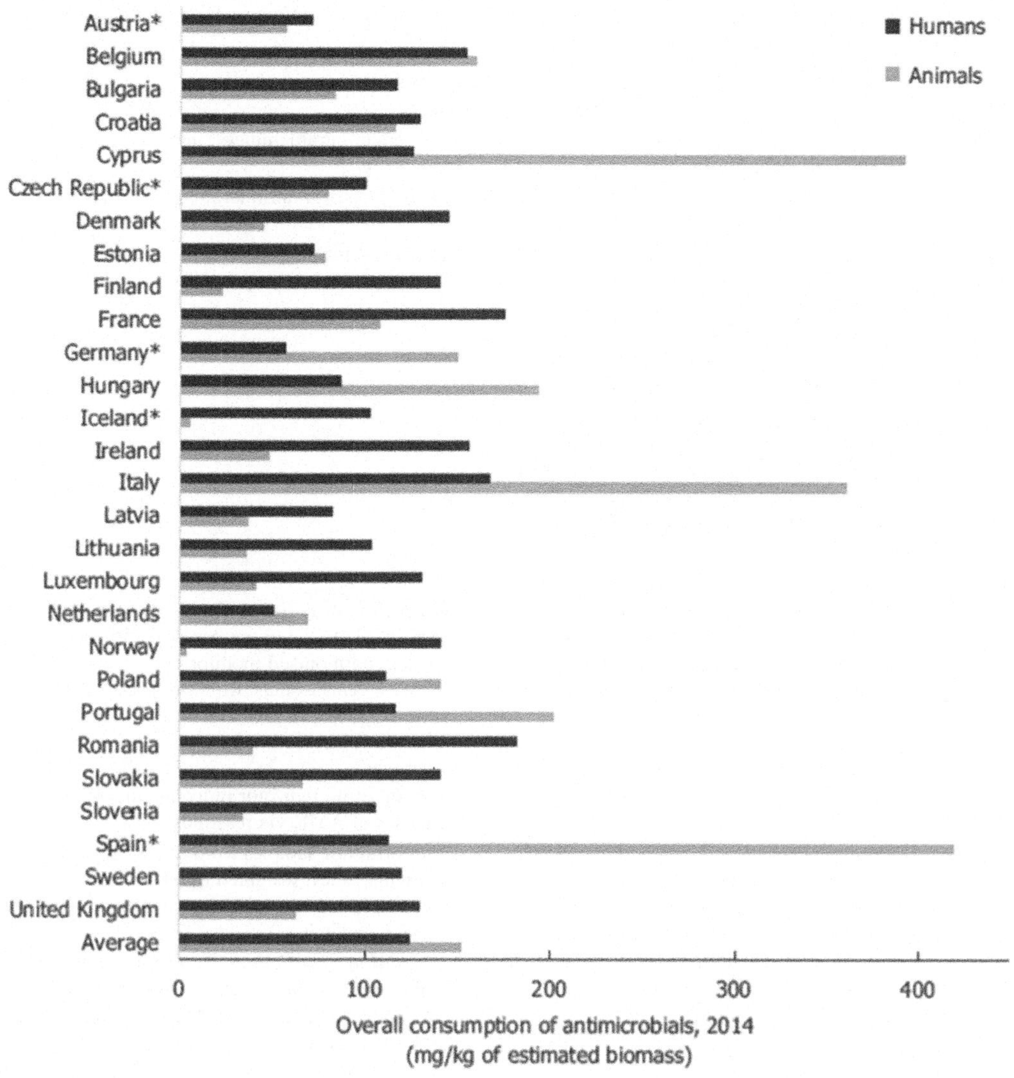

Asterisk (*) denotes that only community consumption was provided for human medicine. The population-weighted mean proportion (%) of the hospital sector AMC of the 2014 total national AMC for EU/EEA MSs that provided data for both sectors is 10%.

Note: 1) The estimates presented are crude and must be interpreted with caution. For limitations that hamper the comparison of consumption of antimicrobials in humans and animals, please see Section 14.

2) The average figure represents the population-weighted mean of data from included countries.

Fig. 4 Comparison of biomass-corrected consumption of antimicrobials (mg/kg of estimated biomass) in humans and food-producing animals by country, EU/EEA MSs, 2014 (JIACRA 2017)

for trend analysis, but also for comparison between countries (if full coverage, gained via methodologies, that allows comparisons); use data are also suitable for benchmarking within the species specific sectors (pig farms, poultry flocks, beef calves, dairy cows etc.)—for that purpose can be used, e.g. species-specific defined daily doses.

A specific situation arises when data on antimicrobial use is intended for the calculation of correlations with antimicrobial resistance selection and spread. The most appropriate is to assess

Table 4 Outcome indicators for the surveillance of antimicrobial consumption and antimicrobial resistance in humans and food-producing animals (ECDC, EFSA 2017)

	Human	Veterinary
Antimicrobial consumption	**Consumption** (systemic antibacterials only) **[DDD[a]/1000 inhabitants/day]**	**Sales** (systemic, intramammary, intrauterine) **[mg/PCU[b]]**
Primary indicators	Overall consumption	Overall sales of antimicrobials
Secondary indicators	*Ratio of community consumption* of certain classes of *broad-spectrum* penicillins, cephalosporins, macrolides (except erythromycin) and fluoroquinolones *narrow-spectrum* penicillins, cephalosporins and erythromycin	third- and fourth-generation cephalosporins, fluoroquinolones, polymyxins
	Proportion of total hospital consumption: glycopeptides, third- and fourth-generation cephalosporins, monobactams, carbapenems, fluoroquinolones, polymyxins, piperacillin and enzyme inhibitor, linezolid, tedizolid and daptomycin.	
Antimicrobial resistance		
Primary indicators	Proportion of methicillin-resistant *Staphylococcus aureus (MRSA)* and third-generation cephalosporin-resistant *Escherichia coli*, as major public health concern pathogens. Expressed as two individual numbers	Proportion of *indicator E. coli* that are fully susceptible to the entire panel of antimicrobials used in harmonised monitoring. Those *E. coli* isolates from *broilers, fattening turkeys, fattening pigs and calves* weighted by the size (expressed in PCU) of the four animal populations. Representing indicator reflecting use as well as transfer of AMR via plasmids.
Secondary indicators	Proportion of *Klebsiella pneumoniae* with combined resistance to aminoglycosides, fluoroquinolones and third-generation cephalosporins, representing hospitals.	Proportion of samples from the above four animal species, weighted by PCU, that are identified as positive for presumptive *ESBL-/AmpC-producing indicator E. coli.*
	Proportion of penicillin-resistant and macrolide resistant *Streptococcus pneumoniae*, representing community.	Proportion of *indicator E. coli* isolates from the same four animal species, weighted by PCU, that are *resistant to at least three antimicrobials* from different classes from the predefined panel of antimicrobials.
	Proportion of carbapenem resistant *Klebsiella pneumoniae*, representing emerging threat.	Proportion of *indicator E. coli* isolates from the four species, weighted by PCU, that are microbiologically resistant to *ciprofloxacin*, representing fluoroquinolones.

[a]*DDD* defined daily dose; the assumed average maintenance dose per day for a drug used for its main indication in adults (ref: https://www.whocc.no/filearchive/publications/2019_guidelines_web.pdf)
[b]*PCU* population correction unit (ref: https://www.ema.europa.eu/en/documents/report/trends-sales-veterinary-antimicrobial-agents-nine-european-countries_en.pdf

the correlation at farm level, especially in "closed" farms (farms, which do not import animals, as imports could lead to "imported resistance"). Ideally you need also to consider the same time period for collecting data on antimicrobial use (including dose and duration of the treatment) and resistance, as well as having a robust denominator (e.g. number of animals including their weight, if different categories or biomass in kilograms). Data on antimicrobial resistance needs to include information on indicator bacteria (such as intestinal commensal *E. coli* isolated from healthy animals, which is considered as one of the best indicators at least for Gram-negatives, and/or staphylococci and enterococci representing Gram-positives) and data on antimicrobial resistance/susceptibility in target pathogens (isolated from diseased animals). Despite having all the above-mentioned datasets available at farm level, the final precise analysis of the data is quite challenging as usually multifactorial analysis considering also other factors is hard to be performed in the most complex way. There should be involved other factors as possible influence of AMR (e.g. co-selection of AMR by disinfectants or heavy metals—Zn and Cu), input of AMR from other sources (water and feed; flies, rodents in the cases of insufficient barriers and deratisation; farmers/vets as carriers; etc.) influencing the correlation. Even taking into account the complexity mentioned above the exposure to antimicrobials can be considered as a main driver for antimicrobial resistance selection and spread also at the farm level.

Utilisation of the number of packages used (and calculating amount of the active substance) is one of the most used variables when expressed numerator, which can serve as a background for further calculations. Different concepts exist, see Fig. 5—for example they can be based on the "Defined Daily Dose/Animal Daily Dose", the "Defined Course Dose" or the treatment incidence (Animal Treatment Index). There are also systems (e.g. in Denmark) that use, especially for benchmark purposes coefficients for multiplication according to the type of antimicrobial and its importance (like CIAs (e.g. third- and fourth-generation cephalosporins) have higher coefficient than, e.g. narrow spectrum penicillins).

4 Qualitative Aspects of Antimicrobials Used

Most classes of antimicrobials are used both for humans and animals. But among the groups used there is a difference, especially considering impact on antimicrobial resistance ;its transfer, risks from public health perspective, importance of certain antimicrobials used in both human and veterinary medicine as life-saving drugs or the only/few alternative for the treatment of human infection as well as the intensity of selective pressure not only for the antimicrobials from the same pharmacological group, but also for the others—resistance co-selection.

4.1 Which Antimicrobials Are Critically Important

Antimicrobials have been classified based on the importance for human and animal medicine by several institutions or committees. The most famous one is the classification done by the World Health Organization (WHO) which classified them on the basis of their importance for human health. The World Animal Health Organisation (OIE) did a similar classification but on the basis of their importance for animal health. On a European level, policy makers mostly base their decision on the Antimicrobial Advice ad hoc Expert Group (AMEG) by the European Medicines Agency, which created a third list, but based on the importance of the antimicrobials both for the human and animal sector considering the European situation and risks for resistance transmission.

4.2 WHO List of Critically Important Antimicrobials for Human Medicine (WHO CIA List)

The WHO has categorised all antimicrobial classes into three groups: critically important (CIAs), highly important (HIAs) and important antimicrobials for human health as declared in WHO list of critically important antimicrobials 6th revision (WHO 2019). The WHO list takes into account the following two criteria to define higher risk: a/the antimicrobial class is the sole or one of few alternatives for the treatment of serious bacterial infections in humans, b/the class is used to treat diseases caused by bacteria that may be transmitted to humans from nonhuman sources or bacteria that may acquire resistance genes from nonhuman sources. The first list was published in 2005, and the list has been revised six times since with the latest revision in 2019. The WHO

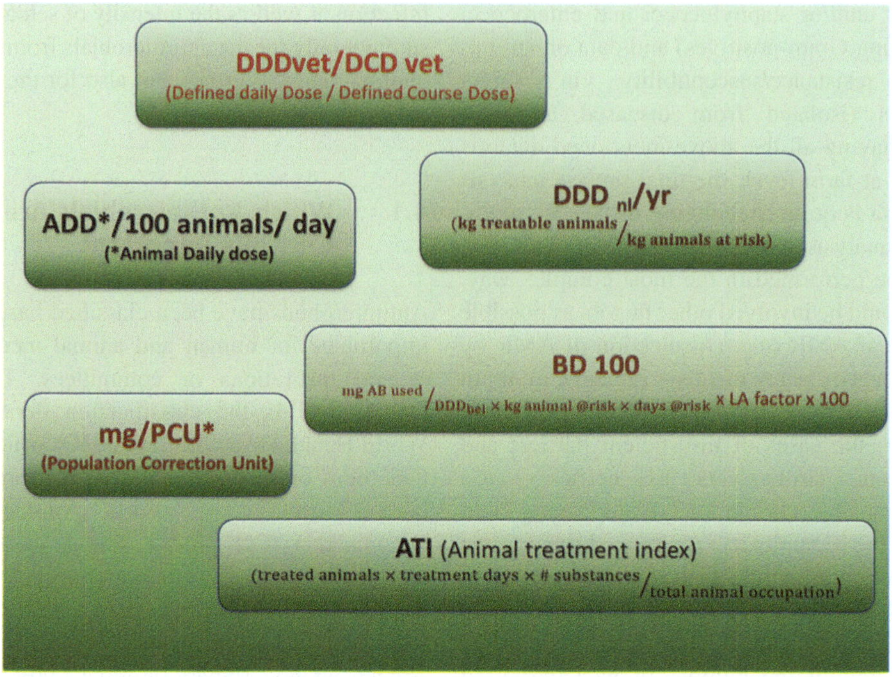

Fig. 5 Examples of the analysis/calculations of raw numerator/denominator data to express use of ATM. Figure explanatory notes: *DDDvet* = assumed average daily dose or *DCDvet* = treatment course dose of active substance (both taking into account of differences in dosing, pharmaceutical form and route of administration used in the different species (EMA 2016a); *ADD/100 animals/day* and *BD 100* both = number of treatment days out of 100 days that an animal was present at the farm, whereas the *DDD nl/yr* = number of treatment days in a year that an animal was present at the farm. *ATI (also called TF = treatment frequency)* = for how many days, on average, an animal in the observed population is treated within a given time period (e.g. how many single doses were administered to one animal on average within the observation period); *mg/PCU* = amount of active substance(s) (numerator) per consensual technical unit expressing population/biomass of animals in the given territory, when numerator was counted (EMA 2017)

classified as highest priority "Critically Important Antimicrobials": cephalosporins (third to fifth generations), quinolones (including fluoroquinolones), aminoglycosides, glycopeptides, macrolides, ketolides and polymyxins. Some of these antimicrobial classes have been used in some countries in the EU frequently to treat a variety of infections in veterinary medicine.

4.3 OIE List of Antimicrobial Agents of Veterinary Importance

Also the OIE categorised all antimicrobial classes into three groups; although slightly differently named than WHO. They differentiate between Veterinary Critically Important Antimicrobial Agents (VCIA), Veterinary Highly Important Antimicrobial Agents (VHIA) and Veterinary Important Antimicrobial Agents (VIA). The OIE list takes into account the following two criteria to define higher risk: a/in a members survey the majority of the respondents (more than 50%) identified the importance of the antimicrobial class and b/the class was identified as essential against specific infections and there was a lack of sufficient therapeutic alternatives. The first list was endorsed by all OIE Member Countries in 2007 and has been revised twice since with the latest revision in 2018 (OIE 2019).

The OIE highest priority list of Veterinary Critically Important Antimicrobial Agents (VCIA) include: aminoglycosides, amphenicols, cephalosporins (third to fifth generations), macrolides, penicillins, fluoroquinolones, sulfonamides and tetracyclines.

4.4 AMEG Categorisation of Antimicrobials: Combining Human and Veterinary Critical Antimicrobials in a European Context

In 2013, the Antimicrobial Advice ad hoc Expert Group (AMEG) of the European Medicines Agency made a first classification of antimicrobials taking into account the World Health Organization list (WHO), the hazards of zoonotic relevance in Europe, the use of antimicrobials in veterinary medicine, and the risk of resistance transfer to humans. AMEG uses another sort of listing, distinguishing Category 1, 2 and 3 classes of antimicrobials taking into account the following criteria: a/their need in human medicine and b/the risk for spread of resistance from animals to humans.

Category 1 Antimicrobials that are authorised for use in veterinary medicine and the risk for public health is estimated as low or limited.

Category 2 Antimicrobials that are authorised for use in veterinary medicine but the risk for public health is estimated higher. This category includes fluoroquinolones, third- and fourth-generation cephalosporins and polymyxins (colistin). Category 2 antimicrobials should only be used in veterinary medicine when there is no alternative available (EMA 2014).

Category 3 Antimicrobials that are not approved for use in veterinary medicine. Consequently, Macrolides are not included in Category 2 of the AMEG categorisation.

Risk profiling for aminoglycosides was finalised in 2018 (reflection papers on aminoglycosides (EMA 2018b) and they were found to have a higher risk to public health, but lower than the classes currently included in category 2. Risk profiling for certain penicillins (with extended spectrum) has been finalised in 2019 (EMA (2018c)). It recognises that in accordance with the categorisation criteria in the first AMEG report, all veterinary authorised aminoglycosides and amoxicillin–clavulanate combinations would be placed in Category 2. However, as the use of these antimicrobials in veterinary medicine was considered to present a lower risk to human health compared to quinolones and third- and fourth-generation cephalosporins, the Committee for Veterinary Medicinal Products recommended not include too many antimicrobials being placed in the higher risk category and therefore proposed more detailed stratification of categories.

In July 2017, the European Commission requested EMA to update the AMEG categorisation of antimicrobials and to further elaborate on the proposed early hazard characterisation. This new categorisation was adopted in December 2019 as document EMA/CVMP/CHMP/682198/2017 (EMA 2019a, b) EMA classified antimicrobials in four different categories, from A to D. For communication purposes, key action words have been attributed for each category. They also made a nice infographic to summarise this information. You could take parts of the Infograph to add as picture in the text. See here: https://www.ema.europa.eu/en/documents/report/categorisation-antibiotics-use-animals-prudent-responsible-use_en.pdf The following categories are included:

Category A = "Avoid" (previously adequate to Category 3), includes antimicrobial classes not currently authorised in veterinary medicine in the EU. In the absence of established maximum residue limits for foodstuff of animal origin, use of these classes of AM in food-producing animals is prohibited and they may only be administered to individual companion animals exceptionally, in compliance with the prescribing "cascade".

Category B = "Restrict" (previously adequate to Category 2), includes the substances listed as highest priority CIAs (HPCIAs) by the WHO (except of macrolides and those classes included in Category A). In fact includes quinolones, third- and fourth-generation cephalosporins and polymyxins. For these antimicrobials, the risk to public health resulting from veterinary use needs to be mitigated by specific restrictions as use for the treatment of clinical conditions when there are no alternative antimicrobials in a lower category that could be effective; use should be based on the results of antimicrobial susceptibility testing, whenever possible.

Category C = "Caution", new intermediate category, taking account of the considerations above (mainly as for aminoglycosides and aminopenicillins/clavulanic acid combination); categories included are individual antimicrobial classes listed in different WHO categories (also macrolides). In general, those antimicrobials are considered to have alternatives in human medicine in the EU but there are few alternatives in veterinary medicine for certain indications. There are included also antimicrobial classes that may co-select for resistance to a substance in Category A through specific multiresistance genes. Warnings related use when there is no substance in Category D that would be effective.

Category D = "Prudence", considered as the lowest risk category. While the risk to public health associated with the use in veterinary medicine of substances included in this category is considered low, some of them are listed as WHO CIAs (aminopenicillins and natural penicillins). It is acknowledged that these antimicrobials are not devoid of negative impact on resistance development and spread, in particular through co-selection. A general recommendation that prudent use principles should be adhered to in everyday practice to keep the risk from use of these classes as low as possible should be applied. Unnecessary use and unnecessarily long treatment periods should be avoided and group treatment should be restricted to situations where individual treatment is not feasible.

This AMEG categorisation is supposed to be used considering Regulation on veterinary medicinal products, especially in relation to use of antimicrobials for prophylaxis, metaphylaxis and under the "cascade" conditions. This categorisation can be used as a tool by those preparing treatment guidelines (Fig. 6).

At least with the EU according to the current level of knowledge about authorised VMPs, only a few classes such as carbapenems, lipopeptides, oxazolidinones are used (authorised) in humans, and some classes used in animals in a great extent are traditionally rarely used in humans (e.g. ionophores).

The importance of (classes of) antibiotics can however change over time, as exemplified by the case of colistin. Due to the rising problem with multidrug-resistant infectious agents, a reintroduction of colistin came to human medicine. Despite its neuro- and nephrotoxicity for humans, for certain healthcare-associated infections caused by multidrug resistant Gram-negative bacteria it became the only life-saving antimicrobial. That fact, together with discovery of horizontal transfer of resistance to colistin (*mcr* genes) lead to big pressure to decrease use of colistin containing veterinary medicinal products in animals (EMA 2016a). Due to the risks recognised, European targets were decided to decrease colistin consumption. The vast majority of colistin containing veterinary medicinal products are orally administered either via medicated feed or medicated drinking water and as indicated by ESVAC figures, in some countries the use for herd/flock medications was high. Therefore, for "high and moderate consumers" the targets and desirable levels were set at 5 mg/PCU and 1 or below 1 mg/PCU, respectively (EMA 2016a). The achievement of the target 5 (or 1) mg/PCU levels by the countries can be also facilitated by the outcomes of the referral, in which prophylaxis was deleted and the therapeutic indication for use was narrowed (salmonella claim deleted), treatment duration were shortened (to maximum 7 days) and prudent use warning in all the VMPs product texts were introduced (reference referral). Also all authorisations containing colistin in fixed combination with other antimicrobials were withdrawn. Comparison (EMA 2019a) of the data in recent years (data 2010 with the data 2017) shows that many countries have managed to decrease consumption

Fig. 6 Comparison listing of CIAs by WHO (2019), OIE (2018b) and AMEG—EMA (2019a) (as valid 2020)

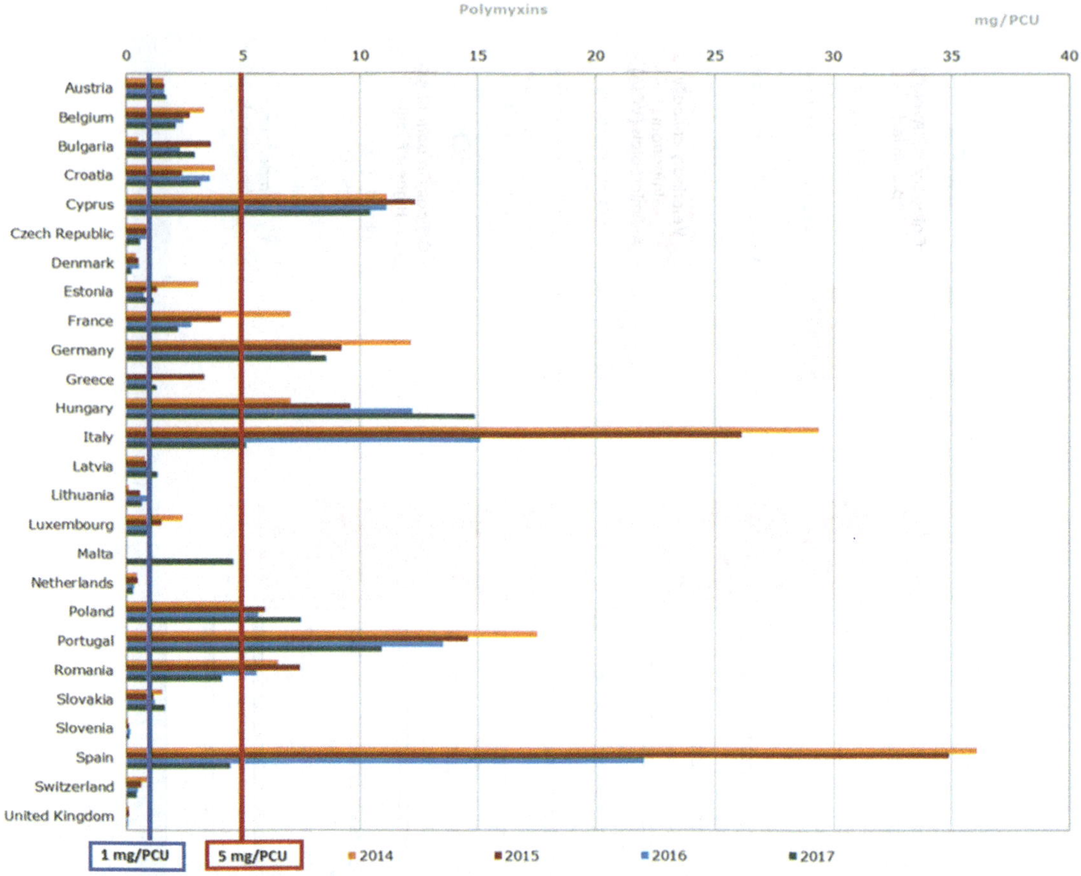

Fig. 7 Changes in sales of polymyxins for food-producing species, in mg/PCU, by 30 European countries, from 2010 to 2017 (EMA 2019b - modified). Note: (1) No sales of polymyxins in Finland, Iceland and Norway for any of the years. (2) For reasons of commercial confidentiality, sales of polymyxins in Ireland and Sweden (≤0.1 mg/PCU in any of the years) are not included in this figure

of polymyxins, from which majority accounts for colistin (Fig. 7). From those countries with initial big consumption, huge decrease of sales was reported by Spain (33.5 mg/PCU in 2011 to 4.4 in 2017) and Italy (30.7 mg/PCU in 2011 to 5.2 mg/PCU in 2017).

As can be seen from the above, the goals and methods of the various categorisation schemes differ, resulting in final categorisations that are not identical. Although harmonising these schemes would bring clarity to antimicrobial resistance discussions and policy, it has the disadvantage of not accounting for regional antimicrobial resistance and use. This could potentially remove effective medicines from clinical use in animals for situations, where they are wholly appropriate. Antimicrobials should be categorised in a One Health manner, where both physician and veterinarian share the responsibility for antimicrobial use (Watts et al. 2020).

It is worth to highlight however that while proposed categorisation by WHO, OIE and AMEG subjects to regular review and update according to new scientific evidence that comes to light and current needs, it should be expected that most of the highest priority antimicrobials for humans will continue to be of critical importance for animal health as well. That underlines the need to emphasise on responsible use of those classes of antibiotics in both sectors, e.g. need for examination, establishment of proper diagnosis and antimicrobial susceptibility testing before their prescription.

Also situations in different parts of the world and under different regulatory surroundings may vary. The Food and Drug Administration (US) divided antimicrobials into two groups: medically important (MI) and non-medically important (NMI). In the group of non-medically important antimicrobials listed are aminocoumarins (novobiocin), glycolipids (bambermycins), ionophores (e.g. lasalocid, monensin, narasin, salinomycin), orthosomycins (avilamycin), pleuromutilin (tiamulin), polypeptides (bacitracin) and quinoxalines (carbadox). While in some groups cross-resistance or co-resistance to antimicrobials used in human medicine seems to be limited, e.g. ionophores or quinoxalines (Ministry of Agriculture and Forestry of New Zealand 2011), in others e.g. pleuromutilin or polypeptides, the cross-resistance within some of the other pharmacological groups is of concern (pleuromutilin—retapamulin used in treatment of Staphylococci resistant to methicillin in human) and bacitracin and issue of co-selection of resistance to colistin (interference with *mcr* genes) (Van Duijkeren et al. 2014; Xu et al. 2018).

5 Use of Antimicrobials in Europe in Food-Producing Animals

5.1 Use Per Species and Region

Within the European Union, the main food-producing animals in terms of biomass meat are cattle (around 89 million heads), pigs (148 million heads), poultry (892 million heads), sheep (85 million heads), goats (12 million heads) and fish (2.3 million tonnes) (Eurostat 2016; FEAP 2016). In smaller proportion, horses, rabbits, turkeys, ducks are kept.

The main diseases/syndromes to use antibiotics per species in Europe are known. Most European countries have taken actions to promote responsible and prudent use of antibiotics in animals.

The use in other animals, such as dogs and cats, in tonnes is unknown in most countries and can only be roughly estimated combining data on tablets and certain injectables. Currently, all species data remain merely estimates, with a lot of work ongoing currently to create reliable and practical systems to measure accurately antimicrobial use per species.

Within Europe, species kept for food production as well as the husbandry and management conditions of how these animals are reared differs greatly between countries. This also applies to the relative proportion of the various animal species/subspecies/category of animals, the climate, epizootiology, intensity of production, infectious disease status and the availability of veterinary antimicrobial products and alternatives. There is no one European reality. As a result, indications to prescribe antimicrobials for and amounts used per species vary greatly per species and per country.

There is also considerable variation between countries in terms of the availability of number of authorised veterinary medicines including vaccines; from 296 products in Iceland to 2944 products in France (EPEC 2011), what is also one factor impacting the use of antimicrobials. Some EU countries tend to have fewer authorised veterinary medicinal products, due to having a smaller market which is less attractive for industry and depending on the presence or absence of local pharmaceutical companies and nationally authorised VMPs. Figure 8 shows the results of analysis of marketing authorisations demonstrating also percentage of authorisations for certain species of animals (i.e. counting with fact, that individual VMP can be authorised for several target animal species) for the 15 countries that provided the necessary data. Please note that not only antimicrobial VMPs were taken into account (EPEC 2011). In the EU/EEA countries in 2016 number of presentations (by product name, form, strength and pack size) of veterinary medicinal products (VMPs) containing antimicrobials differed greatly among countries (lowest 34, highest 723; with average

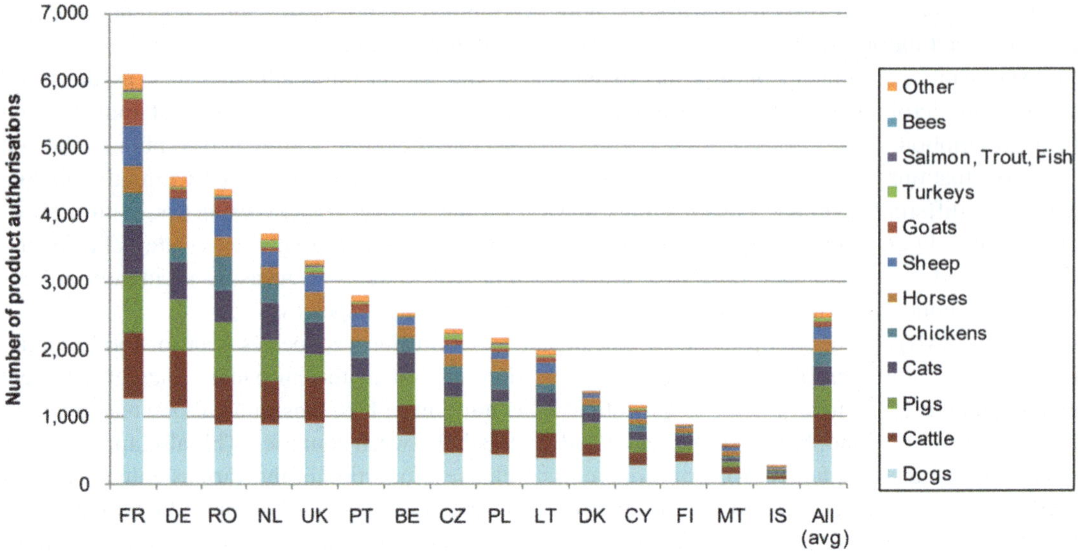

Fig. 8 Number of products authorised for use with each of the 11 species (+ category other) as indicated by 15 participating EU countries (EPEC 2011). Note: The number authorisations (*all* VMPs considered) for use with selected species, as in May 2010. Note that each veterinary medicinal product can have multiple target species, meaning that the *total number* of authorised veterinary medicinal products per country is lower comparing to this chart. The intention of the chart is to show that comparing to dogs, cattle, pigs, cats and chickens the number of authorisations for some other species as horses, sheep, goats, turkeys, fish and bees seems to be significantly lower, with great variability across countries and influence on availability especially on small markets.

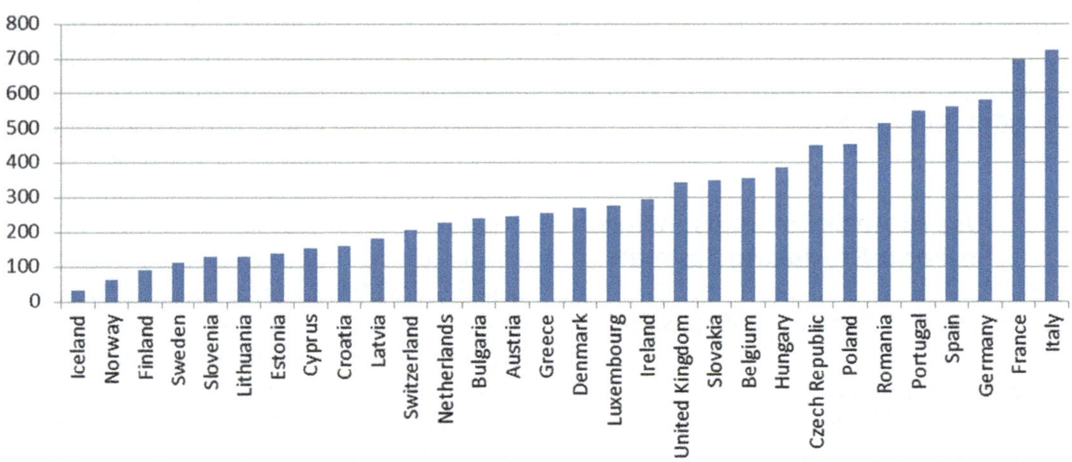

Fig. 9 Number of presentations (by product name, form, strength and pack size) of veterinary medicinal products containing antimicrobials, by country, for 2016 (tablets excluded from the data) (based on EMA (2018) data)

300 VMPs containing antimicrobials) as shown in Fig. 9.

Veterinary medicinal products in the EU can be authorised **centrally** through the procedure coordinated by European Medicines Agency (EMA) that gives the product a single marketing authorisation for the whole of the EU. Valid **national** authorisations of the veterinary medicinal product can be granted via national (just for single one Member State market), decentralised (DCP) or for those already existing at national level can be spread to other Member States via mutual recognition procedure (MRP). After successful DCP or MRP authorisation the VMP can be placed on the market in several so-called Concerned Member States.

It is noted that a high proportion of centrally authorised products while having a single marketing authorisation for all EU countries are not placed by industry on the market of smaller countries as it is seen as not profitable. Especially for minor species such as rabbits, turkeys, sheep, goats and fish, there are very few available authorised veterinary medicinal products. In addition, the spectrum of antimicrobials available is limited. This fact also makes it more difficult to create appropriate treatment guidelines and recommendations for prudent use of antimicrobials based on authorised veterinary medicinal products. Off-label use, mainly using a product authorised for another indication or another species or in another country (through "Cascade prescription"), is often the only solution to treat these animals and is therefore essential. For the future, research should be done to identify new antimicrobials, but also revisiting "legacy molecules" (old molecules) to gain valid data (e.g. on pharmacokinetic) for minor species would be useful.

5.2 Use of Antimicrobials Per Species

According to Federation of Veterinarians of Europe (FVE) research based among the other data also on surveys performed among veterinarians in the European countries in 2012 and 2016 (FVE 2016), the main specific disorders of certain food-producing species, that are targeted within this publication (cattle, pigs, poultry and horses) are addressed below in individual subchapters.

5.2.1 Cattle

Cattle are kept in Europe for dairy products, veal and beef. Group treatment is relatively rare in cattle, except in intensively raised veal calves. According to FVE research in 2012 and 2016 (FVE 2016), the main specific disorders of cattle leading to antimicrobial use, divided by the production categories, are:

Dairy Cattle

- Mastitis (especially the dry cow prevention/treatment):
  - e.g. *Staphylococcus aureus, Streptococcus agalactiae*
- Lameness/foot disease:
  - polymicrobial, e.g. anaerobes
- Urogenital tract disorders
  - e.g. metritis caused by *E. coli or Trueperella pyogenes*
- Surgery

Calves and Veal

- Respiratory diseases
  - e.g. *Mannheimia haemolytica, Pasteurella multocida* and *Histophilus somni*
- Diarrhoea
  - e.g. *E. Coli, Salmonella* spp.

Adult Beef

- Respiratory diseases
  - Mainly at the beginning of the fattening period—e.g. *Mannheimia haemolytica, Pasteurella multocida* and *Histophilus somni*
- Locomotory diseases
  - Lameness, arthritis

CIAs are mostly used to prevent and treat diarrhoea in calves (colistin, fluoroquinolones) and to treat respiratory disorders (fluoroquinolones, macrolides). Cephalosporins of the third and fourth generations are used amongst other reasons to prevent and treat mastitis (intramammary VMPs), uterine and locomotion disorders (from CIAs mainly injectable VMPs) (De Briyne N et al. 2014). Despite the fact that it is highly recommended to perform susceptibility testing in CIAs, in some diseases and their etiological agents (e.g. foot rot, anaerobes like *Fusobacterium necrophorum, Dichelobacter nodosus, Bacteroides melaninogenicus, Porphyromonas asaccharolytica*) routine susceptibility testing is not performed due to technical difficulties of in vitro culturing as well as due to unavailability of the interpretive criteria. Therefore, further investigations are needed to find appropriate methods for susceptibility testing that can be used for practice.

Dairy production systems are present in every European country. Udder health problems are the major indication for antimicrobial use in dairy cattle (Oliver et al. 2011; Stevens et al. 2016). The use of dry cow management, giving cows a long acting antimicrobial in the four quarters of its udder before the dry period to prevent new infections, is much debated and variably done in different countries (Bradley et al. 2018). While in some countries by routine all cows are still being treated with an antibiotic dry cow treatment, other countries moved towards only a selective treatment of cows or herds at high risk of a new infection. It is now seen as good practice and prudent use to avoid routine prevention/treatment with antibiotics of cows at drying-off, by implementing proper hygiene, drying off good practices, using teat sealants and/or vaccination and testing in advance to identify the causative bacterium before treatment (FVE 2019).

Calves, being young animals, are especially sensitive for respiratory and gastrointestinal infections. Calves in many countries are high users of antimicrobials, including critically important antimicrobials (CIAs). Antimicrobial resistance levels for commensals, pathogens and zoonotic agents are found to be high in the specialised veal husbandry. Transport and re-grouping of young calves is a major risk factor. Calves aged 2–4 weeks, originating from a multitude of herds of origin (including from other countries) arrive at large fattening farms (150–2000 animals), where they are fattened in, typically, an 8 months period. Predominantly (96%) oral, group antimicrobial treatments (metaphylaxis) in the milk or milk replacer are used (FVE 2016; Pardon et al. 2012).

Many infections in young calves can be prevented by improving the housing and management of mothers and calves, using vaccination (both of mother and calves) and good colostrum management (using own herd colostrum and making sure all calves receive enough). Calves should never be fed with waste milk from cows that have been treated with antimicrobials. Such practices are considered as risky from the perspective of selective pressure or mutation selection window promoting spread of antimicrobial resistance.

In older beef cattle the main reason for antimicrobial use is bovine respiratory disease (60%). Intensive rearing husbandry systems are most at risk. Bovine respiratory disease (BRD) is the most difficult disease to control on cattle farms because of the multitude of pathogens involved (viruses and bacteria) and the multiple risk factors.

The precise extent of transmission of antimicrobial resistance (either as genetical determinants or via bacteria as carrier) between cattle, cattle products and human is unknown. Cattle to human transfer has been reported in literature such as of *Staphylococcus aureus* complex, CC97 causing mastitis in cows and it is recognised that cattle—as all livestock—can present a reservoir for the emergence of new human-pathogenic clones (Spoor et al. 2013). However, research also demonstrates that the bacterium and its resistance genes are largely maintained within animal and human populations separately and that there is only limited transmission, in either direction (Mather et al. 2013). Current status of scientific knowledge brings the evidence that there are genes, that are of human origin (and some of them remain with the human population only), in animals and again remains within niche of staphylococci causing infections or colonising animals only, but that there is also a big gene pool, which origin is hardly to be decoded as these genes are present in both human and animals (Schwarz et al. 2018).

5.2.2 Pigs

Pigs are kept for meat in most European countries. In total, around 148 million pigs are kept in Europe, with the largest populations in Germany (27 million), France (13 million), Denmark (12 million), the Netherlands (12 million) and Poland (12 million) (European Commission, Eurostat (2019)). Pigs are kept mostly indoors, in large herds in confined space allowance. Most treatment with antimicrobials is group treatment, although moves are seen to go as much as possible to individual treatment of older animals.

According to FVE research in 2012 and 2016, the main specific disorders of pigs leading to antimicrobial use, divided by the production categories, are:

Piglets

- Neonatal diarrhoea
  - *Escherichia coli*
- Diarrhoea and respiratory diseases (especially around weaning)
  - e.g. *Mycoplasma hyorhinis*, *Lawsonia intracellularis*, *E. coli*
- Neurological disorders
  - e.g. *Streptococcus suis*, *Haemophilus parasuis*

Fatteners

- Respiratory
  - e.g. Porcine Respiratory Disease Complex (PRDC)
- Digestive disorders
  - Proliferative Enteropathy (PE): *Lawsonia intracellularis*
  - Swine dysentery: *Brachyspira* spp.
  - Ileitis
  - *Salmonella* spp.

Sows

- Urogenital disorders with Leptospirosis
- Postpartum dysgalactia syndrome (PPDS)
- Pleuropneumonia: *Actinobacillus pleuropneumoniae* (APP) in gilts.

CIAs are mostly used to prevent and treat weaning diarrhoea (colistin, fluoroquinolones and macrolides) and to treat Postpartum Dysgalactia Syndrome (fluoroquinolones, third- and fourth-generation cephalosporins) in sows.

The most critical period is around weaning, when the piglets are very sensitive for gastrointestinal diseases (Moeser et al. 2017). Secondly, antimicrobials are used for respiratory problems and *Streptococcus suis* infections.

Husbandry and management factors such as high stocking density, bad ventilation, not enough roughage, lack of water, poor biosecurity, early weaning and the need to use foster moms because of too large litter size are all contributors to lower immunity and higher dissemination of undesirable microorganisms on pig farms. Mixing batches is especially risky.

In some countries, the use of high concentrations of zinc oxide (ZnO) has traditionally been used for treatment and/or prevention and control of diarrhoea in post-weaning pigs. In 2017, European policy makers took the decision to withdraw high doses of zinc oxide by 2022 due to environmental risks and risks for co-resistance with antimicrobials. Alternative, but more costly, preventive measures such as using vaccination will need to be used to avoid increase of use of antimicrobials (Collineau 2016), especially CIAs (colistin, fluoroquinolones, aminoglycosides).

Antimicrobial treatment during the finishing period increases the risk of transmission of *Salmonella* spp. (Fosse et al. 2009) as well as general AMR development (Holman and Chénier 2015).

In respect to pig–human transfer of resistance, *livestock-associated methicillin-resistant Staphylococcus aureus CC398* is one of the most important pathogens, being discovered in animals (not only pigs, but also other animals on the farm), livestock farmers and retail meat. MRSA colonisation has been identified in people working with pigs, raising concerns about the role of pigs as reservoirs of MRSA for human infection. Studies show that livestock-associated MRSA has a high prevalence in people with direct contact with animals, but that at this moment, the spread from the farms into the community is less obvious (Van den Broek et al. 2009; Van Cleef et al. 2010). A Danish study looked at risk for MRSA transfer to visitors of pig stables, but concluded that for healthy individuals the risk to cause secondary transmissions of MRSA is most likely negligible due to the observed decline to unquantifiable levels in 95% of the nasal samples already after 2 h (Angen et al. 2017).

The results of another study demonstrated that the increasing prevalence of LA-MRSA CC398 in Danish pigs and patients was caused by clonal expansion of three dominant lineages. At the same time this study warned, that livestock-associated methicillin-resistant *Staphylococcus aureus* clonal complex CC398 (LA-MRSA CC398) is resistant to nearly all β-lactams and several non-β-lactam antimicrobials. Over the last decade, it has become widespread in pig farms across Europe and is now an important cause of human infections in several countries, as e.g. NL and DK. This study demonstrates that pig movements between farms in combination with increased bacterial resistance to certain antimicrobials and heavy metals were important drivers of the rapid spread of the clonal complex (Sieber et al. 2018). As MRSA likely spreads between animal species, humans and the pig barn environment, it is important to accurately implement control practices, in which not only pigs should be targeted, but also all other animal species present on farms (Pletinckx et al. 2013).

Streptococcus suis is another important zoonotic pathogen in the porcine industry causing septicaemia, meningitis and arthritis. It can also cause infections in humans. During the last decade, the number of reported human cases due to *S. suis* has dramatically increased, and while most sporadic human cases of infection appear to be due to close occupational contact with pigs/pork products, particularly in Western countries (farmers, veterinarians, butchers, food processing workers, etc.), in Asia the number of human cases have increased and are thought to also possible endanger the general population (Goyette-Desjardins et al. 2014). Studies also show that in some countries resistance levels among isolates are increasing for antimicrobials including aminoglycosides, cephalosporins, fluoroquinolones, pleuromutilins, potentiated sulphonamides and tetracyclines (Hernandez-Garcia et al. 2017).

Similarly as in cattle also in pigs are some frequently occurred diseases, for which antimicrobials are used with limited possibility of in vitro susceptibility testing and with no interpretive criteria (e.g. *Lawsonia intracellularis* and *Mycoplasma* spp.).

5.2.3 Poultry

Poultry is kept for eggs and meat (mostly called broilers) in most European countries. In total, around 6 billion broilers, 350 million laying hens, but also turkeys, ducks, guinea fowls and geese are kept in Europe (Eurostat 2016), with the largest populations of poultry in Poland (16.8%), the United Kingdom (12.9%), France (11.4%), Spain (10.7%), Germany (10.4%) and Italy (8.5%) as indicated by EUROSTAT statistics 2018. Farms with more than 5000 broilers represent barely 1% of the total number of broiler farms in Europe, but they account for above 93% of broilers. European production of broiler chicken meat is the third biggest in the world (the United States, Brazil, the EU).

Poultry kept for meat production are kept mostly indoors, typically in professional farms of more than 5000 broilers in confined space allowance. Although poultry farms with more than 100,000 heads account for less than 1% of total poultry holdings in the EU, Eurostat data shows that these holdings account for 38% of total poultry numbers. The main producing countries tend to adhere to stocking densities of max 33 kg/m^2, but some are of higher densities—maximum of 39 kg/m^2 is permitted if the owner complies with certain environment parameters, 42 kg/m^2 in exceptional cases (please refer to Table 5 below).

Different production systems exist in the EU, also within intensive farming, but also in the farming with improved animal welfare (e.g. lower density). Mostly used are broilers of fast-growing genotypes to produce poultry meat but increasingly gaining attention in many EU countries is the use of slower growing genotypes (slaughter age in the range of 35–45 days). Most treatments with antimicrobials are group treatments. Recently, it has been demonstrated that high levels of ammonia can damage the birds' immune system, hence increasing the birds' vulnerability to bacterial diseases, especially *Escherichia coli* infections (ASOA 2017).

Table 5 Stocking densities (broiler farming) in countries with biggest production rates (EU, 2017—modified—Agra CEAS Consulting (Food Chain Evaluation Consortium), 2017)

Farm characteristics	PL (kg/m^2)	UK (kg/m^2)	DE (kg/m^2)	FR (kg/m^2)	NL (kg/m^2)
Stocking density (majority)	33	38	39	39–42	42

Laying hens are kept in different housing systems from enriched cages, to free range or barn, in large-scale professional farms, small or backyard farms, conventional or organic farming. Much lower amounts of antimicrobials are authorised and used to treat laying hens producing eggs for human consumption, in part due to the effects of withdrawal periods on eggs.

Routine group medication in poultry is also often being done immediately before or after transport of 1-day-old chicks to address perceived potential losses of productivity.

Main pathogens/disorders for which antimicrobials are most frequently used in poultry include:

Broilers

- Gastrointestinal disorders
  - (such as coccidiosis, necrotic enteritis, dysbacteriosis)
- Respiratory diseases
  - (including infections as infectious bronchitis, Newcastle disease, infectious laryngotracheitis that are often followed by secondary bacterial infection—*E. coli*)
- Locomotion-related diseases
  - (bacterial arthritis—due to e.g. *E. coli, Staphylococcus aureus* or *Enterococcus* spp., and secondary bacterial infections connected with tenosynovitis, necrosis of the femur head)
- Septicaemia, omphalitis

Laying Hens

- Gastrointestinal disorders
  - Enteritis caused by *E. coli*, avian intestinal spirochaetosis and *Clostridium perfringens*
  - *Eimeria* spp. infections
- Respiratory and locomotion-related diseases
  - (caused by *E. coli* and *Mycoplasma* spp.)
- Secondary bacterial infections connected, e.g. concomitant with red mite infestation
- Taeniosis (in free-range production systems)

Within the poultry sector turkeys are specific. Except the fact that the period of fattening is longer, they are also prone to certain diseases and use of antimicrobials can be quite high in intensive production systems.

Turkeys

- Respiratory diseases (caused by *Ornithobacterium rhinotracheale* infection).
- Besides avian pathogenic *Escherichia coli, Staphylococcus aureus* and *Ornithobacterium rhinotracheale*, also *Mycoplasma* spp. can be of concern.
- Gastrointestinal disorders (caused by coccidiosis mostly *Eimeria* spp.).

In some countries the use of fluoroquinolones is prohibited and cephalosporins are not authorised for poultry in most European countries. In other countries, such as the United Kingdom, the sector has voluntarily committed to reduce the prophylactic use of fluoroquinolones in 1-day-old broilers (RUMA 2018).

In some EU countries, the poultry sector reduced antibiotic use drastically. A 2019 report by Wageningen Economic Research, at the request of the Dutch Ministry of Agriculture, Nature and Food Quality, concluded that the reduction of veterinary antibiotics sales by 63% in the Netherlands between 2009 and 2017 showed no evidence of a negative effect on either average production or economic results for broiler farmers (Bondt and Korstee 2016). The evaluation was performed for sows and broilers. A set of factors leading to improvement was evaluated directly at broiler farms and following mostly used interventions to improve health of animals that declared by farmers were: cleaning and disinfection of unoccupied stables, avoiding routine use of antibiotics, "all in- all out" system, clean drinking water, improvement of climate control, handling/care of 1-day-old chicks, change to slower growing broilers and preventive vaccination. It has been also recognised that in both sectors more modern buildings are associated with lower animal health costs (Wageningen 2019).

Many authorised poultry vaccines exist such as for Marek's disease, Newcastle/infectious bronchitis, Chicken Infectious Anaemia, Salmonellosis etc. For turkeys, the main vaccines are against Rhinotracheitis and Haemorrhagic Enteritis. Also veterinary autogenous vaccines could be an option. Poultry (broiler chickens and turkeys) are considered as one of the sources of zoonotic bacteria (*Campylobacter coli, Campylobacter jejuni, E.coli, Salmonella* spp.) that can not only cause food-borne alimentary diseases in human population, but can be also a source of resistance genes to be transferred to human (EFSA 2019). In the study by Kittl et al. (2013) population structures of 730 *C. jejuni* and *C. coli* from human cases, 610 chicken, 159 dog, 360 pig and 23 cattle isolates collected between 2001 and 2012 in Switzerland was compared. Based on MLST, human campylobacteriosis was attributed to chicken in 70.9% of cases, 19.3% to cattle, 8.6% to dogs and 1.2% to pigs. Furthermore a host-independent association between sequence type (ST) and quinolone resistance was proven. Interestingly, in both *C. jejuni* and *C. coli* the odds of quinolone resistance were highest in isolates from humans (with suggested reasons as travellers imports, antibiotic treatment prior sampling of human, strains resistant to fluoroquinolones more pathogenic to humans (Kittl et al. 2013)).

5.2.4 Horses

There are around 7 million equines in the European Union. France, the United Kingdom and Romania have the largest equine populations. The issue with horses is more complex as some horses are kept as food-producing animals, some as companion animals and some for sport only. Whether a horse is considered as food-producing animal greatly influences the treatment options. Horses kept as companion animals and declared as not for food production can be treated with a much wider range of veterinary medicinal products as medicine residues in food are not of concern.

Main pathogens/disorders for which antimicrobials are most frequently used in horses include (FVE 2016; De Briyne et al. 2014; Scicluna et al. 2013):

- Respiratory infections—especially for young animals and stables/studs with large numbers of horses or high throughput of horses and horses travelling frequently to competitions
- Wounds and other skin diseases
- Reproductive disorders—mostly in studs, broodmares being treated for in/hypo-fertility or for intrauterine treatment
- Perioperative use of antimicrobials

Some practices can lead to the spread of the diseases as mixing large numbers of horses, especially vulnerable ones such as young horses or horses being transported long distances, and therefore should be avoided.

Guidelines for horses in regard to using antimicrobials for treating wounds, other skin diseases, reproductive disorders and perioperative, can greatly reduce antimicrobial use and the use of critically important antimicrobials, but gaining an evidence for the recommendation for such guidance documents is really difficult, because of a lack of data.

Vaccination is especially indicated against *Equine Influenza* (already widely in use), *EHV1,4* and *Streptococcus equi var equi.*

Transfer from animals to human or vice versa has been proved for MRSA. Studies concluded that horse owners, horse handlers, horse vets and all others in close contact are having a higher risk for transfer of MRSA from horses. Horses are mostly infected by animal-associated isolates that differ from the traditional ST398 LA-MRSA with the particular prevalence of the CC398-IV MRSA clone belonging to the spa-type t011. They also conclude that MRSA can also spread from the people to the horses (Abdelbary et al. 2014; Haenni et al. 2017).

Considering the nosocomial and zoonotic potential of MRSA isolated from horses, equine veterinarians should pay specific attention to both antibiotic treatments and hygiene measures, to limit MRSA selection and transmission. The rate of MRSA carriage in healthy horses, as well as the potential human-to-animal or animal-to-human transmission in both veterinary clinics and in equestrian centres was studied on certain occasions and certainly deserve further investigation (Van Duijkeren et al. 2010; Cuny et al. 2016; Koop 2016).

6 Conclusion

Use of antimicrobials in animals in current days, at least in European and North-American region, has started to change.Less progress is seen in other parts of the world, such as in the Latin American & Caribbean and Sub-SaharanAfrica regions (European Commission, DG Health and Food Safety (2017). It is clearly seen from the ESVAC reports figures that show constant decrease in most countries over the last years. However, there can be expected further changes and movements, not only due to the new legislation, making rules for use and authorisation of antimicrobials more strict and prudent, but also due to the National action plans established in many European countries (European Commission 2016). The EU 'Farm to Fork strategy' released on 20 May 2020 includes the aim to reduce by 50% thesales of antimicrobials for farmed animals and in aquaculture by 2030. These plans introduce health and welfare measures that should lead to a decreased need for use of antimicrobials and establish exact targets for drop down of use of antimicrobials. Europe, therefore, stands in front of a big challenge and at the same time a big opportunity to significantly lower the contribution of animal sector not only to the total use of antimicrobials, but also to minimise the impact on public health and environment from the use of antimicrobials in animals.

References

AACTING (2019a) Antimicrobial usage at herd level and analysis, communication and benchmarking to improve responsible usage: overview of farm-level AMU monitoring systems. http://www.aacting.org/monitoring-systems/. Accessed 4 June 2019

AACTING (2019b) Antimicrobial usage at herd level and analysis, communication and benchmarking to improve responsible usage: guidelines. http://www.aacting.org/guidelines/. Accessed 4 June 2019

Aarestrup FM, Kruse H, Tast E, Hammerum AM, Jensen LB (2000) Associations between the use of antimicrobial agents for growth promotion and the occurrence of resistance among *Enterococcus faecium* from broilers and pigs in Denmark, Finland, and Norway. Microb Drug Resist 6(1):63–70

AB Register vzw (2019). https://www.abregister.be/HOME.php. Accessed 4 June 2019

Abdelbary MM, Wittenberg A, Cuny C, Layer F, Kurt K, Wieler LH, Walther B, Skov R, Larsen J, Hasman H, Fitzgerald JR, Smith TC, Wagenaar JA, Pantosti A, Hallin M, Struelens MJ, Edwards G, Böse R, Nübel U, Witte W (2014) Phylogenetic analysis of *Staphylococcus aureus* CC398 reveals a sub-lineage epidemiologically associated with infections in horses. PLoS One 9 (2)

Alban L, Dahl J, Andreasen M, Petersen JV, Sandberg M (2013) Possible impact of the "yellow card" antimicrobial scheme on meat inspection lesions in Danish finisher pigs. Prev Vet Med 108:334–341

Angen Ø, Feld L, Larsen J, Larsen AR, Rostgaard K, Skov R, Madsen AM (2017) Transmission of MRSA to human volunteers visiting a swine farm. Appl Environ Microbiol 83(23):e01489

ASOA (2017) Alliance to save our antibiotics: real farming solutions to antibiotic misuse: what farmers and supermarkets must do. http://www.saveourantibiotics.org/media/1777/asoa-report-real-farming-solutions-to-antibiotic-misues-what-farmers-and-supermarkets-must-do.pdf. Accessed 4 June 2019

AVMA (2017) American veterinary medical association: veterinary feed directive (VFD) basics. https://www.avma.org. Accessed 4 June 2019

Baertlein M (2015) McDonald's USA to phase out human antibiotics from chicken supply. https://www.reuters.com/article/us-usa-mcdonalds-antibiotics/mcdonalds-usa-to-phase-out-human-antibiotics-from-chicken-supply-idUSKBN0M01L520150304. Accessed 4 June 2019

Bondt N, Kortstee H (2016) Good practices - use of antibiotics. Lei/Wageningen, Lei.library.nl, 2016-030

Bos MEH, Taverne FJ, van Geijlswijk IM, Mouton JW, Mevius DJ, Heederik DJJ (2013) Consumption of antimicrobials in pigs, veal calves, and broilers in The Netherlands: quantitative results of nationwide collection of data in 2011. PLoS One 8(10). https://doi.org/10.1371/journal.pone.0077525. Accessed 4 June 2019

Bradley A, De Vliegher S, Farre M, Jimenez LM, Peters T, de Leemput ES, van Werven T (2018) Pan-European agreement on dry cow therapy. Vet Rec 182:637

Chantziaras I, Boyen F, Callens B, Dewulf J (2014) Correlation between veterinary antimicrobial use and antimicrobial resistance in food-producing animals: a report on seven countries. J Antimicrob Chemother 69:827–834

CIPARS. Canadian integrated program for antimicrobial resistance surveillance. https://www.canada.ca/en/public-health/services/surveillance/canadian-integrated-program-antimicrobial-resistance-surveillance-cipars.html. Accessed 4 June 2019

Collineau L (2016) Quantify, explain and reduce antimicrobial usage in pig production in Europe, PhD Thesis. https://www.theses.fr/2016ONIR091F.pdf. Accessed 4 June 2019

Cuny C, Abdelbary MMH, Kock R, Layer F, Scheidemann W, Werner G, Witte W (2016) Methicillin-resistant *Staphylococcus aureus* from infections in horses in Germany are frequent colonizers of veterinarians but rare among MRSA from infections in humans. One Health 2:11–17

DANMAP. Danish integrated antimicrobial resistance monitoring and research programme. https://www.danmap.org/. Accessed 4 June 2019

De Briyne N, Atkinson J, Pokludová L, Borriello SP (2014) Antibiotics used most commonly to treat animals in Europe. Vet Rec 175(13):325

EAGRI-CZ (2018) Announcement of the rules for certification of the products in quality regimen Q CZ in producers of poultry and processing plants of poultry products, (Vestnik_MZE_03_2018): 86–94, p 92 (Czech language). http://eagri.cz/public/web/file/612588/Vestnik_MZE_03_2018_3.pdf. Accessed 28 June 2019

ECDC/EFSA BIOHAZ/CVMP Panel (2017) European centre for disease prevention and control/European food safety authority panel on biological hazards and/EMA committee for medicinal products for veterinary use: ECDC, EFSA and EMA joint scientific opinion on a list of outcome indicators as regards surveillance of antimicrobial resistance and antimicrobial consumption in humans and food-producing animals. EFSA J 15(10):5017

EFSA (2019) European Food Safety Authority and ECDC (European Centre for Disease Prevention and Control), The European Union summary report on antimicrobial resistance in zoonotic and indicator bacteria from humans, animals and food in 2017. EFSA J 17 (2):5598. https://doi.org/10.2903/j.efsa.2019.5598

EMA (2014) European Medicines Agency: answers to the requests for scientific advice on the impact on public health and animal health of the use of antibiotics in animals (EMA/381884/2014). https://www.ema.europa.eu/en/documents/other/answers-requests-scientific-advice-impact-public-health-animal-health-use-antibiotics-animals_en.pdf. Accessed 4 June 2019

EMA (2016a) Updated advice on the use of colistin products in animals within the European Union: development of resistance and possible impact on human and animal health EMA/231573/2016. https://www.ema.europa.eu/en/documents/scientific-guideline/updated-advice-use-colistin-products-animals-within-european-union-development-resistance-possible_en-0.pdf. Accessed 4 June 2019

EMA (2016b) Defined daily doses for animals (DDDvet) and defined course doses for animals (DCDvet) EMA/224954/2016. https://www.ema.europa.eu/en/documents/other/defined-daily-doses-animals-dddvet-defined-course-doses-animals-dcdvet-european-surveillance_en.pdf. Accessed 4 June 2019

EMA (2017) European Medicines Agency, European Surveillance of Veterinary Antimicrobial Consumption: sales of veterinary antimicrobial agents in 30 European countries in 2015 (EMA/184855/2017). https://www.ema.europa.eu/en/documents/report/seventh-esvac-report-sales-veterinary-antimicrobial-agents-30-european-countries-2015_en.pdf. Accessed 4 June 2019

EMA (2018) European Medicines Agency, European Surveillance of Veterinary Antimicrobial Consumption: Sales of veterinary antimicrobial agents in 30 European countries in 2016 (EMA/275982/2018). https://www.ema.europa.eu/en/documents/report/sales-veterinary-antimicrobial-agents-30-european-countries-2016-trends-2010-2016-eighth-esvac_en.pdf. Accessed 6 October 2020

EMA (2018a) European Medicines Agency: guidance on collection and provision of national data on antimicrobial use by animal species/categories (EMA/489035/2016). https://www.ema.europa.eu/en/documents/scientific-guideline/guidance-collection-provision-national-data-antimicrobial-use-animal-species/categories_en.pdf. Accessed 4 June 2019

EMA (2018b) European Medicines Agency: use of aminoglycosides in animals in the European Union: development of resistance and impact on human and animal health (EMA/CVMP/AWP/721118/2014). https://www.ema.europa.eu/en/use-aminoglycosides-animals-european-union-development-resistance-impact-human-animal-health. Accessed 4 June 2019

EMA (2018c) European Medicines Agency: use of aminopenicillins and their beta-lactamase inhibitor combinations in animals in the European Union: development of resistance and impact on human and animal health (EMA/CVMP/AWP/842786/2015). https://www.ema.europa.eu/en/use-aminopenicillins-their-beta-lactamase-inhibitor-combinations-animals-european-union-development. Accessed 4 June 2019

EMA (2019a) European Medicines Agency. Categorisation of antibiotics in the European Union. December 2019. EMA/CVMP/CHMP/682198/2017. https://www.ema.europa.eu/en/documents/report/categorisation-antibiotics-european-union-answer-request-european-commission-updating-scientific_en.pdf

EMA (2019b) European Medicines Agency, European Surveillance of Veterinary Antimicrobial Consumption, 2019. 'Sales of veterinary antimicrobial agents in 31 European countries in 2017'. (EMA/294674/2019). Accessed 10 Nov 2019

EMA and EFSA (2017) European Medicines Agency and European Food Safety Authority: joint scientific opinion on measures to reduce the need to use antimicrobial agents in animal husbandry in the European Union, and the resulting impacts on food safety (RONAFA). EFSA J 15(1):4666

EPEC (2011) European Policy Evaluation Consortium: assessment of the impact of the revision of veterinary pharmaceutical legislation. https://ec.europa.eu/health//sites/health/files/files/veterinary/11-07-2011_final_report_.pdf. Accessed 4 June 2019

EPRUMA (2008) European platform for responsible use of medicines in animals: best-practice framework for the use of antimicrobials in food-producing animals in the EU. http://www.vif.dk/VIFDokumenter/Epruma.pdf. Accessed 4 June 2019

European Centre for Disease Prevention and Control, European Food Safety Authority, European Medicines Agency (2017) ECDC/EFSA/EMA second joint report on the integrated analysis of the consumption of antimicrobial agents and occurrence of antimicrobial resistance in bacteria from humans and food-producing animals – joint interagency antimicrobial consumption and resistance analysis (JIACRA) report. EFSA J 15 (7):4872

European Commission (2016) Press Release Brussels 16 June 2016 IP-16-2229. http://europa.eu/rapid/press-release_IP-16-2229_en.htm. Accessed 4 June 2019

European Commission (2019a) Regulation (EU) 2019/4 on the manufacture, placing on the market and use of medicated feed. OJ L20, Vol 62

European Commission (2019b) Regulation (EU) 2019/6 of the European Parliament and of the Council of 11 December 2018 on veterinary medicinal products and repealing Directive 2001/82/EC. OJ L20, Vol 62

European Commission, DG Health and Food Safety (2017) Overview report on non-EU countries' national policies and measures on antimicrobial resistance. https://publications.europa.eu/en/publication-detail/-/publication/8f6927e2-6ded-11e8-9483-01aa75ed71a1/language-en. Accessed 4 June 2019

European Commission, Eurostat (2019) Pig Population/Livestock Survey—Annual Data. https://ec.europa.eu/agriculture/sites/agriculture/files/market-observatory/meat/pigmeat/doc/pig-population-survey_en.pdf. Accessed 4 June 2019

European Court of Auditors (2019). Special report on addressing antimicrobial resistance: progress in the animal sector. Doi: 10.2865/487837

European Union (2017) The European Union summary report on antimicrobial resistance in zoonotic and indicator bacteria from humans, animals and food in 2017. EFSA J 17(2):5598

European Union, Health and Food Safety Directorate-General (2018) Overview report measures to tackle antimicrobial resistance through the prudent use of antimicrobials in animals. https://publications.europa.eu/en/publication-detail/-/publication/aa676ddd-2d87-11e8-b5fe-01aa75ed71a1. Accessed 4 June 2019

European Union (2020) Communication: a farm to fork strategy for a fair, healthy and environmentally-friendly food system. COM(2020) 381 final. 20 May 2020. https://ec.europa.eu/info/sites/info/files/communication-annex-farm-fork-green-deal_en.pdf

Eurostat (2016) Agriculture, forestry and fishery statistics. https://ec.europa.eu/eurostat/web/products-statistical-books/-/KS-FK-16-001. Accessed 4 June 2019

FAO (2010) Food and Agriculture Organization of the United Nations: Food and Agriculture data. http://faostat.fao.org/. Accessed 4 June 2019

FAO (2018) Food and Agriculture Organization of the United Nations: antimicrobial resistance policy review and development framework - a regional guide for governments in Asia and the Pacific to review, update and develop policy to address antimicrobial resistance and antimicrobial use in animal production. http://www.fao.org/3/ca1486en/CA1486EN.pdf. Accessed 4 June 2019

FEAP (2016) Federation of European Aquaculture Producers: information for 2016 on fish aquaculture in Europe. http://feap.info/index.php/data/. Accessed 4 June 2019

Federation of Veterinarians of Europe (2016a) Antimicrobial use in food-producing animals, FVE input to RONAFA report. EFSA J 15(1)

Federation of Veterinarians of Europe (2016b) Relationship between animal welfare and the use of antibiotics in food animals. https://www.fve.org/cms/wp-content/uploads/063-FVE_AWW-Position-on-resistance-and-animal-welfare_final.pdf. Accessed 4 June 2019

Federation of Veterinarians of Europe (2018) Monitoring of farm animal welfare using animal indicators. https://www.fve.org/cms/wp-content/uploads/058_AWIndicatorsPaper_finaldraft18sept2018_GA_adopted.pdf. Accessed 4 June 2019

Fine Chemicals (1951) Continuous high level feeding of Aureomycin (advertisement billboard). https://www.propublica.org/article/a-history-of-fda-inaction-on-animal-antibiotics. Accessed 4 June 2019

Fosse J, Seegers H, Magras C (2009) Prevalence and risk factors for bacterial food-borne. Zoonotic hazards in slaughter pigs: a review. Zoonoses and Public Health 56(8):429–454

Federation of Veterinarians of Europe (2016) Antimicrobial use in food-producing animals: replies to EFSA/EMA questions on the use of antimicrobials in food-producing animals in EU and possible measures to reduce antimicrobial use, FVE/016/DOC/O10

Federation of Veterinarians of Europe (2019) FVE position on selective dry cow treatment in dairy cows, FVE/doc/047. https://www.fve.org/cms/wp-content/

uploads/047-Position_dry_cow_treatment_adopted.pdf. Accessed 4 June 2019

Goyette-Desjardins G, Auger JP, Xu J, Segura M, Gottschalk M (2014) *Streptococcus suis*, an important pig pathogen and emerging zoonotic agent-an update on the worldwide distribution based on serotyping and sequence typing. Emerg Microbes Infect 3:1–20

Haenni M, Châtre P, Dupieux-Chabert C, Métayer V, Bes M, Madec JY, Laurent F (2017) Molecular epidemiology of methicillin-resistant *Staphylococcus aureus* in horses, cats, and dogs over a 5-year period in France. Front Microbiol 8:2493

Hernandez-Garcia J, Wang J, Restif O, Holmes MA, Mather AE, Weinert LA, Wileman TM, Thomson JR, Langford PR, Wren BW, Rycroft A, Maskell DJ, Tucker AW, BRADP1T Consortium (2017) Patterns of antimicrobial resistance in *Streptococcus suis* isolates from pigs with or without streptococcal disease in England between 2009 and 2014. Vet Microbiol 207:117–124

Holman DB, Chénier MR (2015) Antimicrobial use in swine production and its effect on the swine gut microbiota and antimicrobial resistance. Can J Microbiol 61:785–798

Ikerd J (2008) Crisis and opportunity: sustainability in American agriculture. University of Nebraska Press, Lincoln, NE

JVARM. Japanese Veterinary Antimicrobial Resistance Monitoring System. http://www.maff.go.jp/nval/tyosa_kenkyu/taiseiki/monitor/e_index.html. Accessed 4 June 2019

Kirchhelle C (2019) Pyrrhic progress. Antibiotics in Anglo-American food production 1935-2013. Newark University Press, Rutgers

Kittl S, Heckel G, Korczak BM, Kuhnert P (2013) Source attribution of human campylobacter isolates by MLST and Fla-typing and association of genotypes with quinolone resistance. PLoS One 8(11):e81796. https://doi.org/10.1371/journal.pone.0081796. Accessed 4 June 2019

Koop G (2016) MRSA transmission between horses and vets: who's doing the infecting? Vet Rec 178:471–472

Lin D, Chen K, Li R, Liu L, Guo J, Yao W, Chen S (2014) Selection of target mutation in rat gastrointestinal tract *E. coli* by minute dosage of enrofloxacin. Front Microbiol 5:468

Maas L (2014) Selective dry cow therapy: the effect of different selection criteria on costs, antibiotic use and mastitis infections at Dutch dairy farms. http://edepot.wur.nl/309122. Accessed 4 June 2019

MacKenzie S (2015) A brief history of agriculture and food production: the rise of "Industrial Agriculture", Johns Hopkins Center for a Livable Future. https://www.saylor.org/site/wp-content/uploads/2015/07/ENVS203-7.3.1-ShawnMackenzie-ABriefHistoryOfAgricultureandFoodProduction-CCBYNCSA.pdf. Accessed 4 June 2019

Mather AE, Reid SW, Maskell DJ, Parkhill J, Fookes MC, Harris SR, Brown DJ, Coia JE, Mulvey MR, Gilmour MW, Petrovska L, de Pinna E, Kuroda M, Akiba M, Izumiya H, Connor TR, Suchard MA, Lemey P, Mellor DJ, Haydon DT, Thomson NR (2013) Distinguishable epidemics of multidrug-resistant Salmonella Typhimurium DT104 in different hosts. Science 341 (6153):1514–1517

Ministry of Agriculture and Forestry of New Zealand (2011) Antibiotic resistance review and update on New Zealand regulatory control of antimicrobial agricultural compounds with regard to antimicrobial resistance. https://www.mpi.govt.nz/dmsdocument/26383/send. Accessed 4 June 2019

Ministry of Environment and Food of Denmark (2016) Low use of antibiotic in Denmark. https://www.foedevarestyrelsen.dk/english/Animal/MRSA/Pages/Low_use_of_antibiotic_in_Denmark.aspx. Accessed 4 June 2019

Moeser AJ, Pohl CS, Rajput M (2017) Weaning stress and gastrointestinal barrier development: implications for lifelong gut health in pigs. Anim Nutr 3(4):313–321

Nienhoff HJ (2019) QS-Report –Meat- and Meat-Products-1-2019-complete-V2. QS-Qualität uns Sicherheit GmbH. https://www.q-s.de/flip/QS-Report-Meat-and-Meat-Products-1-2019-Complete-V2/8/. Accessed 22 June 2019

O'Neil J (2016) The review on antimicrobial resistance. https://amr-review.org/sites/default/files/160518_Final%20paper_with%20cover.pdf. Accessed 4 June 2019

OIE (2017) World Organisation for Animal Health: OIE Annual report on the use of antimicrobial agents in animals. http://www.oie.int/fileadmin/Home/eng/Our_scientific_expertise/docs/pdf/AMR/Annual_Report_AMR_2.pdf. Accessed 4 June 2019

OIE (2018a) World Organisation for Animal Health: OIE Annual report on the use of antimicrobial agents in animals. http://www.oie.int/fileadmin/Home/eng/Our_scientific_expertise/docs/pdf/AMR/Annual_Report_AMR_3.pdf. Accessed 4 June 2019

OIE (2018b) World Organisation for Animal Health: OIE List of antimicrobial agents of veterinary importance. https://www.oie.int/doc/ged/D9840.PDF. Accessed 4 June 2019

Oliver SP, Murinda SE, Jayarao BM (2011) Impact of antibiotic use in adult dairy cows on antimicrobial resistance of veterinary and human pathogens: a comprehensive review. Foodborne Pathog Dis 8 (3):337–355

Pardon B, Catry B, Dewulf J, Persoons D, Hostens M, De Bleecker K, Deprez P (2012) Prospective study on quantitative and qualitative antimicrobial and anti-inflammatory drug use in white veal calves. J Antimicrob Chemother 67(4):1027–1038

Pletinckx LJ, Verhegghe M, Crombé F, Dewulf J, De Bleecker Y, Rasschaert G, Butaye P, Goddeeris BM, De Man I (2013) Evidence of possible methicillin-resistant *Staphylococcus aureus* ST398 spread between pigs and other animals and people residing on the same farm. Pre Vet Med 109(3–4):293–303

Rabiee AR, Lean IJ (2013) The effect of internal teat sealant products (Teatseal and OrbeSeal) on intramammary infection, clinical mastitis, and somatic cell counts in lactating dairy cows: a meta-analysis. J Dairy Sci 96:1–17

Rees VE, Bulitta JB, Nation RL, Tsuji BT, Sörgel F, Landersdorfer CB (2015) Shape does matter: short high-concentration exposure minimizes resistance emergence for fluoroquinolones in *Pseudomonas aeruginosa*. J Antimicrob Chemother 70:818–826

RUMA (2018) Report summarising the progress against antibiotic use targets identified by the UK livestock industry's Targets Task Force in October 2017. https://www.ruma.org.uk/wp-content/uploads/2018/11/RUMA-TTF-1-year-on-Full-Report-FINAL.pdf. Accessed 4 June 2019

Schwarz S, Feßler AT, Loncaric I, Wu C, Kadlec K, Wang Y, Shen J (2018) Antimicrobial resistance among Staphylococci of animal origin. Microbiol Spectr 6(4)

Scicluna C, Gronlund U, Ricardi G, Slater J (2013) Factors influencing the antimicrobial choice of equine veterinarians in Europe 13th WEVA congress Budapest October 2013

Sieber RN, Skov RL, Nielsen J, Schulz J, Price LB, Aarestrup FM, Larsen AR, Stegger M, Larsen J (2018) Drivers and dynamics of methicillin-resistant livestock-associated *Staphylococcus aureus* CC398 in pigs and humans in Denmark. MBio 9(6)

Spoor LE, McAdam PR, Weinert LA, Rambaut A, Hasman H, Aarestrup FM, Kearns AM, Larsen AR, Skov RL, Fitzgerald JR (2013) Livestock origin for a human pandemic clone of community-associated methicillin-resistant *Staphylococcus aureus*. MBio 13 (4)

Stevens M, Piepers S, Supré K, Dewulf J, De Vliegher S (2016) Quantification of antimicrobial consumption in adult cattle on dairy herds in Flanders, Belgium, and associations with udder health, milk quality, and production performance. J Dairy Sci 99(3):2118–2130

Swann MM, Baxter KL, Field HI (1969) Report of the joint committee on the use of antibiotics in animal husbandry and veterinary medicine. HMSO, London

The European Union Summary Report on Antimicrobial Resistance in zoonotic and indicator bacteria from humans, animals and food in 2017/2018. EFSA Journal 18(3)

Van Boeckel TP, Brower C, Gilbert M, Grenfell BT, Levin SA, Robinson TP, Teillant A, Laxminarayan R (2015) Global trends in antimicrobial use in food animals. Proc Natl Acad Sci U S A 112(18):5649–5654

Van Boeckel TP, Glennon EE, Chen D, Gilbert M, Robinson TP, Grenfell BT, Levin SA, Bonhoeffer S, Laxminarayan R (2017) Reducing antimicrobial use in food animals. Science 29(357):1350–1352

Van Cleef BA, Verkade EJ, Wulf MW, Buiting AG, Voss A, Huijsdens XW, van Pelt W, Mulders MN, Kluytmans JA (2010) Prevalence of livestock-associated MRSA in communities with high pig-densities in The Netherlands. PLoS One 5(2)

Van den Broek IV, van Cleef BA, Haenen A, Broens EM, van der Wolf PJ, van den Broek MJ, Huijsdens JA (2009) Methicillin- resistant *Staphylococcus aureus* in people living and working in pig farms. Epidemiol Infect 407(137):700–708

Van Duijkeren E, Moleman M, Sloet van Oldruitenborgh-Oosterbaan MM, Multem J, Troelstra A, Fluit AC, van Wamel WJ, Houwers DJ, de Neeling AJ, Wagenaar JA (2010) Methicillin-resistant *Staphylococcus aureus* in horses and horse personnel: an investigation of several outbreaks. Vet Microbiol 141:96–102

Van Duijkeren E, Greko C, Pringle M, Baptiste KE, Catry B, Jukes H, Moreno MA, Pomba MC, Pyörälä S, Rantala M, Ružauskas M, Sanders P, Teale C, Threlfall EJ, Torren-Edo J, Törneke K (2014) Pleuromutilins: use in food-producing animals in the European Union, development of resistance and impact on human and animal health. J Antimicrob Chemother 69(8):2022–2031

Wageningen Economic Research (2019) The impact of antibiotic reduction on economic results of pig and broiler farms in the Netherlands. https://edepot.wur.nl/475403. Accessed 4 June 2019

Wageningen University & Research (2018). https://www.wur.nl/en/newsarticle/Europe-and-China-join-forces-for-healthy-livestock.htm. Accessed 4 June 2019

Watts J, Sweeney MV, Lubbers B (2020) Current and future perspectives on the categorization of antimicrobials used in veterinary medicine. J Vet Pharmacol Therap:1–8

WHO (2019) World Health Organisation: WHO list of critically important antimicrobials 6th revision. https://www.who.int/foodsafety/publications/antimicrobials-sixth/en/. Accessed 28 May 2020

World organisation of animal health (OIE). OIE list of antimicrobial agents of veterinary importance revision July 2019. https://www.oie.int/fileadmin/Home/eng/Our_scientific_expertise/docs/pdf/AMR/A_OIE_List_antimicrobials_July2019.pdf

Xu F, Zeng X, Hinenoya A, Lin J (2018) MCR-1 confers cross-resistance to bacitracin, a widely used in-feed antibiotic. mSphere 3(5):411–418

Considerations Reflecting Possible Risks from Use of Antimicrobials

Lucie Pokludová and Leona Nepejchalová

Abstract

Whenever antimicrobials are used and despite the undoubtful benefits of their existence with respect to the treatment of the infections of human, animals and in some cases also plant protection, the risks associated with their use should be taken into account. Considering the risks associated with use of antimicrobials is quite complex. Despite the fact that currently mainly risks related to antimicrobial resistance are mentioned at least in the scientific bibliography, further risks cannot be forgotten. One of those also mentioned frequently, especially when antimicrobials are used in animals are risks for consumers (i.e. possible residues of antimicrobials in food of animal origin). It should be noted that well-established rules and system of surveillance of residues (including antimicrobials) have been established within EU decades ago. Also risks for users (veterinarians, farmers, staff in medicated feed mills) are considered when authorising the veterinary medicinal products containing antimicrobials and relevant risk mitigation measures are proposed to be followed in practice. Last but not least should be mentioned the possible risks for the environment, that can come not only from the use of antimicrobials both in human and animals, but can be also linked with pharmaceutical industry production, especially in third countries.

Keywords

Antimicrobial risks · Food safety · User exposure · Occupational exposure · Environment

1 Considerations Reflecting Possible Risks from Use of Antimicrobials in Food-Producing Animals

Lucie Pokludová

Use of antimicrobials in animals, including those called food-producing, pose benefits coming from treatment of animal diseases and due to this ensuring to gain food from healthy animals at the point of their slaughter or harvesting the milk and eggs from them. On the other hand, use of antimicrobials pose also hazards of transmission of resistant bacteria or their resistance determinants (indirectly by food of animal origin or directly by contact with animals), hazards of direct exposure to antimicrobials (handling either with VMP at point of pre-administration, administration, or post-administration), hazards from occupational exposure of staff in pharmaceutical companies or feed mills as well as hazards of AMR coming from the environment (that is partly

L. Pokludová (✉) · L. Nepejchalová
Institute for State Control of Veterinary Biologicals and Medicines, Brno, Czech Republic
e-mail: pokludova@uskvbl.cz; nepejchalova@uskvbl.cz

L. Pokludová (ed.), *Antimicrobials in Livestock 1: Regulation, Science, Practice*,
https://doi.org/10.1007/978-3-030-46721-0_5

loaded by AMR coming from animal husbandries). Exact risks of adverse health effects in human can be identified being exposed to those hazards. Among such adverse health effects can be included some with immediate impact as illness due to presence of organisms resistant to antimicrobials in food, or acquired through direct animal contact, as well as hypersensitivity reactions. From the long-time perspective impacts increased frequency of infections, treatment failures, loss of treatment options and increased severity of disease manifested by prolonged duration of disease, increased hospitalisation, disability and mortality, are of concern.

Risk estimation is based on the following cornerstones (EMA 2018a):

Hazard identification: Identification of resistant bacteria/determinants that are selected by use of the antibiotic in target animal species and may be associated with human illness.

Exposure assessment: Pathways necessary for exposure of humans to resistant bacteria/determinants following from the point of release from the animal to the point of food consumption or direct contact and an estimation of the probability of its occurring should be known.

Release assessment: Pathways and probability that resistant bacteria are present in the animal as a result of use of the VMP at the time of "release" (slaughter, collection of food produce, or via direct animal contact).

Consequence assessment: The probability and severity of adverse human health effects following exposure to resistant bacteria/determinants originating from treated animals and colonisation and infection of human.

There is existing draft of the Guideline on the assessment of the risk to public health from antimicrobial resistance due to the use of an antimicrobial veterinary medicinal product in food-producing animals (EMA/CVMP/AWP/706442/2013) elaborated by EMA-CVMP Antimicrobial Working party (EMA 2018a) that sets the methodology for the risk assessment considering the views and approaches used by the internationally agreed documents—Risk Analysis for Antimicrobial Resistance Arising from the Use of Antimicrobial Agents in Terrestrial and Aquatic Animals by OIE (OIE 2018a; OIE 2018b); Vose et al. (2001); considering also Codex (Guidelines for risk analysis of food-borne antimicrobial resistance, CAC/GL 77-2011) and the requirements in place in other jurisdictions—FDA (2003); Health Canada (2007), APVMA (2014).

2 Impacts on Food Chain Safety: Residues of Antimicrobials

One of the concerns that coming more and more urgent with raising awareness that antimicrobials not provide us with benefits only, is food safety perspective, considering all parts of the whole food chain and its critical points. Many of the drugs, among them antimicrobials, were approved for use in livestock production and one of the parts of the assessment during the authorisation process is assessment of the residues and their possible impacts from the perspective of the acute and chronical effects on human health. Specific assessment is performed in the case of antimicrobials—especially in the phase of the establishment of the so-called Maximum Residue Limits (MRLs). According to the valid legislative (Regulation (EC) 470/2009) MRL is defined as the maximum concentration of residue of (antimicrobial or any other) pharmacologically active substance which may be permitted in food of animal origin. Once the residues drop below this threshold, the food is considered as safe from the perspective of not having harmful effect on health of possible consumers. There should be also commented that not all antimicrobial active substances, that were assessed, were finally approved for use in food-producing animals or in some of them at least risk mitigation measures (including very low MRLs levels approved) were established to minimise the negative impacts on consumer's health. Another important thing is that MRLs have not been established yet for all food-producing species/commodities, but most of them are set for major food-producing species and in well-defined cases (mostly where pharmacokinetic data indicate such possibility) they can be extrapolated to cover also minor food-producing species (e.g. MRLs set for bovine can be extrapolated under defined circumstances to ovine, MRLs defined for *Gallus gallus* can be extrapolated under defined circumstances to other poultry species).

Derivation of MRLs for antimicrobials requires complex assessment of toxicological, pharmacological as well as microbiological data (Baynes et al. 2016). One of the key assessed factors is Acceptable Daily intake (ADI), which is based among the other safety studies in the case of antimicrobials also on assessment of microbiological parameters (please refer to the next subchapter for further details). According to Article 6 of the Regulation (EC) 470/2009 the scientific risk assessment shall consider the metabolism and depletion of pharmacologically active substances in relevant animals species, the type of residues and the amount thereof that may be ingested by human beings over a lifetime without an appreciable health risk expressed in terms of acceptable daily intake (ADI). Alternative approaches to ADI may be used.

The scientific risk assessment shall concern:

1. The type and amount of residue considered not to present a safety concern for human health
2. The risk of toxicological, pharmacological or microbiological effects in human beings
3. Residues that occur in food of plant origin or that which comes from the environment

If the metabolism and depletion of the substance cannot be assessed, the scientific risk assessment may take into account monitoring data or exposure data.

To establish MRLs for a given drug, following data are requested to be provided to the regulatory body:

- **Dosing schedule data** (amount of the active substance, dose interval, total duration of administration)
- **Administration route**
- **Pharmacodynamics data** (including mechanism of action with implications, e.g. for mADI)
- **Pharmacokinetic data** (including metabolic fate)—laboratory as well as target food-producing animals
- **Toxicity data,** effect on reproduction (developmental effects, mutagenicity, carcinogenicity, neurotoxicity, immunotoxicity)
- **Data on microbiological properties of residues** (disruption of colonisation barrier, increase of population of resistant bacteria, potential effect on microorganisms used in industrial processing of foodstuffs)
- **Depletion of residues data using a radiolabeled drug** for the edible tissues for which MRLs is to be established (usually main edible tissues as muscle, fat, liver, kidney AND, where relevant, milk, eggs, honey) and for each (major) target species—for minor species, based on certain data, extrapolations can be indicated to be possible
- **Validated analytical method** for residue detection and quantification (including methods appropriate for monitoring purposes)
- **Identified "marker residue" data a and main "target tissues"**
- **Data defining the effect of residues on food processing**

MRLs are set for pharmacologically active substances and are specific for certain target tissues representing edible meat and offals: muscle (*in fish: muscle + skin*), liver, kidney, skin (*in pigs and poultry: skin + fat in natural proportions*) and for food commodities as milk, eggs and honey. The above briefly described concept also count with the so-called standard food basket for individual commodities and groups of animal species (see Table 1). The concentrations of the MRLs for one active substance may differ both for the individual target tissues/commodities as well as among the species of animals. For exact classification regarding maximum residue limits (i.e. levels of MRLs) of individual pharmacologically active substances should be referred to the Regulation (EU) 37/2010 (2009) as amended.

Other provisions/restrictions can also be listed according to Article 14 (7) of Regulation (EC) 470/2009, e.g. in the cases where no

Table 1 Standard food basket according to EMA-CVMP/WHO (Gupta, 2018)

Mammals		Poultry		Fish		Bees	
Muscle	0.300 kg	Muscle	0.300 kg	Muscle and skin in natural proportion	0.300 kg	Honey	0.20 kg
Fat	0.050 kg[a]	Fat and skin in natural proportion	0.090 kg				
Liver	0.100 kg	Liver	0.100 kg				
Kidney	0.050 kg	Kidney	0.010 kg				
Milk	1.500 kg	Eggs	0.100 kg				

[a]Fat and skin in natural proportions

MRLs are established for eggs/milk, the active substance is not intended to be used in animals from which eggs/milk is produced for human consumption (see also "Other provisions" and "List of antimicrobials" in Table 2). For honey, no EU MRLs are currently defined for any antibiotics.

Maximum residue limits are also considered points of reference for the establishment of withdrawal periods, in accordance with the respective legal provisions including rules for marketing authorisation of veterinary medicinal products, and are to be used in food-producing animals as well as for the control of residues in food of animal origin.

Withdrawal periods (WPs) are indicated for drugs used in different food-producing species of animals as the period of time post-administration of such drugs that must elapse before the edible tissues/commodities are considered safe for direct human consumption or food processing (i.e. when the residue levels are equal/below the MRLs). The withdrawal period is established for each individual veterinary medicinal product that is intended to be used for food-producing animals. In the process of establishing WP, thorough assessment is performed of the relevant and sufficiently robust studies that are submitted to confirm residue depletion after the last administration of the veterinary medicinal product in the dose, dosing schedule (frequency of dosing), route of administration and total duration of treatment. The so-called worst case scenario (highest dose, most frequent administration and longest duration of treatment proposed to be authorised) is considered to establish safe withdrawal period for any veterinary medicinal product (including those with antimicrobials) and in fact all residues in edible tissues should be safely below the established MRLs of individual target tissues/commodities.

One could ask what to do in those cases where no authorised veterinary medicinal product with approved withdrawal period is established for certain animal species or indications. As antimicrobials are prescription-only medicines, it is allowed according to the valid legislation in necessary cases, especially avoiding unacceptable suffering of animals to prescribe certain VMP under the so-called "cascade" principles. It is under the veterinarian discretion and responsibility to prescribe for species, for which the product is not licensed, or for the indication not licensed, or go for, e.g. higher dosage, but in such cases veterinarian should specify the withdrawal period to be set with certain minimum required withdrawal period to be used (please refer to the part related to "off-label" use of antimicrobials in Chap. 7, where more information is given).

From the above text it can be summarised that for the whole process key moments are establishment of the correct ADI and correct MRLs derived from them. It is also taken into consideration in the proportion of the ADI that can be allocated to the use of the active substance in veterinary medicinal products (considering that there can also be other use like, e.g. pesticide). The exposure of even a small amount of residues can cause adverse effects in humans like hypersensitivity (allergic) reactions, disruption of the normal intestinal human microbiome, antimicrobial resistance selection, blood dyscrasias, carcinogenicity effects, mutagenicity, teratogenicity effects, etc. Considering the adverse effects of antimicrobials on human health, it should be distinguished which level of residues can be dangerous from the perspective of the acute or chronic exposure. As an example (Table 3), the groups of the most frequently used antimicrobials in the EU (tetracyclines, penicillins, sulphonamides) are commented briefly on the main concerns related to the active substances belonging to this groups that were considered from the perspective of safety assessment, including MRLs assessment.

THINK ABOUT: Revising of the Methodological Concept of Microbiological ADI

One part of the current concept of microbiological ADIs is based on phenotypic methods of susceptibility testing and therefore on microbes that are cultivable in vitro. For the nearest future it might be of importance to investigate gut microbiome additionally via molecular biology methods like whole genome sequencing to describe changes in composition of bacterial populations (both cultivable/non-cultivable) as well as changes in resistance after exposure to different concentration residues of antimicrobials.

The above-mentioned example of penicillins' and milk's very low MRL (EU) and even zero tolerance in the United States show that for substances with certain concerns, if the risky and safe levels can be quantified, the solution on how to establish safe MRLs can be found. Despite that, from the big portfolio of antimicrobials there were identified those pharmacologically active substances, for which due to the possible serious adverse effect on human health (mutagenicity, carcinogenicity, toxicity) the safe MRLs cannot be established. They are listed in Table II of the Regulation 37/2010 as "prohibited substances" and are as follows: chloramphenicol, dapsone, metronidazole, dimetridazole, ronidazole and nitrofurans. As mentioned above, despite those substances being banned for use in food-producing animals in the EU, the reasons are connected neither with antimicrobial properties nor with antimicrobial resistance.

Reading the text of the above paragraph you might be interested in the level of harmonisation of the safety of residues assessment worldwide. Risk assessment of residues coming from the food of animal origin due to the use of veterinary medicinal products follows worldwide similar essential principles, despite that methodologies can be slightly different. In the European Union the responsible body (based on the rules as stipulated by the Regulation 726/2004 as amended by Regulation EC 5/2019) is the European Medicines Agency (EMA), which, with active participation of the experts coming from individual Member States, perform relevant assessment leading to the establishment and publication of the maximum residue limits for substances intended to be used in veterinary medicinal products authorised in the European Union. It should also be noted that for certain antimicrobials (e.g. ionophoric anticoccidials) that are used in the European Union under the feed additive legislation the MRLs are set by the European Food Safety Authority. In the United States, the Food and Drug Administration (FDA) is the responsible regulatory body establishing the maximum permitted concentrations for veterinary drug residues known as tolerances. Another independent body that recommend the MRLs is the Joint Food and Agricultural Organisation/World Health Organisation Expert Committee on Food Additives (Baynes et al. 2016). It can be said that worldwide is mostly agreed using of the Codex Alimentarius' MRLs (also EU accept them without requiring additional application and reassessment since 2009).

2.1 How the Issue of Possible Residues of Antimicrobials is Tackled in the EU

Leona Nepejchalová

When antimicrobial substances are planned to be used in veterinary medicinal products (VLPs) in

Table 2 Restrictions for commodities: milk and eggs for pharmacologically active substances (antibiotics) in relation to species of food-producing animals (according to Regulation (EC) 470/2009 as specified in Regulation (EC) 37/2010)

Other provisions	Pharmacologically active substances
"Not for use in animals from which milk is produced for human consumption" *According to Article 14 (7) of Regulation (EC) 470/2009*	Apramycin (BO, OV) Difloxacin (BO, OV, CA) Doxycycline (all FPS) Florfenicol (BO, OV, CA) Gamithromycin (all ruminants except bovine) Oxolinic acid (all FPS) Paromomycin (all FPS) Tildipirosin (BO, CA) Tulathromycin (BO, OV, CA)
"Not for use in animals from which eggs are produced for human consumption" *According to Article 14 (7) of Regulation (EC) 470/2009*	Amoxicillin (all FPS) Ampicillin (all FPS, except fin fish) Apramycin (GA) Avilamycin (POU) Benzylpenicillin (all FPS) Cloxacillin (all FPS) Danofloxacin (POU) Dicloxacillin (all FPS) Doxycycline (POU) Enrofloxacin (POU) Florfenicol (POU) Flumequine (POU) Kanamycin (all FPS, except fin fish) Oxacillin Oxolinic acid (all FPS) Paromomycin (all FPS) Sarafloxacin (GA) Spectinomycin (all FPS) Spiramycin (GA) Sulfonamides (all FPS) Thiamphenicol (all FPS) Tilmicosin (GA) Trimethoprim (all FPS) Virginiamycin (GA)

BO bovine, *CA* caprine, *FPS* food-producing species, *GA Gallus gallus*, *OV* ovine, *POU* poultry
Note: Regulation (EC) 37/2010 is subject of updates

food-producing animal species, the acceptable consumer safety level must be assessed and established according to the current legislation. In the EU, an application for the establishment of maximum residue limits (MRLs) needs to be submitted to the European Medicines Agency (EMA). The following assessment critically monitors how the residues of a substance (or metabolite) taken from food can affect human health. The microbiological acceptable daily intake (ADI), based on a concentration that does not adversely affect the gastrointestinal microflora, is decisive for the determination of maximum residue limits (MRLs) for most antimicrobials.

Subsequent use of medicinal products for treatment in veterinary practice and the possible presence of residues of pharmacologically active substances or their metabolites in tissues and products of animal origin are regularly monitored in Europe. EU legislation incorporates the responsibilities of individual Member States to carry out this monitoring on a planned basis. Therefore, the plans are being prepared annually and the results of the national monitoring are being processed and reported to the Commission. Food Safety Agency (EFSA) is summarising the data received from each member state and is publishing a public report.

Table 3 Examples of main concerns related to the active substances considered from the perspective of safety assessment, including MRLs assessment

Group of antimicrobials	Main concerns	Comment	Note
Tetracyclines	Possible influence of human intestine microbiome	• MRLs set based on the microbiological ADI • In the period of EMA assessment, it was concluded that there is no induction of resistant enterobacteria at the dose 2 mg per person per day—on the other hand, in an in vitro study to assess the impact of tetracycline on the human intestinal microbiome, there was screened the variability of the presence of *tet* genes after exposure of low concentrations 0.15, 1.5, 15 and 150 μg/ml of tetracycline, after 24 h and 40 days and variable to slight increase of the tetracycline gene copies occurred.	Chlortetracycline, Oxytetracycline, Doxycycline have same MRLs for edible tissues with exception to the fact that no MRLs have been set for eggs/milk for doxycycline
Penicillins	Hypersensitivity reactions	• 10% of the human population is believed to be allergic[a] • Association with IgE-mediated allergic anaphylaxis[b]	Same MRLs covering the penicillin group: Penicillins: mainly benzyl PNC Aminopenicillins: Amoxicillin, Ampicillin Izoxazolyl penicillins: Oxacillin, Cloxacillin, Dicloxacillin
	Anaphylaxis	• Human reaction based on penicilloyated (amoxicilloyated) residues in milk and meat[c] • Amoxicillin (AX), with or without clavulanic acid, is the most common elicitor of allergy. • Very low levels (6 μg/L) can cause this reaction; therefore, especially for milk low MRLs (4 μg/kg) were established for the group of penicillins by EMA and JECFA (Codex). • USA—zero tolerance for residues in milk	
	Influence of starter cultures in food processing	• Sufficient evidence that consumption of beef or pork containing residues of penicillins exceeding MRLs causing anaphylactic reactions[d,e]	
Sulphonamides	Skin reactions	• Mild rash to severe toxidermia are some of the skin reactions following human exposure to sulphonamide[f]	No studies directly confirmed the effect of residues consumption of products of food origin and of the mentioned adverse reactions
	Hypersensitivity reactions	• Contact sensitisation confirmed for topical medicinal products	
	Blood dyscrasias	• Haemolytic anaemia, neutropenia, thrombocytopenia and pancytopenia	
	Cancerogenicity (thyroid)	• Sulfamethazine dose-dependent increase in follicular cells adenomas of thyroid gland	

[a]Solensky and Khan (2014); [b] Patterson and Stankewicz (2019) [c] Torres et al. (2017) [d] Raison-Peyron et al. (2001); [e] Demoly and Gomes (2005); [f] Choquet-Kastylevsky et al. (2002)

2.2 EU System

To ensure consumer safety, a system of legislative standards is developed in the EU that requires an assessment of the pharmacological activity of each substance that is planned for use in the treatment of food-producing animal species. The appropriate application and request, as described in the Commission implementing regulation (EU) 2017/12, needs to be submitted to European Medicines Agency (EMA) where it is assessed by Committee for Medicinal Products for Veterinary Use (CVMP). The opinion of EMA presented to the Commission consists of a scientific risk assessment and risk management recommendations. Only substances with favourable results of assessment are allowed to be used in practice. Where necessary for the protection of human health, the maximum residue limits (MRLs) need to be established, i.e. maximally permitted acceptable concentrations of residues in food of animal origin. MRLs are important points of reference for calculation of withdrawal period and are used for control purposes in EU member states as well.

The basic legislative standard is Regulation (EC) No 470/2009 that sets the framework for this assessment and its regulatory rules. The methodological principles for the risk assessment and risk management recommendations referred in Reg. (EC) No 470/2009 are specified in Commission regulation (EU) 2018/782. Within the context of previously mentioned legislation, another Commission regulation (EU) 2017/880 lays down rules for extrapolation of MRLs or of classification of assessed pharmacologically active substance between specific species and derived foodstuff that needs to be considered by EMA during the scientific assessment to ensure better conditions for greater possibility to support the availability of authorised veterinary medicinal products.

Results of evaluations undertaken by other scientific bodies, such as the Joint Food and Agriculture Organisation (FAO)/World Health Organisation (WHO) Expert Committee on Food Additives (JECFA), the European Food Safety Authority (EFSA) and European Chemicals Agency (ECHA) are considered during assessment as well.

All these regulatory acts and rules in them point to the obligation of estimation of consumer exposure based on assumed acceptable daily intake (ADI) and derived relevant maximum residue limits (MRLs) at levels that ensure that the total amount of residues from all sources likely to be ingested do not exceed the ADI. Some of the substances are not only used as veterinary medicines; in case of substances used for example as pesticides the portion of 45% should be reserved for veterinary use.

2.3 ADI (Focused on Microbiological ADI)

It is necessary to clearly and enough conservatively estimate the possible residual load during the lifetime (chronic exposure) on the basis of the data provided. This estimation is done by setting the so-called Acceptable Daily Intake (ADI). Appropriate values of NO(A)EL(s) (LO(A)EL (s)) or BMDL(s) obtained from pharmacological, toxicological and where suitable from microbiological studies performed based on valid VICH guidelines (VICH programme harmonising technical requirements for veterinary products authorisation between the EU, Japan, and the United States) are utilised for subsequent derivation of the ADI, together with the use of the justified uncertainty safety factor. The formulas predefined in guidelines are applied.

Antimicrobial effects of substances with appropriate effect on the human intestinal flora may occur at very low doses that are below those incurring toxicity in the toxicity tests. In the case of such substances we can actually talk about the pharmacodynamic effect of residues, meaning pharmacodynamic effect on microorganisms. The principles for establishment of microbiological ADI are described in VICH GL 36 (EMA/CVM/VICH 2012 and EMA/CVMP/VICH rev 2019); this guideline is required to address two aspects of antimicrobial activity of residues:

Disruption of the colonisation barrier (MIC, NOAEC)

and

Increase of the population of resistant bacteria (NOAEC, NOAEL)

The assessment and consideration of these aspects are reliable in cases where residues reach the human colon and remain microbiologically active. Based on the above-mentioned guideline, for determining the need for a microbiological ADI, the following sequence of steps is recommended and described in this guideline. The data may be obtained experimentally or from other appropriate sources such as scientific literature.

Step 1. Are residues of the drug, and (or) its metabolites, microbiologically active against representatives of the human intestinal flora?

Recommended data:

MIC data, obtained by standard test methods, from the following relevant genera of intestinal bacteria (*E. coli*, and species of *Bacteroides, Bifidobacterium, Clostridium, Enterococcus, Eubacterium (Collinsella), Fusobacterium, Lactobacillus, Peptostreptococcus/Peptococcus*).

It is recognised that the understanding of the relative importance of these microorganisms is incomplete and that the taxonomic status of these organisms can change. The selection of organisms should take into account current scientific knowledge.

If no information is available, assume that the compound and (or) its metabolites are microbiologically active.

Step 2. Do residues enter the human colon?

Recommended data:

Absorption, distribution, metabolism, excretion (ADME), bioavailability or similar data may provide information on the percentage of the ingested residue that enters the colon.

If no information is available in humans, use appropriate animal data. If there is no information available, assume that 100% of the ingested residue enters the colon.

Step 3. Do the residues entering the human colon remain microbiologically active?

Recommended data:

Data demonstrating loss of microbiological activity from in vitro inactivation studies of the drug incubated with faeces or data from in vivo studies evaluating the drug's microbiological activity in faeces or colon content of animals.

If the answer to any of the questions in steps 1, 2 or 3 is "no", then the ADI will not be based on microbiological endpoints and the remaining steps need not be addressed.

Step 4. Assess whether there is any scientific justification to eliminate the need for testing either one or both endpoints of concern. Take into account available information regarding colonisation barrier disruption and resistance emergence for the drug. If a decision cannot be made based on the available information, both endpoints need to be examined.

Step 5. Determine the NOAECs/NOAELs for the endpoint(s) of concern as established in step 4. The most appropriate NOAEC/NOAEL is used to determine the microbiological ADI.

The guideline is clear that remains open in case of requirements for specific studies and allows alternative approaches to be used. The text of guideline emphasises that further research is needed to confirm the reliability and validity of all test systems and does not recommend any one particular system for use in regulatory decision-making. Instead, recommendations are provided for a harmonised approach to establish a microbiological ADI and test options are offered rather than specification of a testing regimen.

For calculation of ADI data obtained from studies detecting disruption of the colonisation barrier and studies indicating the changes in resistant bacteria population should be taken into consideration. Afterward the most appropriate values should be chosen for use in calculations via following formulas:

1. **Disruption of the colonisation barrier:**
   (a) Derivation from MIC data:

$$ADI = \frac{MIC_{calc} * \text{volume of colon content } (500\ \text{mL/day})}{\text{fraction of oral dose avalable to microorganisms} * 60\ \text{kg person}}$$

(b) Derivation from results of other in vitro tests (NOAEC):

$$ADI = \frac{NOAEC * \text{volume of colon content } (500\ \text{mL/day})}{\text{fraction of oral dose avalable to microorganisms} * 60\ \text{kg person}}$$

MIC_{calc}—is derived from the lower 90% confidence limit for the mean MIC50 of the relevant genera described in Step 1 for which the drug is active.

NOAEC—is derived from the lower 90% confidence limit for the mean NOAEC from in vitro systems.

Volume of colon content of 500 ml is used in guideline with implementation in August 2019 and was estimated from three-dimensional abdominal magnetic resonance imaging measurement. Version of guideline valid until August is working with the constant value of mass of colon content of 220 g in formula for calculation. This value is based on the colon content measured from accident victims. Experts included in work on the amendment of value used for colon content in VICH guideline have referred to the report of Joint FAO/WHO Expert Committee on Food Additives No. 1008 (JECFA 2018). And they have taken into account the studies based on current imaging technology showing that the hydrated colon of healthy individuals is larger than the 220 g (220 mL) estimate. Based on available information the expert group concluded that the more appropriate value for the colon volume is 500 mL.

1. **Increase in the population(s) of resistant bacteria:**
   (a) Derivation from in vitro data:

$$ADI = \frac{NOAEC * \text{volume of colon content } (500\ \text{mL/day})}{\text{fraction of oral dose avalable to microorganisms} * 60\ \text{kg person}}$$

   (b) Derivation from in vivo data:

This derivation uses the NOAEL divided by the uncertainty factor for estimation of microbiological ADI.

Regulatory rules allow generally to refer to the other alternative toxicological reference value based on valid justification. For example, discussion on the traditional concept of ADI was triggered by the international bodies and concept of Reference Dose (RfD) was developed in the 1980s by USEPA and was used in guideline for pesticides.

In contrast to the Acceptable Daily Intake (ADI) that presents chronic exposure to residues only, the Reference Dose (RfD) assessment is possible for both acute and chronic exposure. On EU level, the use of Acute Reference dose was considered as one of the possible options at a time when the problem of MRLs setting and residues in the site of injection needed to be solved. Finally, the EU experts decided to prosper from "unused" part of ADI.

Question of use of ARfD is still on the table, mainly in such cases when ADI may not be the appropriate value for quantification of the level above which exposure after a single meal or over 1 day can produce acute adverse effects (VICH GL54 2017). The debate on the possible use of the ARfD has currently shifted to evaluation of possible acute effects of residues of antimicrobials (residues of metabolites or parent substances originating from veterinary medicinal products) on the human intestinal microbiota following acute human exposure.

The establishment of the ADI value remains to be the main approach in the EU. The ADI value is compared to the quantified Theoretical Maximum Daily Intake (TMDI) which estimation is based on the MRLs set for each edible tissue (meat, fat, kidney, liver) or relevant product (milk, eggs, honey) and amount of each commodity into the standard consumer basket (see Table 1 above). However, it should be noted that for some "old" substances, such as penicillins, the ADI value has never been set, even we have sets of MRLs established. Furthermore, the Regulation (EC) No. 1950/2006 (amended by Reg. No. 122/2013) allows to use medicines for treatment of horses containing substances listed on the List of substances essential for the treatment of equidae (i.e. substances never been assessed having regard to the consumer safety and the ADI). Consumer safety is guaranteed for them if the withdrawal period of not less than 6 months is applied after treatment.

2.4 MRLs in Relation to VMPs Authorisation

The established MRLs are important for calculation of withdrawal period that is the time necessary between the last administration of the veterinary medicinal product to animals under normal conditions of use and the production of foodstuffs from such animals, in order to ensure that such foodstuffs do not contain residues in quantities in excess of the maximum limits or ADI in case of substance with "No MRL required" status. The proper classification of substances and MRLs can be found in the Regulation (EU) No. 37/2010. In Table I of annex of this regulation, a dependency between values and specific tissues may be noticed; the higher the MRL for a given tissue or commodity, the higher the residual load of this tissue. These tissues are usually organs of metabolism or excretion like the liver or kidneys. This dependency is mainly visible in newly assessed substances. For example, penicillins, cephalosporins, tetracyclines are reaching very high concentrations in kidneys based on the results from residue depletion studies. Fluoroquinolones and pleuromutilins are concentrated in liver.

Results from residues depletion studies are used for calculation of withdrawal period. Circulation (ADME—absorption, distribution, metabolism and elimination) of substance in animal body is affected by many factors, for example by the dose administered, route of administration (in case of injections, the administered volume plays an important role as well) and physiological status of animal. It should be noted that residues depletion studies are performed on healthy, in many cases young, animals. It can be said that the withdrawal period is set under ideal conditions. On other hand, to cover the differences in the population of treated animals, computational softwares work with confidence intervals and provide with high probability a sufficiently long withdrawal period.

> **THINK ABOUT:**
> However, it should be borne in mind that if we treat very ill animals, very young or very old animals, or use combinations of different medicinal products (which can interact with each other, e.g. slow down the activity of liver enzymes), we must always look at all conditions and rather extend withdrawal period. This also applies to the other occurrences of the off-label use of medicine within the cascade.

2.5 Residues Monitoring and Results Available for the EU Area

Based on Council Directive 96/23/EC (on measures to monitor certain substances and residues thereof in live animals and animal products) each EU member state has to prepare, adopt and implement on national level a monitoring plan for residues of specific groups of substances. The annual results need to be submitted to the Commission by the end of March of the following year at the latest. The Commission reports the results to the European Parliament and the Council. Possible presence of residues of substances contained in veterinary medicinal products and of other specific substances (unauthorised or prohibited substances and chemical contaminants) is observed in live animals and the edible products obtained from them. Specific sampling levels and frequency is described for each animal species (bovines, pigs, sheep, goats, equidae, poultry and aquaculture), as well as the group of substances are prescribed by the above-mentioned directive. This directive is amended by Commission Decision 97/747/EC that adds rules for levels and frequencies of sampling for milk, eggs, honey, rabbit meat and game animals (wild and farmed). The concentrations of substances found in the tissues are compared with the limits and evaluated according to the rules described in the following legislative standards:

- Regulation (EU) No. 37/2010 on pharmacologically active substances and their classification regarding maximum residue limits in foodstuffs of animal origin
- Regulation (EC) No. 396/2005 on maximum residue levels of pesticides in or on food and feed of plant and animal origin
- Regulation (EC) 1881/2006 lays down the maximum limits setting maximum levels for certain contaminants in foodstuffs
- Directive 96/22/EC concerning the prohibition on the use in stock-farming of certain substances having a hormonal or thyrostatic action and of β-agonists prohibiting the use of certain substances for specific purposes in food-producing animal
- Regulation (EU) 2018/470 on detailed rules on the maximum residue limit to be considered for control purposes for foodstuffs derived from animals which have been treated in the EU under Article 11 of Directive 2001/82/EC (Text with EEA relevance)

Targeted sampling is recommended and needs to be planned with regard to animals (species, age, gender), farming and fattening systems and available information on medicines' misuse or abuse.

MRLs are used in monitoring as a reference value for all substances classified based on reg. 470/2009. Because MRLs are not set for all tissues or commodities, the Commission has published detail rules on MRLs consideration for control purposes (reg. 2018/470). For purpose of extrapolation of MRLs values used for control in national reference laboratories, target species are considered as to be related or more closely related to each other and the rules for equation of tissues are established.

Illegal use of prohibited substances or unauthorised substances is controlled as well and therefore, the values of Reference points for action (RPA) were set up for them (chloramphenicol, malachite green) and used for monitoring. The RPA needs to be reviewed regularly in the light of new scientific knowledge. In Table 4 overview of non-compliant results for prohibited substances is given (data 2017)

Methods used for purpose of monitoring and control are mainly based on liquid chromatography (HPLC or LC, or MS/MS) with different final detection and need to be properly validated. Reference laboratories are using multianalyte (multiclass) methods that are fast and sensitive.

Table 4 Overview on the non-compliant results for prohibited substances from EFSA report 2017 (EFSA 2019)

Substance	Species/ products	Number of non-compliant results	Member States reporting non-compliant results
Chloramphenicol			
Chloramphenicol	Bovines	1	Poland
	Milk	1	Croatia
	Pigs	4	The Czech Republic, Germany
	Poultry	2	Cyprus, Poland
Nitrofurans			
AHD (1-aminohydantion)	Milk	1	Crotia
AMOZ (5-methylmorpholino-3-amino-2-oxazolidone)	Aquaculture	1	Greece
	Milk	1	Crotia
	Poultry	7	Belgium
AOZ (3-amino-2-oxazolidone)	Honey	1	Germany
	Milk	1	Croatia
SEM (semicarbazide)	Honey	1	Finland
	Milk	1	Croatia
	Pigs	1	Italy
	Poultry	3	Cyprus, the Netherlands
Nitroimidazoles			
Metronidazole	Pigs	1	France
	Poultry	1	Belgium

3 Food Safety from Resistance Perspective

Lucie Pokludová

In the previous parts of the chapter perspectives of food safety understood as "chemical safety" for exact chemical compounds of antimicrobial nature presented as residues have been taken into account. Another perspective of great importance, speaking about antimicrobials and risks for consumers of food of animal origin, is transmission of the microorganisms bearing resistance or transfer of genetic determinants of resistance. Due to use of antimicrobials either to treat food-producing animals or to prevent the diseases of those animals, pressure to select resistant bacteria that can be present in the food of animal origin belongs to the significant risks. Antimicrobial resistance genes transfer can occur in the favourable conditions of the gastrointestinal tract between bacterial members of the gastrointestinal microbiota (Haug et al. 2011) and thus present a public health risk (Jans et al. 2018). Transmission among food and humans occurs during handling of raw commodities (meat and offal, milk, eggs) as well as cross- and re-contamination between different food products at production, distribution and household levels.

The risk for consumers is multifactorial and depends also on food hygiene along the whole food chain, but also on consumption habits. As raw meat/eggs products mostly undergo a cooking step or in the case of milk pasteurisation prior to consumption, the level of risks from ingestion of (resistant) bacteria decrease, but the level of risk of resistance genes transfer remains still significant. In contrast to raw meat or other food commodities undergoing a cooking step, fermented products are generally consumed without a prior heating step that leads to high-level colony counts of technologically relevant bacteria and indicator bacteria per gram of product (Ross et al. 2002) and this can also pose higher risks for the transfer of resistance. As campylobacteriosis and non-typhoidal salmonellosis belong to the predominating zoonotic food-borne infections and 96 and 79 million infection cases are reported worldwide each year (Havelaar et al. 2015), these microorganisms are monitored for the presence in food, but also for resistance (EFSA and ECDC 2018). From the above listed commodities, raw meat products represent a major factor for cross-contamination of bacteria in the kitchen or at the table to infect humans and therefore likely also for the transfer of resistant food-borne bacteria to human (Jans et al. 2018). Considering also other zoonotic bacteria, as well as indicator bacteria signalising the resistance patterns/the emerging resistance, further data in existing European reports and datasets are focused mainly on Gram-negative (mainly *Salmonella* (all serotypes), *E. coli, Campylobacter* spp.) and Gram-positive (*Staphylococcus, Enterococcus*) food-borne pathogens and indicator organisms in meat products (EFSA and ECDC 2018). Surveillance data from individual European countries are further compiled under the umbrella of the European Centre for Disease Control (ECDC) and European Food Safety Authority (EFSA) to prepare comprehensive annual reports on AMR prevalence dynamics in key bacteria groups in animals, food (mainly raw meat) and human medicine as well as joint report including also the data on antimicrobials consumption both in human and veterinary sectors (ECDC/EFSA/EMA 2015).

Only limited knowledge as indicated in the Jans et al. (2018) systematic review is available regarding other resistant bacteria (e.g. *Clostridium difficile*), resistant bacteria from other food commodities (milk, eggs) as well as on resistance in technologically important bacteria including starter cultures, e.g. *Bifidobacterium* spp., *Lactobacillus* spp., *Lactococcus* spp., *Leuconostoc* spp., *Pediococcus* spp., *Streptococcus* spp. and *Weissella* spp. Therefore the determination of the potential for resistance genes transfer between technologically important (e.g. starter culture bacteria as lactic acid bacteria), as well as to gut commensals, opportunistic, and obligate pathogens in order to estimate the associated public health risk is identified as the gap in the knowledge related to food safety with respect to resistance to antimicrobials.

Despite the fact that data coming from molecular biology investigations is still limited for the whole spectrum of relevance for public health, some information on the food-borne bacteria of importance/food commodity and resistance determinants known can be summarised (Table 5). The work of Jans et al. (2018) using a systematic literature review of data published between 1996 and 2016 can be considered as very comprehensive and shows the analysis of the data from 313 out of 9473 collected studies (122,438 food samples; 38,362 bacteria isolates of which 30,092 samples and 8799 isolates were AMR positive). A median prevalence of bacteria with resistance >50% was observed for meat and seafood harbouring *Campylobacter* spp., *Enterococcus* spp., *Salmonella* spp., *Escherichia coli, Listeria* spp., and *Vibrio* spp., lower prevalence for milk products harbouring starter culture bacteria. Aminoglycosides, cephalosporins, fluoroquinolones, penicillins, sulphonamides, and tetracyclines phenotypic profiles (in some studies confirmed genetically) were determined in Gram-negatives. Exposures scores of levels 1 (medium) and 2 (high) were determined as for AMR for *Campylobacter* spp., *Salmonella* spp., *E. coli* in meat as well as *Vibrio* spp. and *E. coli* in seafood. Glycoproteins, lincosamides, macrolides and nitrofurans were recognised of importance in Gram-positives—mainly *Staphylococcus* spp. and *Enterococcus* spp. in meat sources, *Staphylococcus* spp. in seafood as well as *Enterococcus* spp. and technologically important bacteria (including starters) in fermented or processed dairy products.

There is also increasing level of knowledge coming from the European surveillance system known as harmonised monitoring of AMR in zoonotic and indicator bacteria from animals and food which is legally based on Directive 2003/99/EC and Commission Implementing Decision 2013/652/EU. The monitoring of antimicrobial resistance is performed considering a public health perspective, there is defined and prioritised list of combinations of bacterial species, antimicrobials tested, food-producing animal populations and foodstuffs, which are monitored in harmonised manner and results are regularly

Table 5 Examples of food-borne bacteria of importance/food commodity and resistance determinants (Jans et al. 2018) modified, amended as listed below in the table)

Bacteria	Food commodity of animal origin/food product, where predominantly identified	Examples of determinants of resistance	Resistance identified for groups of antimicrobials	Notes
Campylobacter spp.[a] *Campylobacter coli* *Campylobacter jejuni*	MEAT: beef, lamb, pork, poultry, fish	*Tet (O)* *gyr*A 23SrRNA, *erm*B *aph*A-7	Tetracyclines Fluoroquinolones Macrolides Aminoglycosides	*C.coli* infections 9.2% of human *Campylobacter* infections in Europe (EFSA and ECDC 2018b). *C. coli* is the predominant species in pigs, may also be quite prevalent in broilers or in turkeys in some countries (EFSA and ECDC 2016, 2018a, b).
	Raw, also frozen	*cme*ABC, *cfr*C	Multidrug efflux pump	
Salmonella spp. *Salmonella enterica* subsp. *enterica*/main five serovars: Derby, Enteritidis, Infantis, Stanley, Typhimurium	MEAT: Poultry, pork, beef, fish, seafood, eggs	*aad*A, *aph*A, *strA/B* bla_{CMY}, bla_{CTX}, and bla_{TEM} *qnr*S, *oxx*A/B *AmpC* *cat*A, *flo*R, *cml*A *sul*1/2/3 *dfr*A *tet*A/B/C/G *mcr* bla_{OXA}, bla_{VIM}, bla_{NDM}	Aminoglycosides Cephalosporins Fluoroquinolones Penicillins Phenicols Sulphonamides Trimethoprim Tetracyclines Colistin Carbapenems[b]	Recent publications show that *S. typhimurium* strains can harbor 20–22 genes of resistance Multidrug-resistant non-typhoidal *Salmonella* infections may have more serious human health implications compared to those of pan-susceptible strains (Parisi et al. 2018).
E. coli	MEAT (including fish)	*strA/B* bla_{CMY}, bla_{CTX}, bla_{OXA}, bla_{TEM}	Aminoglycosides Cephalosporins	
	MILK	AmpC sul1/2/3 dfrA tet mcr	Penicllins Sulfonamides Trimethoprim Tetracyclines Colistin	
	CHEESE	bla_{OXA}, bla_{VIM}, bla_{NDM}	Carbapenems[b]	
Enterobacteriaceae	MEAT : beef, poultry predominant	*Aac* bla_{ACC}, bla_{CMY}, bla_{CTX}, bla_{DHA}, bla_{FOX}, bla_{OXA}, bla_{OXY}, bla_{SHV}, or bla_{TEM}	Aminoglycosides Cephalosporins	
Citrobacter, Cronobacter, Enterobacter, Klebsiella, Kluyvera, Pantoea, Shigella	MILK CHEESE SEAFOOD	*AmpC* *mcr* bla_{OXA}, bla_{NDM}	Penicllins Kolistin Carbapenems[b]	

(continued)

Table 5 (continued)

Bacteria	Food commodity of animal origin/food product, where predominantly identified	Examples of determinants of resistance	Resistance identified for groups of antimicrobials	Notes
Non- *Enterobacteriaceae* indicator	MEAT (including fish)		Aminoglycosides, Cephalosporins Penicillins	
Acinetobacter, Aeromonas, Pseudoalteromonas, Pseudomonas.	MILK		Lower prevalence: Fluoroquinolones, Macrolides, Phenicols, Polypeptids, Rifamycins	
	CHEESE			
Yersinia entrocolitica	MEAT : poultry, pork Raw	No certain genes reported	Penicillins Cephalosporins Macrolides	
Staphylococcus spp.	MEAT	mecA; bla l/R/Z *erm*A/B/C/D/F/G/J/K *tet*A/K/L/M *aac*A-*aph*D, *aad*D	Penicillins Macrolides Tetracyclines Aminoglycosides Fluoroquinolones Glycopeptides Lincosamides	
S. aureus, S. epidermidis, S. carnosus, S. equorum, and *S. saphrophyticus*.	MILK	*vga*A/C		
	CHEESE			
Enterococcus	MEAT (including fish)	*aac*(6′)-Ie-*aph*(2″)-Ia *aph*(3″)-IIIa *aph*(2″)-Ib, *aph*(2″)-Ic, *aph*(2″)-Id	Aminoglycosides	A total of 390 samples of ready-to-eat meat products were investigated. Genes of resistance as listed in the third column of this table predominant.
E.faecalis, E. faecium. E. avium, E. durans, E.hirae or E. casseliflavus	MILK	*erm*A, *erm*B	Fluoroquinolones Glycopeptides Lincosamides Macrolides Penicillins Tetracyclines	Nearly half of the isolates contained a conjugative transposon of the Tn*916*/Tn*1545* family. Many isolates were antibiotic resistant and carry transferable resistance genes.[c]
	CHEESE (soft, semi-hard, hard)	*tet*M/L/K/W/O	Rifampicin Low prevalence: Carbapenems Cephalosporins	

Listeria monocytogenes	MEAT (including fish, seafood)		Lincosamides Penicillins	The presence of *cad*A (heavy metal transporting efflux system), *ebr*B, and *qac* (efflux transport and resistance to quaternary ammonium compounds) in food isolates likely promotes survival in food and in the food-processing environment by mitigating cell exposure to harmful chemicals[e]
	MILK		Aminoglycosides	
	CHEESE		Fluoroquinolones Cephalosporins	
Clostridium difficile	MEAT (including fish, seafood)		Carbapenems Cephalosporins Fluroquinolones	RT078 and RT027 strains have been isolated from pork, beef, and chickens in the Northern Hemisphere
		*erm*B	Lincosamides	Numerous transposons (rather than other mobile elements) are associated with antimicrobial resistance in *C. difficile*, including Tn*916* and Tn*5397* with *tet*M, Tn*4453a/b* with chloramphenicol R, TnB*1230* *tet*W and Tn*5398* with two copies of *erm*B encodes a 23S rRNA methylase conferring the MLS (macrolide-lincosamide-streptogramin B) phenotype, the most common resistance type in *C. difficile*.[d]
			Penicillins	
		*tet*M/W	Tetracyclines	
Technologically important (lactic, starter cultures, food processing, probiotic)			Aminoglycosides Fluoroquinolones Glycopeptides Penicillins Sulfonamides Tetracyclines	
Bifidobacterium spp., *Lactobacillus* spp., *Lactococcus* spp., *Leuconostoc* spp., *Pediococcus* spp., *Streptococcus* spp., *Weissella* spp.				
Other Gram-positive (*Bacillus* spp.)	MEAT (including fish) CHEESE		Aminoglycosides Lincosamides Macrolides Penicillins Phenicols Tetracyclines	
Other Gram-negative bacteria	MEAT (including fish)	*bla*$_{CTX-M1}$, bla$_{CTX-M14}$, *bla*$_{OXA-48}$	Aminoglycosides Cephalosporins	
	CHEESE		Penicillins	

[a] Wieczorek and Osek (2013); [b] Madec et al. (2017); [c] Chajęcka-Wierzchowska et al. (2016); [d] Knight et al. (2015); [e] Pirone-Davies et al. (2018)

reported as analysed data on AMR for food-producing animals and food. Due time course not only phenotypic methods, but more in-depth sample analysing by genetic methods is performed, analysed and reported (EFSA and ECDC 2018). Despite this achievement, it seems still not fully sufficient data package on all bacterial species (especially those not easily culturable) as well as on resistance gene prevalence and on sufficiently precise elucidation of genetic linkages between resistance genes and mobile genetic elements such as plasmids, transposons, integrons and gene cassettes that have been still rarely reported (Lanza et al. 2015; Martínez et al. 2017). This unfortunately hinders better estimations on AMR gene prevalence and transferability. Of particular relevance is the assessment of transferability of AMR genes located on chromosomes versus those on mobile genetic elements in terms of clonal expansion versus horizontal transfer as exemplified for *Campylobacter* spp. and *Enterobacteriaceae* (EFSA and ECDC 2018). There can be mentioned an example of colistin, where *mcr* genes were not known and horizontal transfer of resistance genes (later recognised as *mcr*) was identified as risky especially in *Enterobacteriaceae* (EFSA and ECDC 2016; Florez-Cuadrado et al. 2016). Another case is the example of *Campylobacter* spp. horizontal gene transfer of erythromycin resistance was associated with *erm*B being located on a chromosomal multidrug-resistant genomic island, which rendered all recipients also resistant to lincosamides and aminoglycosides (Wang et al. 2014).

Muloi et al. (2018) performed a systematic review to explore the evidence that food animals are responsible for the transfer of AMR *E. coli* and their AMR determinants to humans. Larger number of studies involved in review did not suggest providing an evidence of transmission in certain direction. Big variability in sampling methodologies and antibiotics tested may have affected the conclusions made regarding the epidemiological connection between food animals and humans. Also molecular techniques, such as MLST and PCR, used in most studies in this review, are considered as limited in resolution (Didelot and Gardy 2014). The demonstration of overlapping patterns should be interpreted with care as the direction of transmission is difficult to infer, and co-colonisation from a shared source is also possible. As pointed out by Grad and Lipsitch (2014), demonstrating the direction of transmission and thus the epidemiological history of pathogens and their determinants requires a quantitative description of relatedness, including phylogenetic analysis.

Studies of the genetic organisation of AMR genes are of high importance for an AMR risk assessment and should be more systematically implemented. Future studies in the area of sources and epidemiological links of genes of resistance should benefit from combining phylogeographic methods with new methods of WGS (including NGS), which allow to proof of quantitative hypothesis for inferring pathogen movement between host populations. New methodologies as Multiple-Locus Variable number tandem repeat Analysis (MLVA) and in silico typing based on WGS—Single Nucleotide Polymorphism (SNP) typing, Multi Locus Sequence Typing (MLST) and "next-generation" sequencing (NGS) methods such as the Roche454 and Illumina methods currently offer the highest level of bacterial strain discrimination and are powerful tools for studying transmission events and deep insight into bacterial genomes. Once the results/metadata gained by these advanced methods will be analysed by sophisticated electronical tools, but also adequately interpreted, it can be expected to elucidate and describe the links and causality among source-vector/vehiculum-host/patient (Knight et al. 2015). All these methods can help to gain sufficiently robust data on different food-borne pathogenic, zoonotic, indicator and commensal bacteria, that can be present in the food (including that of animal origin) as a source of genes of antimicrobial resistance transferable to human.

Gaining evidence-based knowledge is of importance, but also raising awareness and spreading the information on situation can help further promote responsible use of antimicrobials in food-producing animals. From this perspective can be mentioned the new infographic tool released by the EFSA and ECDC in 2019 (please see Fig. 1).

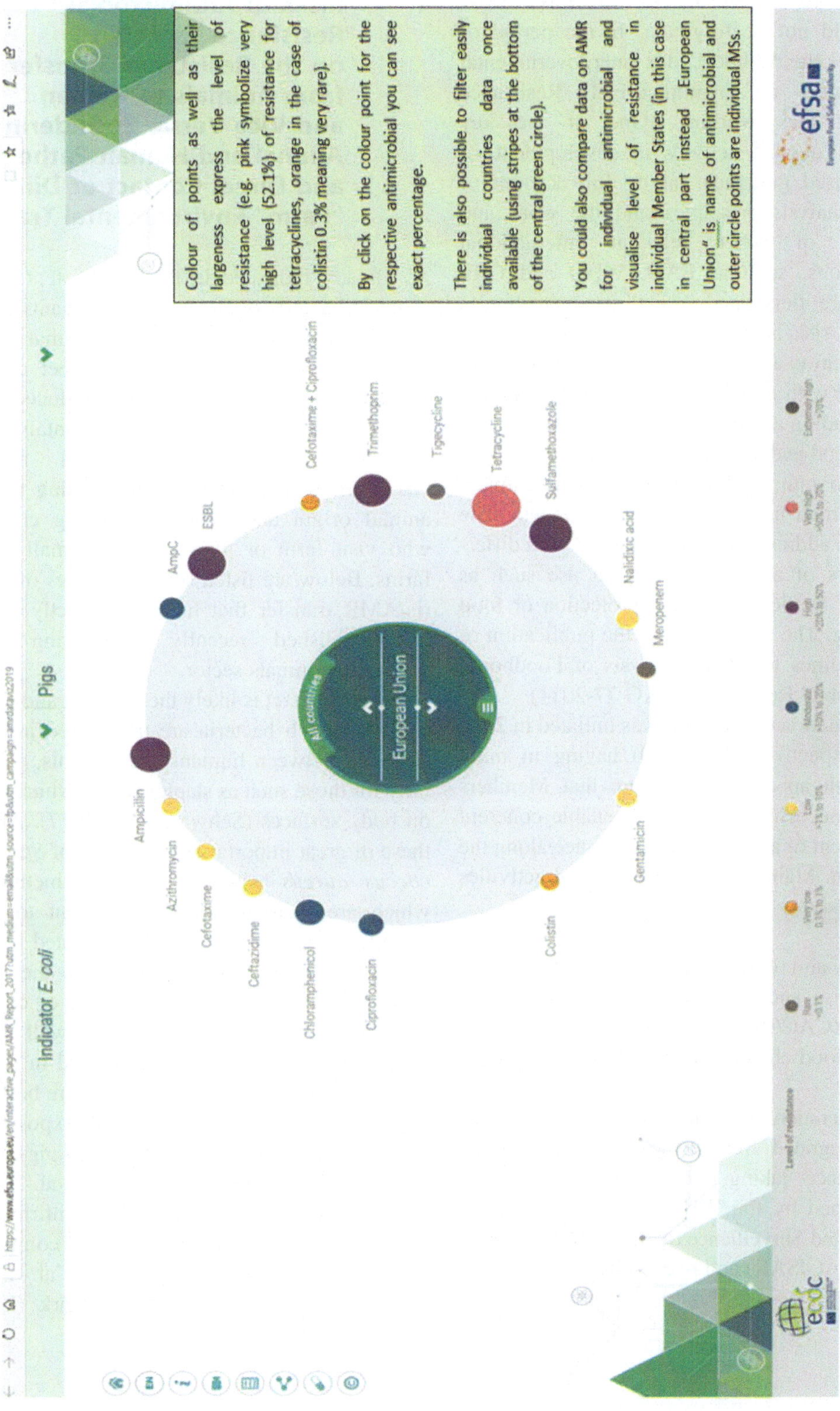

Fig. 1 Infographic figure shows the situation in AMR for all countries within the EU reporting results for indicator *E. coli* isolated from healthy pigs and the antimicrobials tested (as appeared on https://www.efsa.europa.eu/en/interactive-pages/AMR-Report-2017, amended by comment on right side)

Global activities related to food-borne AMR also should not be forgotten. In the period of 2007–2011 the Ad hoc Codex Intergovernmental Task Force on Antimicrobial Resistance (TFAMR) was working. One of the key objectives of this activity is development of science-based guidance, taking full account of its risk analysis principles and the work and standards of other relevant international organisations, such as FAO, WHO and OIE. There were developed guidelines providing a structured risk analysis framework (see Fig. 2) with the aim to address the risks to human health associated with the presence in food and animal feed, including aquaculture, and the transmission through food and animal feed, of microorganisms resistant to antimicrobials or determinants linked to non-human use of antimicrobial agents. The guidelines address the risk associated with different sectors of antimicrobial agent use such as veterinary applications, plant protection or food processing. The activity led to the publication of the Guidelines for Risk Analysis of Foodborne Antimicrobial Resistance (CXG 77-2011).

Continuation of the work was initiated in 2016 (with perspective 2017–2020) having in mind One-Health approach, to ensure that Members have the necessary guidance to enable coherent management of antimicrobial resistance along the food chain. Main working streams and activities are targeted on:

- Review and revision of the Code of Practice to Minimise and Contain Antimicrobial Resistance (CAC/RCP 61-2005) to address the entire food chain, in line with the mandate of Codex.
- Consideration of the development of Guideline on Integrated Surveillance of Antimicrobial Resistance, taking into account the guidance developed by the WHO Advisory Group on Integrated Surveillance of Antimicrobial Resistance (AGISAR) and relevant OIE documents.

4 Risks of Antimicrobial Resistance Targeted on the Resistance Transfer from Animals to Human and Vice Versa, Considering Animal and Human Pathogens and Direct Contact or Direct "Farm" Environmental Transfer

As already indicated in the subchapters above, resistant bacteria or resistant determinants can be transmitted through foodstuff of animal origin. Another way of transfer is via direct contact, e.g. handling animals or animal products or via inhalation of dust and aerosols that contain bacteria by farm workers, animal owners, veterinarians, abattoir workers handling food of animal origin and people (including children) who visit farm or living, e.g. on small family farms. Below are listed some examples of bacteria/AMR transfer that has been exactly proved and published recently considering food-producing animals sector.

Direct contact is likely the quickest and easiest way by which bacteria are transferred in either direction between humans and animals, particularly for those such as staphylococci which reside on body surfaces (Schwarz et al. 2017). Among these of great importance are strains of *Staphylococcus aureus* with resistance to methicillin, which are many times also resistant to other antimicrobials. This livestock associated (LA)—MRSA (in Europe mostly belong to the clonal lineage CC398, MLST ST 398) are of concern considering both direct contact as well as the exposure to farm dust as documented in several studies and different animal sources are best visible in Fig. 3. The occupational exposure of persons at farm level to MRSA from pigs, cattle or poultry is very frequent (Goerge et al. 2017).

Within Europe most reports that confirm transfer of MRSA from animal to human come from pig farms in Spain (Reynaga et al. 2016), Germany (Alt et al. 2011), Denmark (Larsen

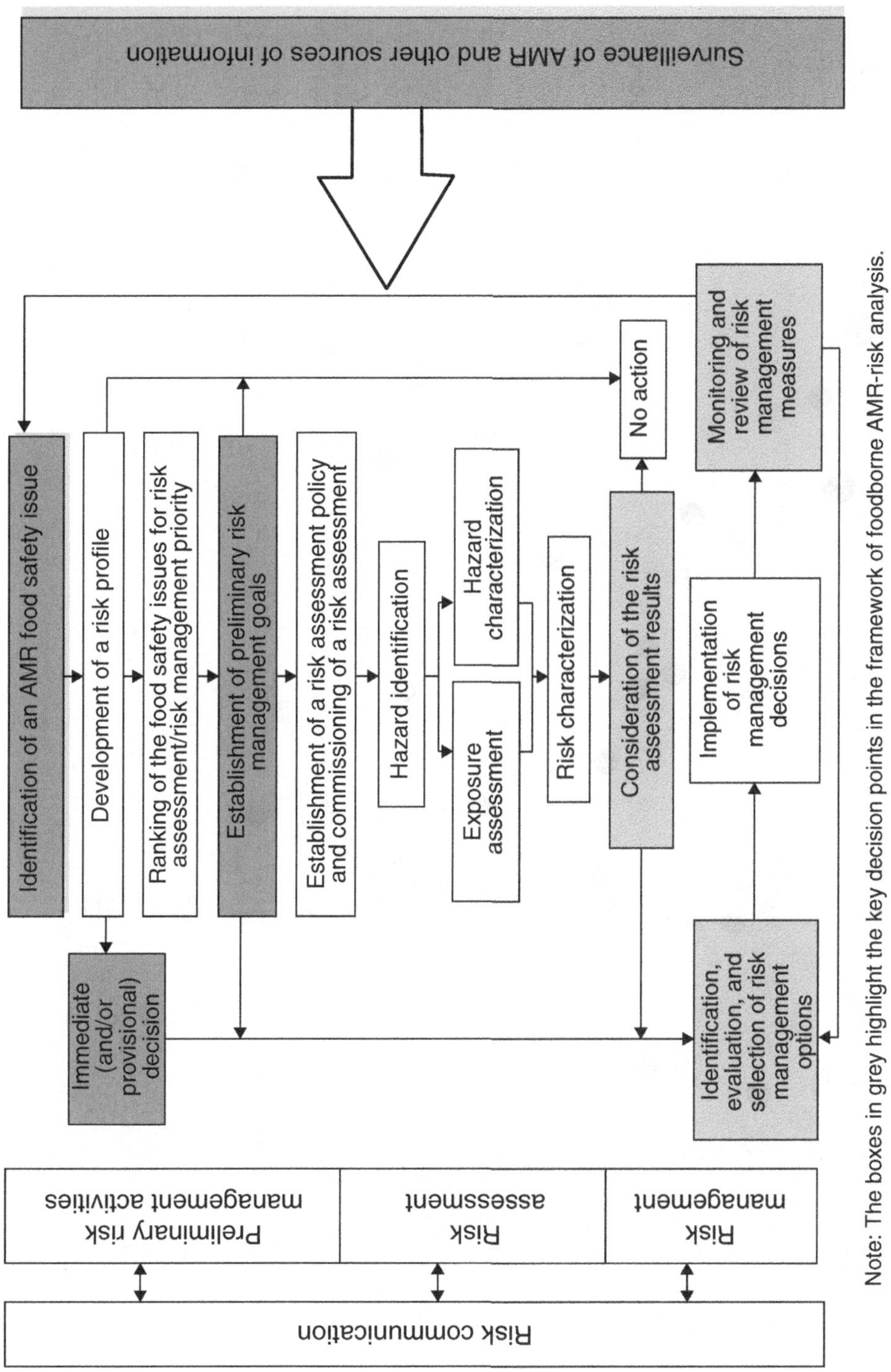

Fig. 2 Food-borne Risk analysis framework targeted on AMR food safety issues (CAC/GL_CXG 77-2011)

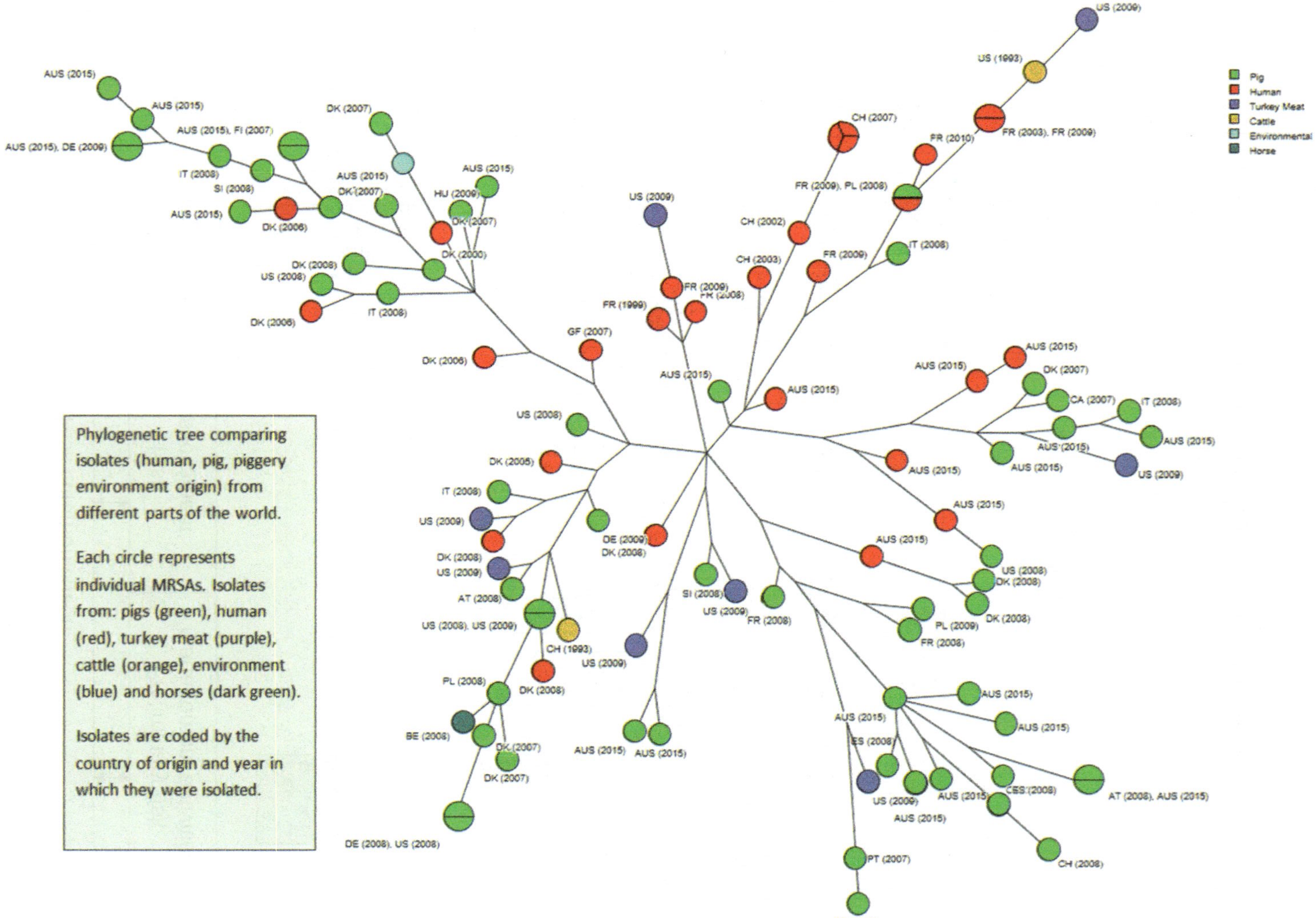

Fig. 3 Phylogenetic tree constructed by core genome SNPs of MRSA ST398 isolated from humans, pigs and piggery environment comparing isolates from different parts of the world. Each circle represent individual MRSAs, animal species, sources of isolates are differentiated by colours (Sahibzada et al. 2017)

et al. 2017), the Netherlands (where according to Cuny et al. 2009, 45% of veterinarians attending pig farms were MRSA-positive), Belgium (Crombé et al. 2013), Portugal (Conceição et al. 2017) and Poland (Mroczkowska et al. 2017). A study by Voss et al. (2005) demonstrated a 760-fold higher MRSA carriage rate among pig farmers compared to the general Dutch population. Duration and intensity of animal contact and the number of MRSA-positive animals on a farm have been linked with human CC398 colonisation and infection (Köck et al. 2013; Graveland et al. 2011). Furthermore, as CC398 has the ability to survive in the environment, environmental contamination may contribute towards further dissemination. German study reported that MRSA ST398 that carried SCCmec IV or V, accessory gene regulator type I and capsule type 5 in pig primary production herds was higher in their country (45–70%) than in the rest of the European Union (Alt et al. 2011; Köck et al. 2009). In 2018, the Danish Veterinary and Food Administration carried out a study on the situation regarding methicillin-resistant *Staphylococcus aureus* (MRSA) on conventional, outdoor and organic pig farms, dairy livestock, turkeys, laying hens, mink and horses. In the case of conventional pig farms, the study was performed on 130 farms, and 116 of them were positive for MRSA, this entailing 89% of the farms, this value being similar to a study carried out in 2015. Regarding outdoor and organic farms, of the 104 farms tested, 20%, i.e. 21 farms, were positive (Pig Health 2019).

Except evidence of MRSA carriage and transfer on pig farms, to date, a great number of studies have reported the isolation (with variable frequencies) of MRSA from livestock, wild animals and derived foods, both raw and ready-to-eat, as well as from professionals working in the animal husbandry or the food production chain settings (Chon et al. 2017).

A study involving 26 dairy farms in the Netherlands revealed that the same LA-MRSA types, based on pulsed-field gel electrophoresis (PFGE) type, spa type and resistance patterns, were detected not only among dairy cattle and their contact personnel (e.g. milkers), but occasionally also among other animals living on the same farm (Fessler et al. 2012).

Another study from Italy reported that the sheep farm casuistic proved that the MRSA isolates from human and animal sources within the farm were same. Moreover, after 2 years from the first isolation, the authors confirmed the presence of the same multidrug-resistant strain of MRSA sequence type (ST)1, clonal complex (CC)1, spa-type t127, staphylococcal cassette chromosome mec (SCCmec) type IVa (Carfora et al. 2016).

From poultry sources, Kraushaar et al. (2017) determined that there are other poultry (broiler chickens and turkey) associated clones of MRSA (mainly CC9 and CC5) besides the predominant CC398. The study also shows the presence of genes *erm*(C), *aac*A-*aph*D and *tet*(K); therefore, resistance to macrolides/lincosamides/streptogramins; aminoglycosides and tetracycline can be expressed phenotypically. Genes *qac*C conferring resistance to quaternary ammonium compound disinfectants was also found. Within the isolates tested MRSA harbouring classical enterotoxin genes (*sea, seb, sed*) were identified.

Also studies bringing evidence of transfer of MRSA in equine practices are available. Results of investigation of a total of 272 methicillin-resistant MRSA from equine infections (17 equine hospitals and 39 veterinary practices and 67 isolates from personnel working at equine clinics in Germany) can be given as an example. Samples were subjected to molecular typing—the majority of isolates from horses was attributed to clonal complex (CC) 398 (82.7%). Nasal MRSA colonisation was found in 19.5% of veterinary personnel with occupational exposure to horses. As the proportion of isolates exhibiting characteristics of MRSA from equine medicine when searched human isolates database in Germany was very low (<0.5%) it is supposed that threat from MRSA coming from equine practices will be negligible (Cuny et al. 2016).

The presence of MRSA in airborne dust from pig farms in Denmark indicates that dust might be an important vehicle for transmission of LA-MRSA that was found to survive well in farm dust with half-lives of 5 days. Dependent on the initial concentration they could be found in farm dust for weeks (the 99.9% die-off rate was 66 days for LA-MRSA). Thus, farm dust can pose an exposure risk for humans in the farm environment, but also when transported to other environments (Feld et al. 2018).

A study conducted in Germany showed that 85.8%, i.e. 97 of 113 swine farmers but only 4.3%, i.e. 5 of their 116 family members were positive for LA-MRSA. Likewise, 44.9%, i.e. 22 of 49 swine veterinarians but only 9.1%, i.e. 4 of their 44 family members were positive for LA-MRSA in another report. These observations suggest that the human-to-human transfer of LA-MRSA occurs distinctly more rarely than the animal-to-human transfer.

As an example of Gram-negative germs to be livestock associated are well-known representatives of the Enterobacterales as well as, e.g. *Pseudomonas* spp. and related bacterial species. Initially, ESBL/AmpC-producing bacteria were only observed in human medical practice, but during the last decade it has been recognised, first in companion animals and later on also increasingly in livestock due to start of European harmonised monitoring studies concentrated on major species of food-producing animals. The danger connected with *bla*ESBL and *bla*AmpC genes from *Enterobacterales* is that they can be spread by horizontal transfer being often associated with mobile genetic elements, like transposons (integrons), and insertion sequences. Thus, such gene transfer by mobilisation or conjugation has a major impact on the dissemination of β-lactam resistance among bacteria of different origin (Liebana et al. 2013). As indicated by Ewers et al. (2012) a similar distribution of major ESBL/AmpC types was apparent only in human isolates, regardless of their geographical origin from Europe, Asia, or the Americas, whereas in animals this varied extensively between animal groups and across different geographical areas. Even though exposure to LA-MRSA and risks associated with it was recognised as significant, evidence for a direct transfer of ESBL/AmpC-producing bacteria from animals to humans through close contacts is limited. Nonetheless, the size of the commensal ESBL/AmpC reservoir in non-human sources is dramatically rising. This may constitute an indirect risk to public health by increasing the gene pool from which pathogenic bacteria can pick up ESBL/AmpC/carbapenemase genes (Madec et al. 2017). Some further recent works elucidated that some predominant ESBL/AmpC genes were identified in human, animal and also environmental reservoirs. However, proportional similarity indices (PSIs) and principal component analyses (PCAs) revealed close human–animal ESBL/AmpC gene similarity between human farming communities and their animals in the case of broilers and pigs. Another research brought the evidence that isolates from people in the general population had higher similarities to those from human clinical settings, surface and sewage water and interestingly also with wild birds (0.7–0.8), while similarities to livestock or food reservoirs were lower (0.3–0.6) (Dorado-García et al. 2018). One of the most recent references (Ceccarelli et al. 2019) brings very comprehensive results from Dutch isolates coming from period 2007–2017 gained both on selective and non-selective culturing methodologies. A collection of 2304 extended-spectrum cephalosporin-resistant (ESC-R) *E. coli* isolated from faeces of broilers, dairy cattle, slaughter pigs, turkeys, ducks and veal calves was investigated and ESBL/pAmpC genes were determined. In 473 *E. coli* isolates was determined and by typing of plasmids identified 22 different ESBL/pAmpC genes with *bla*CTX-M-1 being the most prevalent gene in livestock (43.7%)—what is in line with (EFSA/ECDC 2017) followed by *bla*CMY-2 and *bla*SHV-12, independent of the animal source. Prevalence of typically human-associated *bla*CTX-M-15 was highest in cattle. Cecarelli and her team (2019) also analyse where the genes were localised, what is of importance from the epidemiological perspective, considering speed of spread and transfer of antimicrobial resistance. Majority (92%) of ESBL/AmpC genes were plasmid located and mostly on Inc. plasmids. The most represented plasmid family in isolates from all animals was IncI1α (86%). In veal calves, dairy cattle and slaughter pigs, as the second most prevalent was IncF, followed by

IncK detected in broilers and laying hens and also IncX1 in broilers. Emerging IncX3 was identified in broilers and dairy cattle. As also commented by Ceccarelli et al. (2019) IncI1 plasmids encoding $bla_{CTX\text{-}M\text{-}1}$ or $bla_{TEM\text{-}52c}$ were recognised as the most prevalent gene–plasmid combinations in *Enterobacterales* from slaughter pigs worldwide (Geser et al. 2011; Randall et al. 2014; Biasino et al. 2018; Dang et al. 2018). The IncI plasmids harbour several genes for formatting type IV pili. Besides motility and mating, type IV pili contributes to adhesion and invasion of *E. coli* (STEC) and other Gram-negative pathogenic bacteria.

Another study (Irrgang et al. 2018), performed by German scientists, was based on the screening of 2256 food samples (from poultry, pork, beef, milk, cheese and vegetables) for cephalosporin-resistant *E. coli*. A total of 437 phenotypically resistant isolates were obtained. An ESBL or AmpC genotype was confirmed for 404 of these isolates. The majority ($n = 212$) of them harboured a CTX-M-1 β-lactamases, from those 89 isolates were characterised more in detail. Fifty-one different ST-types were detected. The most abundant type was ST117 ($n = 11$) followed by ST88 ($n = 6$) and ST10 ($n = 5$), what is in line also other study (Day et al. 2016).

Among vehicles transporting resistant bacteria, including resistance genes, can be considered dust present on a farm. Hoffmann (2014) proved the association of CTX-M-1-positive dust samples obtained from pig farms with positive faecal samples from farm workers. The positivity of CTX-M-1 in pig and human faeces was significantly associated. Therefore, he hypothesised a possible transmission of CTX-M-1 subtypes to humans via inhalation of contaminated dust particles during exposure in the stable environment.

As a summary it can be concluded that a significant amount of studies gives the evidence of possibility of contact transfer and spread of resistance among animals and human and vice versa for at least MRSA and ESBL/AmpC E. coli. Considering mostly Enterobacterales the list of risky resistance profiles, that can be spread cannot be limited just to ESBL and AmpC, but should be amended also by epidemiologically most important Carbapenemases: metalo-beta-lactamases (MBLs) such as VIM (Verona integron-encoded), IMP (imipenemase) and NDM (New Delhi MBL), KPC (K. pneumoniae carbapenemase), OXA (Carbapenem-hydrolysing oxacillinase). Another horizontally transferable, plasmid harboured genes are *mcr* (currently known *mcr*-1 to *mcr*-9) that confer resistance to colistin, the last resort live saving antimicrobial in human medicine. Those genes were proven (Liu et al. 2017) to be able to transfer from *E. coli* to other species of bacteria (e.g. *Cronobacter sakazakii*). Other genes of importance in Gram-negatives are *qnr* genes (resistance to quinolones), PMQR efflux pump genes (*qep*A and *oqx*AB) affecting quinolones as well as encoding acetyltransferase and causing resistance to aminoglycosides. Of course pool of other genes is also detectable in some spectrum of *E. coli* and other members of order *Enterobacterales*, but they do not cause resistance to antimicrobials considered of critical importance for human medicine (those are, e.g. *cat*A1, *flo*R and *cml*A1 (amfenicol resistance), *sul* genes (sulphonamides), *dfr* genes (trimethoprim) and *tet* genes (tetracyclines)).

It can be summarised that not only staphylococci and bacteria from order Enterobacterales, even the resistance genes, mechanism of transfer and casuistics might be the best investigated, are of importance. Direct contact or indirect exposure within farm environment can cause transfer of other important bacteria that could harbour resistance and can cause serious diseases, mainly described as associated with hospital care. Therefore, other bacterial species from the group of pathogens known under the abbreviation ESC(K)APE (*Enterococcus faecium, Staphylococcus aureus, Clostridium difficile (Klebsiella pneumoniae), Acinetobacter baumannii, Pseudomonas aeruginosa,* and bacteria from order *Enterobacterales)* waiting for further in-depth studies. Especially *Acinetobacter baumannii* seems to be of grooving importance (Van der Kolk et al. 2018; Wareth et al. 2019) (not only isolates from companion animals, but also from livestock animals and food—mainly poultry meat). Table 6 gives only some

Table 6 Selected resistance of importance in *Clostridium difficile, Enterococcus* spp., *Acinetobacter baumanii* and *Pseudomonas aeruginosa* linked to livestock animal sources (modified from Argudín et al. 2017, amended by Van der Kolk et al. 2018; Wareth et al. 2019; Ma et al. 2019)

Bacterial species	Resistance
Clostridium difficile	**MLS$_B$** macrolide-lincosamide-streptogramin
	PhLOPSA$_A$ phenicols, lincosamides, oxazolidinones, pleuromutilins and streptogramin A
	Tetracycline
Enterococccus spp.	**MLS$_B$** macrolide-lincosamide-streptogramin
	PhLOPSA$_A$ phenicols, lincosamides, oxazolidinones, pleuromutilins and streptogramin A
	Oxazolidinones, Glycopeptides, Chloramphenicol, Tetracyclines, Trimethoprim
	Co-resistance—plasmids *optr*A (ABC transporter) and *erm* (MLS$_B$) and *fex*A (phenicols)
Acinetobacter baumanii[a]	**Metalo-beta-lactamases** *bla*$_{NDM-1}$, *bla*$_{OXA-23*, 58*, 497}$, **AmpC cephalosporinases**
	Polymyxins *pmr*A, *pmr*B *or pmr*C; *mcr 4.3—isolate* pig faeces
	Efflux systems resistance to β-lactams, aminoglycosides, fluoroquinolones, tetracyclines
	Sulphonamides, Tetracyclines, Phenicols, Macrolides
	QRDR quinolone resistance determining region
	Aminoglycosides *str*A, *str*B, *aad*A, *aph, aac*
	Glycocyclines
Pseudomonas aeruginosa	**Metalo-beta-lactamases** *bla*$_{IMP-4}$, *bla*$_{VIM-2}$,
	Quinolones e.g. *gyr*A, *gyr*B, *par*C and/or *par*E genes
	Aminoglycosides e.g. *aac*A4 and *aad*A6

[a]human - > companion animal transfer of AMR; some strains are pathogenic also for animals, *bla*$_{OXA-23*, 58*}$ naturally occurring in *A. baumannii; mcr-4.3—plasmid mediated* (Ma et al. 2019)

illustrative, but non-exhaustive examples of their resistances of public health importance that was detected in animal source isolates.

Despite the fact of growing evidence during the last years supported by the latest analytical and genetic methods, that allow us to gain more information and stronger proof of evidence on exact pathways there is still a lot of work to be done. Urgent need to investigate how, in a most effective way, to interrupt the way of resistance transfer and spread and minimise the impact of spread infections, zoonotic agents, resistance to antimicrobials and the harmful effect on human and animal health is knocking on researchers doors.

5 Risks from Occupational Exposure Related to Workers at Slaughter, Food Processing and Retail

Several studies in European abattoirs report a very high prevalence of nasal MRSA carriage in slaughter pigs—in Germany slaughter pigs were determined to be MRSA-positive up to 70.8%, of 99% slaughter batches found to be positive for MRSA in the Netherlands (Dierikx et al. 2016). The study of Normanno et al. (2015) shows the great genetic diversity of MRSA strains in slaughtered pigs and in abattoir employees in Italy, where the MRSA prevalence among pigs at slaughter was 38%. The main path of contamination is caused by smear infections due to improper removal of the intestines, especially the intestinal package.

Broens et al. (2011) published study about transmission of methicillin-resistant *Staphylococcus aureus* among pigs during transportation from farm to abattoir (Broens et al. 2011). All pigs ($n = 117$) tested MRSA-negative before transportation. On arrival at the abattoir, 10.3% pigs (12/117) in two batches tested MRSA-positive. In lorries that tested positive after transportation, the prevalence of MRSA-positive pigs was 21.1%, whereas no MRSA was detected in pigs that had been transported in lorries that tested negative after transportation. At stunning, all batches and 59.8% pigs (70/117) tested MRSA-positive. Pigs can become MRSA-positive in the short period of time during transportation from the farm to stunning at the abattoir. In all parts of this chain (farm employees, lorry drivers, abattoir workers, further food processing workers) humans are at risk of colonisation by MRSA. A contamination of meat with ESBL-producing *E. coli* and MRSA was confirmed in the study of Petternel et al. (2014). The large diversity of ESBL-producing *E. coli* could indicate a growing dissemination of ESBL genes in *E. coli* found in meat products from porcine and bovine origin. This can be considered as risky not only from end-consumer and food processing at individual home-kitchens, but also for all parts of further food chain and people that would be in contact with raw meat.

Humans involved in food processing chain need to be considered as very important intermediate reservoirs and vectors for ESBL genes. Food dealers/retailers especially may represent a reservoir for ESBL genes, because while working with contaminated food they are at a greater risk for infection with ESBL-producing bacteria. Lavilla et al. (2008) found that 27.5% of food retailers (372 tested persons) are colonised with ESBL-producing microorganisms.

5.1 Risks from Occupational Exposure to Antimicrobials

Adverse health effects of direct occupational exposure to antimicrobial veterinary medicinal products of different users (e.g. a veterinarian's assistant, a farmer, a bystander, a breeder, a miller incorporating a medicated premix into a finished feed) can also be taken into account, when speaking about risks from use of antimicrobials, especially considering "One Health" approach. Weak and moderate effects include hypersensitivity, allergic skin reactions and respiratory symptoms as well as more severe reactions as e.g. anaphylactic shock can be exhibited either in short- or long-term exposure. From the long-term perspective also influence on the antimicrobial resistance as well as microbiome of the professions frequently handling with antimicrobials is of concern.

5.2 Persons Handling Antimicrobials on Farm Level

Veterinarians, farmers, technicians, breeders as well as the technical staff performing on farm mixing using oral powders intended for use either in drinking water or as "top dressings" or powders that can be mixed via on farm mixing technologies are in the highest exposure risk. Examples of tasks and situations that may lead to exposure and possible consequences are listed in Table 7. Despite the fact that protective equipment should be used to minimise the risks, workers handling antibiotics report that they smell the drugs, have a bitter taste in their mouth, and sometimes also observe splashes and leakages during the preparation and handling with antimicrobials (Sessink 2018).

Veterinary medicinal products (including those containing antimicrobials) undergo assessment of the "User Safety" during the marketing authorisation procedure. Such assessment results, except the definition of risks and proposal for risk mitigations, in approval of the Summary of Product Characteristics (SPC) and Product Leaflet that contain warnings and guidance which equipment should be used to minimise harmful effect on the user. Below are listed some examples (parts of SPCs referring to user safety) that can be considered as representing the current internationally authorised veterinary medicinal products of different pharmaceutical forms. As very specific can be considered example of injectable tilmicosin, where serious concerns, other than hypersensitisation or AMR, are of importance in the case of accidental self-injection (the cardiovascular system is the target of toxicity, and this toxicity may be due to calcium channel blockade). Therefore, for the injectable products containing tilmicosin very detail special warnings, advices and contact point in national toxicology centre are included in the product texts and on packages to be easily accesible to vets exposed to this compound.

Table 7 Exposure to antimicrobials in different phases of use of antimicrobial veterinary medicinal products, including the description of situations, main routes of exposure and types of users exposed (in line with Guideline on user safety for pharmaceutical veterinary medicinal products, EMA (2010), modified/amended by references as cited in the table)

Phase of handling with antimicrobial	Task and situation	Type of user exposed	Main routes of exposure	Examples of possible consequences	Comments/examples
Pre–administration phase	Opening or accessing the product (taking product out of packaging)	• Miller incorporated a medicated premix • Veterinarian • Farmer • Breeder	Dermal	Hypersensitivity, allergic skin reactions Influence of microbiome	**Influence on microbiome of skin** • e.g. carriage of MRSA/ CoaNS) can be supported by dermal exposure to selected antimicrobials • Isolation of multidrug-resistant coagulase-negative staphylococci from pharmaceutical workers as a result of occupational exposure (Haddadin et al. 2013) **Respiratory tract/oral cavity** • Microbiome susceptibility/resistance • Research by Hamscher et al. (2003) demonstrated that 90% of the dust samples collected during two decades from a swine production facility exhibited antibiotics, including, e.g. tylosin, and various tetracyclines, sulfamethazine, in total amounts up to 12.5 mg kg − 1 dust. It can lead to the exposure even not linked with direct handling of antimicrobials
			Inhalation	Respiratory symptoms Influence of microbiome	
			Ocular	Hypersensitivity, tearing	
			Parenteral	Self-injection risk (tilmicosin)	
	Mixing and/or diluting of concentrates (mixing with feed or water)	• Miller incorporated a medicated premix • Farmer • Breeder	Dermal	Hypersensitivity, allergic skin reactions Influence of microbiome	
			Inhalation	Respiratory symptoms Influence of microbiome	
			Ocular	Hypersensitivity, tearing	
	Loading application apparatus or system: (drinking water equipment, dosing gun)	• Veterinarian • Farmer • Breeder	Dermal	Hypersensitivity, allergic skin reactions Influence of microbiome	
			Parenteral	Self-injection risk	
Phase of administration	Administration to the animal	• Veterinarian • Farmer • Breeder	Dermal	Hypersensitivity, allergic skin reactions Influence of microbiome	
			Parenteral	Self-injection risk	
Post-administration phase	Cleaning equipment and preparation area and disposal activities	• Technical staff • Farmer • Breeder	Dermal	Hypersensitivity, allergic skin reactions Influence of microbiome	
			Inhalation	Respiratory symptoms Influence of microbiome	
			Ocular	Hypersensitivity, tearing	
			Parenteral	Self-injection risk (tilmicosin)	

MRSA = *Staphylococcus aureus* resistant to methicillin CoaNS = coagulase negative staphylococci

Special precautions to be taken by the person administering the veterinary medicinal product containing antimicrobials to animals, examples from recent marketing authorisations of VMPs are as described in the following paragraphs.

Example of Intramammary VMP (Fixed Combination VMP: Penicillin, Penethamate and Neomycin)

Persons administering the product should avoid contact with this preparation as occasionally skin allergy may occur.

Penicillins and cephalosporins may cause sensitisation following injection, inhalation, ingestion or skin contact. Sensitivity to penicillins may lead to cross sensitivity to cephalosporins and vice versa. Allergic reactions to these substances may occasionally be serious.

Do not handle this product if you know that you are sensitised or if you have been advised not to work with such preparations.

If you develop symptoms such as a skin rash following exposure, seek medical advice and show this warning to the doctor. Swelling of the face, lips or eyes, or difficulty with breathing are more serious symptoms and require urgent medical attention.

Example of Orally Administered VMP (Chlortetracycline Oral Powder or Similarly Premix)

Handle this product with care to avoid exposure when adding to feed (water) and administering medicated feed (water) to the animals.

Take adequate measures to avoid dust formation when adding the product to feed.

Those handling the product should do so in a mechanically ventilated area.

Wear either a disposable half-mask respirator conforming to European Standard EN149 or a non-disposable respirator to European Standard EN 140 with a filter to EN 143 when mixing and handling the product.

Direct contact of the product with the skin, eyes and mucous membranes should be avoided.

Wear protective gloves, overalls and approved safety glasses.

In case of accidental exposure, wash area immediately with water.

Hands and exposed skin should be washed thoroughly after use.

Do not smoke, eat or drink when handling the product.

Example of Injectable VMP (Amoxicillin)

Care should be taken to avoid accidental self-injection. In the case of accidental self-injection, seek medical advice immediately.

Penicillins and cephalosporins may cause hypersensitivity (allergy) following injection, inhalation, ingestion or skin contact. Hypersensitivity to penicillins may lead to cross-reactions to cephalosporins and vice versa. Allergic reactions to these substances may occasionally be serious.

Do not handle this product if you know you are sensitised, or if you have been advised not to work with such preparations.

Handle this product with great care to avoid exposure, taking all recommended precautions.

If you develop symptoms following exposure such as skin rash, you should seek medical advice and show the doctor this warning. Swelling of the face, lips or eyes or difficulty with breathing, are more serious symptoms and require urgent medical attention.

Wash hands after use.

5.3 Pharmaceutical Industry/Feed Mills Workers

High exposure to different chemicals, including antimicrobials can occur in pharmaceutical industry, especially in the grinding, sieving, compression, granulation, mixing, filling and packing steps during the manufacture of medicines. The effect of health of workers due to occupational exposure to antibiotics has been studied by several authors. Asthma, other respiratory diseases (D'ıaz Angulo et al. 2011), dermatitis, and allergies (Møller et al. 1990; Møller and Würden 1992; Rebandel and Rudzki 1990; Stejskal et al. 1987; Rudzki et al. 1986) have been reported to be associated with exposure to antibiotics, mainly penicillin. However, most studies investigate AMR as a result of dermal contact to antibiotics and not airborne exposure to antibiotics (Moore and Nygren 2004). Few studies reported quantitative monitoring of penicillin dust and AMR in pharmaceutical industries (Moore and Nygren 2004).

Most of the current studies come from countries, in which pharmaceutical industry targeted on manufacturing of the active product ingredient, including antimicrobials, is settled. Similarly to those manufacturing plants also staff of the feed mills, especially those with not very advanced technologies is in danger of the contact dermal or inhalation exposure, that can lead to either acute or chronical adverse effects, including influence of the staff microbiome and resistance of bacteria involved in it.

Sarker et al. (2014) assessed the level of antibiotic resistance among occupationally exposed and compared the degree of bacterial resistance between pharmaceutical workers ($n = 20$_ and non-pharmaceutical workers ($n = 20$) in Bangladesh. Results indicated that all of the isolated species of bacteria showed a significant AMR in pharmaceutical workers compared with non-pharmaceutical subjects. Another study (Farshad et al. 2016) results indicate that the percentage of penicillin resistance was nearly 93% in pharmaceutical workers of penicillin production line and 71.4% among food industry workers, indicating a high level of resistance in both groups (in comparison a surveillance by WHO on drug resistance has reported 33.9% *S. pneumoniae* resistance or non-susceptibility to penicillin within the population of the country, where occupational study was performed). The results of studies listed above show that especially the manufacturing processes of the active product ingredients of antimicrobial nature can pose significant occupational risks for workers. Special attention should be therefore paid to manufacturing plants and technologies to strictly follow good manufacturing practices and additionally use all means possible to minimise exposure of workers as well as rules for using of the personal protective equipment (gloves, respirators, special air supplied suits) (Binks 2003). It has also been proven that sufficient staff trainings have to be performed to avoid inappropriate use or poor fit of the rules (Burgess and Mashingaidze 1999).

6 Risks for the Environment

Last part of this chapter is targeted on the considerations of the impact(s) on ecosystems, animal and human health from the presence of antimicrobial residues and/or pathogens and commensals harbouring antimicrobial resistance genes in the environment resulting from the use of (antimicrobial) veterinary medicinal products. An increasing body of evidence indicate that environment is the "cross-road" of exchange and transfer of antimicrobial resistance among different resistomes coming from human/animal/natural environment. Sewage, waste water treatment plants, agricultural and veterinary hospital effluents, drinking water (consumed either by humans or livestock/companion animals), recreational water, airborne aerosols, dust, wildlife fauna and contaminated food from agriculture or aquaculture are vectors enabling the potential transmission of bacteria and resistance determinants between hosts through the environment (EMA 2018b). Despite the prevalence of antibiotic residues and antibiotic resistance is assumed to be significantly influenced by the inputs from agricultural animal husbandry (Westphal-Settele et al. 2018), there has been some limitations of spread recognised. Bengtsson-Palme et colleagues (2018) summarised that the transfer of resistance genetic elements to animal and human bacteria from environmental bacteria, which are often less phylogenetically related, would likely be less common, but not necessarily insignificant as environmental stressors may induce horizontal genes transfer to and from (opportunistic) human pathogens in environmental settings. Remaining question is what level of stressor (e.g. residual concentration of antimicrobial or co-selector is needed to promote this transfer). Most of the resistance genes can be horizontally transferred by transduction, which is considered as the main mechanism conferring drug resistance in drinking water, surface water and wastewater (Lupo et al. 2012; Schwartz et al. 2003) but also limit the scale of transfer.

Considering the use of antimicrobials in animals, the residues of parent substances as well as microbiologically active metabolites are excreted to the environment via faeces, urine, and other products (discarded milk, blood) being considered as additional load to soil, water and sediment. Also bacteria from animal microbiome as well as pathogens are released to the environment altogether with their genetic elements (extra- and intra-chromosomal), among which resistance genes are present. Therefore, environment loaded by both residues and microbes/genes acts like a "mixer" of mobile genetic elements that interact, disperse and move to other ecological niches like human and wild/companion/livestock animals. There is growing evidence that (multi)resistant pathogens have developed through these pathways. Co-acting with the other conditions, especially those favourable for pathogenic strains able to cause infection diseases of human and animals, it is becoming common threat that can significantly limit future success of antimicrobial treatment of bacterial diseases in human and veterinary medicine.

What essential factors to be considered thinking about environment and AMR:

- Natural resistome of the environment due to the intrinsic resistance of bacterial environmental pool (being aware of many natural producers of antimicrobials, e.g. in soil)
- Selective pressure in nature due to physicochemical factors that can cause changes in microbial genomes (e.g. mutations with impact on intra- and extra- chromosomal genetic information of microbes)
- Co-resistance (bypass of different antimicrobial targets via linked resistance determinants)
- Cross-resistance (bypass of same antimicrobial targets via the same resistance determinant)
- Use of antimicrobials (but also other co-selectors like biocides, heavy metals (mostly known Zn and Cu), plant protective substances, other pharmacologically active medicinal substances):
  - In human medicine
  - In veterinary medicine
  - In plant protection
  - In food processing (including e.g. disinfectants, antimicrobial preservatives)
  - In industry (pharmaceutical, cosmetic, but also all industrial branches producing heavy metal pollution and other physicochemical load with possible influence)
  - In house holding (cleaning and disinfection products, cosmetics)
- Direct or indirect transfer of bacteria, genes, infectious diseases across geographical areas:
  - Trade with animals, plants, food, other goods that can serve as vehicle:
    - Directly the "subjects" of trade
    - Means of transport (e.g. lorries transporting animals)
  - Transfer of people (travelling, tourism, business trips, "surgery tourism"):
    - Directly persons (sick or as carriers of pathogens/commensals)
    - Means of transport (e.g. international transportation—airplanes)

Coming back to the use of antimicrobials in livestock animals and considering "*chemical pathway of the (antimicrobial) active substance environmental influence*" a range of rates of excretion and degradation, and possible transformation events, are seen which are dependent on the individual active substance. For example, results obtained by several authors indicated that tetracyclines have the highest concentrations and are most frequently reported antibiotic residues in manure (Pan et al. 2011; Chen et al. 2012; Massé et al. 2014); another study (Sengeløv et al. 2003) also indicate that tetracycline resistance levels in soil are temporarily influenced by the addition of pig manure slurry. The results indicate also that increased amount of pig manure slurry amendment may result in increased levels of tetracycline resistance in the soil. Other groups of antibiotics with considerable concentrations in manure are fluoroquinolones (Zhao et al. 2010; Van Doorslaer et al. 2014) and sulphonamides (Martínez-Carballo et al. 2007). Among the macrolide antibiotics, the highest concentration in manure was measured for tylosin (Dolliver et al. 2008). Compared to manure, biosolids contain much lower amounts of antibiotics (Jones-Lepp and Stevens 2007).

Since antibiotic substances and AMR genes have different rates of depletion/degradation in the body of the treated animal and the environment and also to be considered, e.g. further food processing, the "hot spots" for resistance development and spread may not be exactly those, where antibiotic substance consumption is the highest. Properties of antibiotics; especially in terms of their stability, sorption, physicochemical properties (e.g. molecular structure, size, shape, solubility, hydrophobicity, reactivity etc.) and persistence characteristics, partitioning to soil or water compartments as well as other "substrates and surfaces" present in the environment should be thoroughly considered in establishment of hazards and risks. Also the characterisation of the environmental conditions plays a significant role (climatic conditions—including rainfalls, type of soil, hydrogeological conditions). Degradation processes can be influenced by biotic (bacteria, yeasts, fungi and plants) as well as abiotic conditions (oxidation, reduction, hydrolysis and complexing) influenced by moisture, temperature, pH and other physico-chemical properties. Biggest differences can be considered among use in terrestrial animals (and therefore primarily "terrestrial conditions") and in aquatic animals (primarily "aquatic conditions" with direct influence of the surface water and indirect also on ground water).

Accurate quantification of antibiotics and their transformation products in the soil is of utmost importance and requires advanced analytical methods, such as high-performance liquid chromatography with tandem mass spectrometry (HPLC/MS) (Aga et al. 2016). Also other advanced technologies that have been currently developed can facilitate monitoring of residues of antimicrobials as well as presence of resistant microbes and genes of resistance in different environmental matrices (Charmaine and Yew-Hoong Gin 2019)—one of the examples for detection of resistant bacteria as well as genes of resistance is called OMIC approaches.

Except traditional quantitative polymerase chain reaction, the more recent high-throughput qPCR (HT-qPCR) platform with capabilities of detection of ~200 different antimicrobial resistance genes and mobile genetic elements is used to compare relative concentrations of AMR contamination across a variety of aquatic environments including water treatment plants (Liu et al. 2018; Zhu et al. 2017, Muziasari et al. 2017; Xu et al. 2016; An et al. 2018, Karkman et al. 2016).

OMIC approaches such as metagenomics are able to provide a holistic picture of the diversity of ARGs, MGE and vectors (e.g., integrons, plasmids) that assist horizontal gene transfer, and the overall microbial community structure (bacteria, viruses) in environmental systems and wastewaters (Bondarczuk and Piotrowska-Segat 2018; Chu et al. 2018; Gupta et al. 2018; Ng et al. 2017; Guo et al. 2017). Other OMIC approaches, such as metatranscriptomics, enable the identification of active microbial members within a community. In the context of AMR it also enables the measurement of transcription activity of bacteria resistant to antimicrobials through resistance genes expression (Rowe et al. 2017). ResCap, a targeted capture platform (TCP) designed to analyse ~78,000 ARGs, metal resistance, and plasmid markers is a targeted metagenomics approach for qualitative and quantitative resistome analysis (Lanza et al. 2018) .

Considering "chemical pathway of the active substance(s)" once used in livestock animals, could be considered from different perspectives:

Routes of administration:

- Oral administration (terrestrial vs aquatic animals, "cross" contamination of the environment)
- Parenteral administration (individualisation, pathways through animal body)
- Intramammary, Intrauterine, topical administrations

Release to the environment:

- Unused drinking water/waste water from farm
- Biofilms on/in equipment of the farm (e.g. drinking/medicated water distributing equipment with algae/biofilms can be reservoir of the antimicrobials/resistant bacteria)
- Direct excretion from animals (pasture)
- Excretion in housings and follow-up application of animal (un)processed manure(s) or slurry to areas of agricultural use as fertilisers
- Discharge of effluents from animal production units (husbandry—discarded milk and slaughter houses) to surface waters and soils, including aquaculture

Once the microbiologically active residue is released to the environment it can act and create selective pressure. This selective pressure can be considered from the perspective of influence on "naturally occurred environmental microbiome" and its gene pool, but is considered as very risky from the perspective of selected bacterial species with human health importance. Among those important belong *Aeromonas* spp., *Acinetobacter* spp., *Bacillus* spp. (mainly *cereus*), *Burkholderia cepacia*, *Pseudomonas* spp., *Serratia* spp., *Stenotrophomonas maltophilia*, and newly recognised Corynebacteria that can grow outside the animal/human body and use the environment as an alternative or main habitat (Raphael and Riley 2017; Tsai et al. 2018; Jumat et al. 2018). Pressure can lead to mobilisation of environmental resistance genes and their spread as well as development of new resistances (mutations/ recombinations), leading to shift within the ecosystem from less resistance prevalence to more resistance prevalence. Contaminated slurry was presented as the major emission source for ESBL/AmpC-producing *E. coli* in pig fattening farms (Dohmen et al. 2017). An increase in the prevalence of resistant clones of bacteria including enterococci, *E. coli* and *Acinetobacter* spp., after wastewater treatment has been observed by several authors, despite a reduction of bacterial load in treated wastewater compared to the raw wastewater (Ferreira da Silva et al. 2007; Łuczkiewicz et al. 2010; Zhang et al. 2009). However, a possible link between the prevalence of ESBL Enterobacteria in hospitals and other sources such as local food, water or animal sources has not been identified (Moore et al. 2010) or is difficult to interpret what the initial source was of either resistant bacteria or genes of resistance in certain cases.

Selective pressure among other factors depends on the concentration of antimicrobial (s) and possible co-selectors. It is important how much lower the real concentration is compared to the minimum inhibitory concentration, but also on the fact, if even very low concentration can pose ecological advantage of certain bacteria within bacterial community. For example, Gullberg et al. (2011) demonstrated resistance at levels as low as ng/l for ciprofloxacin (microbiologically active metabolite of enrofloxacin—one of the most used fluoroquinolones in livestock animals).

It should be noted that specific procedures of treatment of (drinking) water, sewage and other contaminated residues can reduce concentrations of certain classes of antibiotics, but invariably, a fraction of antibiotics remains after treatment (Watkinson et al. 2007). Water chlorination helps to degrade antibiotics such as beta-lactams and trimethoprim (Dodd and Huang 2004; Li et al. 2008). Traditional methods for wastewater treatment can eliminate up to 80% of fluoroquinolones and tetracyclines but they are less efficient in the removal of macrolides (Gulkowska et al. 2008; Shellie et al. 2002; Sukul and Spiteller 2007).

Despite all facts summarised above the issue is of even larger complexity. It means that not only selective pressure, but also capability of some soil microbiome bacteria that degrade antibiotics (and have been isolated from antibiotics-contaminated soils) should be taken into account (Cycoń et al. 2019). For example, strains belonging to the genera *Microbacterium* (Topp et al. 2013), *Burkholderia* (Zhang and Dick 2014), *Stenotrophomonas* (Leng et al. 2016), *Labrys* (Mulla et al. 2018), *Ochrobactrum* (Zhang et al. 2017; Mulla et al. 2018), and *Escherichia* (Mulla et al. 2018; Wen et al. 2018) were capable of degrading sulfamethazine, penicillin G, tetracycline, erythromycin and doxycycline in liquid cultures, respectively. Other bacteria belonging to the genera *Acinetobacter, Escherichia* (Zhang et al. 2012), *Klebsiella* (Xin et al. 2012), *Microbacterium* (Kim et al. 2011), *Labrys* (Amorim et al. 2014) and *Bacillus* (Rafii et al. 2009; Erickson et al. 2014) that were capable of degrading chloramphenicol, sulphapyridine, sulphamethazine, ciprofloxacin, norfloxacin and ceftiofur have been isolated from patients, sediments, sludge, animal faeces and seawater.

Moreover, Heinemann et al. (2017) give the evidence of increasing bacterial loads after cleaning and disinfection, which could lead to a vertical transfer of pathogens to newly arriving pigs. They evaluate methods for cleaning performances in pig stables as an important factor to minimise the risk of spread of bacteria as well as potential genes of resistance.

Another aspect is that any use of antimicrobial substance (but also any contact with co-selector substance) creates certain level of selective pressure both on animal microbiome as well as on pathogenic bacteria. From this perspective especially gastrointestinal tract and the microbiome settled here can be considered as the hotspot for mixing microbial genetic elements. With the excreta/faeces huge amount of bacteria and genetic elements are released into environment. Other reservoirs of the animal body can also be considered of importance—upper respiratory tract, urogenitary tract and skin. In relation of use of antimicrobials in animals we should therefore consider "*microbial and genetic elements pathways of environmental influence*". Transfer of genes between bacteria can in theory occur anywhere (Bengtsson-Palme et al. 2018), but transfer is more likely to occur between phylogenetically closely related bacteria (Philippot et al. 2010) and especially if the host and receiving bacteria share the same ecological niche, at least temporarily (Wiedenbeck and Cohan 2011). There will probably be higher probability of horizontal gene transfer among identical or closely related bacteria specific for one host, i.e. human to human and animal to animal, but considering that some bacterial species are zoonotic and some of lineages were proven to colonise both hosts the probability of resistance genes horizontal transfer is growing.

6.1 Terrestrial Animal

As for the administration of antimicrobials to terrestrial animals, the biggest part of the risks can be allocated to mass medication and excretion and following it release of microbial and genetic elements to manure or slurry. Manure/slurry is often stored prior to its land application. Degradation of active substances or metabolites occurs for some antibiotics (to different extents according to the bacterial species) but not others. Exchange of the genetic elements and survival of live bacteria differs. Therefore, research is needed to investigate the best methods of storing/treatment of manure and slurry to reduce the levels of residual antibiotics (especially those with long persistence), resistance genetic elements and those species of bacteria with the biggest potential for transfer and spread of antimicrobial resistance. This task is extremely difficult, because several experiences used, e.g. in waste water plants or sewage cleaning tanks have been proven to rather create environment for further selection of resistance. The highest risks are therefore related to the area with high density of farming, with big farms with high use of antimicrobials, but also high use of co-selectors (Zn, disinfectants and generally biocides). From this perspective, regions with the highest livestock density, such as Brittany (France), Po Valley (Italy) and most of Denmark, Belgium and particularly the Netherlands, can be considered (Bos et al. 2013). Doubling pig, cattle and veal calf densities per municipality increased the odds of LA-MRSA carriage over carriage of other types of MRSA by 24.7% (95% CI: 0.9%–54.2%), 76.9% (95% CI: 11.3%–81.3%) and 24.1% (95% CI: 5.5%–45.9%), respectively, after adjusting for direct animal contact, living in a rural area, and the probable source of MRSA carriage. Controlling the spread of LA-MRSA thus requires giving attention to community members in animal-dense regions who are unaffiliated with livestock farming (Feingold et al. 2012).

The emissions of ESBL-producing *E. coli* from pig farms to the surrounding environment, faecal and environmental samples from six pig farms were collected. In total, 119 ESBL-producing *E. coli* were isolated from faeces, air samples, water, sludge and soil samples. Antibiotic susceptibility testing showed that the ESBL-producing isolates were resistant to multiple antibiotics and isolates of different origin within the same farm showed similar resistance phenotypes (Gao et al. 2012). Environmental samples on selected pig and poultry Dutch farms (dust, animal feed, manure) and samples from pigs and farmers were contaminated with MRSA (Pletinckx et al. 2011). The work of Hartmann et al. (2012) has proven long-term survival of CTX-M isolates in soil; isolates were from soil that had been treated with manure 1 year before sampling. Von Salviati et al. (2015) detected high prevalence of ESBLs-positive *E. coli* in manure and proved the emission potential via manure and transmission via flies in pig farms and their surroundings in selected parts of Germany. In the vicinity of the pig barns ESBL/AmpC-producing *E. coli* were detected in 16.1% (14/87) of the examined boot swab samples taken from various ground surfaces and in 6% (2/36) of ambient air samples. The majority of slurry samples (82.4%; 14/17) and three of four samples of digestate from biogas plants were also tested positive for these resistant bacteria. In total 274 *E. coli* isolates were further analysed by phenotypical and genotypic methods. The authors of the above study summarised that contaminated slurry presented the major emission source for ESBL/AmpC-producing *E. coli* in the pig fattening farms (von Salviati et al. 2015).

Fertilisers (e.g. manure) used in **plant production** serve as vehicle of chemical (residues), microbial and genetic elements and therefore having influence on crop. Since fresh vegetables are often consumed raw, consumption may result in the ingestion of resistant bacteria and genetic elements that, depending on the bacterial species, are able to colonise the gut or pass through the intestine, thus posing a potential public health risk (FAO 2016). Vital et al. (2018) proved that multidrug-resistant isolates were observed in irrigation water, soil, and vegetables in urban farms that were most prevalent in water (25.3%) compared to soil (2.8%) and vegetable (8.4%) isolates, indicating that water serves as a possible route for a wide distribution across all kinds of borders.

6.2 Use in Aquacultures

Antimicrobials are also used in aquaculture where they are generally used as in-feed medication in Europe, but only a percentage is assumed to be absorbed by the fish. Ultimately, antimicrobials can reach various external environmental compartments such as rivers, lakes and soils (Kümmerer 2009; Martínez-Carballo et al. 2007; Sukul and Spiteller 2007) where they can continue to exert their effects. Rigos et al. (2004) estimated that 60–73% of oxytetracycline administered to sea bass on Greek farms is released to the environment. As for the marine environment, even bigger knowledge gaps were identified. However, available studies indicate potential ecological risks. In the cases of treatment of furunculosis in salmon, oxytetracycline and florfenicol have been used. High concentrations of those antibiotics were shown to inhibit growth of algae.

As the water ecosystems are prone to quick exchange of genetic elements, each input by any substance with potential to (co)select resistance should be thoroughly considered. Also there emerges the need to understand the effects of chronic, low-level exposure to antimicrobials and other AMR co-selectors (including the combinations) in wild species (Pittenger et al. 2007). In the past, antimicrobials were used much more liberally in aquaculture. In response to growing awareness and stricter regulations on their use, they are now generally used in less extent, at least in many EU countries. Improvements in farming practices have led to improved animal health and have reduced the need for the use of antimicrobials Commission Notice (2015). Moreover, the development and use of vaccines is also a key factor in reducing antibiotic use in aquaculture as was shown by Norwegian example.

There should be noted not just release of the risky elements from the livestock animal sector, but also inputs to this sector as, e.g. high bacterial loads in animal drinkers coming from poor quality water contaminated by bacteria and resistance genes.

More highly resistant Gram-negative strains were found in waste water treatment plants with urban/clinical influence than in waste waters with rural influence (Müller et al. 2018). Diallo et al. (2013) identified a significantly higher prevalence of ESBL-producing E. coli from municipal waste water (8.4%), compared to slaughterhouse waste water (1.2%).

There is also evidence that especially in waste water coming from international airports as well places as harbours and docks that can be considered as cross-roads big portfolio of resistant genes and bacteria harbouring resistance is present (Berendonk 2018).

7 Need for "One Earth" Approach

Thinking from global perspective "One Earth" approach should be considered as new concept, promoting "One health" perspective together with the environmental aspects but throughout the whole Earth. There should be considered all inputs and outputs as well as interactions in their complexity. We should not forget the Earth perspective, especially considering the need for environmental balance. Huge amount of antibiotics is not used in veterinary medicine only, but vast majority are the same substances or at least classes of antibiotics, used also in human medicine and emissions from industrial sites can be considerable, especially in developing countries (Larsson 2014). As vast majority of currently manufactured active product ingredients of medicinal products coming from those countries, we should help to improve technologies to minimise risks for the environment. Antibiotics are also used in culture medium for the production of biological pharmaceuticals, but emissions from this industry seem to be negligible compared to classical pharmaceuticals/antimicrobials active product ingredient manufacturing.

8 Conclusion

Significant gaps in our knowledge around the specific mechanisms and pathways of AMR spreading and associated risks for human, animals and environment are still there. Among the other knowledge missed information whether putative changes induced in communities of bacteria, naturally present in the environment, may affect the emergence and spread of AMR in bacteria of clinical relevance for human or animals (EMA 2018b).

The situation is even more complex due to possibility to not only select, but co-select resistance. This is one of the reasons why many resistance genes persist for long periods in the absence of antibiotics. It was proven that in absence of antibiotics in *Escherichia coli* (study with nine plasmids from six major incompatibility groups and mixed populations carrying multiple plasmids) there is still sufficient extent of conjugation to maintain resistance in the population. Authors of this study predict that combining conjugation inhibition and promoting plasmid loss would be an effective strategy to limit conjugation-assisted persistence of antibiotic resistance. Results of this study suggest that reducing antibiotic use alone is likely insufficient for reversing resistance (Lopatkin et al. 2017).

Moreover, not only bacteria, but also consumers are under the pressure both of individual substances concentration of residues (even all below MRLs) and mixtures of chemical residues from different sources. Despite the fact that the European concept of the residue/consumer safety is considered as very conservative, counting with safety factors and precautionary principle, there are still existing unknown risks of the mixture of below MRLs residues and their pharmacological, toxicological or hypersensitivisation effects.

While there are still important knowledge gaps, all the above considerations and evidences bringing information on risks associated with use of antimicrobials in animal, human and plant sector are therefore leading to the conclusion that minimising of the use of antimicrobials and co-selectors of AMR is one of the most important ways to mitigate these risks.

References

Aga DS, Lenczewski M, Snow D, Muurinen J, Sallach JB, Wallace JS (2016) Challenges in the measurement of antibiotics and in evaluating their impacts in agroecosystems: a critical review. J Environ Qual 45 (2):407–419

Alt K, Fetsch A, Schroeter A, Guerra B, Hammerl JA, Hertwig S, Senkov N, Geinets A, Mueller-Graf CH, Braeunig J, Kaesbohrer A, Appel B, Hensel A, Tenhagen BA (2011) Factors associated with the occurrence of MRSA CC398 in herds of fattening pigs in Germany. BMC Vet Res 7:69

Amorim CL, Moreira IS, Maia AS, Tiritan ME, Castro PM (2014) Biodegradation of ofloxacin, norfloxacin, and ciprofloxacin as single and mixed substrates by *Labrys portucalensis* F11. Appl Microbiol Biotechnol 98 (7):3181–3190

An XL, Su JQ, Li B, Ouyang WY, Zhao Y, Chen QL, Cui L, Chen H, Gillings MR, Zhang T et al (2018) Tracking antibiotic reistome during wastewater treatment using high throughput quantitative PCR. Environ Int 117:146–153

APVMA (2014) Australian Government: Australian pesticides and veterinary medicines authority antibiotic resistance risk assessments. https://apvma.gov.au/node/1018. Accessed 20 June 2019

Argudín M, Deplano A, Meghraoui A, Dodémont M, Heinrichs A, Denis O, Nonhoff C, Roisin S (2017) Bacteria from animals as a pool of antimicrobial resistance genes. Antibiotics 6(2):12

Baynes RE, Dedonder K, Kissell L, Mzyk D, Marmulak T, Smith G, Tell L, Gehring R, Davis J, Riviere JE (2016) Health concerns and management of select veterinary drug residues. Food Chem Toxicol 88:112–122

Bengtsson-Palme J, Kristiansson E, Larsson DGJ (2018) Environmental factors influencing the development and spread of antibiotic resistance. FEMS Microbiol Rev 42(1):fux053

Berendonk T (2018) Sewage from airports exhibits high abundance and diversity of antibiotic resistance genes. Environ Sci Technol 53(23):13898–13905

Biasino W, De Zutter L, Garcia-Graells C, Uyttendaele M, Botteldoorn N, Gowda T, Van Damme I (2018) Quantification, distribution and diversity of ESBL/AmpC-producing *Escherichia coli* on freshly slaughtered pig carcasses. Int J Food Microbiol 281:32–35

Binks SP (2003) Occupational toxicology and the control of exposure to pharmaceutical agents at work. Occup Med 53:363–370

Bondarczuk K, Piotrowska-Segat Z (2018) Microbial diversity and antibiotic resistance in a final effluent-receiving lake. Sci Total Environ 650:2951–2961

Bos JFFP, Smit AL, Schroder JJ (2013) Is agricultural intensification in the Netherlands running up to its limits? NJAS Wagen J Life Sci 66:65–73

Broens EM, Graat EA, Van der Wolf PJ, Van de Giessen AW, De Jong MC (2011) Transmission of methicillin resistant *Staphylococcus aureus* among pigs during transportation from farm to abattoir. Vet J 2011 (189):302–305

Burgess GL, Mashingaidze MT (1999) Respirator leakage in the pharmaceutical industry of Northwest England. Ann Occup Hyg 43:513–517

Carfora V, Giacinti G, Sagrafoli D, Marri N, Giangolini G, Alba P, Feltrin F, Sorbara L, Amoruso R, Caprioli A, Amatiste S, Battisti A (2016) Methicillin-resistant and methicillin-susceptible *Staphylococcus aureus* in dairy sheep and in-contact humans: An intra-farm study. J Dairy Sci 99(6):4251–4258

Ceccarelli D, Kant A, van Essen-Zandbergen A, Dierikx C, Hordijk J, Wit B, Mevius DJ, Veldman KT (2019) Diversity of plasmids and genes encoding resistance to extended Spectrum Cephalosporins in commensal *Escherichia coli* from Dutch livestock in 2007-2017. Front Microbiol 10:76

Chajęcka-Wierzchowska W, Zadernowska A, Łaniewska-Trokenheim Ł (2016) Diversity of antibiotic resistance genes in Enterococcus strains isolated from ready-to-eat meat products. J Food Sci 81:2799–2807

Charmaine NG, Yew-Hoong Gin K (2019) Monitoring Antimicrobial Resistance Dissemination in Aquatic Systems, Special edition. Water 11(1):71

Chen YS, Zhang HB, Luo YM, Song J (2012) Occurrence and assessment of veterinary antibiotics in swine manures: a case study in East China. Chin Sci Bull 57:606–614

Chon J, Sung K, Khan S (2017) Methicillin-resistant *Staphylococcus aureus* (MRSA) in food-producing and companion animals and food products. In:

Frontiers in *Staphylococcus aureus*. In Tech, London, pp 48–101

Chu BTT, Petrovich ML, Chaudhary A, Wright D, Murphy B, Wells G, Poretsky R (2018) Metagenomics reveals the impact of wastewater treatment plants on the dispersal of microorganisms and genes in aquatic sediments. Appl Environ Microbiol 84(5):e02168–e02117

Codex Alimentarius Commission (2011) Guidelines for risk analysis of foodborne antimicrobial resistance (CAC/GL 77–2011). www.fao.org/input/download/standards/11776/CXG_077e.pdf. Accessed 20 June 2019

Commission Notice (2015) Guidelines for the prudent use of antimicrobials in veterinary medicine (2015/C 299/04), Official Journal of the European Union, C299/7, https://ec.europa.eu/health/sites/health/files/antimicrobial_resistance/docs/2015_prudent_use_guidelines_en.pdf. Accessed 6 October 2020

Conceição T, de Lencastre H, Aires-de-Sousa M (2017) Frequent isolation of methicillin resistant *Staphylococcus aureus* (MRSA) ST398 among healthy pigs in Portugal. PLoS One 12(4):e0175340

Choquet-Kastylevsky G, Vial T, Descotes J (2002) Allergic adverse reactions to sulfonamides. Curr Allergy Asthma Rep 2(1):16–25

Crombé F, Argudín MA, Vanderhaeghen W, Hermans K, Haesebrouck F, Butaye P (2013) Transmission dynamics of methicillin-resistant *Staphylococcus aureus* in pigs. Front Microbiol 4:57

Cuny C, Nathaus R, Layer F, Strommenger B, Altmann D, Witte W (2009) Nasal colonization of humans with methicillin-resistant *Staphylococcus aureus* (MRSA) CC398 with and without exposure to pigs. PLoS One 4(8):e6800

Cuny C, Abdelbary MMH, Köck R, Layer F, Scheidemann W, Werner G (2016) Methicillin-resistant *Staphylococcus aureus* from infections in horses in Germany are frequent colonizers of veterinarians but rare among MRSA from infections in humans. One Health 2:11e7

Cycoń M, Mrozik A, Piotrowska-Seget Z (2019) Antibiotics in the soil environment—degradation and their impact on microbial activity and diversity. Front Microbiol 10:338

Dang STT, Bortolaia V, Tran NT, Le HQ, Dalsgaard A (2018) Cephalosporin-resistant *Escherichia coli* isolated from farm workers and pigs in northern Vietnam. Tropical Med Int Health 23(4):415–424

Day MJ, Rodríguez I, van Essen-Zandbergen A, Dierikx C, Kadlec K, Schink AK, Wu G, Chattaway MA, DoNascimento V, Wain J, Helmuth R, Guerra B, Schwarz S, Threlfall J, Woodward MJ, Coldham N, Mevius D, Woodford N (2016) Diversity of STs, plasmids and ESBL genes among from humans, animals and food in Germany, the Netherlands and the UK. J Antimicrob Chemother 71(5):1178–1182

Demoly P, Gomes ER (2005) Drug hypersensitivities: definition, epidemiology and risk factors. Eur Ann Allergy Clin Immunol 37(6):202–206

Diallo AA, Brugère H, Kérourédan M, Dupouy V, Toutain PL, Bousquet-Mélou A, Oswald E, Bibbal D (2013) Persistence and prevalence of pathogenic and extended-spectrum beta-lactamase-producing *Escherichia coli* in municipal wastewater treatment plant receiving slaughterhouse wastewater. Water Res 47(13):4719–4729

Didelot X, Gardy J (2014) Bayesian inference of infectious disease transmission from whole-genome sequence data. Mol Biol Evol 31:1869–1879

D'ıaz Angulo S, Szram J, Welch J, Cannon J, Cullinan P (2011) Occupational asthma in antibiotic manufacturing workers: case reports and systematic review. J Allergy 2011:1

Dierikx CM, Hengeveld PD, Veldman KT, de Haan A, van der Voorde S, Dop PY, Bosch T, van Duijkeren E (2016) Ten years later: still a high prevalence of MRSA in slaughter pigs despite a significant reduction in antimicrobial usage in pigs the Netherlands. J Antimicrob Chemother 71(9):2414–2418

Dodd CM, Huang CH (2004) Transformation of the antibacterial agent sulfamethoxazole in reactions with chlorine: kinetics, mechanisms, and pathways. Environ Sci Technol 38(21):5607–5615

Dohmen W, Schmitt H, Bonten M, Heederik D (2017) Air exposure as a possible route for ESBL in pig farmers. Environ Res 155:359–364

Dolliver H, Gupta S, Noll S (2008) Antibiotic degradation during manure composting. J Environ Qual 37 (3):1245–1253

Dorado-García A, Smid JH, van Pelt W, Bonten MJM, Fluit AC, van den Bunt G, Wagenaar JA, Hordijk J, Dierikx CM, Veldman KT, de Koeijer A, Dohmen W, Schmitt H, Liakopoulos A, Pacholewicz E, Lam TJGM, Velthuis AG, Heuvelink A, Gonggrijp MA, van Duijkeren E, van Hoek AHAM, de Roda Husman AM, Blaak H, Havelaar AH, Mevius DJ, Heederik DJJ (2018) Molecular relatedness of ESBL/AmpC-producing *Escherichia coli* from humans, animals, food and the environment: a pooled analysis. J Antimicrob Chemother 73:339–347

ECDC/EFSA/EMA (2015) European Centre for Disease Prevention and Control/European food safety authority/European medicine agency: ECDC/EFSA/EMA first joint report on the integrated analysis of the consumption of antimicrobial agents and occurrence of antimicrobial resistance in bacteria from humans and food-producing animals. EFSA J 13:4006

EFSA (2019) European food safety authority: Ttechnical report: report for 2017 on the results from the monitoring of veterinary medicinal product residues and other substances in live animals and animal products. EFSA J 16(5):1578E. https://doi.org/10.2903/sp.efsa.2019.EN-1578

EFSA and ECDC (2016) The European union summary report on antimicrobial resistance in zoonotic and indicator bacteria from humans, animals and food in 2014. EFSA J 14:4380. https://doi.org/10.2903/j.efsa.2016.4380

EFSA, ECDC (2018) European Food Safety Authority, European Centre for Disease Prevention and Control: The European Union summary report on antimicrobial resistance in zoonotic and indicator bacteria from humans, animals and food in 2016. EFSA J 16(2):5182

EFSA, ECDC (2019) European Food Safety Authority, European Centre for Disease Prevention and Control: The European Union summary report on antimicrobial resistance in zoonotic and indicator bacteria from humans, animals and food in 2017. EFSA J 17(2):5598

EMA (2010), European Medicines Agency: guideline on user safety for pharmaceutical veterinary medicinal products EMA/CVMP/543/03-Rev.1. https://www.ema.europa.eu/en/documents/scientific-guideline/guideline-user-safety-pharmaceutical-veterinary-medicinal-products_en.pdf. Accessed 20 June 2019

EMA (2018a) European medicines agency: guideline on the assessment of the risk to public health from antimicrobial resistance due to the use of an antimicrobial veterinary medicinal product in food-producing animals (EMA/CVMP/AWP/706442/2013). https://www.ema.europa.eu/en/documents/scientific-guideline/second-draft-guideline-assessment-risk-public-health-antimicrobial-resistance-due-use-antimicrobial_en.pdf. Accessed 20 June 2019

EMA (2018b) European medicines agency: reflection paper on antimicrobial resistance in the environment: considerations for current and future risk assessment of veterinary medicinal products (EMA/CVMP/ERA/632109/2014). https://www.ema.europa.eu/en/documents/scientific-guideline/draft-reflection-paper-antimicrobial-resistance-environment-considerations-current-future-risk_en.pdf. Accessed 20 June 2019

EMA/CVM/VICH (2012) VICH GL36 (R) Studies to evaluate the safety of residues of veterinary drugs in human food: General approach to establish a microbiological ADI(EMA/CVMP/VICH/467/2003). https://www.ema.europa.eu/documents/scientific-guideline/vich-gl36r-studies-evaluate-safety-residues-veterinary-drugs-human-food-general-approach-establish_en.pdf. Accessed 20 June 2019

EMA/CVM/VICH rev (2019) VICH GL36 (R2): studies to evaluate the safety of residues of veterinary drugs in human food: general approach to establish a microbiological ADI - revision 2. https://www.ema.europa.eu/documents/scientific-guideline/vich-gl36r2-studies-evaluate-safety-residues-veterinary-drugs-human-food-general-approach-establish_en.pdf. Accessed 20 June 2019

Erickson BD, Elkins CA, Mullis LB, Heinze TM, Wagner RD, Cerniglia CE (2014) A metallo-β-lactamase is responsible for the degradation of ceftiofur by the bovine intestinal bacterium *Bacillus cereus* P41. Vet Microbiol 172(3–4):499–504

European Commission (2009) Commission regulation EU 37/2010 of 22 December 2009 on pharmacologically active substances and their classification regarding maximum residue limits in foodstuffs of animal origin (Text with EEA relevance). https://ec.europa.eu/health/sites/health/files/files/eudralex/vol-5/reg_2010_37/reg_2010_37_en.pdf. Accessed 20 June 2019

European Parliament and the Council of the European Union (2009) Regulation (EC) No 470/2009 of the European Parliament and of the Council of 6 May 2009 laying down Community procedures for the establishment of residue limits of pharmacologically active substances in foodstuffs of animal origin, repealing Council Regulation (EEC) No 2377/90 and amending Directive 2001/82/EC of the European Parliament and of the Council and Regulation (EC) No 726/2004 of the European Parliament and of the Council (Text with EEA relevance). Off J Eur Commun 152:11

Ewers C, Bethe A, Semmler T, Guenther S, Wieler LH (2012) Extended-spectrum β-lactamase-producing and AmpC-producing *Escherichia coli* from livestock and companion animals, and their putative impact on public health: a global perspective. Clin Microbiol Infect 18(7):646–655

FAO (2016) Drivers, dynamics and epidemiology of antimicrobial resistance in animal production. http://www.fao.org/3/a-i6209e.pdf. Accessed May 2020

Farshad AA, Enferadi M, Bakand S, Orak RJ, Mirkazemi R (2016) Penicillin dust exposure and penicillin resistance among pharmaceutical workers in Tehran, Iran. Int J Occup Environ Health 22(3):218–223

FDA (2003) U.S. Food and drug Administration guidance for industry #152: evaluating the safety of antimicrobial new animal drugs with regard to their microbiological effects on bacteria of human health concern. https://www.fda.gov/regulatory-information/search-fda-guidance-documents/cvm-gfi-152-evaluating-safety-antimicrobial-new-animal-drugs-regard-their-microbiological-effects. Accessed 20 June 2019

Feingold BJ, Silbergeld EK, Curriero FC, van Cleef BAGL, Heck MEOC, Kluytmans JAJW (2012) Livestock density as risk factor for livestock-associated methicillin-resistant *Staphylococcus aureus*, the Netherlands. Emerg Infect Dis 18(11):1841–1849

Feld L, Bay H, Angen Ø, Larsen AR, Madsen AM (2018) Survival of LA-MRSA in dust from swine farms. Ann Work Expo Health 62(2):147–156

Ferreira da Silva M, Vaz-Moreira I, Gonzalez-Pajuelo M, Nunes OC, Manaia CM (2007) Antimicrobial resistance patterns in *Enterobacteriaceae* isolated from an urban wastewater treatment plant. FEMS Microbiol Ecol 60(1):166–176

Fessler AT, Olde Riekerink RG, Rothkamp A, Kadlec K, Sampimon OC, Lam TJ, Schwarz S (2012) Characterization of 971 methicillin-resistant *Staphylococcus aureus* CC398 obtained from humans and 972 animals on dairy farms. Vet Microbiol 160:77–84

Florez-Cuadrado D, Ugarte-Ruiz M, Quesada A, Palomo G, Dominguez L, Porrero MC (2016)

Description of an erm(B)-carrying campylobacter coli isolate in Europe. J Antimicrob Chemother 71:841–843
Gao P, Munir M, Xagoraraki I (2012) Correlation of tetracycline and sulfonamide antibiotics with corresponding resistance genes and resistant bacteria in a conventional municipal wastewater treatment plant. Sci Total Environ 421–422:173–183
Geser N, Stephan R, Kuhnert P, Zbinden R, Kaeppeli U, Cernela N, Haechler H (2011) Fecal carriage of extended-spectrum beta-lactamase–producing *Enterobacteriaceae* in swine and cattle at slaughter in Switzerland. J Food Prot. 74:446–449
Goerge T, Lorenz MB, Alen S, Hübner NO, Becker K, Köck R (2017) MRSA colonization and infection among persons with occupational livestock exposure in Europe: prevalence, preventive options and evidence. Vet Microbiol 200:6–12
Government of Canada (2007) Guidance for industry preparation of veterinary new drug submissions. https://www.canada.ca/en/health-canada/services/drugs-health-products/veterinary-drugs/legislation-guidelines/guidance-documents/guidance-industry-preparation-veterinary-new-drug-submissions-health-canada-2007.html. Accessed 20 June 2019
Grad YH, Lipsitch M (2014) Epidemiologic data and pathogen genome sequences: a powerful synergy for public health. Genome Biol 15:538
Graveland H, Wagenaar JA, Bergs K, Heesterbeek H, Heederik D (2011) Persistence of livestock associated MRSA CC398 in humans is dependent on intensity of animal contact. PLoS One 6(2):e16830
Gulkowska A, Leung HW, So MK, Taniyasu S, Yamashita N, Yeung LW, Richardson BJ, Lei AP, Giesy JP, Lam PK (2008) Removal of antibiotics from wastewater by sewage treatment facilities in Hong Kong and Shenzhen, China. Water Res 42 (1–2):395–403
Gullberg E, Cao S, Berg OG, Ilbäck C, Sandegren L, Hughes D, Andersson DI (2011) Selection of resistant bacteria at very low antibiotic concentrations. PLoS Pathog 7(7):e1002158
Guo J, Li J, Chen H, Bond PL, Yuan Z (2017) Metagenomic analysis reveals wastewater treatment plants as hotspots of antibiotic resistance genes and mobile genetic elements. Water Res 123:468–478
Gupta RC (2018) Regulatory aspects for the drugs and chemicals used in food-producing animals in the european union. In: Veterinary toxicology: basic and clinical practice, 3rd edn. Academic Press, Cambridge, Massachusetts, p 1238
Gupta SK, Shin H, Han D, Hur HG, Unno T (2018) Metagenomic analysis reveals the prevalence and persistence of antibiotic- and heavy metal-resistance genes in wastewater treatment plant. Journal of Microbiology 56(6):408–415
Haddadin RN, Saleh SA, Ayyash MA, Collier PJ (2013) Occupational exposure of pharmaceutical workers to drug actives and excipients and their effect on Staphylococcus spp. nasal carriage and antibiotic resistance. Int J Occup Environ Health 19(3):207–214
Hamscher G, Pawelzick HT, Sczesny S, Nau H, Hartung J (2003) Antibiotics in dust originating from a pig-fattening farm: a new source of health hazard for farmers? Environ Health Perspect 111(13):1590–1594
Hartmann A, Locatelli A, Amoureux L, Depret G, Jolivet C, Gueneau E, Neuwirth C (2012) Occurrence of CTX-M producing *Escherichia coli* in soils, cattle, and farm environment in France (Burgundy region). Front Microbiol 3:83
Haug MC, Tanner SA, Lacroix C, Stevens MJ, Meile L (2011) Monitoring horizontal antibiotic resistance gene transfer in a colonic fermentation model. FEMS Microbiol Ecol 78(2):210–219
Havelaar AH, Kirk MD, Torgerson PR, Gibb HJ, Hald T, Lake RJ, Praet N, Bellinger DC, de Silva NR, Gargouri N, Speybroeck N, Cawthorne A, Mathers C, Stein C, Angulo FJ, Devleesschauwer B, World Health Organization Foodborne Disease Burden Epidemiology Reference Group (2015) World Health Organization global estimates and regional comparisons of the burden of foodborne disease in 2010. PLoS Med 12 (12):e1001923
Heinemann C, Petersen B, Steinhoff-Wagner J (2017) Evaluation of methods for determining cleaning performance in pig stables. J Anim Sci 95(Supplement 4):48
Hoffmann A (2014) Presence of CTX-M-1 group extended-spectrum-ß-lactamase in dust from Dutch pig farms. In Proceedings of faculty of veterinary medicine, Utrecht
Irrgang A, Hammerl JA, Falgenhauer L, Guiral E, Schmoger S, Imirzalioglu C, Fischer J, Guerra B, Chakraborty T, Käsbohrer A (2018) Diversity of CTX-M-1-producing *E. coli* from German food samples and genetic diversity of the blaCTX-M-1 region on IncI1 ST3 plasmids. Vet Microbiol 221:98–104
Jans C, Sarno E, Collineau L, Meile L, Stärk KDC, Stephan R (2018) Consumer exposure to antimicrobial resistant Bacteria from food at Swiss retail level. Front Microbiol 9:362
JECFA (2018) Evaluation of certain veterinary drug residues in food. Eighty-fifth report of the Joint FAO/WHO Expert Committee on Food Additives). WHO technical report series, No. 1008, 2018; ISBN 978-92-4-121017-1
Jones-Lepp TL, Stevens R (2007) Pharmaceuticals and personal care products in biosolids/sewage sludge: the interface between analytical chemistry and regulation. Anal Bioanal Chem 387(4):1173–1183
Jumat MR, Haroon MF, Al-Jassim N, Cheng H, Hong PY (2018) An increase of abundance and transcriptional activity for Acinetobacter junii post wastewater treatment. Water 10:436
Karkman A, Johnson TA, Lyra C, Stedtfeld RD, Tamminen M, Tiedje JM, Virta M (2016) High-throughput quantification of antibiotic resistance

genes from an urban wastewater treatment plant. FEMS Microbiol Ecol 92(3):fiw014

Kim DW, Heinze TM, Kim BS, Schnackenberg LK, . Woodling KA, Sutherland JB (2011) Modification of Norfloxacin by a microbacterium sp. strain isolated from a wastewater treatment plant. Appl Environ Microbiol 77(17): 6100–6108

Knight DR, Elliott B, Chang BJ, Perkins TT, Riley TV (2015) Diversity and evolution in the genome of *Clostridium difficile*. Clin Microbiol Rev 28(3):721–741

Köck R, Harlizius J, Bressan N, Laerberg R, Wieler LH, Witte W, Deurenberg RH, Voss A, Becker K, Friedrich AW (2009) Prevalence and molecular characteristics of methicillin-resistant *Staphylococcus aureus* (MRSA) among pigs on German farms and import of livestock-related MRSA into hospitals. Eur J Clin Microbiol Infect Dis 28(11):1375–1382

Köck R, Schaumburg F, Mellmann A, Köksal M, Jurke A, Becker K, Friedrich AW (2013) Livestock-associated methicillin-resistant *Staphylococcus aureus* (MRSA) as causes of human infection and colonization in Germany. PLoS One 8(2):e55040

Kraushaar B, Ballhausen B, Leeser D, Tenhagen BA, Käsbohrer A, Fetsch A (2017) Antimicrobial resistances and virulence markers in methicillin-resistant *Staphylococcus aureus* from broiler and Turkey: a molecular view from farm to fork. Vet Microbiol 200:25–32

Kümmerer K (2009) Antibiotics in the aquatic environment--a review--part I. Chemosphere 75(4):417–434

Lanza VF, Tedim AP, Martínez JL, Baquero F, Coque TM (2015) The plasmidome of firmicutes: impact on the emergence and the spread of resistance to antimicrobials. Microbiol Spectr 3(2). https://doi.org/10.1128/microbiolspec.PLAS-0039-2014

Lanza VF, Baquero F, Martinez JL, Ramos-Ruiz R, Gonzalez-Zorn B, Andremont A, Sanchez-Valenzuela-A, Ehrlich SD, Kennedy S, Ruppe E et al (2018) In-depth resistome analysis by targeted metagenomics. Microbiome 6(1):11

Larsen J, Petersen A, Larsen AR, Sieber RN, Stegger M, Koch A, Aarestrup FM, Price LB, Skov RL, and for the Danish MRSA Study Group (2017) Emergence of livestock-associated Methicillin-resistant *Staphylococcus aureus* bloodstream infections in Denmark. https://www.ncbi.nlm.nih.gov/pmc/articles/PMC5850567/. Accessed 20 June 2019

Larsson DGJ (2014) Pollution from drug manufacturing: review and perspectives. Philos Trans R Soc Lond Ser B Biol Sci 369(1656):20130571

Lavilla S, González-López JJ, Sabaté M, García-Fernández A, Larrosa MN, Bartolomé RM, Carattoli A, Prats G (2008) Prevalence of qnr genes among extended-spectrum beta-lactamase-producing enterobacterial isolates in Barcelona, Spain. J Antimicrob Chemother 61(2):291–295

Leng Y, Bao J, Chang G, Zheng H, Li X, Du J, Snow D, Li X (2016) Biotransformation of tetracycline by a novel bacterial strain *Stenotrophomonas maltophilia* DT1. J Hazard Mater 318:125–133

Liebana E, Carattoli A, Coque TM, Hasman H, Magiorakos AP, Mevius D, Peixe L, Poirel L, Schuepbach-Regula G, Torneke K, Torren-Edo J, Torres C, Threlfall J (2013) Public health risks of enterobacterial isolates producing extended-spectrum β-lactamases or AmpC β-lactamases in food and food-producing animals: an EU perspective of epidemiology, analytical methods, risk factors, and control options. Clin Infect Dis 56(7):1030–1037

Li D, Yang M, Hu J, Zhang Y, Chang H, Jin F (2008) Determination of penicillin G and its degradation products in a penicillin production wastewater treatment plant and the receiving river. Water Res 42 (1–2):307–317

Liu BT, Song FJ, Zou M, Hao ZH, Shan H (2017) Emergence of Colistin resistance gene *mcr*-1 in *Cronobacter sakazakii* producing NDM-9 and in *Escherichia coli* from the same animal. Antimicrob Agents Chemother 61(2):1444–1416

Liu L, Su JQ, Guo Y, Wilkinson DM, Liu Z, Zhu YG, Yang J (2018) Large-scale biogeographical patterns of bacterial antibiotic resistome in the waterbodies of China. Environ Int 117:292–299

Lopatkin AJ, Meredith HR, Srimani JK, Pfeifer C, Durret R, Lingchong Y (2017) Persistence and reversal of plasmid-mediated antibiotic resistance. Nat Commun 8:1689. https://doi.org/10.1038/s41467-017-01532-1

Łuczkiewicz A, Jankowska K, Fudala-Książek S, Olańczuk-Neyman K (2010) Antimicrobial resistance of fecal indicators in municipal wastewater treatment plant. Water Res 44(17):5089–5097

Lupo A, Coyne S, Berendonk TU (2012) Origin and evolution of antibiotic resistance: the common mechanisms of emergence and spread in water bodies. Front Microbiol 3:18

Ma F, Shen C, Zheng X, Liu Y, Chen H, Zhong L, Liang Y, Kang L, Xia Y, Tian GB, Yang Y (2019) Identification of a novel plasmid carrying *mcr*-4.3 in *an Acinetobacter baumannii* strain in China. Antimicrob Agents Chemother 63(6). https://doi.org/10.1128/AAC.00133-19

Madec JY, Haenni M, Nordmann P, Poirel L (2017) Extended-spectrum β-lactamase/AmpC- and carbapenemase-producing *Enterobacteriaceae* in animals: a threat for humans? Clin Microbiol Infect 23(11):826–833

Massé DI, Saady NMC, Gilbert Y (2014) Potential of biological processes to eliminate antibiotics in livestock manure: an overview. Animals 4:146–163

Martínez-Carballo E, González-Barreiro C, Scharf S, Gans O (2007) Environmental monitoring study of selected veterinary antibiotics in animal manure and soils in Austria. Environ Pollut 148(2):570–579

Martínez JL, Coque TM, Lanza VF, de la Cruz F, Baquero F (2017) Genomic and metagenomic technologies to explore the antibiotic resistance mobilome. Ann N Y Acad Sci 1388(1):26–41

Møller NE, Nielsen B, von Würden K (1990) Changes in penicillin contamination and allergy in factory workers. Contact Dermatitis 22:106–107

Møller NE, Würden K (1992) Hypersensitivity to semi-synthetic penicillins and cross-reactivity with penicillin. Contact Dermatitis 26(5):351–352

Moore GA, Nygren O (2004) The Nordic expert group for criteria documentation of health risks from chemicals: 134. Penicillins. http://www.inchem.org/documents/kemi/kemi/ah2004_06.pdf. Accessed 20 June 2019

Moore JE, Watabe M, Millar BC, Loughrey A, McCalmont M, Goldsmith CE, Heaney JC, Buckley T, Egan C, McDowell DA, McMahon MA, Dooley JS, Xu J, Rooney PJ (2010) Screening of clinical, food, water and animal isolates of *Escherichia coli* for the presence of blaCTX-M extended spectrum beta-lactamase (ESBL) antibiotic resistance gene loci. Ulster Med J 79(2):85–88

Mroczkowska A, Żmudzki J, Marszałek N, Orczykowska-Kotyna M, Komorowska I, Nowak A, Grzesiak A, Czyżewska-Dors E, Dors A, Pejsak Z, Hryniewicz W, Wyszomirski T, Empel J (2017) Livestock-associated *Staphylococcus aureus* on Polish pig farms. PLoS One 12(2):e0170745

Müller H, Sib E, Gajdiss M, Klanke U, Lenz-Plet F, Barabasch V, Albert C, Schallenberg A, Timm C, Zacharias N, Schmithausen RM, Engelhart S, Exner M, Parcina M, Schreiber C, Bierbaum G (2018) Dissemination of multi-resistant gram-negative bacteria into German wastewater and surface waters. FEMS Microbiol Ecol 94(5):15–25

Mulla SI, Hu A, Sun Q, Li J, Suanon F, Ashfaq M, Yu CP (2018) Biodegradation of sulfamethoxazole in bacteria from three different origins. J Environ Manage 206:93–102

Muloi D, Ward MJ, Pedersen AB, Fèvre EM, Woolhouse MEJ, van Bunnik B (2018) Are food animals responsible for transfer of antimicrobial-resistant *Escherichia coli* or their resistance determinants to human populations? A systematic review. Foodborne Pathog Dis 15(8):467–474

Muziasari WI, Pitkanen LK, Sorum H, Stedtfeld RD, Tiedje JM, Virta M (2017) The Resistome of farmed fish feces contributes to the enrichment of antibiotic resistance genes in sediments below Baltic Sea fish farms. Front Microbiol 7:2137. (Including corrigendum Front. Microbiol. Aug 2;8:1491)

Ng C, Tay M, Tan B, Le TH, Haller L, Chen H, Koh TH, Barkham T, Thompson JR, Gin KYH (2017) Characterization of metagenomes in urban aquatic compartments reveals high prevalence of clinically relevant antibiotic resistance genes in wastewaters. Front Microbiol 16(8):2200

Normanno G, Dambrosio A, Lorusso V, Samoilis G, Di Taranto P, Parisi A (2015) Methicillin-resistant *Staphylococcus aureus* (MRSA) in slaughtered pigs and abattoir workers in Italy. Food Microbiol 51:51–56

OIE (2018a) World Organisation for Animal Health: terrestrial animal health code: Chapter 6.11. Risk analysis for antimicrobial resistance arising from the use of antimicrobial agents in animals. http://www.oie.int/index.php?id=169&L=0&htmfile=chapitre_antibio_risk_ass.htm. Accessed 20 June 2019

OIE (2018b) World Organisation for Animal Health: aquatic animal health code: Chapter 6.5. Risk analysis for antimicrobial resistance arising from the use of antimicrobial agents in aquatic animals. http://www.oie.int/index.php?id=171&L=0&htmfile=chapitre_antibio_resp_risk_analysis.htm. Accessed 20 June 2019

Pan X, Qiang Z, Ben W, Chen M (2011) Residual veterinary antibiotics in swine manure from concentrated animal feeding operations in Shandong Province, China. Chemosphere 84:695–700

Parisi A, Crump JA, Glass K, Howden BP, Furuya-Kanamori L, Vilkins S, Gray DJ, Kirk MD (2018) Health outcomes from multidrug-resistant infections in high-income countries: a systematic review and meta-analysis. Foodborne Pathog Dis 15(7):428–436

Patterson RA, Stankewicz HA (2019) . Penicillin Allergy In: StatPearls Publishing; 2019 Jan. https://www.ncbi.nlm.nih.gov/books/NBK459320/. Acessed 20 June 2019

Petternel C, Galler H, Zarfel G, Luxner J, Haas D, Grisold AJ, Reinthaler FF, Feierl G (2014) Isolation and characterization of multidrug-resistant bacteria from minced meat in Austria. Food Microbiol 44:41–46

Philippot L, Andersson SG, Battin TJ, Prosser JI, Schimel JP, Whitman WB, Hallin S (2010) The ecological coherence of high bacterial taxonomic ranks. Nat Rev Microbiol 8(7):523–529

Pig Health (2019) Denmark: high proportion of pig farms with MRSA. https://www.pig333.com/latest_swine_news/denmark-high-proportion-of-pig-farms-with-mrsa_14707/. Accessed 20 June 2019

Pirone-Davies C, Chen Y, Pightling A, Ryan G, Wang Y, Yao K, Allard MW (2018) Genes significantly associated with lineage II food isolates of *Listeria monocytogenes*. BMC Genomics 19(1):708

Pittenger R, Anderson B, Benetti DD, Dayton P et al (2007) Sustainable marine aquaculture: fulfilling thepromise; managing the risks. Marine Aquaculture Task Force, Takoma Park, MD

Pletinckx LJ, Verhegghe M, Dewulf J, Crombé F, De Bleecker Y, Rasschaert G, Goddeeris BM, De Man I (2011) Screening of poultry-pig farms for methicillin-resistant *Staphylococcus aureus*: sampling methodology and within herd prevalence in broiler flocks and pigs. Infect Genet Evol 11(8):2133–2137

Rafii F, Williams AJ, Park M, Sims LM, Heinze TM, Cerniglia CE, Sutherland JB (2009) Isolation of bacterial strains from bovine fecal microflora capable of degradation of ceftiofur. Vet Microbiol 139 (1–2):89–96

Raison-Peyron N, Messaad D, Bousquet J, Demoly P (2001) Anaphylaxis to beef in penicillin-allergic patient. Allergy 56(8):796–797

Randall LP, Lemma F, Rogers JP, Cheney TE, Powell LF, Teale CJ (2014) Prevalence of extended-spectrum-β-lactamase-producing *Escherichia coli* from pigs at slaughter in the UK in 2013. J Antimicrob Chemother 69(11):2947–2950

Raphael E, Riley LW (2017) Infections caused by antimicrobial drug-resistant saprophytic gram-negative Bacteria in the environment. Front Med 4:183

Rebandel P, Rudzki E (1990) Occupational allergy in the production of drugs. Pol Tyg Lek 45:82–84

Reynaga E, Navarro M, Vilamala A, Roure P, Quintana M, Garcia-Nuñez M, Figueras R, Torres C, Lucchetti G, Sabrià M (2016) Prevalence of colonization by methicillin-resistant *Staphylococcus aureus* ST398 in pigs and pig farm workers in an area of Catalonia, Spain. BMC Infect Dis 16(1):716

Rigos G, Nengas I, Alexis M, Troisi GM (2004) Potential drug (oxytetracycline and oxolinic acid) pollution from Mediterranean sparid fish farms. Aquat Toxicol 69:281–288

Ross RP, Morgan S, Hill C (2002) Preservation and fermentation: past, present and future. Int J Food Microbiol 79(1–2):3–16

Rowe WPM, Baker-Austin C, Verner-Jeffreys DW, Ryan JJ, Micallef C, Maskell DJ, Pearce GP (2017) Overexpression of antibiotic resistance genes in hospital effluents over time. J. Antimicrob. Chemother 72:1617–1623

Rudzki E, Rebandel P, Rebandel B (1986) Occupational allergy to antibiotics. Med Pr 37(6):383–387

Sahibzada S, Abraham S, Coombs GW, Pang S, Hernández-Jover M, Jordan D, Heller J (2017) Transmission of highly virulent community-associated MRSA ST93 and livestock-associated MRSA ST398 between humans and pigs in Australia. Scientific Reports 7:5273

Sarker MR, Islam KN, Huri HZ, Rahman M, Imam H, Hosen MB, Mohammad N, Sarker MZ (2014) Studies of the impact of occupational exposure of pharmaceutical workers on the development of antimicrobial drug resistance. J Occup Health 56:260–270

Schwartz T, Kohnen W, Jansen B, Obst U (2003) Detection of antibiotic-resistant bacteria and their resistance genes in wastewater, surface water, and drinking water biofilms. FEMS Microbiol Ecol 43(3):325–335

Schwarz S, Loeffler A, Kadlec K (2017) Bacterial resistance to antimicrobial agents and its impact on veterinary and human medicine. Vet Dermatol 28(1):82–e19

Sengeløv G, Agersø Y, Halling-Sørensen B, Baloda SB, Andersen JS, Jensen LB (2003) Bacterial antibiotic resistance levels in Danish farmland as a result of treatment with pig manure slurry. Environ Int 28 (7):587–595

Sessink PJM (2018) Occupational exposure of healthcare workers to antibiotics. J Infec Dis Treat J Infec Dis Treat 19(3):207–214

Shellie RA, Xie LL, Marriott PJ (2002) Retention time reproducibility in comprehensive two-dimensional gas chromatography using cryogenic modulation an intralaboratory study. J Chromatogr A 968 (1–2):161–170

Solensky R, Khan DA (2014) Evaluation of antibiotic allergy: the role of skin tests and drug challenges. Curr Allergy Asthma Rep 14(9):459

Stejskal VD, Forsbeck M, Olin R (1987) Side chain-specific lymphocyte responses in workers with occupational allergy induced by penicillins. Int Arch Allergy Appl Immunol 82:461–464

Sukul P, Spiteller M (2007) Fluoroquinolone antibiotics in the environment. Rev Environ Contam Toxicol 191:131–162

Topp E, Chapman R, Devers-Lamrani M, Hartmann A, Marti R, Martin-Laurent F, Sabourin L, Scott A, Sumarah M (2013) Accelerated biodegradation of veterinary antibiotics in agricultural soil following long-term exposure, and isolation of a sulfamethazine-degrading sp. J Environ Qual 42(1):173–178

Torres MJ, Romano A, Celik G, Demoly P, Khan DA, Macy E, Park M, Blumenthal K, Aberer W, Castells M, Barbaud A, Mayorga C, Bonadonna P (2017) Approach to the diagnosis of drug hypersensitivity reactions: similarities and differences between Europe and North America. Clin Transl Allergy 7:7

Tsai HC, Chou MY, Shih YJ, Huang TY, Yang PY, Chiu YC, Chen JS, Hsu BM (2018) Distribution and genotyping of aquatic *Acinetobacter baumannii* strains isolated from the Puzi River and its tributaries near areas of livestock farming. Water 10:1374

Van der Kolk JH, Endiminiani A, Graubner C, Gerber V, Perreten V (2018) Acinetobacter in veterinary medicine, with an emphasis on *Acinetobacter baumannii*. J Glob Antimicrob Resist 16:59–71

van Doorslaer X, Dewulf J, Van Langenhove H, Demeestere K (2014) Fluoroquinolone antibiotics: an emerging class of environmental micropollutants. Sci Total Environ 500–501:250–269

VICH GL54 (2017) (Safety) ARfD: Studies to evaluate the safety of residues of veterinary drugs in human food: general approach to establish an acute reference dose (ARfD), implementation by Nov 2017. https://www.vichsec.org/en/component/attachments/attachments/1548.html?task=download. Acessed 20 June 2019

Vital PG, Zara ES, Paraoan CEM, Dimasupil MAZ, Abello JJM, Santos ITG, Rivera WL (2018) Antibiotic resistance and extended-Spectrum Beta-lactamase production of *Escherichia coli* isolated from irrigation waters in selected urban farms in metro Manila, Philippines. Water 10(5):548

von Salviati C, Laube H, Guerra B, Roesler U, Friese A (2015) Emission of ESBL/AmpC-producing *Escherichia coli* from pig fattening farms to surrounding areas. Vet Microbiol 175(1):77–84

Vose D, Acar J, Anthony F, Franklin A, Gupta R, Nicholls T, Tamura Y, Thompson S, Threlfall EJ, van Vuuren M, White DG, Wegener HC, Costarrica ML, Office International des Epizooties Ad hoc Group (2001) Antimicrobial resistance: risk analysis methodology for the potential impact on public health of antimicrobial resistant bacteria of animal origin. Rev Sci Tech 20(3):811–827

Voss A, Loeffen F, Bakker J, Klaassen C, Wulf M (2005) Methicillin-resistant *Staphylococcus aureus* in pig farming. Emerg Infect Dis 11:1965–1966

Wang Y, Zhang M, Deng SZ, Wu C, Zhang J, Zhang Q, Shen J (2014) Emergence of multidrug-resistant campylobacter species isolates with a horizontally acquired rRNA methylase. Antimicrob Agents Chemother 58 (9):5405–5412

Wareth G, Neubauer H, Sprague LD (2019) *Acinetobacter baumannii* - a neglected pathogen in veterinary and environmental health in Germany. Vet Res Commun 43(1):1–6

Watkinson AJ, Murby EJ, Costanzo SD (2007) Removal of antibiotics in conventional and advanced wastewater treatment: implications for environmental discharge and wastewater recycling. Water Res 41 (18):4164–4176

Wen X, Wang Y, Zou Y, Ma B, Wu Y (2018) No evidential correlation between veterinary antibiotic degradation ability and resistance genes in microorganisms during the biodegradation of doxycycline. Ecotoxicol Environ Saf 147:759–766

Westphal-Settele K, Konradi S, Balzer F, Schönfeld J, Schmithausen R (2018) The environment as a reservoir for antimicrobial resistance : a growing problem for public health? Bundesgesundheitsblatt Gesundheitsforschung Gesundheitsschutz 61(5):533–542

Wieczorek K, Osek J (2013) Antimicrobial resistance mechanisms among campylobacter. Biomed Res Int 2013:340605

Wiedenbeck J, Cohan FM (2011) Origins of bacterial diversity through horizontal genetic transfer and adaptation to new ecological niches. FEMS Microbiol Rev 35(5):957–976

Xin Z, Fengwei T, Gang W, Xiaoming L, Qiuxiang Z, Hao Z, Wei C (2012) Isolation, identification and characterization of human intestinal bacteria with the ability to utilize chloramphenicol as the sole source of carbon and energy. FEMS Microbiol Ecol 82 (3):703–712

Xu L, Ouyang W, Qian Y, Su C, Su J, Chen H (2016) High-throughput profiling of antibiotic resistance genes in drinking water treatment plants and distribution systems. Environ Pollut 213:119–126

Zhang Y, Marrs CF, Simon C, Xi C (2009) Wastewater treatment contributes to selective increase of antibiotic resistance among Acinetobacter spp. Sci Total Environ 407(12):3702–3706

Zhang WW, Wen YY, Niu ZL, Yin K, Xu DX, Chen LX (2012) Isolation and characterization of sulfonamide-degrading bacteria Escherichia sp. HS21 and Acinetobacter sp. HS51. World J Microbiol Biotechnol 28(2):447–452

Zhang Q, Dick WA (2014) Growth of soil bacteria, on penicillin and neomycin, not previously exposed to these antibiotics. Sci Total Environ 493:445–453

Zhang W, Qiu L, Gong A, Yuan X (2017) Isolation and characterization of a high-efficiency erythromycin A-degrading *Ochrobactrum* sp. strain. Mar Pollut Bull 114(2):896–902

Zhao L, Dong YH, Wang H (2010) Residues of veterinary antibiotics in manures from feedlot livestock in eight provinces of China. Sci Total Environ 408 (5):1069–1075

Zhu YG, Zhao Y, Li B, Huang CL, Zhang SY, Yu S, Chen YS, Zhang T, Gillings MR, Su JQ (2017) Continental-scale pollution of estuaries with antibiotic resistance genes. Nat Microbiol 2(16):270

COMMISSION NOTICE (2015) Guidelines for the prudent use of antimicrobials in veterinary medicine (2015/C 299/04), Official Journal of the European Union, C299/7, https://ec.europa.eu/health/sites/health/files/antimicrobial_resistance/docs/2015_prudent_use_guidelines_en.pdf Accessed 6 October 2020

Prevention Is Better Than Cure

Lucie Pokludová

Abstract

Preventive measures and health programmes should help significantly to keep animals healthy. If animal welfare principles and good animal husbandry practices are also followed, minimal or no use of antimicrobials can be, with high probability, achieved. Setting priorities in biosecurity, which fits exact conditions of farm/husbandry is vital. Thorough mechanic cleaning, rational use of disinfection, disinsection and deratisation, proper ventilation and keeping the proper temperature and humidity contribute to keep good environment both in old stables and hi-tech husbandries. Health programmes, including vaccination tailored for local conditions, animal species and technologies used in the respective husbandry should be defined by educated veterinarians, specialised not only on treatment, but also on preventive medicine, use of alternatives to antimicrobials and management. Close cooperation of vets, farmers and people taking immediate care of animals and facilities is the basic prerequisite of the effectivity of such system. Therefore, tools for motivation and socio-economical aspects also belong among the key elements for effective preventive measures, which finally can help to minimise or skip the use of antimicrobials and help to combat antimicrobial resistance.

Keywords

Preventive medicine · Health programmes · Vaccination · Biosecurity · Hygiene · Good husbandry practices · Animal welfare · Disinfection · Disinsection · Deratisation · Alternatives to antimicrobials · Socio-economical aspects of prevention

1 Prevention Is Better Than Cure

The concept that "prevention is better than cure" (P>C) in veterinary medicine or agricultural sector generally should be understood in its complexity. Even holistic and integrated approach is necessary for the performance of this "P>C concept", primary target is to keep animals in good health and welfare status. As pointed out by the European Commission (2015) Guidelines for the prudent use of antimicrobials, preventing infections in the first instance is the best way to achieve reduction of consumption of antimicrobials, through minimising the need to use antimicrobials, as reducing the number of infections reduces the number of treatments needed. The "P>C concept" is fully in line with the new Animal Health Strategy according to European Commission (2014).

L. Pokludová (✉)
Institute for State Control of Veterinary Biologicals and Medicines, Brno, Czech Republic
e-mail: pokludova@uskvbl.cz

L. Pokludová (ed.), *Antimicrobials in Livestock 1: Regulation, Science, Practice*,
https://doi.org/10.1007/978-3-030-46721-0_6

The objective of reducing the use of antimicrobials is also in line with animal welfare, aims to reduce the density of the farm animal population and might be to do "step back" in intensity of the farming production, going more close to natural biological characteristics of the animals. That can be seen on the first side as step back also in economical parameters, but it could be compensated by lower costs of veterinary care on diseased animals as well as higher price purchase of better quality products from the healthy and sustainable husbandries. Overstocking is believed to be one of the major risk factors in the emergence and spread of infections that require the use of antimicrobials to reduce the suffering of sick animals (European Commission 2015).

There is a broad spectrum of medical and non-medical factors and drivers influencing antimicrobial usage in food-producing animals. One of other key factors, that can be considered across different livestock sectors, is the herd health status that significantly influences the need for antimicrobial treatments. Considering the real farm practice view, the observation of changes in feed and water intake by the animals as well as clinical signs was reported as the main driver for farmers to ask veterinarian to initiate an antimicrobial treatment (Friedman et al. 2007). Gastrointestinal disorders in young animals can be considered as one of the most frequent reasons for prescribing antimicrobials (for group/herd medication) among different sectors (piglets in the weaning period, broiler chickens and calves). Among other very frequent reasons belong respiratory clinical signs (beef calves/cattle and pig fatteners and turkeys, in less scale in broiler chickens). Reproductive clinical signs in sows and in cows are also of importance when considering frequency of antimicrobial prescription (De Bryine et al. 2014; FVE 2016a, b; Jensen et al. 2014; van Rennings et al. 2015).

There are also non-medical drivers. Those considered technical as herd characteristics, such as farm size, production type or management (e.g. farrowing rhythm, calves housing, oestrus synchronisation), biosecurity level were shown to be significantly associated with the amount of antimicrobials used in a herd (Fertner et al. 2015; Postma et al. 2016). However, social and behavioural factors, as interplay, cooperation, teamwork and professional relationship among farmers and veterinarians, including attitudes and habits towards responsible use of antimicrobials as well as economical drivers also plays a significant role (Coyne et al. 2016, 2018; Moreno 2014). Once framers are couched and start to believe they could use other means than antimicrobials to keep animals healthy and that the farms are still doing well and are in good productivity condition, they will not go back to the practices of overuse of antimicrobials. Educated farmers aware of antibiotic policies as well as those perceived higher risk of using antimicrobials also had lower actual antimicrobial usage (Visschers et al. 2016). The results of Scherpenzeel et al. (2017) show that it is more likely that broiler farmers who use a low amount of antibiotics for their broilers perceive more control and less risk and uncertainty with regard to the reduction of antibiotics use than high users. Uncertainty as driver for antibiotics use is mentioned as well in dairy farming with regard to the prevention and treatment of mastitis (Swinkels et al. 2015). Trujillo-Barrera et al. (2016) found that perceived risk appeared to be a barrier to the adoption of sustainable practices, while risk tolerance appeared to be a positive moderator of the relationship between economic rewards and adoption.

Animal husbandry and disease prevention measures that can be implemented to improve animal health and welfare can be divided into three main categories [modified according RONAFA report (EMA and EFSA 2017)]:

Primary prevention category includes practices reducing the introduction and spread of microorganisms/diseases between farms. There should be implemented principles of external biosecurity, performed compartmentalisation and eradication measures.

Secondary prevention category includes practices to reduce transmission or spread of microorganisms/diseases within a farm. There should be implemented principles of internal biosecurity and properly performed groupings of production. Appropriate housing design, building and maintenance.

Tertiary prevention category should be targeted on increase of the ability of animals to cope with the infectious disease causative pathogens: proper housing, well-balanced nutrition and improved feed strategies, reduction of stress, farm health plans including vaccination programmes tailored for the farm conditions and epidemiological situations, early diagnostics including tools like smart farming as well as genetic selection are necessary prerequisites to keep good health status of the herd/flock. Care should be taken both at individual, and also group of animals/herd/flock level.

Despite the fact that the antimicrobials seem to be still the only powerful tool for the treatment of acute diseases, there are some alternatives available to antimicrobials (*Note: author think that should be rather called "alternative tools"*), that could be used as preventive care/treatment before onset of clinical signs, or, e.g. in cases where it is expected that some stressful conditions make animals prone to be sick. Among these alternative tools (Table 5) belong (non-exhaustive list): prebiotics, probiotics, "competitive exclusion" products (i.e. excluding pathogenic bacteria from the host by competition with innocuous bacteria) (Callaway et al. 2008), bacteriophages, immunomodulators, phytotherapeuticals (including etheric oils and mixtures of extracts, tinctures and others), organic acids, vitamins, minerals although data are generally lacking about their efficacy, feasibility and return on investment (Collineau 2016; EMA and EFSA 2017).

An expert elicitation conducted among 111 European pig health experts identified improved internal biosecurity, external biosecurity and housing conditions as the alternative measures with the highest perceived effectiveness, whereas increased vaccination, increased use of anti-inflammatory products and improved water quality were reported as having the highest feasibility (Collineau 2016). The highest perceived return-on-investment was reported to be associated with improved internal biosecurity, use of zinc/metals and increased diagnostics to develop disease control action plans for pig farms (Postma et al. 2016). In other sectors (poultry, cattle), other factors can be considered of bigger importance and once improved—e.g. in the poultry sector 1-day-old chickens' quality, appropriate temperature and humidity, slower growing of chickens, lower density (decrease of ammonium)—can help significantly to reduce the use of antimicrobials (ASOA 2017). In dairy cattle sector EFSA (2009) in its scientific opinion on farming systems has stated that "the farming system by itself is a major factor determining the health problems of dairy cattle" and "the genetic component underlying milk yield has also been found to be positively correlated with the incidence of lameness, mastitis, reproductive disorders and metabolic disorders" and in beef cattle/veal with high use of antimicrobials according to Catry et al. (2016) were respiratory diseases linked to overstocking, inadequate ventilation, mixing of animals and failure of early diagnosis and treatment. So solving these issues in dairy and beef cattle can be a proactive solution instead of reactive approach with high usage of antimicrobials.

If the vast majority of the above-mentioned solution approaches is implemented and on the other hand wrong practices and (extremely) high intensity of farming are avoided, the animals will be less stressed, well cared and there can be expected increase of ability of an animal's immune system to respond appropriately to an infectious challenge helping to keep animals healthy.

2 Biosecurity and Hygiene

According to the "Animal Health Law" (Regulation (EU) No 2016/429) biosecurity is defined as the sum of management and physical measures designed to reduce the risk of the introduction, development and spread of diseases to, from and within an animal population or an establishment zone, compartment, means of transport, premises or location.

Locally tailored biosecurity plans that identify potential pathways for the introduction and spread of disease in a zone or compartment, and describe plan and performance of the measures that are being or will be applied to mitigate the disease risks should be in current time common practice of each farm. Biosecurity plans are also in accordance with the recommendations in the OIE Terrestrial Code (OIE 2013).

There are two cornerstones of the biosecurity: EXTERNAL pertains the introduction of the pathogen to the population of animals kept in some unit (can be understood as, e.g. country or farm) and INTERNAL targets reduction of the spread of pathogen within the population of animals kept in the respective unit.

FAO (2010) considers three main elements of biosecurity:

1. Segregation
   The creation and maintenance of barriers to limit the potential opportunities for infected animals and contaminated materials to enter an uninfected site. When properly applied, this step will prevent most contamination and infection.
2. Cleaning Materials
   (e.g. vehicles, equipment) that have to enter (or leave) a site must be thoroughly cleaned to remove visible dirt. This will also remove most of the pathogens that contaminate the materials. Mechanical cleaning (water and brush as well as high-pressure washer (110–130 bar, or where appropriate with hot water/steam) are efficient tools not introducing any chemical co-selectors of resistance.
3. Disinfection, e.g. according to the OIE Terrestrial code (OIE 2014)
   When properly applied and proper actives selected, including rotation to ensure efficacy and minimise/avoid resistance, disinfection will inactivate any pathogen that is present on materials that have already been thoroughly cleaned.

Speaking about biosecurity in connection with the reduction of need to use antimicrobials, mostly the unit is understood as farm (herd/flock) and the farm owner or manager is responsible for the setting of a biosecurity system.

For epidemic notifiable infections, as considered also by Animal Health law, the competent authority of the respective country/or e.g. the EU is in charge of the necessary biosecurity measures nationally and internationally, and outbreak measures. Those measures include banning import and other protective measures to avoid introduction of any animals/products that can cause introduction/spreading of the epidemic agent such as checking of imported animals and sources they are coming from, inspections, surveillance, check of the animals/products on the market and at the slaughterhouse level.

There should also be considered different routes of transmission (see Table 1) and their "rate of importance" in the transmission of the exact diseases altogether with combination of the probability of transmission and frequency of occurrence of transmission routes taking into account also certain pathogens/diseases incidence in the source of transmission.

Table 1 The key routes of disease transmission that can be mitigated by biosecurity and hygienic measures

Route of transmission	Examples of vehicles and vectors	Example of pathogens	Notes
Direct contact among animals and their secretions	Body fluids (urine, faeces, secrets/saliva, milk, blood)	All pathogens, of high importance e.g. MRSA (*Staphylococcus aureus* resistant to methicillin)	Not only clinically diseased animals to be considered as source, use screen methods for carriers investigation.
	Breeding (mother to piglets)		Entry into the susceptible animal through contact with the mucous membranes, such as the eyes, nose or mouth.
	Perinatal (mother giving birth)		
			Isolation/quarantine whenever possible.
			Frequent disinfection, enforced hygienic measures.
	Tissues (e.g. skin lesions)		Open wounds or breaks in the skin due to injuries, biting or rubbing against each other.
Aerogenous transmission (longer distance)	Small dust particles, air flow	*Mycoplasma pneumoniae*	High level of risk even for distance about 4 kilometres, high biosecurity level farms with air filters/ conditioning
Aerogenous transmission (within the housing)	Coughing, sneezing, aerosol droplets, small dust particles	*Actinobacillus pleuropneumoniae*	Most microorganisms are not able to survive for extended time periods - close proximity of infected and susceptible animals is required for disease transmission. Aerosol transmission can also occur when infected droplets from urine, faeces, or birthing material get stirred up from contaminated surfaces/ dust and inhaled.
			Adequate ventilation, ammonium decrease, dust minimisation
Staff		All bacteria with zoonotic potential as well as other just as by contaminated clothes/hands/or as healthy carriers	All hygienic measures to be thoroughly followed
			Also check for carriers
Vehicles, equipment	Needles, balling guns	*Actinobacillus pleuropneumoniae, Streptococcus suis*	Proper cleaning, disinfection, where possible use e.g. needle free delivery devices
	– Buckets (feed/ water)		
	– Bedding, shovels		
	– Vehicles, trailers		
	– Boots/clothing		
Feed		*Salmonella* spp.	
Water		All bacteria, yeasts, algae participating in biofilm	

(continued)

Table 1 (continued)

Route of transmission	Examples of vehicles and vectors	Example of pathogens	Notes
Manure and bedding		*Salmonella* spp., *Campylobacter* spp., *Enterobacteria* (including ESBLs/AmpC, *mcr-* genes carriers), Enterococci, *Mycobacterium (tuberculosis, avium), Brachyspira* spp.	Potential to survive differs across different pathogens and is highly dependent on the environmental conditions (rather cold/vet support survival, drying/hot do not support)
Rodents, birds, bats, wild animals, domestic animals	Most frequently birds	*Bordetella* spp., erysipelas, avian tuberculosis	
		Viruses: Classical Swine Fever, PRRS, influenza and Transmissible gastroeneteritis	
	Rodents	Atrophic rhinitis *Pasteurella multocida* alone or in combination with *Bordetella bronchiseptica*, E. *coli* diarrhoea, *Streptococcus suis*, leptospirosis, salmonellosis, *Brachyspira* spp., rotaviral diarrhoea, PRRS	
	Wild boars	Classical Swine Fever and African Swine Fever	
Flies and other insect vectors		*Enterobacteria* (including ESBLs/ AmpC, *mcr-* genes carriers), *Streptococcus suis*	
Semen			Most bacterial contaminants of semen are from faecal/environmental contamination.
			Semen collection and distribution and hygiene in all steps seems therefore critical, together with routine screening of breeding females for semen-spread infections.

2.1 External and Internal Biosecurity

2.1.1 External Biosecurity Subcategories

1. Purchasing policy:
    - (a) Animals preferably from one source/ same supplier:
        - (i) Avoid mixing of sources
        - (ii) If necessary, minimum number of sources with same health status/vaccination status is preferred (this can be difficult for some species, e.g. calves)
        - (iii) Reduction of mixing of animals from different batches
    - (b) Semen and embryos from reputable sources/health status declaring
    - (c) Routine cleaning and disinfection of means of transport, prior to transport checks:
        - (i) Boxes/containers, trucks, lorries, boats should be properly cleaned and disinfected (e.g. in pigs evidence that ASF, *Actinobacillus pleuropneumoniae*, TGE and

Streptococcus suis can be spread by contaminated vehicles).
   (ii) For 1-day-old poultry it is extremely important to check for proper working of the all fans in the transport vehicle/truck, cleaning/changing of filters, checking the heaters and air ventilation/circulation, checking if proper humidity can be ensured during the transport to avoid stress making the poultry prone to disease.
   (iii) Avoid/minimise use of litter, enrichment materials or feed ingredients originating from other farms, as they increase the risk of pathogen transfer.
   (iv) Ensure proper water sources during the transport.
   (v) Avoid any interfering "risky contacts".
 (d) Health status documented
 (e) Quarantine period
   (i) Isolate the animals with signs of sickness.
   (ii) Ensure personnel hygiene and disinfection, changing the boots/clothes.
 (f) Following the rules stipulated by competent authorities of country, regional authorities

2. Rules for removing:
 (a) Animals
 (b) Manure and slurry
 (c) Bedding (Ensure that appropriate disposal of soiled bedding is carried out in order to prevent the spread of diseases)
 (d) Contaminated single use equipment (ensure proper disposal of, e.g. syringes)
 (e) Carcasses (Dispose carcasses as soon as practical, ensure safe disposal—ensure disposed carcasses are not eaten by pest animals)

3. Rules for supply of:
 (a) Feed:
   (i) Avoid feed contamination by the raw materials used, post-production and during transport, or by exposure to rodents and birds on the property.
   (ii) Avoid poor quality or damaged feed, especially once bacteria/moulds are present and can be a concern.
 (b) Water:
   (i) Avoid surface water contamination, if possible (especially animals on the pasture, outside the "closed water systems"—hard to manage the quality).
   (ii) Regularly check water quality and water supply system in the stable/poultry house.
 (c) Bedding:
   (i) Ensure bedding material is fit-for-purpose.
   (ii) Ensure areas where bedding is stored are kept as dry and vermin free as is practically possible.
 (d) Equipment

4. Access check (including visitors book/exclusion rules):
 (a) Full fencing of the farm surrounding area, and where possible minimise the number of entry points and restrict access to the farm.
 (b) Define, and where appropriate signpost, "permitted access areas" for farm contractors (e.g. veterinarians, livestock agents, insemination technicians), delivery and pick-up vehicles (e.g. milk tankers, livestock and feed transporters) and service personnel (e.g. utility company technicians, government officers) and notify relevant operators prior to entry, check also the staff for not keeping animals (e.g. pigs) at backyards/home.
 (c) Availability of hygiene lock.
 (d) Strict separation of dirty and clean area in hygiene lock.

5. Control of communication with outside influencing factors (building design/structure/effective maintenance—doors, walls, screens, meshes, drain covers and all measures designed to prevent access of animals):
 (a) Ventilation, influx of spread agents (e.g. PRRS, *Mycoplasma hyopneumoniae*)

   (b) Control of animal vectors: pest/wild/domestic animals (feral animals; domestic animals, rodents, insects and other invertebrates (such as ticks or mites, rodents, wild birds); e.g. for birds/fly protection install nets
6. Location and environment:
   (a) Pig herds located in an area with high density of pigs (average pig density at municipality level >300 pigs/km^2), spotting of wild boars.
   (b) Air: avoid transmission of pathogens via aerosol or dust, whenever possible.
7. Staff education, training, recording, incentives

2.1.2 Internal Biosecurity Subcategories

1. Management of diseases:
   (a) Train facility staff to recognise and report diseased animals:
      (i) Early disease detection systems to be introduced (including new tools of "smart farming technologies").
      (ii) Frequent visiting (frequency depending on the age/health status of the animals).
      (iii) If detected animal with sign of the disease (either act according the plan or contact veterinarian).
      (iv) Work with a veterinarian to develop treatment protocols and monitor response rates on routine visits to the facility.
   (b) Make available of hospital pens:
      (i) Thoroughly cleaned/disinfected to avoid spread of infection
      (ii) Manage disposing excreta
      (iii) Manage disposing of deadstock animals
   (c) Clean and disinfect all equipment, clothing, boots, etc. that come into contact with ill animals.
   (d) Place a secondary identification on all animals that were treated for illness so they can be rapidly identified and more closely monitored once they have returned to their home pen.
2. Vaccination plan:
   (a) Implement protocols for routine vaccinations.
   (b) Vaccination guns, well cleaned/disinfected/properly changing the needles.
3. Farrowing, suckling period:
   (a) Frequency of cross-fostering of suckling pigs
   (b) Minimise frequency of manipulating (vaccination, castration) suckling pigs
4. Nursery period:
   (a) All-in/all-out-management.
   (b) Avoid mixing of different age groups.
   (c) Minimise stacking and stressful events.
5. Fattening period:
   (a) In pigs, e.g. compartmentalising, working lines, equipment
6. Avoid pets' movement among sections/stables/poultry houses (pigs, poultry, cattle)
7. Tools and equipment (including, e.g. injectors, dosing automats)
8. Cleaning and disinfection:
   (a) Removal of all mechanic dirties
   (b) Cleaning:
      (i) Mechanic
      (ii) Wet-soap use cleaning
      (iii) Water high pressure washer/high temperature cleaning
      (iv) High importance of proper drying off
   (c) Disinfection
      (Effective against target selected microorganisms, preferably non-toxic/irritating, non-corrosive, non-AMR (co) selecting, important to follow thoroughly all phases).
      (i) Application
      (ii) Allowance of contact time
      (iii) Rinse and highly important to let dry again (commercial desiccants, properly cleaned fans and heaters can be used)
   (d) Keep the stable free (the duration depending on consideration of local situation in incidence and type of disease/s in

pervious course of fattening/laying as well as drying of period)
   (e) Special attention to water supply/water medication system:
      (i) Avoid biofilm formation.
      (ii) Avoid residual amounts of antimicrobials.
8. Staff education, training, couching, recording, incentives
   (a) Explain and let follow internal biosecurity plan (e.g. once moving across pens/farm facilities by technical staff, keepers, nutritionists, veterinarians etc.)

3 Vaccination

Vaccines have, from the 1930s, made a major contribution to improving farm animal health, welfare and productivity. They are vital components in preventing a wide variety of diseases (RUMA 2016). A survey conducted among citizens in 2016 showed that 66% of respondents believe pets should be vaccinated, while only 54% think the same applies to farm animals. Over 40% of them replied they did not know that animal vaccination prevents the transfer of infectious disease from animals to humans (EPRUMA 2019).

One of the very important tools to improve health status of the herd/flock except following welfare practices and biosecurity is vaccination. The benefits can be seen from different perspectives, e.g. benefit for the individual animal to induce protection and improve its immunostatus, minimising the risk to get sick and reduce mortality (linked also to reduction of secondary, opportunistic infections), help to keep productivity, help build the "herd/flock immunity" and lower transfer of disease. Public health can also be protected through vaccination and a good example of this is the vaccination against zoonotic diseases of livestock animals. Vaccines and proper vaccination strategy at the herd/region level represent the single most cost-effective medical countermeasure that can be used to confront the threat of antimicrobial resistance. The vaccine effectiveness in preventing diseases has been far-reaching, and could significantly reduce the need and use of antibiotics (OIE 2015). Reduced antimicrobial use was also identified as being strongly associated with vaccination, e.g. against Porcine Circovirus type 2 in finishing pig farms (Raith et al. 2016), but vaccinating against more pathogens did not necessarily lead to lower antimicrobial use (Postma et al. 2016). Therefore, it should be borne in mind that vaccines optimally fulfil their potential when used as part of an overall programme of infection prevention and infection control in animal husbandries. Such a programme would be inclusive of veterinary oversight and care, good biosecurity and husbandry practices including welfare rules, proper quality and nutritional and protective balance of the feed, and improved diagnostics (including antimicrobial susceptibility testing, where appropriate) to help ensure pathogen specific, targeted treatment, including, where necessary, treatment by properly chosen antimicrobials (first choice-narrow spectrum whenever possible). The vaccination schedule should be prepared with high level of knowledge of the herd/flock status and anamnesis, considering also parents herds/flocks and should be tailored thoroughly on the exact farm conditions, using both commercial vaccines (in the case that they are appropriate and available), but also thinking on use of autogenous vaccines (if possible to prepare them in high quality standard, and in a way that fit the current pathogens load of the herd/flock).

THINK ABOUT The Research to Support the Development of Multivalent Vaccines

"A reliable supply of pure, safe, potent, and effective vaccines is essential for maintenance of animal health and the successful operation of animal health programmes"

There is the need for research as well as production of safe and effective multivalent vaccines that potentially cover a broad range of issues and disciplines, including

(continued)

discovery of new aetiological agents for inclusion in such vaccines. To close the diagnostic gap, identification of improved surrogate markers of protective immunity is of importance. It should also include an understanding of the mechanisms of interference and diminished efficacy that can be a consequence of combined vaccines. Encouragingly, new technologies and a major shift on how we approach vaccine discovery research may provide new opportunities for addressing these challenges (OIE 2018b).

In July 2019 (SAPHIR 2019), there was released the announcement related to outcomes of the European project under Horizon 2020 umbrella called SAPHIR, that should bring novel vaccine strategies to the market. SAPHIR has selected two representative pathogens of pigs (Porcine Reproductive and Respiratory Syndrome Virus and *Mycoplasma hyopneumoniae*), chickens (*Eimeria* spp. and *Clostridium perfringens*) and cattle (Bovine Respiratory Syncytial Virus and *Mycoplasma bovis*) to set vaccine development and production approaches applicable to other pathogens.

As many current vaccines fall short of ideal vaccines in one or more respects, promising breakthroughs to overcome these limitations include new biotechnology techniques, new oral vaccine approaches, novel adjuvants, new delivery strategies based on bacterial spores, and live recombinant vectors; they also include new vaccination strategies in ovo, and strategies that simultaneously protect against multiple pathogens. However, translating this research into commercial vaccines that effectively reduce the need for antibiotics will require close collaboration among stakeholders, for instance through public–private partnerships (Hoelzer et al. 2018a, b).

Despite several activities, there exist serious reasons why it is difficult to develop and introduce in practice new (antibacterial) commercial vaccines and those can be summarised as follows:

Biological: as, e.g. different serotypes => cross-protection among serotypes is often poor => not sufficiently protect the animal from infection of other serotypes. Also combination of different serotypes limited by interference in the immune response between different strains and tolerability issues due to increased side effects.

Technical: as, e.g. investment into manufacturing facilities as a limiting factor

Economic: as combinations of serotypes is increasing costs. Also production yields and minimum immunising dose (MID) play a key role for the capacity requirements and cost in commercial manufacturing,

Regulatory: as increasing demands on dossiers to be submitted together with application for new, innovative vaccine.

Development of effective and safe vaccines is still a mid- to long-term objective for many bacterial diseases requiring substantial additional "ground" research.

3.1 Vaccination Targets and Types of Vaccines

Main target of vaccination in general is to create the immunity that can lead to protection against disease/s. Proper vaccination of animals helps to prevent or even eradicate infectious diseases (animal health target: stop the spread of certain diseases and therefore protect animal health) but also ensures public health (human health target: preventing zoonotic, in contact and food-borne diseases) and especially on farm where previously antibiotic was being highly used , switching to proper health programme together with vaccination can significantly decrease use of antimicrobials.

Table 2 Routes of administration of vaccines in cattle, pigs, poultry and horses

Cattle	Pigs	Broilers	Horses
Injection (SC/IM) (needle/needle-free)	Injection (SC/IM) (needle/needle-free)	*In ovo* (hatcheries)	Injection (most common IM, less common SC) (needle/needle-free)
Oral (suspension, e.g. lungworm)	Oral (drinking water)	Aerosols (spray, nebulisation, fogging)	Intranasal (in the nostril)
Intranasal (anti respiratory diseases)	Intranasal	Oral (drinking water, viral, coccidiosis)	
		Intraocular (Eye drop instillation)	
		Nasal drops	
		Injection (SC/IM; wing web) (needle/needle-free)	

Ideally herd or flock vaccination is based on the vaccination plan/schedule that is the part of the animal health plan of the certain herd/flock and considers the animal husbandry system. The vaccination plan/schedule should take into consideration a range of different risk factors related to their age, lifestyle, prevailing disease threats and movement of the animals. Specific marker vaccines (DIVA) enable the differenciation between naturally infected and vaccinated animals, what can be of importance in the cross-border movement (export/import) of the animals. These factors should be discussed with the attending veterinarian with knowledge of the herd to decide on the most appropriate choice of vaccine and vaccination plan/protocol. There are three basic objectives in vaccination to provide immunity (RUMA 2016):

- To the animal or group of animals (active immunity)
- To the offspring of an animal via vaccination of the dam (passive immunity)
- To the animal or group of animals and their offspring (active and passive immunity)

Vaccines used can cause *active* immunisation: vaccine contains either pathogen/s that induce immunity, but not cause the disease, or antigenic components (parts of pathogens). The protective immunity should occur and is based on "immunologic memory", once faced with the same/very closely similar pathogen heightened immunological response will prevent the disease. *Passive* immunisation is based on the antibodies, e.g. also maternally derived antibodies transferred either in perinatal period (across the placenta) or post-natal (via suckling the colostrum). Vaccines, whose effect is based on passive immunisation usually contain antibodies (in the form of immune serum or hyperimmune serum, obtained from animals with very high antibody levels to the infection). Typical example is tetanus antitoxin used in farm animals.

The choice of the type of vaccine developed to target a particular organism is based on the nature of the organism itself, its invasive properties and the immune response the organism generates. Consequently, a range of vaccine types are available including modified live vaccines such as attenuated and recombinant vector vaccines and killed/inactivated vaccines, subunit, conjugate and DNA vaccines. Administration of vaccines can be by a wide variety of mucosal delivery routes: oral, nasal, oro-nasal, conjunctival through water, baits, spray, or using the more classical subcutaneous or intramuscular injection to bypass the difficulties of mucosal immune activation (EMA and EFSA 2017) and is detailed per species in Table 2.

Provoking immunological response aiming to gain immunisation/protection against certain diseases can be reached by (modified according Jorge and Dellagostin 2017):

- **Live—modified**
  Mostly attenuated vaccines (pathogenic agent is modified/weakened not being able to cause infection, but causing immune response).
  Pros: usually requires single dose; usually less likely to produce local reactions
  Cons: depending on exact vaccine—troubles if amount of the dose is too small to be correctly administered; slight signs of animal disease can occur; in sick animals at time of vaccination may prevent or reduce the amount of antibody production as well as cause increased expression of the undercurrent disease; not possible administer concomitantly with antimicrobials; interference with maternal immunity in some vaccines; packages should be used immediately after reconstitution (no preservatives, higher risk of contamination); not mixing vaccinated animals to those not vaccinated.
- **Inactivated vaccines**
  Pathogen is killed/inactivated by physical/chemical factors; often need boosters to maintain immunity (the interval indicated for individual vaccines should be followed); proper adjuvant can help to improve protective immune response; mainly producing IgG response, but some also cell-mediated immunity.
  Pros: vaccine cannot cause disease; no general illness if healthy animals vaccinated; less likely influenced by maternal immunity (some vaccines can); longer shelf life; less prone to contamination, if preservatives used.
  Cons: higher volumes than in live vaccines to be administered (especially older vaccines); the quantity of the immune response is mainly, but not always, dependent on the amount of antigen present in the vaccine; due to adjuvants local reactions could appear (some persisting for long time); twice or more administration—risk of improper compliance; increase stress from animals handling; in killed, whole bacteria containing vaccines, higher risk of induction of autoimmunity troubles via molecular mimicry.
- **Recombinant subunit**
  Pathogen-specific genetic material is processed using genetic engineering technology to produce immunity stimulating proteins, primarily humoral immune response, need of adjuvant.
  Live recombinant vaccines: still able to multiply in cells, usually give an excellent immune response (via simulating natural infection). Disease virulence genes can be altered/removed, thereby safer vaccines can be produced; also genes allowing spread to environment can be removed.
- **Purified protein/subunit vaccines**
  Made up of either monoclonal antibodies or protein molecules (purified antigens) that have been extracted from pathogens or the exotoxins they produce.
- **RNA/DNA based**
  Humoral and cellular immune responses. Challenges in adequate cellular uptake and expression.
  Long-term persistence of immunogen. The risk of integration of nucleic acid or part of it into host genome cannot be completely excluded. Unstable and quite expensive production (for RNA vaccines).
- **Non-vaccine immunomodulators**
  Alter the immune response by augmenting or reducing the ability of the immune system to produce antibodies or sensitised cells that recognise and react with the antigen which initiated their production. The mode of action includes augmentation of the anti-infectious immunity by the cells of the immune system including lymphocyte subsets, macrophages and natural killer cells. Other mechanisms can involve induction or restoration of immune effector functions. Use of immunomodulatory agents seems an attractive approach as an adjunct modality for control of several parasitic diseases. Examples are cytokines, interferons, interleukins and tumour necrosis factors (Ratna and Arora 2018).
- **Combination vaccines**
  An issue of the multistrain/multiserotype as well as combined bacterial/viral infections exists. Those cases are the most

complicated to be prevented by vaccination programmes. For some purposes they fit the combination vaccines, that differ from those "single disease/pathogen" vaccines that provide specific protection for one organism or strain of an organism.

Combination vaccines:

- May present antigens that include different strains of the organism and/or provide protection against a number of microorganisms often of the same type such as in the clostridial vaccines.
- May be designed against different microorganisms which result in the same types of disease (e.g. respiratory disease of calves) and may combine several viral antigens and also in some cases bacterial components.
- May contain completely different antigens (in terms of diseases caused) but they are administered together as they are important causes of infection in the species concerned or they provide required enhanced protection at certain stages in the animal's life such as with sheep (vaccines against clostridial diseases/pasteurellosis).
- Require proper design and formulation to ensure that they are produced so that each component is in sufficient quantity to initiate an immune response in the animal.
- Can pose an advantage of reduced number of vaccination, decreasing risk from potential stress from handling and the administration of the vaccine.
- Can pose another advantage—for the farmer is less demanding to remember as to when to undertake initial vaccination or subsequent boosters.

3.2 Routes of Administration of Vaccines, Main Reasons for Vaccination Failure

For different types of vaccines, different species and different production categories/husbandry systems specific routes of administration are available for the respective veterinary medicinal products. Table 2 gives a brief summarisation of them, showing also that vaccination can be strictly individualised (horses via administration by injection/intranasally) or is available/applicable for groups of animals or even flocks/herds.

In the cases of individualised administration the risks of incorrect dose are minimal as well as no risk, that certain animals will not receive the vaccination. On the other hand, the mass vaccination (e.g. via vaccines administered orally via drinking water) is considered to pose some advantages as improved safety and compliance, and easier manufacturing and administration as well as stimulation of humoral and cellular immune responses at both systemic and mucosal sites to establish broader and long-lasting protection. On the other hand, difficulties especially due to possible damage of the vaccine active parts by gastrointestinal tract conditions and finally delivery of sufficient amount of antigen to provoke adequate immunogenic response cannot be forgotten. Therefore, oral vaccines are required to be designed for successful delivery of the intact and active antigen to the intestine, proper transport across the mucosal barrier and subsequent sufficient activation of antigen-presenting cells. Care should be paid to the proper use and administration of such vaccines to achieve the effect expected.

Essential factors that must be taken into account when using a vaccine consist of proper handling with vaccine, correct administration and use, performed by well-trained staff that follows the approved product texts being aware of the use instructions (route of administration, correct site of administration of injectables) as well as warnings, recording of vaccination is important not only if some troubles occur, but, e.g. having an evidence for the future herd health/vaccination plan(s).

Possible failures (as listed in Table 3) can be influenced by several factors, but some of mistakes can be relatively easily avoided (incorrect handling/use), some need the effort, but could be overcomed (vaccination programme weaknesses, improper quality/suitability of

Table 3 Main possible sources of failures in vaccination

Due to incorrect handling	Due to incorrect use	Weakness of vaccination programme	Vaccine-related issues	Animal related
Incorrect storage (cold chain issues, not exposing sunlight)	Incorrect dose	Unavailability of vaccine (permanent, in the period of need)	Low potency	Poor health/unwell/ stressed animals
Incorrect handling not warmed enough prior to administration (troubles to pass needle, temperature shock in small animals)	Incorrect route of administration	Incorrect timing	Manufacturing quality problems (inter-batches differences, poor batch quality)	Immunological interference (e.g. maternal immunity)
Using improper syringes/ devices (especially in oily suspensions)	Failure in proper administration (e.g. broken needle, due to improper needle size/not sufficiently sharp; SC/IM failures)	Suboptimal schedules (number of doses, intervals among doses)	Serotype not covering the field isolate	Production/age status Young: Immature immune system Mother close to birth: less immune responsive
Improper cleaning/ disinfection/not letting to dry off (especially live vaccines susceptible to the rest of disinfecting agents) – water supply systems – if use automatic syringes	No booster vaccination	Misdiagnosis of disease leading to incorrect vaccination scheme		Immunodeficiency/ immunosuppression Old animals
Partly used bottles—risks of contamination (especially when not aseptically collected from vial/bottle)	Vaccination of unwell animals (mainly risk of adverse vaccine effect)	Interference with other vaccines		Suboptimal immune response
Partly used bottles, damage of the stopper—oxidation damage of the antigen or carrier	Vaccination under concomitant antimicrobial use in live vaccines (Lawsonia vaccines in pigs)			Infection already in incubation period (e.g. viral infections/ mycoplasma infection suppressing immunity response)
Partly used bottles—temperature fluctuation leading to damages, especially in live vaccines longer duration of reconstitution causes death of the vaccination microorganism	Mixing vaccines in one syringe, once not recommended			Nutrition/feed deficiencies—e.g. especially proteins, vitamin A and E, selenium.
Vaccine beyond the expiry date	Injection into same injection site (even in several days) if warned not to do so			Not all animals obtained their dose (e.g. oral vaccines via drinking water)

vaccines), some are animal related and the influencing is hard/need big experience of vet.

3.3 Diseases for Which Vaccines Could Reduce Antimicrobial Use in Animals

Principally, based on consideration of the following factors OIE made the priorisation of the diseases, for which vaccination can reduce need for use of antimicrobials (OIE 2015, 2018a, b):

- Identification of the most prevalent and important bacterial infections
- Identification of common nonbacterial infections (e.g. protozoal, viral) showing clinical signs that trigger empirical antibiotic treatment (e.g. for diarrhoea) and which also result frequently in bacterial co-infection or secondary bacterial infection that need use of antimicrobials
- An assessment of antibiotic use in response to the syndromic indication or diagnosed disease
- The availability of a vaccine(s), and if available, their effectiveness
- The potential for a new or improved vaccine to reduce the need for antibiotic treatment

Proposed lists are for chicken, swine (according diseases/pathogens) and fish (according groups of fish species/pathogens) diseases (OIE 2015), continuing with lists for cattle, sheep and goats (OIE 2018a). Below text also contains other (bacterial diseases) of poultry, pigs and cattle that were not prioritised by OIE, but are also treated or prevented by use of antimicrobials and therefore, vaccination can bring an achievement of better health status as well as minimise use of antimicrobials (at least in the region of Europe). There are also examples of vaccines whose use has led to significant and well-described decline in the use of antimicrobials in different parts of the world and in different types of animal species/types of husbandries. This is the case of the vaccination against furunculosis in fish (salmon) in Norway, followed by the huge decrease use ATM in aquacultures (Midtlyng et al. 2011). In Denmark vaccination against *Lawsonia intracellularis*, causative agent of ileitis in pigs, also helps to decrease the use of certain group of antimicrobials (DANMAP 2014).

3.3.1 Poultry Diseases

Two main bacterial pathogens of *Gallus gallus* (considering broilers, breeders and layers) were identified:

- *Escherichia coli*
  Key diseases to be covered were identified: yolk sac infection (omphalitis), airsacculitis, cellulitis, salpingitis and peritonitis. There were also recognised limitations of the use of vaccines as degree of strain coverage (fully cross-protective), ease of administration (e.g. aerosol), minimal adverse effects. Among other challenges were that vaccine is needed for very early stages (consider possibility to stimulate at maternal level). Within the EU area both live and inactivated vaccines (monovalent or polyvalent) are available, and there is also possibility to detect the *E. coli* strains and profiles and use autogenous vaccines.
- *Clostridium perfringens* (type A)
  Causing necrotic enteritis and high production losts. Therefore, broadly prevented by antimicrobials. The duration of passive immunity induced by toxoid vaccines in layers is short lasting. The need for a vaccine to achieve active immunity, particularly for broilers.

Further examples of bacterial pathogens causing the diseases in *Gallus* including those for which antimicrobials are also frequently used and for which vaccines are available in the territory of Europe are as follows (live and inactivated vaccines (monovalent or polyvalent) for active or passive immunisation are available):

- *Salmonella* spp. (mostly against *enteritidis, typhimurium, gallinarum*)
- *Pasteurella multocida*
- *Ornithobacterium rhinotracheale*
- *Mycoplasma (gallisepticum, synoviae)*

Also coccidial infection predisposes to secondary bacterial infections and improvement in the degree of cross-protection of current vaccines would result in a decrease of secondary bacterial infection. Live vaccines for passive and active immunisation are available in the European territory.

Two key viral infections influencing health and immune status of animals and getting them prone for other bacterial diseases, for which vaccination is available, are Infectious Bursal Disease Virus (IBDV) and Infectious Bronchitis virus.

Those reading this chapter can ask for the information related to vaccination of other bacterial diseases of poultry with increasing importance not only from the perspective of animal health, but having also potential to affect human health (e.g. staphylococcal and enterococcal infections). It should be unfortunately noted that up to this day, despite some attempts, as e.g. study on vaccination of breeder hens with a polyvalent killed vaccine for pathogenic *Enterococcus cecorum*, that shows that does not protect offspring from enterococcal spondylitis (Borst et al. 2019), there is no available commercial vaccine against *Staphylococcus* spp. and *Enterococcus* spp. in poultry.

3.3.2 Swine Diseases

Nine bacterial pathogens and three viral infections (resulting frequently in secondary bacterial infections) were identified by OIE to be considered when setting the priority list. The main issue of currently existing vaccines are the range of pathogen strain coverage and degrees of cross-protection.

Respiratory Tract Infections of Swine Main Causative Agents and the Vaccination Possibilities

- *Pasteurella multocida*
  Associated with pneumonia, an effective toxoid vaccine for atrophic rhinitis exists. Portfolio of live and inactivated vaccines (monovalent or polyvalent) for active or passive immunisation is available within the European territory, autogenous vaccines use also possible.
- *Streptococcus suis*
  Causing infections as meningitis, arthritis, sepsis or infection of soft tissues mainly in post-weaned piglets. In addition to the current most worldwide spread sequence type 2 vaccine will be beneficial to have vaccines protected against other strains (e.g. sequence types ST1, ST 20 described for the EU region and the other STs for other world regions). Due to this variability is hard to find "universal vaccine". Improvement of immunogenicity is needed and maternal antibody interference with the *H. parasuis* vaccine should be solved. Also more studies on maternal antibody interference are necessary in order to determine conclusively whether it is preferable to vaccinate sows or piglets, and when exactly. Actually, autogenous bacterins are discussed as the available option in the field, keeping in mind that improved diagnostic and proper understanding the epidemiology of *S. suis* diseases is essential for using bacterins successfully (Rieckmann et al. 2020). These vaccines are bacterins prepared for a specific farm (need for bacteriological analysis of samples from affected farm). As such, despite the huge variation in *S. suis* infections by region, vaccinated animals are protected from the same strain (s) causing clinical problems within the herd in question.
- *Actinobacillus pleuropneumoniae*
  Causing pleuropneumonia in pigs, live and inactivated vaccines (monovalent or polyvalent) for active or passive immunisation are available. Types of vaccines currently available are bacterins based (washed and killed whole bacteria with serotype-specific protection), mainly autogenous vaccines, with diagnosed and isolated serotype/s from exact farm and purified toxoid-based vaccines (sometimes enriched with surface proteins): mostly Apx I, Apx II and Apx III toxoids are present. There is a strong interference with maternal antibodies: usually, first dose should not be applied before 7–8 weeks of age. Also not sufficient single dose-booster dose needed. Interestingly, also high level of antibodies after infection or vaccination do not eliminate

APP from tonsils of carriers. It is beneficial to use, e.g. new serotyping PCRs (mPCR1 and mPCR2) as tools to identify virulent serotypes for choosing the proper vaccination (by commercial or autogenous vaccine)—one of the most effective (but not routinely available yet) is the whole genome sequencing, that allows also to recognise new serotypes (Bossé et al. 2018).

- *Mycoplasma hyopneumoniae*
  Pathogen with an important role in the porcine respiratory disease complex. The vaccine does not eradicate the pathogen. Live and inactivated vaccines (monovalent or polyvalent) for active or passive immunisation are available in the European territory. Vaccination schemes frequently used in the European region are traditional two-shot formulations, one-shot formulations, and bivalent one-shot formulations containing both *M. hyopneumoniae* and porcine circovirus type 2 (PCV2) antigens. In general, vaccination reduces the occurrence of clinical signs and lung lesions and improves performance, but on the other hand does not prevent colonisation of the respiratory tract epithelia by mycoplasma organisms (Cvjetković et al. 2018).

Further examples of bacterial respiratory pathogens causing the diseases of pigs including those for which antimicrobials are also frequently used and for which vaccines are available in the territory of Europe are as follows (live and inactivated vaccines (monovalent or polyvalent) for active or passive immunisation are available):

- *Haemophilus parasuis*
- *Pasteurella multocida*
- *Bordetella bronchiseptica*

Two key viral infections influencing respiratory tract polyfactorial infections spread, including secondary bacterial infections, for which vaccination is available and could help minimising of use of antimicrobials are:

- Porcine Reproductive and Respiratory Syndrome (PRRS) virus
- Swine Influenza virus (SIV)

Enteric Tract Infections in Swine Main Causative Agents and the Vaccination Possibilities

- *E. coli*
  One of the most important pathogen causing enteritis as well as oedema disease, for which maternal vaccines which provide passive immunity to neonates exist. Despite this for *E. coli* vaccines in weaners/finishers, complications are maternal antibody interference and the relatively short window for induction of immunity. Some new vaccines exist, e.g. live non-pathogenic *Escherichia coli* O141:K94 (F18ac) and O8:K87 (F4ac), as well as combined vaccine for F4 and F18, but further research is needed. Live and inactivated vaccines (monovalent or polyvalent) for active or passive immunisation are available as well as use of autogenous vaccines.
- *Lawsonia intracellularis*
  Porcine proliferative ileitis is a major economic burden for the swine industry, affecting growing pigs and young adult pigs. A modified live-attenuated vaccine has been commercially available since 2001, but due to the live nature of the oral vaccine, concurrent use with antibiotics effective against *L. intracellularis* was not possible. Recently (2018) inactivated injectable vaccine has been introduced, that is intended to be administered to 3-week-old pigs under typical field conditions that can pose protection against ileitis, help reduce bacterial shedding 15-fold and help maintain gut barrier function integrity (Roerink et al. 2018).
- *Brachyspira hyodysenteriae* and *Brachyspira pilosicoli*
  Are considered to be re-emerging issue, not solved by change of husbandry practices, but in most cases sold by repopulation and restructuralisation of the husbandries. As for vaccine, the issue is not easy culturing and work with strains to develop vaccine. No

commercial vaccines are available for prevention of *B. pilosicoli* infections. In pigs studies with autogenous bacterin induced systemic antibody titres, but the vaccinated animals still became colonised and developed diarrhoea (Hampson 2018). Recently, recombinant Bmp72 C-terminus has been shown to give potential to be developed for use as a vaccine component to provide protection against *B. pilosicoli* infections (La et al. 2019b). Currently (2019), an atypical weakly haemolytic strain of *Brachyspira hyodysenteriae* has been described to occur in Europe and Australia, and due to its avirulence it can be used to protect pigs from developing swine dysentery (La et al. 2019a).

Further examples of bacterial enteric pathogens causing the diseases of pigs including those for which antimicrobials are also frequently used and for which vaccines are available in the territory of Europe which are as follows (live and inactivated vaccines (monovalent or polyvalent) for active or passive immunisation are available):

- *Clostridium perfringens*
- *Salmonella* spp.

As for rotavirus infections, influencing the whole pig performance and can be complicated by bacterial enteric pathogens entry, authorised vaccines are available protecting against these infections.

Further vaccination to be considered in pigs are:

- *Erysipelothrix rhusiopathiae*
- PCV2 (viral)

Intestinal infections are a major problem and account for a large proportion of total antimicrobial consumption in Danish pigs. Viral infections—such as swine influenza, PRRS and porcine circovirus type 2 (PCV2)—also increase antimicrobial consumption as they are associated with secondary bacterial infections. Given that many of these diseases can be handled with vaccinations and good management, further reduction in antimicrobial use is possible without compromising animal welfare. The sales of vaccines for pigs increased, according to the figures provide by the Danish Veterinary and Food Administration from 28 million doses in 2009 to 55 million doses in 2017 (FAO 2019).

3.3.3 Cattle Diseases

In cattle, considering young animals (new born calves in which antimicrobials are used in milk replacers) as well as dairy and feedlot cattle use of antimicrobials is of concern. Considering different systems (respiratory, gastrointestinal, urogenitary) as well as specific disease complexes (mastitis, lameness) there were identified diseases caused either primarily by bacterial pathogens (or viral agents with concomitant or secondary bacterial pathogens contributions) that can be prevented by proper vaccination. Both OIE priority lists (OIE 2018a, b) as well as vaccines and other immunological/biological veterinary medicinal products available predominantly in the European region were taken into account where following text was completed.

***Respiratory* Tract Infections of Cattle Main Causative Agents and the Vaccination Possibilities**

The bovine respiratory disease complex (BRD) is a multifactorial disease attracting high levels of antimicrobial use in cattle, especially in feedlots. For vaccine development, a syndromic, multipathogen, approach would be preferable to address all animal health risks (OIE 2018a, b).

The major organisms involved are:

- *Mannheimia haemolytica* (MH)
  Regarded as a primary pathogen and features a lack of cross-protection among different strains. Most vaccines are targeted MH serotype 1, also in combination with HS, or with viral inactivated parts (bovine respiratory syncytial virus), parainfluenza 3 virus and/or bovine viral diarrhoea virus.

- *Pasteurella multocida* (PM)
  Primary and a secondary pathogen causing respiratory diseases. It was recognised that the existing vaccines notably have marginal efficacy and there is a potential lack of cross-protection among PM field isolates (where autogenous vaccines can be of choice).
- *Histophilus somni* (HS)
  Opportunistic pathogen, that is less common, in the EU region commercial vaccines available (including e.g. combined vaccine against HS and MH used in claves from 2 months of age, two doses).
- *Mycoplasma bovis*
  The role in BRD is considered to be lower than for other pathogens, and that although it was found with increasingly higher occurrence, its role as a causal agent in BRD was uncertain.

Combined vaccines for bacterial and viral diseases as well as individual agents covering vaccines are available at least in the European region:

- Bovine viral diarrhoea virus (BVDV): Considered by the group to be the viral pathogen that elicits the most significant use of antimicrobial agents in BRD.
- Parainfluenza virus 3 (PI3), BHV-1 (IBR): Both these viruses were recognised as being lesser contributors to antimicrobial use, and existing vaccines are effective and safe. For IBR, DIVA vaccines have been shown to be useful for eradicating the disease in several countries of Europe.
- Bovine respiratory syncytial virus (BRSV): Adequate vaccines are available.
- Bovine coronavirus: Recognised as an emerging respiratory pathogen. While a vaccine is available, its efficacy is uncertain.

Apart from BRD, the group considered another respiratory disease as within the scope, Contagious Bovine Pleuropneumonia (CBPP, *Mycoplasma mycoides* subsp. *mycoides*). CBPP is one of the most relevant diseases in Africa, where it entails high use of antimicrobial agents, which could lead to establishment of a carrier state.

Enteric **Tract Infections in Cattle Main Causative Agents and the Vaccination Possibilities**

Enteric diseases are an important cause of antimicrobial use, especially in feedlot systems:

- *Fusobacterium necrophorum*
  Entails high use of antimicrobials, especially in feedlots, arising from acidosis. No vaccines are labelled for enteric disease/acidosis/liver abscesses; and off-label use of *F. necrophorum* vaccines designed for other diseases provides limited efficacy.
- Enterotoxigenic *E. coli*
  Provokes a high use of antimicrobials, especially in dairy farms. Effective vaccines do exist, in the European region several vaccines are available as, e.g.:
  - Combined vaccines declaring reduction of severity of diarrhoea caused by *E. coli* F5 (K99), of incidence of scours caused by rotavirus and shedding of virus by calves infected with rotavirus or coronavirus
  - Combined vaccine stimulating serological and colostral antibodies against rotavirus and coronavirus antigens and against *E. coli* K99, Y, 31A and F41 antigens passed to the calf to reduce neonatal diarrhoea infection caused by agents containing these antigens (target categories pregnant cows and heifers, two doses, second at least 2 weeks prior calving)
- *Salmonella enterica*
  Is a notable zoonotic disease. Strains many times have different and sometimes broad portfolio of genes conferring antimicrobial resistance. The disease's greatest effects on animals are in dairy calves soon after birth, which are exposed to the challenge before the onset of immunity that might be derived from vaccination. *Salmonella* spp. vaccines are available to address the prevalent subspecies/serotypes in the various regions (e.g. *S. enterica* serotype Dublin, *S. enterica*

serotype Newport, *S. enterica* serotype Typhimurium). These vaccines are generally used in herd programmes to control the level of *Salmonella* spp. bioburden within the vaccinated herd, leading to lower levels of *Salmonella* spp. exposure to the new animals entering in the herd. This then results in a lower level of the disease.

- *Mycobacterium avium* subsp. *paratuberculosis*
  Causing Johne's disease, often undiagnosed or misdiagnosed, and maybe mistaken for other forms of bacterial enteritis. Vaccine availability is geographically limited, and existing products present several drawbacks.
- Bovine rotavirus and bovine coronavirus are also causal agents of neonatal diarrhoea in calves, which may be treated with antimicrobials because the cause of symptoms is frequently undifferentiated. Rotavirus infections, being more prevalent than coronavirus, are likely to attract higher use of antimicrobials. In both cases, effective vaccines exist, several of them as combined vaccines (together with selected types of *E. coli*)
- *Cryptosporidium parvum* and *Eimeria* spp. no vaccines available for cattle.
- As there was indicated overuse of anthelmintic also provoking anthelmintic resistance, therefore need for vaccines against Helminths research and development is high.

Non-vaccine immunological veterinary medicinal products like bovine concentrated lactoserum also should not be forgotten, that is intended to be used for neonatal calves less than 12 hours of age and which contains high levels of IgG against *E. coli* K99 (indicated to be used for the reduction of mortality caused by enterotoxicosis associated with *E. coli* F5 (K99) in the first days of new born calves).

Mastitis Main Causative Agents and the Vaccination Possibilities

- *Streptococcus agalactiae, Streptococcus uberis,* Coagulase-negative Staphylococci, *Staphylococcus aureus, Escherichia coli* and *Mycoplasma bovis*.
- Antimicrobial use for mastitis is considered to be higher in modern, intensive dairy production located in stables rather than in grass-based production. The occurrence of multiple strains, the lack of cross-protection of available vaccines, and the difficulty of building a specific immune response at the site of infection were identified as current difficulties. Other (less frequent and less demanding antimicrobial use) mastitis causative pathogens are not discussed. Dry cow therapies provide control against a number of different contagious and environmental pathogens. From a herd perspective, development of a vaccine against individual pathogens will not eliminate the need for control of the other pathogens often found in infected cows. Development of combination vaccines that address the common mastitis pathogens would offset this issue, but represents a difficult technical challenge that would require a significant investment in research and development (OIE 2018a, b).
- Combined vaccine declaring reduction of the incidence of subclinical mastitis and the incidence and the severity of the clinical signs of clinical mastitis caused by *Staphylococcus aureus*, coliforms and coagulase-negative staphylococci exist at least in EU region, three injections needed.
- Recently (2018) was also granted the marketing authorisation to another one veterinary medicinal product containing the active substance called biofilm adhesion component including lipoteichoic acid, which is derived from the sticky film produced by *Streptococcus uberis* strain 5616. The product is indicated for an active immunisation of healthy cows and heifers to reduce the incidence of clinical intramammary infections caused by *Streptococcus uberis*, to reduce the somatic cell count in *Streptococcus uberis* positive quarter milk samples.
- Biological, non-vaccine veterinary medicinal product authorised in 2015 in European Union is also further taken into account as solution for injection intended to be used in

dairy cattle and heifers. As for composition, ovine granulocyte colony stimulating factor (bG-CSF) is a modified form of the naturally occurring immunoregulatory cytokine, which is a naturally occurring protein produced by mononuclear leukocytes, endothelial cells and fibroblasts. The immunoregulatory activities of granulocyte colony stimulating factor concerns notably cells of the neutrophilic granulocyte lineage which bear cell surface receptors for the protein. The use of bG-CSF increases the number of circulating neutrophils and enhances myeloperoxidase hydrogen peroxide halide mediated microbiocidal capabilities of neutrophils. Direct or indirect influence on other cells/receptors and cytokine pathways are also predicted.

***Cattle Lameness* Main Causative Agents and the Vaccination Possibilities**

Lameness is a priority issue for the dairy sector, together with mastitis. Interdigital and digital dermatitis as the dominant lameness syndromes attracting antimicrobial use were identified by OIE (2018a, b):

- *Fusobacterium necrophorum*
  Considered as a major pathogen of importance, but in the EU region also *Bacteroides melaninogenicus, Dichelobacter nodosus, Porphyromonas levii, Prevotella melaninogenica, Treponema* spp. *and Trueperella pyogenes* are of concern (Kontturi et al. 2019). For this kind of disease is frequently used also one of the critically important cephalosporin of third generation (strong selector of ESBLs)—ceftiofur. Veterinary medicinal products containing ceftiofur are many times preferred to older molecules due to zero milk withdrawal period in dairy cattle, what promote further use of ceftiofur by vets and farmers.
  Vaccines are not available.

***Systemic Infections of Cattle* Main Causative Agents and the Vaccination Possibilities**

- *Pasteurella multocida*
  as a causative agent of haemorrhagic septicaemia, provokes high use of antimicrobials, even though the existing vaccines appear effective.
- *Leptospira* spp.
  Regional differences in serovars act to limited vaccine availability and use, but combined vaccines (e.g. against six serotypes together currently authorised).
- *Bacillus anthracis*
  Effective vaccines are available.
- *Clostridium* spp.
  Mostly autovaccines: *Clostridium perfringens* (A, B, D), but also for other Clostridia (*C. difficile, C. novyi, C. sordellii* etc.).

***Genitourinary* Tract Infections in Cattle Main Causative Agents and the Vaccination Possibilities**

Metritis/endometritis syndrome associated with *Trueperella pyogenes, E. coli* and *Fusobacterium necrophorum* was considered by OIE experts. No vaccines authorised to cover these metritis pathogens.

3.4 Veterinary Autogenous Vaccines

Another possibility of reducing the need for use of antimicrobials also in the cases where no commercial vaccine is available, or where commercial vaccine covering certain bacterial/viral specific serotypes or combination of both is not accessible, veterinary autogenous vaccines can be the functioning option. One example can be from pig farming, where autogenous vaccination of the sows with a vaccine based on exhB-positive *Staphylococcus hyicus* isolates reduced metaphylactic treatment with antimicrobials as well as the morbidity and mortality rates in weaned pigs compared with pigs from non-vaccinated sow batches (Arsenakis et al. 2018).

Autogenous vaccines can provide an individual solution where licensed commercial vaccines are not available or lack effectiveness due to the antigenic diversity of the causal, bacterial and

Table 4 Examples of the bacteria used in VAV (Hoelzer et al. 2018b, amended)

	Species/genus of bacteria		Virus
Pigs			
Respiratory	*Actinobacillus pleuropneumoniae* *Histophilus somni* *Mannheimia haemolytica* *Mycoplasma hyorhinis* *Pasteurella multocida* *Staphylococcus pk-negative* *Staphylococcus hyicus* *Staphylococcus aureus* *Streptococcus dysgalactiae* *Streptococcus equisimilis* *Streptococcus hyosynoviae* *Streptococcus suis* *Trueperella pyogenes*	Gatrointestinal Other	*Rotavirus* *Porcine rotavirus A* *Corona virus* *Clostridium difficile* *Clostridium novyi* *Clostridium perfringens* *Clostridium perfringens A* *Clostridium perfringens D* *Escherichia coli* *Haemophilus parasuis* *Streptococcus suis*
Cattle			
Respiratory	*Histophilus somni* *Clostridium perfringens* *Moraxella catarrhalis* *Mannheimia haemolytica* *Pasteurella multocida* *Trueperella pyogenes*	Gastrointestinal	*Clostridium perfringens* *Clostridium perfringens A* *Clostridium perfringens D* *Clostridium perfringens B* *Escherichia coli*
Mastitis	*Klebsiella pneumoniae* *Staphylococcus aureus* *Streptococcus uberis*	Keratoconjunctivitis	*Moraxella bovoculi*
Poultry			
Bacterial	*Avibacterium* *Bordetella avium* *Bordetella bronchiseptica* *Clostridium perfringens* *Enterococcus cecorum* *Enterococcus faecalis* *Enterococcus spp.* *Escherichia coli* *Gallibacterium anatis* *Mycoplasma* spp. *Ornithobacterium rhinotracheale* *Pasteurella multocida* *Riemerella anatipestifer* *Staphylococcus aureus*	Viral	*Adenovirus* *Infectious bronchitis virus* *Reovirus* *Rotavirus*

viral, agent. In certain situations of clinical practice, therefore, they represent a unique option for use in addition to commercial vaccines. Examples of autogenous vaccines used in the European region are listed in the Table 4. As veterinary autogenous vaccines are considered inactivated vaccines which are manufactured using a disease agent isolated in a particular epidemiological unit (herd/flock) and are intended and permitted for use only in that unit. This means that a farm may include more than one herd/flock, provided the flows of animals are clear and clearly documented. The new Regulation on Veterinary medicinal products speaking about the veterinary autogenous vaccine indirectly by characterising them under the Scope (Article 2, paragraph 3) as inactivated immunological veterinary medicinal products which are manufactured from pathogens and antigens obtained from an animal or animals in an epidemiological unit and used for the treatment of that animal or those animals in the same epidemiological unit or for the treatment of an animal or animals in a unit having a confirmed epidemiological link. Animal keepers and owners of food-producing animals must keep record of the use of autogenous vaccines.

Active substance (antigen) of the product are inactivated immunogens of the relevant isolates, which activate protective mechanisms (macrophages, opsonins, B and T lymphocytes) and this leads to establishment of the immunity against individual causative agents. The individual components of the vaccine are in the organism gradually degraded and processed by the immune system. This leads to creation of specific antibodies against individual components.

Generally speaking, veterinary autogenous vaccines can be produced for any species of animal. The disease agents isolated from the herd/flock are cultured, identified, thoroughly typed (e.g. serotypes and virulence factors) and selected depending on their virulence factor content and/or immunogenicity traits. Selected pure cultures of the isolates are propagated. Then following several processing steps finally living compartments (bacterial cells/viral particles) are thoroughly inactivated, antigens gained by specific processes are purified and concentrated. The active ingredient of the each specific veterinary autogenous vaccine can represent several compartments as, e.g. toxoids APX I, II and III, outer membrane proteins and lipopolysaccharides of inactivated bacterium of different serotypes (e.g. *Actinobacillus pleuropneumoniae* of serotypes 2 and 9). There is, like in commercial vaccines, also important role of selection of adjuvants, to stimulate immune answer.

The period of use of single veterinary autogenous vaccine is limited, but repeated production of VAV on the basis of disease agents isolated from the herd/flock epidemiologically linked to farm of origin of the "first" isolate is possible. Several companies on the EU market is able to perform all steps of production of the veterinary autogenous vaccines under GMP-compliant conditions.

Because autogenous vaccines take some considerable time to produce they are only useful in case of chronic or recurrent disease/s. In practical conditions several weeks (4–6 weeks for bacterial and in the case of combined viral/bacterial vaccines or e.g. combination of serotypes 6–8 weeks) are usually necessary to produce the vaccine.

Despite the advantages (as e.g. improved efficacy targeted on exact serotypes of pathogens), there are also disadvantages as such as the fact that each vaccine carries the risk of unwanted or adverse effects as autogenous vaccines are not subject of full battery of tests as commercially produced vaccines that are more broadly tested.

3.5 Conclusion to Vaccines

Vaccines have contributed significantly to the control of infectious diseases in production animals, and both commercial as well as autogenous vaccines tailored for epidemiological units can also help to reduce the need for especially prophylactic use of antimicrobials. Despite the use of vaccination, certain infectious diseases still wait for solution. In individual herds other alternative tools may have a more significant effect. Also it should be noted that vaccines against viral diseases can impact through prevention of immune suppression and secondary bacterial infections prevention the overuse of antimicrobials. It should be also highlighted that despite significant technical progress, there is a big challenge for development of the new generation vaccines/or more generally speaking immunologicals, with improved design, ease of administration as well as efficacy. For the cases where commercial vaccines are not an option autogenous vaccines can be a potent tool, but further harmonisation of their manufacturing and conditions in the EU region will be beneficial.

As indicated already in RONAFA report (EMA and EFSA 2017) and later on SAPHIR (2019) revisited, there are main areas where antimicrobials are used and can be solved by vaccination:

- For pigs: post-weaning diarrhoea causative agents, vaccines against Porcine Reproductive and Respiratory Syndrome Virus, *Mycoplasma hyopneumoniae* and *Streptococcus suis* are needed.

- For poultry: vaccines against various types of *E. coli* as well as against coccidia (*Eimeria* spp.) and *Clostridium perfringens*.
- For cattle: mastitis and viral diseases in veal production, but new (or re-emerging) pathogens such as bovine Respiratory Syncytial Virus and *Mycoplasma bovis* are of concern.

THINK ABOUT Mycotoxins and effectivity of vaccination and influence on susceptibility to pathogens causing infectious diseases

There is no question that mycotoxins have an effect on the immune system, but exact mode of such (inter)action continues to be discussed. Numerous studies supporting the fact that mycotoxins have an effect on the efficiency of the immune response—aflatoxins, ochratoxin A, fumonisins, zearalenone, trichothecenes, deoxynivalenol and T-2 toxin belong among mycotoxins that all target the immune system, but have also further toxic effects (Pierron et al. 2016). It is well recognised that in poor quality feed containing moulds or directly mycotoxins produced by them harm the health and immunity of the animals. Aflatoxins interact with the cytokines, a part of the immune system. They also influence the important inflammatory response. In pigs vaccinated with model antigen, an impaired lymphocyte activation in pigs exposed to Aflatoxin B1 was suggested as one of the outcomes (Meissonnier et al. 2008). Ochratoxin A also effects the efficacy of vaccinations. In weaned pigs ingested diet contaminated by ochratoxin A was proved decreased capacity to respond with cytokine expression (mRNA and protein) to ex vivo challenge with lipopolysaccharides (LPS) (Bernardini et al. 2014). Therefore, it is very important to consider quality of feed and possible mycotoxin contamination of it when developing herd management programmes, to ensure maximum efficacy of vaccinations. Influence of *Fusarium* mycotoxins on susceptibility to infectious diseases in pigs via affecting the intestinal health and the innate as well as adaptive immune system has also been described (Antonissen et al. 2014).

4 Other Tools Available to Keep Animals Healthy

4.1 Disease Management Strategies

The **eradication** of a clinically relevant or production-limiting disease can, in well-defined cases/criteria can also help to reduce the use of antimicrobials and bring economic benefits (Sasaki et al. 2015). Those criteria consist of:

- Life cycle and transmission dynamics precise information
- Accessible and robust diagnostic tools (satisfactory sensitivity and specificity)
- Availability of the effective, efficient and practical interventions (e.g. a vaccine) to interrupt the transmission pathways of the agent

Most effective are interventions, where there is only one maintenance host and when the disease cannot/have very limited amplification in the environmental conditions.

Except biological, scientific, technical conditions, also societal and political commitment to perform measures are of critical importance for disease eradication (EMA and EFSA 2017). The use of vaccination programmes or antimicrobial treatment before the introduction of an eradication campaign can help reduce the susceptible population (Dieste-Perez et al. 2016) and so-called marker vaccines, together with a companion Differentiating Infected from Vaccinated Animals (DIVA) test can be particularly supportive.

Certain fundamental principles need to be fulfilled in the preparatory stages and prior to the design for an eradication programme:

- Thorough evaluation of the epidemiological situation in the region/country or area of focus (including identifying and controlling other relevant domestic and wild reservoirs, to minimise the risk of reintroduction of infection).
- Correct choice of the relevant epidemiological unit, e.g. a largely closed integrated poultry company or closed primary pig breeding herds, and sound epidemiological parameters.
- Effective control (decrease in the prevalence of infection) of the disease as a preliminary step towards eradication.
- The authorities and other players involved in the programme need to be clear on their respective responsibilities and an appropriate legislative framework for the programme must be established together with an administrative team with sufficient and proportionate financial resources for the running of the programme.
- Training at the appropriate level is mandatory for all relevant parties (awareness campaign targeted at farmer participants)—all involved stakeholders should clearly understand the tasks and duties involved.

Eradication can be successfully achieved in poultry production systems, as "all-in-all-out" production facilitates a clean break between flocks. For diseases where the risk of transmission between herds is high, control/eradication should preferably be done on an area/region/country-level.

Zoning and compartmentalisation (including SPF) are disease management strategies that pursue essentially the same objective: to establish animal populations with distinct health status, based on effective separation of populations of different status and application of biosecurity measures to prevent the introduction of infection. Zoning relies more heavily on geographic factors, such as natural or human-made barriers, while compartmentalisation focuses more on management and biosecurity within the establishments comprising the compartment, to ensure the maintenance of health status.

The SPF system can be cost-effective: less medication is needed and vaccination costs are reduced. It does depend on strict biosecurity and a closed herd policy or strict sourcing and transportation from controlled herds with similar health status. The impact of the SPF system will be greatest in farms towards the top of the breeding pyramid (for example, grandparent and parent broiler stock) (EMA and EFSA 2017).

As an example can be mentioned Danish system of the special SPF status in pig sector, that ensures the SPF herds are declared for a number of diseases, including mycoplasma, pleuropneumonia, swine dysentery, mange, lice and atrophic rhinitis. SPF herds can only be established by total depopulation, when the previous herd has been slaughtered and the whole unit is cleaned, disinfected and left empty for a specific period, until the introduction of SPF animals. The security of the system is based on a high level of biosecurity and close veterinary supervision. SPF herds may carry one of the above diseases, but, in this event, the herd may remain an SPF herd but with a qualification, e.g. SPF + ms (SPF with mycoplasma).

Currently, around 75% of all sows and 38% of finishers in Denmark have SPF status and 75% of Danish pigs are born in the SPF system (DAFC 2018). Many other herds operate to similar rules and standards, although they do not have the formal SPF accreditation (DAFC 2017).

SPF Health rules for Danish pig farms requirements:

- Protection against infection
  - Receipt and delivery of pigs, distance to neighbouring herds, visitors, deliveries of feed and litter
- Health inspection
  - Daily inspection by personnel
  - Monthly inspection by external veterinaries, reporting of undesirable symptoms
  - Monthly/annual testing of blood samples:
    Analysis of:

(i) Pleuropneumonia (*Actinobacillus pleuropneumoniae*)
(ii) Enzootic pneumonia (*Mycoplasma hyopneumoniae*)
(iii) Dysentery (*Brachyspira hyodysenteriae*)
(iv) Salmonella
(v) Rhinitis
(vi) PRRS

- Purchase of pigs
  - With known health status
- Transport of pigs to herd
  - In approved SPF vehicles which are owned by a carrier approved by SPF-SuS

4.2 Smart and Precise Farming

Modern, computerised and artificial intelligence technologies also are currently coming to the farming. They can be considered as another, recently introduced disease management strategy. Different sensors such as remote sensors: cameras, microphones, thermometers and accelerometers and/or sensors touched to the body of animals like podometers, collars, thermometer ear sensors, vaginal sensors, tail (chip) implants/sensors, detection bolus in rumen, heart function sensors, monitor or capture information such as images, sound, heat or motion from groups or individual animals. The data from the sensors can be either stored (internally/externally, e.g. on clouds) or sent directly to a specific node for further processing. Processing is usually performed as software algorithm used to solve an individual task/issue or cluster of tasks/issues. Then transformation to the "outcome message" that can be translated should be done (e.g. in the case of podometer to evaluate if the cow has troubles with lameness). Computers can evaluate large scale of situations/examples either automatically or semi-automatically and learn from sets of examples comparing them to the measured/existing data. Computerised systems allow to process and analyse large data sets to track variables and produce estimates at a rate that would not be possible for humans or conventional (individual) statistical methods. Integrated, processed and analysed data can provide credible information and alerts regarding animal health, welfare and productivity (Table 5).

4.3 Alternative Tools

Despite the main purpose of this book targeted mainly on the use of antimicrobials in animals, in the era of growing issue of antimicrobial resistance, when conventional treatment strategies using antimicrobials have become ineffective due to the occurrence of drug-resistant bacteria, focus must be shifted towards alternative tools or therapies for prevention and treatment of infection diseases. Also consumers' pressure and worries towards harmful effects of antibiotic use on produced food and the ban of antibiotics in the EU have prompted researchers to think about alternatives to antibiotics (Diara and Malouin 2014; EMA and EFSA 2017, ASOA 2017).

Although some alternatives have been already investigated also for use in food-producing animals, there are challenges the current research, regulatory bodies as well as veterinarians and farmers in practice are faced with and that make hard to implement them successfully in real practice.

Another aspect of alternative tools is lack of the proof of evidence that they are working effectively either as treatment options or prevention tools without any significant harmful effect that can lead to conclusion that risks/uncertainty overweigh the benefits from the alternative use. The majority of the studies that were identified in search by authors of RONAFA report (EMA and EFSA 2017) failed to meet one or more of the inclusion criteria. Most of the articles focussed on the effect of the potential alternative to antimicrobials on performance and did not include health parameters as end points of the studies. In a few studies, a comparison with an antimicrobial treatment was reported, mainly in the context of antimicrobials used as growth promoters and not within the context of the treatment of acute disease outbreaks.

Table 5 Examples of smart/precise farming technologies for pig, dairy cattle and poultry

Animal species	Parameter	Brief description/detecting tools
Pigs	Lameness and reduced mobility	In groups, sows with non-resolved lameness were observed to move and stand less, lie down more, and were in contact with the wall more than healthy control sows These differences in behaviour could be interpreted as signs of pain or as a way of seeking shelter and isolation from the group. *Pressure force plate systems measurement of* pressure distribution of claws, weight distribution on all four legs of sows and leg loading and weight shifting *Pressure mat measurement of* maximum pressure, stride length, stance time, stride time, and activated sensor count per foot in both sows and weaned pigs *Imaging* Motion tracking between frames from video consecutive images, a lateral motion path is calculated and compared to the actual forward movement of each sow *Accelerometers* Devices attached to the leg of sows to detect posture and stepping behaviour, standing duration, latency to lie down after feeding, and step frequency when feeding and from pre-parturient nesting activity, detected the onset of farrowing. Ear detectors can indicate from high activity (distance walked) and rest phases (lying down) signs of lameness.
	Pen-level activity monitoring	For welfare improvement and automate pig health monitoring. *Imaging (including 3D imaging)* Video images to measure pen-level activity such as antagonistic behaviours, chasing, tail and flank biting, fighting, head-to-head knocks between pigs Depth imaging tracking to monitor pig location, eating, drinking and aggression interactions between pigs. 3D cameras and machine learning to detect pig activity and provide an automatic warning of tail biting "outbreaks". *Sensor data differentiation* of lying patterns of pigs (thermal comfort behaviour), standing pigs from moving pigs, and lateral from sternal recumbency.
	Temperature/illness detection	*Infrared thermography* is used at skin measurement sites for pigs, with the highest correlation to body temperature, are the ear base, eyes and udder, can also detect individual illness in groups of piglets. Vaccination is known to initiate high skin temperatures and huddling responses were observed up to 20 hours post vaccination of piglets.
	Sound signals detection	*Microphones/vocal sensors/analysers* To detect heat stress and high frequency "screams" of pain as a consequence of tail biting or fighting. Distinguishing between infectious productive coughs and non-infectious, non-productive coughs (ammonia or dust) from differences in the acoustic variables, enabling treatment for respiratory disease and ventilation changes at a pen level. In noisy barns or with insufficient microphones, detection level is limited.
	Live weight, shape, growth and body composition measurement	Crucial factors in the management of swine production because individual pig weight and growth affects the herd in factors such as barn flow and space allowance, and audit parameters. *Imaging/extracting the 3D shape* of pigs for automatic mass and weight estimations.

(continued)

Table 5 (continued)

Animal species	Parameter	Brief description/detecting tools
Dairy cattle	Measurement of heat in certain body parts	*Infrared Thermography* Non-invasive detecting of dissipation of heat in individual animals or specific regions of the body for the purpose of rapidly detecting diseases such as mastitis, locomotion disorders, and respiratory disease in bovine.
	Body condition scoring	Several parameters can describe health/body condition *3D Imagination* Lower back image (loin, rump, hook, tailhead, fatness)
	Oestrus, feeding and health signs	Activity, behaviour and health signs detection *Accelerometer (3-axial accelerometer)* Different technologies using ears, leg (podometers/accelerometers) or neck detectors (collar located) measuring activity
	Lameness	*Time and pressure sensors* Stepmetric, podometers Platform/mat for gait consistency
	Mating alert	*Heat detector* (Special belt) that detects standing or cow mounts in heat, the device send an SMS to the inseminator and saves records on the web
	Calving process	*Vaginal device with thermistor* Measurement of the vaginal temperature *Tail sensors* (e.g. ring with accelerometer)
	Rumen functions	*Rumen bolus with pH electrode and thermistor* Measurement of pH, temperature, drinking
	Integrated (rumen activity, oestrus, drinking, temperature)	*3-axial accelerometer, thermistor*
Poultry	Welfare evaluation laying hens/chickens	*Image analysis* *Thermometer and relative humidity sensors, carbon dioxide and ammonium detectors, luxmeters* Monitoring temperature, humidity, air speed, CO_2, ammonium and light measurement, to ensure balanced conditions for birds (importance of sensors location at levels of living birds)
	Thermal comfort/heat loss in chickens	*Noise/vocalisation analysis* Correlation between bird grouping pattern and vocalisation during thermal stress exposure *Thermal imaging in chickens/hens* Infrared thermography evaluation of groups of birds
	Chick performance	*Imaging technologies* Captured images were analysed using raster image analysis software to determine the body surface area and a linear equation to estimate weights

Modified according Benjamin and Yik (2019); Lokhorst (2018); Berckmans (2014)

Despite the difficulties identified, from the advancements within biotechnology, genetic engineering and synthetic chemistry, but also research of the substances occurred in natural products, seems that new perspectives are coming to be opened up towards the discovery of the alternatives to antimicrobials and preventive alternative tools. While some promising alternatives appear, thanks to the enormous effort especially in human medicine research, on veterinary side some of the new alternatives are kept out of practice also from the reason of high costs comparing to conventional therapies.

Table 6 Examples to the various existing/promising alternatives or alternative tools that can help prevention or treatment of infectious diseases

Alternative	Comments
Vaccines (examples)	Inactivated, attenuated, recombinant, autogenous—see subchapters above for more details.
Vaccine adjuvants	Adjuvants are crucial components of vaccines as they reduce the amount and number of doses required to elicit effective immunity. Three-component adjuvant contained a Toll-like receptor agonist, either poly:cytosine (poly-c) or CpG oligodeoxynucleotide, a host defense peptide and polyphosphazene.
Probiotics	*Lactobacillus* spp., *Bacillus* spp., *Enterococcus* spp., *Bifidobacterium* spp., *Lactococcus* spp., *Pediococcus* spp., *Streptococcus* spp., *Propionibacterium* spp., *Saccharomyces* spp. are the most commonly used, but some strains under the scrutiny because AMR transfer.
Prebiotics	Galactooligosaccharides
Phytogenic feed additives	Directly inhibit bacterial growth by inhibiting cell membrane functions As an immunostimulants plant polysaccharides (algae polysaccharides, *Astragalus* polysaccharide, chitosan, ganoderan, lentinan, *Polyporus* polysaccharide).
Essential oils	Terpinen-4-ol (from tea tree) inhibits pro-inflammatory cytokine, upregulates anti-inflammatory cytokine expression, and displays tissue healing characteristics in mastitis.
Phyto-extracts	Extracts from *Allium sativum* also exhibits antibacterial, anti-diarrhoeal, anti-inflammatory, and immune-modulatory properties.
Polyphenols	Antioxidant, anti-inflammatory, anticancer and antimicrobial properties of diferuloylmethane, a polyphenol isolated from turmeric (*Curcuma longa*) rhizomes.
Organic acids	Different short-chain fatty acids, medium-chain fatty acids and other organic acids and their salts (e.g. formic acid, acetic acid, lactic acid, propionic acid, butyric acid, sorbic acid, benzoic acid) have been tested in animal nutrition. While the mode of action of organic acids for feed preservation and water hygiene mainly reflects pH reduction, their role in the gut is still not completely elucidated.
Amino acids	As an immunostimulant (arginine, leucine), cytosine-phosphate-guanine (CpG)-based immunostimulant.
CRISPR/Cas9	Clustered regularly interspaced short palindromic repeats-Cas are designed to cleave plasmids carrying AMR and virulence genes (gene silencing) or gene editing (insertion of a new sequence)—working on a selective site and creates a double stranded nick in the DNA, modifying or permanently replacing the target sequence. Can have potential application in controlling AMR at dairy farms, through application of, e.g. sprays/liquids on farm environment, dairy personnel hands.
Bacteriophages	Lytic phages are used, mostly still at research phase except topical products. Genetic engineered phage-based delivery system as an antimicrobial against *Staphylococcus aureus*—able to overcome the current shortcomings in phage-based delivery systems such as inefficient delivery, narrow host range, and potential transfer of virulence genes exist. Purified phage genes products like endolysins are used e.g. against *S. aureus.*
Bacteriocins	Directly inhibit bacterial growth by inhibiting cell membrane functions or inhibiting gene expression and proteosynthesis (nisin and lysostaphin inhibition of *S. aureus*—mastitis supportive therapy/teat sealant inclusion). Lacticin effective against *S. aureus, S. dysgalactiae*, and *Streptococcus uberis* Promising microcins and colicins isolated from specific *E. coli* strain against enteral infections of calves.
Bacterial extracts	As an immunostimulant β-glucan, peptidoglycan, lipopolysaccharides, muroetasin, prodigiosin.
Bacterial predators	Gram-negative bacteria *Bdellovibrio bacteriovorus* and *Micavibrio aeruginosavorus* attack and kill certain pathogenic bacteria: as e.g. multiresistant *E. coli, Klebsiella* spp., *Pseudomonas aeruginosa, Stenotrophomonas maltophilia* (at the stage of in vitro studies).

(continued)

Table 6 (continued)

Alternative	Comments
Engineered peptides	Antibacterial, antifungal, antiviral. Directly inhibit bacterial growth by inhibiting cell membrane functions, the issue of practical application is related to instability, but also toxicity concerns and AMR. Under research and development are structurally nanoengineered antimicrobial peptide polymers could be a low-cost and effective against Gram-negative bacteria.
Nanoparticles	Metal based—blockage of enzyme pathways, alteration of cell wall, and nucleic material pathways Research targeted on being explored as vehicles for delivery of antimicrobial agents. Nitric oxide and tilmicosin-solid lipid nanoparticles as well as silver in nanoparticles were investigated against *S. aureus* and mastitis prevention.
Immunomodulators/ immunostimulants	Directly enhance innate immune responses through the activation of phagocytes, neutrophils, alternative complement system, and increased lysozyme activity. Typical representatives are biological cytokines as interferon, transfer factor, interleukin, immunoglobulins. Chemically synthesised (cimetidine, imiquimod, levomisole, pidotimod, polyinosinic acid, sodium houttuyfonate, tilorone, ubenimex).
Antibodies (IgY)	IgY is a major serum immunoglobulin in birds and is available in high concentration from chicken egg yolk. Promising tools in GIT infections (rotavirus in animals neonates), orally administrated immunoglobulin Y has been used to prevent or treat bacterial and viral diseases in mammalian, avian and aquatic species. The antibodies in the egg or the yolk can be incorporated into diet, prepared in dry form by spray- or freeze-drying, but the best technology proven is encapsulation [nanocomposite matrix]
Vitamins	As an immunostimulant (A,E,C)
Minerals	As an immunostimulant (Se, Zn), new technologies like lipid encapsulated low concentration zinc occured.
Enzymes	Used in feed additives. Carbohydrates (e.g. amylases, glucanases, cellulase, invertase) Lipase, Proteases (bromelain, papain, trypsin, pepsin), oxidoreductases, phosphatase acting not only by affecting the feed and its digestibility, but interacting also with minerals and influencing also the production and secretion of mucin, which influence the organisation of intestinal epithelial surface and eventually microbial composition of the gut. Also can have an impact on microbial population by providing selective nutritional components to specific group of microbes. Also effects on innate immunity.
Quorum sensing inhibitors/ quorum quenchers	Could control virulence of pathogens by inhibiting the binding of auto-inducers to respective receptors. QSIs have been classified into peptide (autoinducing peptide homologs), protein QSIs, and non-peptide small molecules, can interfere with QS signal molecule synthesis or their binding to the receptor.

Modified, amended according: Mehdi et al. (2018), Marquardt and Li (2018), Sharma et al. (2018), Garg et al. (2017), Park et al. (2017), Kovacs-Nolan and Mine (2012), Bikard and Barrangou (2017), Li et al. (2017), Jiang et al. (2017), Hassanein and Soliman (2010), Nava et al. (2009), Bednarczyk et al. (2016), Mohammadagheri et al. (2016), Rasschaert et al. (2016), EMA and EFSA (2017)

Table 6 provides an introduction and also examples to the various existing/ promising approaches that can, either solely, or in certain cases in combination with rational/minimal use of antimicrobials, help to keep animals healthy.

5 How the Welfare Can Help to Keep Animals Healthy and Free from Infectious Disease

As the antimicrobials serve the possibility to treat the diseased animals, it can be anticipated that healthy animals do not need them. The OIE's formal recognition of the scientifically proved critical relationship and connection between animal welfare and animal health, and the resulting development of the international recommendations set out in the OIE Code, provide strong evidence of the growing consensus on the importance of animal welfare standards. There are available scientific papers that give the evidence of compliance with animal welfare standards strengthens both the health of farm animal populations (including their resistance to disease outbreaks) and the quality of animal food products. Animal welfare science identifies a number of common areas of synergy between animal welfare, animal health and productivity (Fraser 2006).

Ensuring Good animal welfare is the farmer/animal owner/vet responsibility that includes consideration of all aspects of animal well-being, including proper housing, management, nutrition, disease prevention and treatment, responsible care, humane handling and, when necessary, humane euthanasia (AVMA 2019).

Animal welfare reflects how well an animal is coping with the conditions in which it lives. Animals have good welfare if they are healthy, comfortable, well-nourished, safe, able to express the natural behaviours of their species and are not suffering from unpleasant states such as pain, fear and distress.

As an essential prerequisite following Five Freedoms rules were defined and well characterised animal welfare principles (Mellor 2016):

1. Freedom from hunger and thirst—by ready access to fresh water and a diet designed to maintain full health and vigour
2. Freedom from discomfort—by the provision of an appropriate environment including shelter and a comfortable resting area
3. Freedom from pain, injury or disease—by prevention or through rapid diagnosis and treatment
4. Freedom to express normal behaviour—by the provision of sufficient space, proper facilities and company of the animal's own kind
5. Freedom from fear and distress—by the assurance of conditions that avoid mental suffering

There has been also released WQP criteria for the assessment of animal welfare as follows (according Vapnek and Chapman 2010):

Criteria

1. Animals should not suffer from prolonged hunger, i.e. they should have a sufficient and appropriate diet.
2. Animals should not suffer from prolonged thirst, i.e. they should have a sufficient and accessible water supply.
3. Animals should have comfort around resting.
4. Animals should have thermal comfort, i.e. they should neither be too hot nor too cold.
5. Animals should have enough space to be able to move around freely.
6. Animals should be free from physical injuries.
7. Animals should be free from disease, i.e. farmers should maintain high standards of hygiene and care.
8. Animals should not suffer pain induced by inappropriate management, handling, slaughter or surgical procedures (e.g. castration, dehorning).
9. Animals should be able to express normal, non-harmful social behaviours (e.g. grooming).
10. Animals should be able to express other normal behaviours, i.e. they should be able to express species-specific natural behaviours such as foraging.
11. Animals should be handled well in all situations, i.e. handlers should promote good human–animal relationships.
12. Negative emotions such as fear, distress, frustration or apathy should be avoided, whereas positive emotions such as security or contentment should be promoted.

As mentioned also in RONAFA (EMA and EFSA 2017) current research focusses on animal-based measures (ABMs) which are directly linked to an animal's response to adverse circumstances (as opposed to non-animal-based measures which describe the situation in which they are kept, and only indirectly represent the effect they may have on the animal) (Welfare Quality 2009). ABMs include physiological parameters (such as the stress hormones—see below for stressors), behavioural parameters (aggression, restlessness, but stereotypies, social withdrawal and apathy), as well as the actual appearance of the animal (skin lesions, clinical signs of illness, overall health status). Of great importance is minimising of stressors that can be listed in relation with currently used (intensive) farming technologies as well as animal movements: heat, cold, improper air ventilation (too much ammonia, hydrogen sulphide, carbon dioxide), lighting, crowding, mixing of animals from different sources, farrowing/hatching/calving, weaning and animals separation (calves/piglets), limit-feeding and water supply, insufficient bedding, improper milking, noise, pests and parasites and restraint. Improvement of human–animal interaction (handling), enrichment (environment that provides complexity, manipulability and cognitive stimulation of animals), prevention of abnormal behaviours (such as tail biting in pigs, prevention of feather pecking and cannibalism), minimising pain and stress in painful procedures (tail docking, teeth clipping or grinding, nose ringing), and proper transportation conditions can significantly improve welfare of animals. Detailed information on animal welfare specified for pigs, chickens, beef cattle, dairy cattle and working equids are described in OIE (2019), Terrestrial Animal Health Code, Section 7. All of these stressors, even not in a same scale, have been shown to alter the immune system of animals.

Despite the commonly accepted fact that ensuring animal welfare can help to keep animals healthy, also another aspect should be taken into account in current days, where, under extreme pressure to minimise the use of antimicrobials or even ban of some antimicrobials, can appear the situation, in which sick animals will not be properly treated, what can hamper animal welfare. Also consumers' increasing demand on raising animals without antibiotic (i.e. antibiotic-free production avoiding any use of them) can provoke the questions on sustainability of such systems as well as impact on animal welfare (Cervantes 2015; Karavoilas et al. 2018). Banning or severely restricting the use of antimicrobials in animals may negatively impact the veterinarian's ability to protect animal health and prevent suffering from disease, which could potentially lead to worsening of welfare as well as finally lower quality of food from food-producing animals. Current works as e.g. Kruse et al. (2019) commented on association between the antimicrobials use and lesions at slaughter that were found. The authors of study concluded that prevalence of lesions was slightly (non-significant) higher in pig herds with no registered antimicrobial use than with prescriptions of antimicrobials. In another study increasing incidence of eye burns, footpad lesions and airsacculitis in chicken broilers has been documented for broiler flocks with no use of antimicrobials (Karavoilas et al. 2018). Therefore, rational, evidence-based approach should be used when handling antimicrobials—i.e. use them only in cases when necessary, but in sufficient dose, intervals and duration.

Summarising above facts, in relation to use of antimicrobials in food-producing animal husbandries, animal welfare improvement can serve as a potent tool for improvement of animal immunity and health, that in the end will lead with no/low need to use antimicrobials. On the other hand too strict restrictions of antimicrobial use, without considering actual situation at each individual farm and condition, can lead to negative impact on animal welfare. Balanced, rational and responsible approach should be therefore chosen by vets in cooperation with farmers.

6 Socio-economical Aspects to Be Considered

Production of the food commodities (mainly meat, eggs and milk) contribute in a large scale to cover the need of proteins and generally food demand by human population. Within the European area, well supplied by food, ethical food consumption strategies raised during last years (Miele and Evans 2010), which are supported by appropriate regulatory and infrastructural regimes (Davies et al. 2013). Currently more and more consumers thinking on animal welfare and are aware of the resistance issue, also environmental questions are raised with increased frequency. Call for changes or improvement of the farming sector to sustainable and nature-friendly production, that ensure safe food has become urgent.

Previous chapters have shown that antimicrobial use in food-producing animals technologies of production are influenced by several types of drivers (Collineau 2016); these include not only technical drivers, e.g. 1-day-old chicks and piglets health status (van Rennings et al. 2015, prudent use GL), vaccination schemes or biosecurity level (Rojo-Gimeno et al. 2016), but also psychosocial drivers that are related, among others, to farmers' and veterinarians' attitudes and habits towards antimicrobial usage (Visschers et al. 2016; Coyne et al. 2016). The relative importance of technical versus psychosocial drivers seems to be critical for setting of the well balanced policies.

Views on the aspects/drivers influencing the use of antimicrobials to be considered can be from different perspectives. One of the examples can be RESET Mindset Model (Lam et al. 2017) that contains the most important cues to change human behaviour, being Rules and regulations, Education and information, Social pressure, Economics, and Tools.

6.1 Regulatory/Legal/Rules Setting

Non-binding documents: Different documents can be adopted globally—usually as resolutions, guidelines and recommendations (e.g. Codex Alimentarius level, OIE level), creating global consensus, but should be implemented voluntarily. Similarly that can be done at the European level (e.g. Prudent Use Guidelines or AMR resolutions) or at National level (National Action plans—depending on exact way of approval and making certain steps obligatory by, e.g. national legal provisions). The main issue with such documents is that they create rather "general policies", partly raising awareness and make global societal pressure, but, according to the book author opinion, are not leading directly to change of the behaviour in everyday practices of farming. Despite this, without global consensus and recognising AMR as a threat and releasing the signal of importance to find solution, further more pragmatic steps could remain isolated.

Legally binding documents: Depending on the cultural habits and traditions, such setting of obligatory and many times also restrictive or at least prescriptive rules could be helpful. At the European level—new regulation on veterinary medicinal products, at national level several examples can be listed: national law on mandatory susceptibility testing (mainly prior critically important antimicrobials to be used); national law banning the profit from sales of antimicrobial veterinary medicinal products by vets; national law setting the specific taxes on antimicrobials; national law on benchmarking (and further measures mainly targeted on those exceeding the thresholds) etc. Setting of such legal rules should be accompanied with both educational/motivation/stimulation activities as well as checking activities (unfortunately not only supervision, but also system of penalties/punishments).

6.2 Social Pressure

From public health/one health perspective social pressure can increase due to the raising of awareness of the whole society, but also is more targeted on smaller group of interested parties. Among powerful tools belong campaigns that show current antimicrobial consumption at least per animal production sector and also set the targets for (rational) reduction.

With respect to responsible use of antimicrobials, vets and other consultants play an important role in shaping this societal frame of reference, because they have a strong influence on farmers' opinions about animal health. In

Table 7 Characteristics of attitudes of the sow farmers who use less antibiotics (according Policy paper, Economics of antibiotic use, Bergevoet et al. 2019)

Have a higher intention to get or keep the usage of antibiotics under the target value and are more positive about it.
Think that less usage of antibiotics increases work pleasure and is good for animal health, animal welfare and human health.
Think to a lesser extent that farm results will get worse if they reduce the usage of antibiotics
Perceive less risk and uncertainty.
Perceive to have enough knowledge, time (and money) available to keep or get antibiotic usage under the target value.
Think that they use less antibiotics and that the health status of their farm is better when they compare themselves with other farms.
Are less negative about policymakers.
Think to a greater extent that other pig farmers, customers, the government, their partner and their neighbour find it important that antibiotic usage is low.

many cases, veterinarians decide whether to treat an animal or not with antimicrobials, select the antimicrobial to be used, as well as define the dosage and route of administration. Veterinarians also advise farmers on animal health (including vaccination programmes), biosecurity and production management issues that can strongly influence animal health and finally also the need for use of antimicrobials and the transmission of resistant bacteria. Considering vets role, of importance are setting, e.g. recommendation for preventive health programmes (e.g. rules for setting of husbandry tailored vaccination schedules based on epidemiologic evidence) and also first, second and last choice of antimicrobials, altogether with availability of benchmarking that can strengthen the stimulation "not to be worse than my neighbours". Farmers are also a key stakeholders from the farm management, biosecurity, animal health, and welfare perspective, based on everyday experience and contact with animals, they can identify risk factors (and consequently intervene) associated with issues of their animals and husbandry issues and finally also influence the need for use of antimicrobials in livestock. Example of specific attitudes and characteristics of sow farmers, who use less antibiotics (Bergevoet et al. 2019) is given in Table 7.

Applying social psychology to Antimicrobial prescribing practices several models have been described yet. Using of the TPB theory of planned behaviour (Ajzen 1991) to intensive farming of food-producing animals, that identified three most important determinants of behaviour, was one of them:

ATTITUDE	*Is the behaviour good to do?*
SUBJECTIVE NORMS	*What do other actors expect me to do, and do I care what they think?*
PERCEIVED BEHAVOURAL CONTROLS	*How easy or difficult is the behaviour for me to do?*

Change of behavioural models create the up to date approach, how to influence animal health and welfare and finally change/decrease antimicrobials use patterns. There has been recognised steps to be followed for behaviour change and coaching approach has been applied in certain studies yet. Model applicable for farmer/vet is indicated in Fig. 1.

One of the change management models already well incorporated in corporate business is the ADKAR® model. This model identifies the five different elements essential for the successful implementation of change: Awareness, Desire, Knowledge, Ability and Reinforcement and allow the scoring of the individual model parts. Very recent study presented by Caebeke (2019) shows that after a first coaching session, the average ADKAR scores increased, meaning that the farmers changed behaviour to more prudent antimicrobial use after only 6 months. It should be noted that those farms already achieved large reduction in the amount of antimicrobials used, but this specific action plan can further help in reducing the use of antimicrobials by

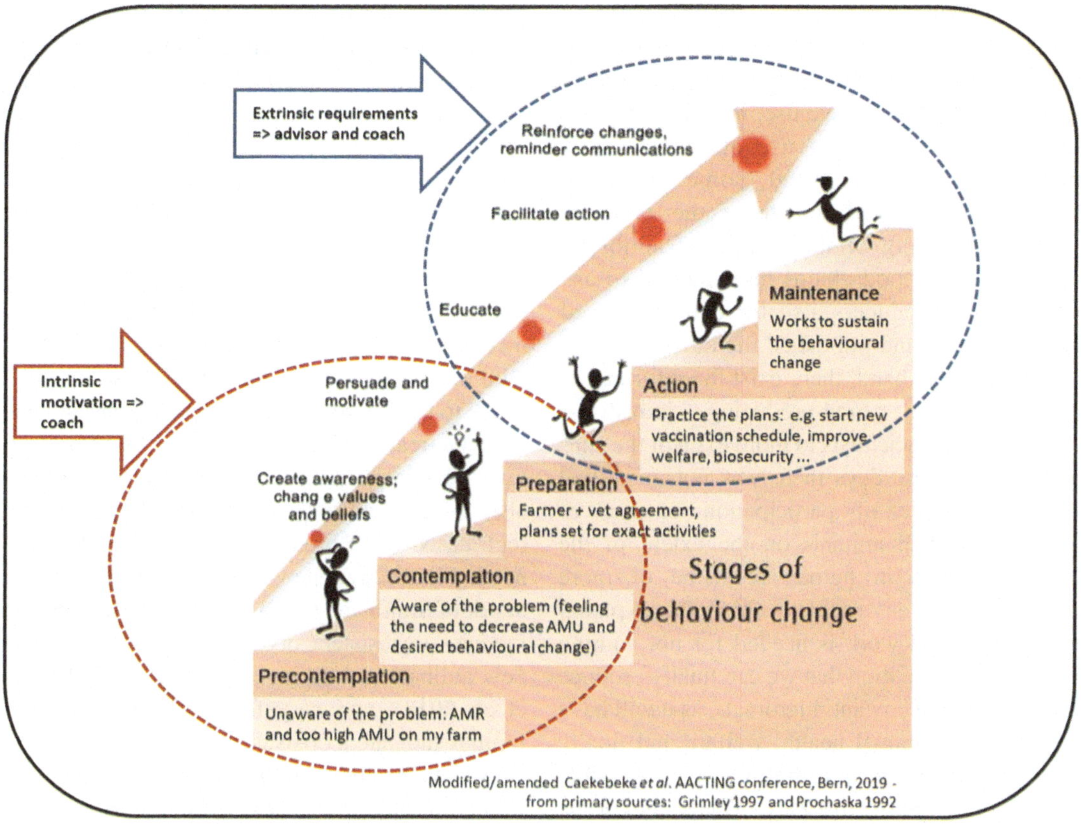

Fig. 1 Model of stages of behaviour change linked to reduction of use of antimicrobials

focusing on improvements in biosecurity and the use of alternative tools.

6.3 Customers' Demand

With raising awareness of the general public, related to risks associated with use of antimicrobials in food-producing animals there have also started to raise consumers' worries related to residues of antimicrobials in food as well as antimicrobial resistance and pollution of the environment due to extensive use of veterinary antimicrobials. Also use of antimicrobials as growth promoters, still being administered to huge amount of animals in certain part of the world (except the EU) is of great consumer concern. For many years, many supermarkets failed to take the issue seriously. In the past couple of years, in particular emphasised also by the publication of the government-commissioned Antimicrobial Resistance Review—the "O'Neill Report" (O'Neill 2015), there has been a considerable amount of work happening behind the scenes in certain countries. One of the good examples to be mentioned is the activity of the head offices of the UK's largest supermarkets. Agricultural teams, Corporate Social Responsibility teams and antimicrobial-resistance experts have been working together in a variety of ways to devise new policies to reduce antibiotic use in their supply chains. In more integrated supply chains (the poultry industry), it has been easier to get rapid shifts in practices, but for more fragmented sectors (cattle, pig, sheep), supermarkets need to deal with a much wider range of suppliers. Nevertheless, with some supermarkets we are starting to see new policies being introduced across all species (ASOA 2017). From certain period some of the supply chains introduced the policy of requirement on primary producers as, e.g. they have banned to have in their nets products from husbandries with routine preventative use, introduce restriction policies as for the use of the critically important antibiotics, and some of them announced that will not trade the products from farms using of the last-resort antibiotic colistin. Some food chains have started to ask the primary producers also for data on use of antimicrobials and published the targets.

On the other hand, there exist the information "gaps" among food-producing industry—consumers and scientists to more explain the situation. Until these days there was not quantified the exact scale of participation of use of antimicrobials in animals on the extent of the issue of AMR in human medicine or more broadly in the environment. Also the explanatory/educational work is needed for not gaining the general perception that we can finally produce all food without any antimicrobials, as it will have an impact on animal health, welfare and finally also on human health through, e.g. possible spread of zoonotic diseases. There cannot be introduced false perception, that any animal will never get sick in the flock/herd and that the spread of infection in the groups or big herds/flocks on the farm can be easily blocked without any use of antimicrobials. Instead of it, as rational and responsible as possible approaches should be investigated, communicated with professionals (vets/farmers), explained to public and especially applied for using antimicrobials in animal sector on everyday basis.

6.4 Economy Pressure

The possible benefits from the use of antimicrobials in animals need to be balanced against their cost and the costs of application and the costs in humans and animals caused by increased risk of emergence of resistance, but also other risks (see also the Chap. 5). The picture should be viewed not only from the short-term perspective (i.e. one treatment course, or one course of fattening gains), but long-term perspective should be considered. For example, whilst antimicrobials may enhance the growth and efficiency of livestock, it could well lead over time to the emergence of resistance to antimicrobials and any outbreaks of disease of organisms with resistance genes would require the use of more expensive antimicrobials (either directly in animals, or consequently after spread of resistant microbes also in human). Conversely not using antimicrobial prophylactically may increase feed costs and perhaps costs associated with disease and death loss, but diseases are less likely to be caused by resistant pathogens and can often be treated with less expensive first-line antimicrobial drugs (Rushton 2015). There should be also counted not only direct costs, but also indirect costs (e.g. broad and extensive research needed to discover new effective antimicrobials). Work of Smith and Coast (2013) indicated that an increase in resistant organisms coupled with no new antibacterial discovered since 1987 (Davies et al. 2013), and very few antivirals and antifungals indicate a crisis.

Speaking about long perspective, there is clearly visible need for reduction of use of antimicrobials (to the minimum level needed for the treatment/justified metaphylaxis) that can be promoted also via economical tools. Some tools have been developed and already used, e.g. decoupling of veterinarians incomes and avoiding profit from prescription and selling of antimicrobials, or impose tax either on specific pharmaceutical form of antimicrobials (e.g. in Germany on medicated feed) or on certain group/s of antimicrobials (e.g. colistin or selected critically important antimicrobials as e.g. cephalosporins of third and fourth generation or fluoroquinolones). Also tax "incentives", e.g. for vaccination, can be the option. More tricky is discovery and introduction to practice of effective and safe alternatives, that will be economically attractive. Also presentation of success in reduction of the use of antimicrobials as well as use of alternatives can be sources of the knowledge, but full information should be presented and explained and also equally highlighted. For example the policy paper issued by the Wageningen University (Bergevoet et al.

2019) declaring and highlighting "The reduction in antibiotic usage on broiler and pig farms in the Netherlands from 2009 to 2017 did not result in a deviation from the long-term trend in average production and economic results in these sectors. To improve animal health, which made a reduction in antibiotic usage possible, farmers used a variety of relatively easy and cheap measures, such as more attention to hygiene, use of pain killers and anti-inflammatory agents or more preventive vaccinations." should be more discussed considering all information meant in individual part of the report—as costs of enlarged vaccination programmes, cost of individualisation of care in sows, costs of modernisation/newly build premises (with improved air conditioning, water supply/water pipe systems of clean water), creating of separate sickbays for piglets and separate for sows, injection without needles, keeping breeding sows in quarantine, that are, as "proactive approach", in complexity more expensive than "reactive approach", i.e. use of antimicrobials. Also should be considered the resources (national government or EU interventions/incentives), e.g. building/reconstruction of new modern facilities.

In summary, some qualitative research showed that some pig farms managed to have simultaneously low antimicrobial use and high technical performance (Fertner et al. 2015); it would be interesting to explore how these "top farms" differ from the others in terms of health status, farm management practices and herd characteristics as well as for real economy parameters. This should contribute to better inform and target future risk mitigation strategies and accompany them with exact measures for practice. From the economical perspective the overview of measures and actions preventing/replacing the use of antimicrobials evaluated for their real costs will be beneficial. Performing thorough analysis of such data and finally providing the real picture for the economists as well as farmers will allow to choose farmers the most economically sustainable way of reduction of use of antimicrobials.

6.5 Education

There is the need for targeted education for professionals veterinarians/farmers/zootechnicians/feed mills/laboratory staff and other professional stakeholders. Education tailored for each level and role of individual professions in the system, that can influence level of awareness of AMR issue, set adequate level of knowledge of preventive programmes for animal health and welfare. For veterinarians specific set of trainings targeted on the continuous improvement of knowledge on disease prevention, vaccination programmes, diagnostics (including e.g. sampling techniques, precise farming technologies), advances in treatment options as well as recommendations for judicious and locally tailored evidence based use of antimicrobials seems to be beneficial. Also models targeted on change of behaviour frameworks and negotiation techniques (improving relationships/trust especially among farmers and vets) could help.

Awareness campaigns targeted on food processors/retailers but also on general public (including consumers explanatory campaigns) can bring improvement of common understanding of the issue.

Last but not least as for importance—involvement of the politicians and policy/decision makers influencing the legal and regulatory rules as well as socio-economical surrounding is necessary.

7 Concluding Remarks

Use of antimicrobials either via prophylaxis, metaphylaxis or treatment to solve the animal infectious diseases issues is reactive approach/solution. It has great benefits, considering especially acute outbreaks of diseases, where causative agent is still susceptible to antimicrobial of choice and animal/s can be successfully treated and protected from unnecessary suffering. In such case, once proper dose, duration and frequency of treatment is chosen and the treatment start in the early stage of disease, we can, with the high probability say, that no alternatives, of the same potency, same effectivity and also same price are currently available. On the other hand, use of antimicrobials also pose significant risks (see Chap. 5) which cannot be overlooked. Rising awareness of antimicrobial resistance as the real threat that, in some cases, have already caused treatment failure both in human and animals lead the global society to ask for decreasing the use of antimicrobials and saving them for life-threatening infections of people. Therefore, increased need for tools of preventive medicine, need for establishing tailored herd/flock health programmes, that include also vaccination plans, ask for more strict following biosecurity and hygiene principles altogether with good husbandry practices keeping animals under good animal welfare seems to be vital. Big challenge is in front of research and development as for new alternatives to antimicrobials.

References

Ajzen I (1991) The theory of planned behavior. Organ Behav Hum Decis Process 50(2):179–211

Antonissen G, Martel A, Pasmans F, Ducatelle R, Verbrugghe E, Vandenbroucke V, Li S, Haesebrouck F, Van Immerseel F, Croubels S (2014) The impact of Fusarium mycotoxins on human and animal host susceptibility to infectious diseases. Toxins 6(2):430–452

Arsenakis I, Boyen F, Haesebrouck F, Maes DGD (2018) Autogenous vaccination reduces antimicrobial usage and mortality rates in a herd facing severe exudative epidermitis outbreaks in weaned pigs. Vet Rec 182:744

ASOA (2017) Alliance to save our antibiotics: real farming solutions to antibiotic misuse, what farmers and supermarkets must do. Briefing 28 p. https://www.soilassociation.org/media/14072/asoa-real-farming-solutions-to-antibiotic-misuses-what-farmers-supermarkets-must-do-091117pdf. Accessed 22 July 2019

AVMA (2019) American Veterinary Medical Association: Animal welfare: seeing the forest and the trees. AVMA 6002482862. https://www.avma.org/KB/Resources/Reference/AnimalWelfare/Documents/animal_welfare_brochurepdf. Accessed 22 July 2019

Bednarczyk M, Stadnicka K, Kozlowska I, Abiuso C, Tavaniello S, Dankowiakowska A et al (2016) Influence of different prebiotics and mode of their administration on broiler chicken performance. Animal 10:1271–1279

Benjamin M, Yik S (2019) Precision livestock farming in swine welfare: a review for swine practitioners. Animals 9:133

Berckmans D (2014) Precision livestock farming technologies for welfare management in intensive livestock systems. Scientific and Technical review of the Office International des Epizooties 33(1):189–196

Bergevoet R, van Asseldonk M, Bondt N, van Horne P, Hoste R, de Lauwere C, Puister-Jansen L (2019) Wageningen economic research. Policy paper, Economics of antibiotic use, 2019-026. https://edepot.wur.nl/475403. Accessed 22 July 2019

Bernardini C, Grilli E, Duvigneau JC, Zannoni A, Tugnoli B, Gentilini F (2014) Cellular stress marker alteration and inflammatory response in pigs fed with an ochratoxin contaminated diet. Res Vet Sci 97:244–250

Bikard D, Barrangou R (2017) Using CRISPR-cas systems as antimicrobials. Curr Opin Microbiol 37:155–160

Borst LB, Mitsu Suyemoto M, Chen LR, Barnes HJ (2019) Vaccination of breeder hens with a polyvalent killed vaccine for pathogenic *Enterococcus cecorum* does not protect offspring from enterococcal spondylitis. Avian Pathol 48(1):17–24

Bossé JT, Li Y, Sárközi R, Fodor L, Lacouture S, Gottschalk M, Casas Amoribieta M, Angen Ø, Nedbalcova K, Holden MTG, Maskell DJ, Tucker AW, Wren BW, Rycroft AN, Langford PR, BRaDP1T Consortium (2018) Proposal of serovars 17 and 18 of *Actinobacillus pleuropneumoniae* based on serological and genotypic analysis. Vet Microbiol 217:1–6

Caekebeke N (2019) Use of a livestock-adapted ADKAR ® change management model for reducing AMU. AACTING Bern, p 1–31. https://aacting.org/swfiles/files/AACTING_Bern_Caekebeke_71.pdf. Assessed May 2020

Callaway TR, Edrington TS, Anderson RC, Harvey RB, Genovese KJ, Kennedy CN, Venn DW, Nisbet DJ (2008) Probiotics, prebiotics and competitive exclusion for prophylaxis against bacterial disease. Anim Health Res Rev 9:217–225

Catry B, Dewulf J, Maes D, Pardon B, Callens B, Vanrobaeys M et al (2016) Effect of antimicrobial consumption and production type on antibacterial resistance in the bovine respiratory and digestive tract. PLoS One 11(1)

Cervantes HM (2015) Antibiotic-free poultry production: is it sustainable? J Appl Poultry Res 24(1):91–97

Collineau L (2016) Quantify, explain and reduce antimicrobial usage in pig production in Europe. PhD Thesis. https://www.theses.fr/2016ONIR091F.pdf. Accessed 22 July 2019

Coyne LA, Latham SM, Williams NJ, Dawson S, Donald IJ, Pearson RB, Pinchbeck GL (2016) Understanding the culture of antimicrobial prescribing in agriculture: a qualitative study of UK pig veterinary surgeons. J Antimicrob Chemother 71(11):3300–3312. https://doi.org/10.1093/jac/dkw300

Coyne LA, Latham SM, Dawson S, Donald IJ, Pearson RB, Smith RF, Williams NJ, Pinchbeck GL (2018) Antimicrobial use practices, attitudes and responsibilities in UK farm animal veterinary surgeons. Prev Vet Med 161:115–126

Cvjetković V, Sipos S, Szabó I, Sipos W (2018) Clinical efficacy of two vaccination strategies against *Mycoplasma hyopneumoniae* in a pig herd suffering from respiratory disease. Porcine Health Manag 4:19

DAFC (2017) Pig industry quality manual, 5th edn, 1st issue. Danish Agriculture & Food Council, Copenhagen. https://www.lf.dk/~/media/lf/aktuelt/publikationer/svinekod/2018/qsg-english-2017.pdf?la=da. Accessed 22 July 2019

DAFC (2018) SPF system Denmark, CHR lookup. http://spfsus.dk/en. Accessed 22 July 2019

DANMAP (2014). http://www.danmap.org/Downloads/Reports.aspx. Accessed 22 July 2019

Davies DS, Grant J, Catchpole M (2013) The drugs don't work. A global threat. Penguin, London

De Briyne N, Atkinson J, Pokludova L, Borriello S (2014) Paper: Antibiotics used most commonly to treat animals in Europe. Vet Rec 175:325

Diarra MS, Malouin F (2014) Antibiotics in Canadian poultry productions and anticipated alternatives. Front Microbiol 5:282

Dieste-Perez L, Frankena K, Blasco J, Muñoz P, de Jong M (2016) Efficacy of antibiotic treatment and test based culling strategies for eradicating brucellosis in commercial swine herds. Prev Vet Med 126:105–110

EFSA (2009) European Food Safety Authority: Panel on Animal Health and Animal Welfare, Scientific Opinion on the overall effects of farming systems on dairy cow welfare and disease. EFSA J 1143:1–38

EMA, EFSA (2017) European Medicines Agency and European Food Safety Authority: Joint Scientific Opinion on measures to reduce the need to use antimicrobial agents in animal husbandry in the European Union, and the resulting impacts on food safety (RONAFA). EFSA J 15(1):4666

EPRUMA (2019) European platform for responsible use of medicines in animals: best-practice framework for the use of vaccines in animals, 8 p. https://www.eprumaeu/wp-content/uploads/2019/04/Best-practice-framework-on-vaccines_23-APRIL-2019pdf. Accessed 22 July 2019

European Commission (2014) Discussion paper on Progress under the Animal Health Strategy for the European Union (2007–2013) where "Prevention is better than cure" and possible future steps. Directorate G – Veterinary and International Affairs Unit G2 – Animal health. https://ec.europa.eu/food/animals/health/strategy2007-2013_en. Accessed 22 July 2019

European Commission (2015) Guidelines for the prudent use of antimicrobials in veterinary medicine. Eur Official J, 2015/C 299/04. https://ec.europa.eu/health//sites/health/files/antimicrobial_resistance/docs/2015_prudent_use_guidelines_en.pdf. Accessed 22 July 2019

FAO (2010) Food and Agriculture Organization of the United Nations/World Organisation for Animal Health/World Bank. Good practices for biosecurity in the pig sector – Issues and options in developing and transition countries. FAO Animal Production and Health Paper No. 169, Rome

FAO (2019) FAO and Denmark Ministry of Environment and Food – Danish Veterinary and Food Administration: Tackling antimicrobial use and resistance in pig production: lessons learned from Denmark. Rome. 52 pp. Licence: CC BY-NC-SA 3.0 IGO

Federation of Veterinarians of Europe (2016a) Antimicrobial use in food-producing animals, FVE input to RONAFA report. EFSA J 15(1)

Federation of Veterinarians of Europe (2016b) Relationship between animal welfare and the use of antibiotics in food animals. https://www.fve.org/cms/wp-content/uploads/063-FVE_AWW-Position-on-resistance-and-animal-welfare_final.pdf. Accessed 4 June 2019

Fertner M, Sanchez J, Boklund A, Stryhn H, Dupont N, Toft N (2015) Persistent spatial clusters of prescribed antimicrobials among Danish pig farms – a register-based study. PLoS One 10:e0136834

Fraser D (2006) Animal welfare assurance programs in food production: a framework for assessing the options. Anim Welf 15:93–104

Friedman D, Kanwat C, Headric M, Patterson N, Neely J, Smith L (2007) Importance of prudent antibiotic use on dairy farms in South Carolina: a pilot project on farmers' knowledge, attitudes and practices. Zoonoses Public Health 54:366–375

Garg R, Babiuk L, van Drunen Littel-van den Hurk S, Gerdts V (2017) A novel combination adjuvant platform for human and animal vaccines. Vaccine 35 (Pt A):4486–4489

Hampson DJ (2018) The spirochete *Brachyspira pilosicoli*, enteric pathogen of animals and humans. Clin Microbiol Rev 31:e00087-17. https://doi.org/10.1128/CMR.00087-17. Accessed 22 June 2019

Hassanein SM, Soliman NK (2010) Effect of probiotic (*Saccharomyces Cerevisiae*) adding to diets on intestinal microflora and performance of hy-line layers hens. J Am Sci 6

Hoelzer K, Bielke L, Blake DP, Cox E, Cutting SM, Devriendt B, Erlacher-Vindel E, Goossens E, Karaca K, Lemiere S, Metzner M, Raicek M, Collell Suriñach M, Wong NM, Gay C, Van Immerseel F (2018a) Vaccines as alternatives to antibiotics for food producing animals. Part 1: Challenges and needs. Vet Res 49(1):64

Hoelzer K, Bielke L, Blake DP, Cox E, Cutting SM, Devriendt B, Erlacher-Vindel E, Goossens E, Karaca K, Lemiere S, Metzner M, Raicek M, Collell Suriñach M, Wong NM, Gay C, Van Immerseel F (2018b) Vaccines as alternatives to antibiotics for food producing animals. Part 2: New approaches and potential solutions. Vet Res 49(1):70

Jensen VF, de Knegt L, Andersen VD, Wingstrand A (2014) Temporal relationship between decrease in antimicrobial prescription for Danish pigs and the "Yellow Card" legal intervention directed at reduction of antimicrobial use. Prev Vet Med 117:554–564

Jiang Y, Zheng W, Kuang L, Ma H, Liang H (2017) Hydrophilic phage-mimicking membrane active antimicrobials reveal nanostructure-dependent activity and selectivity. ACS Infect Dis 3:676–687

Jorge S, Dellagostin OA (2017) The development of veterinary vaccines: a review of traditional methods and modern biotechnology approaches. Biotechnol Res Innov 1:6–13

Karavolias J, Salois MJ, Baker KT, Watkins K (2018) Raised without antibiotics: impact on animal welfare and implications for food policy. Transl Anim Sci 2 (4):337–334

Kontturi M, Junni R, Simojoki H, Malinen E, Seuna E, Klitgaard K, Kujala-Wirth M, Soveri T, Pelkonen S (2019) Bacterial species associated with interdigital phlegmon outbreaks in Finnish dairy herds. BMC Vet Res 15(1):44

Kovacs-Nolan J, Mine Y (2012) Egg Yolk Antibodies for Passive Immunity. Annu Rev Food Sci Technol 3 (1):163–182

Kruse AB, Kristensen CS, Lavlund U, Stege H (2019) Antimicrobial prescription data in Danish national database validated against treatment records in organic pig farms and analysed for associations with lesions found at slaughter. BMC Vet Res 15:218

La T, Phillips ND, Coiacetto F, Hampson DJ (2019a) An atypical weakly haemolytic strain of *Brachyspira hyodysenteriae* is avirulent and can be used to protect pigs from developing swine dysentery. Vet Res 50 (1):47. https://doi.org/10.1186/s13567-019-0668-5

La T, Phillips ND, Hampson DJ (2019b) Vaccination of chickens with the 34 kDa carboxy-terminus of Bpmp72 reduces colonization with Brachyspira pilosicoli following experimental infection. Avian Pathol 48(1):80–85

Lam TJGM, Jansen J, Wessels RJ (2017) The RESET Mindset Model applied on decreasing antibiotic usage in dairy cattle in the Netherlands. Ir Vet J 70:5

Li J, Koh JJ, Liu S, Lakshminarayanan R, Verma CS, Beuerman RW (2017) Membrane active antimicrobial peptides: translating mechanistic insights to design

Lokhorst C (2018) An introduction to smart dairy farming. Van Hall Larenstein University of Applied Sciences, 108 p. https://www.google.com/url?sa=t&rct=j&q=&esrc=s&source=web&cd=13&ved=2ahUKEwiOw4mO6sjjAhXww8QBHdzhCNcQFjAMegQIAxAC&url=https%3A%2F%2Fwww.greeni.nl%2Fwebopac%2FMetaDataEditDownload.csp%3Ffile%3D2%3A144032%3A1&usg=AOvVaw2rlyOpFA2ejzJnEHagR97S. Accessed 28 June 2019

Marquardt RR, Li S (2018) Antimicrobial resistance in livestock: advances and alternatives to antibiotics. Anim Front 8(2):30–37

Mehdi Y, Létourneau-Montminy MP, Gaucher ML, Chorfi Y, Gayatri S, Rouissi T et al (2018) Use of antibiotics in broiler production: global impacts and alternatives. Anim Nutr 4:170–178. https://doi.org/10.1016/j.aninu.2018.03.002

Meissonnier GM, Pinton P, Laffitte J, Cossalter AM, Gong YY, Wild CP (2008) Immunotoxicity of aflatoxin B1: impairment of the cell-mediated response to vaccine antigen and modulation of cytokine expression. Toxicol Appl Pharmacol 231:142–149

Mellor DJ (2016) Updating animal welfare thinking: moving beyond the "Five Freedoms" towards "A Life Worth Living". Animals (Basel) 6(3):21

Midtlyng PJ, Grave K, Horsberg HE (2011) What has been done to minimize the use of antibacterial and antiparasitic drugs in Norwegian aquaculture? Aquac Res 42:28–34

Miele M, Evans A (2010) When foods become animals, ruminations on ethics and responsibility in care-full spaces of consumption. Ethics, Place and Environment 13(2):1–20

Mohammadagheri N, Najafi R, Najafi G (2016) Effects of dietary supplementation of organic acids and phytase on performance and intestinal histomorphology of broilers. Vet Res Forum 7:189–195

Moreno MA (2014) Opinions of Spanish pig producers on the role, the level and the risk to public health of antimicrobial use in pigs. Res Vet Sci 97:26–31

Nava GM, Attene-Ramos MS, Gaskins HR, Richards JD (2009) Molecular analysis of microbial community structure in the chicken ileum following organic acid supplementation. Vet Microbiol 137:345–353

O'Neill J (2015) Securing new drugs for future generations: the pipeline of antibiotics. Wellcome Trust, London. https://amr-review.org/sites/default/files/160518_Final%20paper_with%20coverpdf. Accessed 22 July 2019

OIE (2013) Terrestrial Animal Health Code: Biosecurity procedures in poultry production. http://www.oieint/indexphp?id=169&L=0&htmfile=chapitre_biosecu_poul_productionhtm. Accessed 22 July 2019

OIE (2014) Terrestrial Animal Health Code: General recommendations on disinfection and disinsection. http://www.oieint/indexphp?id=169&L=0&htmfile=chapitre_disinfect_disinsecthtm. Accessed 22 July 2019

OIE (2015) Report of the meeting of the OIE ad Hoc group on priorisation of diseases for which vaccines could reduce antimicrobial use in animals. http://www.oieint//fileadmin/SST/adhocreports/Diseases%20for%20which%20Vaccines%20could%20reduce%20Antimicrobial%20Use/AN/AHG_AMUR_Vaccines_Apr2015pdf. Accessed 22 July 2019

OIE (2018a) Report of the meeting of the OIE ad Hoc group on priorisation of diseases for which vaccines could reduce antimicrobial use in cattle, sheep and goats. http://www.oieint//fileadmin/SST/adhocreports/Diseases%20for%20which%20Vaccines%20could%20reduce%20Antimicrobial%20Use/AN/AHG_AMUR_Vaccines_ruminants_May2018pdf. Accessed 22 July 2019

OIE (2018b) Terrestrial Animal Health Code: Vaccination. http://www.oieint/indexphp?id=169&L=0&htmfile=chapitre_vaccinationhtm. Accessed 22 July 2019

OIE (2019) Terrestrial Animal Health Code: Animal welfare. http://www.oieint/indexphp?id=169&L=0&htmfile=titre_17htm. Accessed 22 July 2019

Park JY, Moon BY, Park JW, Thornton JA, Park YH, Seo KS (2017) Genetic engineering of a temperate phage-based delivery system for CRISPR/Cas9 antimicrobials against *Staphylococcus aureus*. Sci Rep 7:44929. https://doi.org/10.1038/srep44929

Pierron A, Alassane-Kpembi I, Oswald IP (2016) Impact of mycotoxin on immune response and consequences for pig health. Anim Nutr (Zhongguo xu mu shou yi xue hui) 2(2):63–68

Postma M, Backhans A, Collineau L, Loesken S, Sjölund M, Belloc C, Emanuelson U, Grosse Beilage E, Stärk KDC, Dewulf J (2016) The biosecurity status and its associations with production and management characteristics in farrow-to-finish pig herds. Animal 10:478–489

Raith J, Trauffler M, Firth CL, Lebl K, Schleicher C, Köfer J (2016) Influence of porcine circovirus type 2 vaccination on the level of antimicrobial consumption on 65 Austrian pig farms. Vet Rec 178:504

Rasschaert G, Michiels J, Tagliabue M, Missotten J, De Smet S, Heyndrickx M (2016) Effect of Organic Acids on Salmonella Shedding and Colonization in Pigs on a Farm with High Salmonella Prevalence. J Food Prot 79:51–55

Ratna A, Arora SK (2018) Immunomodulators as therapeutic option in parasitic infections. J Bacteriol Vaccin Res 1(1):1002

Rieckmann K, Pendzialek SM, Vahlenkamp T, Baums CG (2020) A critical review speculating on the protective efficacies of autogenous *Streptococcus suis* bacterins as used in Europe. Porcine Health Management 6:12

Roerink F, Morgan CL, Knetter SM, Passat MH, Archibald AL, Ait-Ali T, Strait EL (2018) A novel inactivated vaccine against *Lawsonia intracellularis* induces rapid induction of humoral immunity, reduction of bacterial shedding and provides robust gut barrier function. Vaccine 36(11):1500–1508

Rojo-Gimeno C, Postma M, Dewulf J, Hogeveen H, Lauwers L, Wauters E (2016) Farm-economic analysis of reducing antimicrobial use whilst adopting improved management strategies on farrow-to-finish pig farms. Prev Vet Med 129:74–87

RUMA (2016) RUMA Guidelines: Responsible use of vaccines and vaccinations in farm animal production. https://www.ruma.org.uk/wp-content/uploads/2014/09/farm-vaccine-long.pdf

Rushton J (2015) Anti-microbial use in animals: how to assess the trade-offs. Zoonoses Public Health 62(Suppl 1):10–21

SAPHIR (2019) EC-CORDIS: Strengthening animal production and health through the immune response: Fact sheet. https://cordis.europa.eu/project/rcn/193183/factsheet/en. Accessed 22 July 2019

Sasaki Y, Sekiguchi S, Uemura R, Sueyoshi M (2015) The effect of depopulation and restocking on reproductive and growth performances on Japanese commercialswine farms. J Vet Med Sci 78:333–335

Scherpenzeel CGM, Santman-Berends IMGA, Lam TGJM (2017) Veterinarians' attitudes toward antimicrobial use and selective dry cow treatment in the Netherlands. J Dairy Sci 101:1–10

Sharma C, Rokana N, Chandra M, Singh BP, Gulhane RD, Singh Gill JP, Pallab R, Puniya AK, Panwar H (2018) Antimicrobial resistance: its surveillance, impact, and alternative management strategies in dairy animals. Front Vet Sci 08:fvets.2017.00237

Smith R, Coast J (2013) The true cost of antimicrobial resistance. BMJ 346:f1493

Swinkels JM, Hilkens A, Zoche-Golob V, Krömker V, Buddiger J, Jansen J, Lam TJGM (2015) Social influences on the duration of antibiotic treatment of clinical mastitis in dairy cows. J Dairy Sci 98:2369–2380

Trujillo-Barrera A, Pennings JME, Hofenk D (2016) Understanding producers' motives for adopting sustainable practises: the role of expected rewards, risk perception and risk tolerance. Eur Rev Agric Econ 43:359–382

van Rennings L, von Munchhausen C, Ottilie H, Hartmann M, Merle R, Honscha W, Kasbohrer A, Kreienbrock L (2015) Cross-sectional study on antibiotic usage in pigs in Germany. PLoS One 10:e0119114

Vapnek J, Chapman M (2010) for the Development Law service Food and Agriculture Organization of the United Nations Legal Office: Legislative and regulatory options for animal welfare. Legislative Study 104. http://www.fao.org/3/i1907e/i1907e00.pdf. Accessed 22 July 2019

Visschers VHM, Backhans A, Collineau L, Loesken S, Nielsen EO, Postma M, Belloc C, Dewulf J, Emanuelson U, Grosse Beilage E, Siegrist M, Sjölund M, Stärk KDC (2016) A comparison of pig farmers' and veterinarians' perceptions and intentions to reduce antimicrobial usage in six European countries. Zoonoses Public Health 63:534–544

Welfare Quality (2009) Aims and objectives Welfare Quality project. www.welfarequality.net. Accessed 22 July 2019

Mass Medications: Prophylaxis and Metaphylaxis, Cascade and Off-label Use, Treatment Guidelines and Antimicrobial Stewardship

Keith Edward Baptiste and Lucie Pokludová

Abstract

Antimicrobials are used in animals not only for treatment, but as recognized by the European Surveillance of Veterinary Antimicrobial Consumption report (data 2016), more than 90% are used for mass medication of groups of animals, including healthy animals, mostly orally, especially in pigs and poultry and can reasonably be expected as used for either prophylaxis or metaphylaxis. It is unclear if this annual tonnage of consumption of antimicrobials for prophylaxis/metaphylaxis is necessary. There can be a need to use prophylaxis in just exceptional cases, for individual animals or for a well-defined restricted group of animals where it is known that the risk of development or spread of infectious disease is very high or the consequences of the infection are likely to be severe. Prophylactic use should always be under the responsibility of the attending veterinarian and based on epidemiological and clinical knowledge as well as the justification for such use should be documented. Also, metaphylactic use should be minimized. Growth promotion utilizing antimicrobials should remain banned in EU, and other countries.

Exceptionally when authorized veterinary medicinal products are not available, off-label use which fits the needs of the treatment and to avoid suffering of animals can be used based on the responsible decision of veterinarians. Treatment according to approved veterinary medicinal product texts (on the label) as well as exceptional off-label use should be evidence based. Proper and timely diagnosis, knowledge of pharmacokinetic and pharmacodynamic data and setting the appropriate route of administration and dosing schedule (frequency, duration, and dose) seem key factors for effective treatment. Principles of antimicrobial stewardship should be promoted and become standard for everyday practices.

Keywords

Treatment · Prophylaxis · Metaphylaxis · Growth promoters · Off-label use · Antimicrobials · Responsible use · Diagnosis

1 Mass Medications: Prophylaxis, Prevention, and Metaphylaxis

Different terms describe "mass medication" practices in food animals, such as prophylaxis, prevention, control, group treatment, and metaphylaxis. Prophylaxis and prevention are

K. E. Baptiste (✉)
Danish Medicines Agency, Copenhagen, Denmark
e-mail: keb@dkma.dk

L. Pokludová
Institute for State Control of Veterinary Biologicals and Medicines, Brno, Czech Republic
e-mail: pokludova@uskvbl.cz

L. Pokludová (ed.), *Antimicrobials in Livestock 1: Regulation, Science, Practice*,
https://doi.org/10.1007/978-3-030-46721-0_7

defined similarly as the administration of a veterinary medicinal product (VMP) to healthy animals to prevent infection/s based on a risk/s or possible consequences (ECDC/EFSA/EMA 2015; EMA 2016). Typically, the risk is neither clearly defined, standardized, nor quantified. Examples of common risk factors promoting routine prophylaxis/metaphylaxis use of antimicrobials are listed in Table 1. Originally, metaphylaxis, sometimes referred to as control or group medication, was defined similarly to prophylaxis, the difference being that prophylaxis is applied to individuals and metaphylaxis encompasses whole groups/flocks/herds (Urban-Chmiel and Grooms 2012). In Europe, "metaphylaxis" is redefined as the mass medication of healthy animals when the disease-of-interest is present within the group/flock/herd (EMA 2016). More precise criteria as to when to apply antimicrobial metaphylaxis are rarely discussed. According to Edwards (2010) and Smith et al. (2001), the criterion for antimicrobial metaphylaxis occurs when the morbidity (i.e., attack rate) exceeds 10% for two to three consecutive days. Other criterion used includes the presence of fever in part of the group or just "in-contact" with animals displaying visual clinical signs.

As for animal categories within the major food-producing species, it can be summarized that calves and piglets, at the neonatal stage and later after weaning as well as 1 day to 1 week of age in broiler chicks are of major concern for routine dosing of antimicrobials for mass medications. A common link leading to routine antimicrobial mass medication is the phenomenon that within 2–4 weeks after establishing a group following a "stressor" (e.g., weaning, transport of feedlot/veal calves, newly established all-in-all-out batch group) the majority of clinical cases are identified. Thus, alternative-to-antimicrobial products (e.g. probiotics), as well as alternative animal husbandry management, could serve an important niche to prevent disease and supporting immunity during this "transition" window/s of opportunistic infectious disease scenarios. An OIE symposium in 2012 identified five categories of potential new medical alternatives to antimicrobials, including (1) gene-encoded natural antibiotics including host-derived antimicrobial peptides, such as defensins and cathelicidins, (2) prebiotics and probiotics, (3) bacteriophages, (4) recombinant synthesized enzymes, such as phytases and carbohydrases, and (5) natural phytogenic feed additives. Other alternatives include vaccination, animal welfare and biosecurity measures, animal nutrition, and animal genetic measures (Seal et al. 2013). At the individual animal level, the resilience and tolerance, or immune response of an animal as it adapts to environmental/management changes is a key factor in disease prevention that can be improved through improved housing, biosecurity, appropriate nutrition, stress reduction, vaccination, and genetic selection (Ziping 2018).

Antimicrobials were approved for growth promotion in the United States since 1949 and since 1953 in the United Kingdom (Swann et al. 1969). The current OIE (OIE 2018a) report provides a certain level of precision, and an overview of the current use of antimicrobials for growth promotion. The data published within this third report shows that a total of 110 (n = 155; 71%) responding countries did not use any antimicrobial agents for growth promotion in animals in their countries during the period when data were collected. The 45 remaining countries (n = 155; 29%) reported use of antimicrobials for growth promotion, either with direct authorization of some compounds, or because the country (27 cases, n = 45, 60%) had no regulatory framework on this issue (Table 2).

Mounting evidence about antimicrobial-resistant bacteria and transference of resistance genes from animal to human microbiota led to a full withdrawal of antimicrobial growth promoters in the European Union, since 2006 (Regulation 1831/2003), followed later by, e.g., Canada, United States, South Korea, and other countries. The example commonly cited as justification is the emergence of vancomycin-resistant Enterococci (VREs), especially VanA-VRE, in food animal production systems attributed to the widespread use of sub-therapeutic avoparcin (a glycopeptide for animal use) for growth

Table 1 Examples of occasions, where prophylaxis/metaphylaxis use of antimicrobials frequently occur

Risks	Examples of measures to prevent	Comments
Weaning period	• Vaccination of mothers • Proper suckling to reach immunological status • Postponed weaning • Starter feed of appropriate quality • "Alternatives" (prebiotics, probiotics, symbiotics, organic acids, immunomodulators, and clay minerals) • Biosecurity (intrinsic)	Of the major food-producing species, piglets and calves are most prone to get sick at this time, especially under insufficient care and management Antimicrobials are used for prevention/ metaphylaxis, especially gastrointestinal infections in both piglets and calves, and respiratory infections in calves
Transport stress	• Reducing "middlemen" in the supply chain for animals • Purchasing from dealers with fully disclosed health records • Ensure proper pre-transport management, proper cleaning, and disinfection of vehicles, minimize noise, vibration, novelty, social regrouping, and crowding • Ensure proper transport conditions (provide feed, water, controlled climatic conditions, and avoid mixing and crowding) • Avoid long distance/duration of transportation	All animals, especially weaned calves destined for feedlots or veal calf facilities Antimicrobials used for prevention/ metaphylaxis of different infectious bacterial diseases
One–day-old chicks poor quality	• Improving hatchery hygiene • Vaccination and good health status of parents´ flocks • Avoid mixing of birds from different sources • Scoring the quality of arriving chicks, take samples for bacteriology at arrival • Ensure proper temperature and humidity, ensure sufficient approach to clean water, proper lighting, and ventilation	In some non-EU countries, there is still practices that within vaccination (Marek disease), concomitant administration of 3rd generation cephalosporins (ceftiofur) at early stage of chicks (Saraiva et al. 2018) Especially in 0–7 days old chicks are dosed as prophylaxis/metaphylaxis with antimicrobials (including CIAs - e.g. enrofloxacin)
Wrong input conditions/ feeding of newly housed/weaned animals	• Sufficient cleaning/disinfection of housings with drying off after these procedures • Good quality of feed (incl., e.g., avoiding feeding with feed contaminated by pathogens/mycotoxins)	Gastrointestinal disorders—e.g., Clostridial enteritis due to imbalanced feed (improper fatty acids in diet)—many times given prophylactic antimicrobials Once routine use of (subtherapeutic levels) of antimicrobials is removed, then mycotoxins become more important because they can impair animal health and performance, disrupt the gut barrier and worsen vaccine effectiveness. Therefore, the quality of feed as for no/low level of mycotoxins is of high importance to avoid the use of AGP or routine prophylaxis with antimicrobials
Environmental stress	• Ensure proper conditions of the environment in the stable/hall adequate to age/production category of the animals	For example, especially in broiler chicks, environmental factors (e.g. air quality, lightening, water supply) are of high importance: Leg diseases due to extensive/ rapid growth can be once recognized early, it can be minimized by regulation of lightening/diet balance instead of prophylactic use of antimicrobials

Table 2 List of antimicrobials, including categorization to pharmacological class and importance according to WHO and number of countries reporting their use as antimicrobial growth promotion in different species (data modified from OIE third report (2018a)

Importance according WHO classification	Pharmacological class	Antimimicrobial	No of countries reporting use as AGP (OIE statistics based) / species where mainly used (different references)
CIA- Highest Priority	Fluoroquinolones	Enrofloxacin	1 / poultry
CIA- Highest Priority	Quinolones and quinoxalines	Quinocetone	1 / pigs, chickens, carps
CIA- Highest Priority	Polymyxins	Bacitracin (IA) Colistin (CIA)	18 / poultry, pigs, rabbits, beef cattle 12 / poultry, pigs, calves
CIA-Highest Priority	Macrolides	Tylosin Kitasamycin Erythromycin Tilmicosin Spiramycin Josamycin	17 / pigs 5 / poultry, pigs 2 / poultry, pigs, calves 1 / poultry, pigs, rabbit 2 / poultry, pigs, calves 1 / poultry, pigs, calves
CIA- High Priority	Penicillins	Amoxicillin Benzylpenicillin procaine	3 / poultry, pigs 2 / poultry, pigs
CIA- High Priority	Aminoglycosides	Apramycin Neomycin Streptomycin Kanamycin	1 / pigs 3 / poultry, pigs, cattle 2 / nf 1/ nf
HIA	Streptogramins	Virginiamycin	15 / poultry, pigs, cattle
HIA	Lincosamides	Lincomycin	7 / poultry, pigs
HIA	Tetracyclines	Oxytetracycline Chlortetracycline Tetracycline Doxycycline	12 / poultry, pigs, cattle 9 / poultry, pigs, cattle 1 / poultry, pigs, cattle 1 / nf
HIA	Sulphonamides	Sulfamethazine Sulfachlorpyridazine	2 / pigs 1 / chickens
HIA	Amphenicols	Florfenicol Chloramphenicol	1 / pigs, chickens 2 / nf
IA	Pleuromutilins	Tiamulin	4 / poultry, pigs
IA	Aminocyclitols	Spectinomycin	1 / nf
NCY	Ionophores[1]	Total Monensin Haloquinol	14 8 / cattle 8 / pigs
NCY	Glycophospholipids	Flavofosfolipol	17 / poultry, pigs, cattle
NCY	Orthosomycines	Avilamycin	16 / poultry, pigs
NCY	Other	Enramycin Fosfomycin Bicozamycin Nosiheptide Efrotomycin	13 / poultry, pigs 2 / poultry, pigs 1 / chicken, pigs 2 / pigs 1 / pigs

Antimicrobials: CIA = Critically Important Antimicrobial; HIA = Highly Important Antimicrobial, IA = Important Antimicrobial; NCY= Not Classified Yet at EU level, in some countries classified as "non medically important"

[1]Note: For ionophores, as they are frequently used as feed additives/anticoccidials, not all cases are reported/known as antimicrobial growth promoters in the OIE list

promotion throughout Europe and other countries in the mid-1990s (Klare et al. 1995a, b). On the contrary, other countries like Canada and the United States never approved avoparcin and did not report VRE in animals until 2008, and only in rare occasions (Nilsson 2012). Avoparcin, as a growth promoter, was banned in the EU in 1997 as detailed in the Commission Directive 97/6/EC. New regulations on veterinary medicinal products (Regulation 6/2019/EC) and on medicated feed (4/2019/EC) published in Official Journal at the beginning of January 2019 maintains the ban of antimicrobial use as growth promoters in the European Union. The new rules also propose making EU standards reciprocal for imported foodstuffs and animals. In other words, trading partners will need to respect the ban on antimicrobials for growth promotion, as well as the restriction on antimicrobials reserved for use in human medicine only. On the other hand, the new EU regulations on veterinary medicinal products and medicated feed now bring more restrictive rules for the EU agriculture and use of antimicrobials (see main changes for prophylaxis and metaphylaxis in Table 3).

Other countries still allow routine subtherapeutic antimicrobial administration in some capacity in food animal industries. For example, subtherapeutic chlortetracycline and tylosin are given to feedlot beef cattle to prevent liver abscessation in the United States and Canada and also in Brazil, Mexico, and Australia.

Since the ban on antimicrobial growth promoters, then "mass medications" (prophylaxis and metaphylaxis) has become the most common use in healthy food animals. Despite modern advances in animal production systems, both antimicrobial prophylaxis and metaphylaxis persist typically regarded as herd/flock management measures designed to maintain health (individual or group of animals) and prevent disease. Ancillary benefits of antimicrobial mass medications include better average daily weight gains (i.e., growth promotion) and a belief that less antimicrobial treatments are required later in the production cycle (i.e., reduced labor costs). Both prophylaxis and metaphylaxis leads to substantial antimicrobial consumption since "healthy" individuals will always outnumber "sick" individuals in any given infectious disease scenario that could require antimicrobials. In some EU member states, specific legislation prohibits mass medication of food animals for prevention/ prophylaxis purposes (e.g., Denmark, Netherlands).

Exact consumption figures for antimicrobial mass medications (e.g., prophylaxis and metaphylaxis) in food animals are relatively unknown. However, the sales of different types of a veterinary medicinal product (VMP) antimicrobial formulations can provide an important surrogate measure of the extent of mass medications. For example, two additional types of food animal antimicrobial VMP formulations are designed for easy mass medication, for dissemination in either bulk animal feed or common drinking water supply (e.g., premixes, oral powders/granules/solutions for drinking water) that all animals receive. Since it is considered "inconvenient" or impractical to separate diseased from healthy animals, under intensive livestock production conditions, then these two types of antimicrobial VMP formulations (e.g., premixes, oral powders/granules/solutions for drinking water) are administered commonly. European data (30 countries for 2016: overall sales = 7787.1 tonnes of active ingredient of antimicrobials) reveals just over 90% of antimicrobial sales reported, in mg/PCU (Population Correction Unit, in 1000 tonnes: The estimated weight at the treatment of livestock and slaughter animals.), where these two types of VMP formulations (EMA 2018). In the United States, approximately 95% of the medically important antimicrobials approved for use in U.S. food-producing animals by volume are sold as additives to animal feed or drinking water as mass medications, for routine disease prevention purposes (FDA 2016). This is a consistent annual phenomenon for those countries that collect data on total veterinary antimicrobial sales.

Furthermore, injectable antimicrobials are also used for prophylaxis/metaphylaxis, as utilized in the feedlot/veal calf industries. This reveals that

Table 3 Main changes for prophylaxis and metaphylaxis according to the Regulation 6/2019/EC[a]

	New rules for prophylaxis[2]	**New rules for metaphylaxis[2]**
Regulation 6/2019/EC -VMP		
Definitions	**Article 4/16** 'prophylaxis' means the administration of a medicinal product to an animal or group of animals before clinical signs of a disease, in order to prevent the occurrence of disease or infection	**Article 4/16** 'metaphylaxis' means the administration of a medicinal product to a group of animals after a diagnosis of clinical disease in part of the group has been established, with the aim of treating the clinically sick animals and controlling the spread of the disease to animals in close contact and at risk and which may already be subclinically infected
Preamble explanations		
Avoiding routine prophylaxis and metaphylaxis	**Preamble 41** Requirement for urgent and coordinated intersectoral action in accordance with the 'One Health' approach. Such action includes strengthening of the prudent use of antimicrobials, avoiding their routine prophylactic and metaphylactic use, actions to restrict the use in animals of antimicrobials that are of critical importance for preventing or treating life-threatening infections in humans and encouraging and incentivising the development of new antimicrobials.	
Prophylaxis *Antimicrobials:* Only well-defined cases, individual/restricted number of animals, very high risk of infection and serious consequences. *Antibiotics:* Exceptional cases, only individual animals **Metaphylaxis (groups)** High risk of spread of infection/disease, no alternatives available.	**Preamble 44** Antimicrobial medicinal products should not be used for prophylaxis other than in well-defined cases for the administration to an individual animal or restricted number of animals when the risk for infection is very high or its consequences are likely to be severe. Antibiot ic medicinal products should not be used for prophylaxis other than in exceptional cases only for the administration to an individual animal.	**Preamble 44** Antimicrobial medicinal products should be used for metaphylaxis only when the risk of spread of an in fection or of an infectious disease in a group of animals is high and where no appropriate alternatives are available. Such restrictions should allow the decrease of prophylactic and metaphylactic use in animals towards representing a smaller proportion of the total use of antimicrobials in animals.
Veterinary prescription		
Justification of prescription	**Article 105/3** The veterinarian shall be able to provide justification for a veterinary prescription of antimicrobial medicinal products, in particular for metaphylaxis and for prophylaxis.	
Quantity for one treatment/ Limitation of duration covering period of risk	**Article 105/6** The quantity of the medicinal products prescribed shall be limited to the amount required for the treatment or therapy concerned. As regards antimicrobial medicinal products for metaphylaxis or prophylaxis, they shall be prescribed only for a limited duration to cover the period of risk.	

	New rules for prophylaxis[2]	New rules for metaphylaxis[2]
Use of antimicrobial VMPs		
No routine use of antimicrobials No compensation of poor hygiene/wrong husbandry management or lack of care by antimicrobials.	**Article 107/1** Antimicrobial medicinal products shall not be applied routinely nor used to compensate for poor hygiene, inadequate animal husbandry or lack of care or to compensate for poor farm management	
Ban of antimicrobial growth promoters.	**Article 107/2** Antimicrobial medicinal products shall not be used in animals for the purpose of promoting growth nor to increase yield.	
Regulation 4/2019/EC -MF		
Preamble (27)	**Prophylaxis or use of medicated feed to enhance the performance of animals should not be allowed**, except, in certain cases, as regards medicated feed containing antiparasitics and immunological veterinary medicinal products.	**The use of medicated feed containing antimicrobials for metaphylaxis should only be allowed when the risk of spread of an infection or of an infectious disease is high**, in accordance with Regulation 2019/6.
Article 17/3	**Medicated feed containing antimicrobial veterinary medicinal products** shall be used in accordance with Article 107 of Regulation (EU) 2019/6, except as regards paragraph 3 thereof, and **shall not be used for prophylaxis.**	

Note: *the table above and rules for prophylaxis and metaphylaxis should be read in the context of both regulations and especially considering context of any rules touching regulation, authorization, use, distribution and manufacturing of antimicrobials/antibiotics as stipulated by both regulation on VMP and regulation on MF.*

[1] Please see the Chapter 2.1 for further rules as stipulated by Regulations 6/2019 an 4/2019 as for antimicrobials and AMR

[2] Citation word by word is used to exactly express wording of the legal provision

antimicrobial mass medications are both common and routine in food animal production. Individual scientific investigations have found a range of total percentage of farm-level antimicrobial prophylaxis/metaphylaxis. In the United States, 59% of all feedlot cattle are given prophylactic antimicrobials upon arrival (USDA 2011). Proprietary beef feedlot data shows that the extent of prophylaxis/metaphylaxis depends on cattle weight class equating to 86.85% of cattle between 550 and 625 lb., 23.10% of 626–775 lb., 3.59% of 776–925 lb., and 26.00% of all cattle placed (Dennis et al. 2018). These estimates are higher than official reports by the US *National Animal Health Monitoring System (NAHMS)* of 68.01%, 18.01%, 2.81%, and 20.50%, respectively, for each of the three placement weight categories and overall cattle treatment.

In a Belgian study focused on veal calves, approximately 13.0% of antimicrobials were used preventively (immediately after arrival on farm) and 87.0% for metaphylactic use or as a curative measure (Pardon et al. 2012). Another Belgian survey concerning antimicrobial drug consumption in pigs, injectable antimicrobial drugs were found to be mainly administered for preventive treatments at birth and castration and included broad-spectrum penicillins and cephalosporins. Metaphylaxis were mainly for diarrhea, using fluoroquinolones, aminoglycosides, and polymyxin E (colistin) (Timmerman et al. 2006). A later Belgian survey (Callens et al. 2012) identified 93% use for prophylaxis/metaphylaxis and often lacked a precise diagnosis. The most frequently used antimicrobials at the oral group level were colistin (30.7%), mainly to prevent post-weaning *Escherichia coli* infections, and amoxicillin

(30.0%), as prevention against streptococcal infections. Of concern was a shift from oral mass medications toward the use of long-acting injectable formulations, some of which included amphenicols, third- and fourth-generation cephalosporins. Farmers of large production facilities often consider antimicrobial prophylaxis, despite the cost, in order to achieve lower morbidity/mortality, better production, as well as less labor costs further in the production system (Callens et al. 2012). A survey of Spanish farrow-to-finish farms Spain showed that antimicrobial prophylaxis/prevention occurred on 96% of farms during the 6-month period of the survey, with digestive and respiratory disorders being the most common reasons. Antimicrobial metaphylaxis was associated with 65.8% of records (Moreno 2012, 2014).

Antimicrobial prophylactic/metaphylactic "blanket" dry cow therapy has been a standard since the 1950s (e.g., Five-point Mastitis Control Plan). The goal is to treat and prevent intramammary infections. In a survey of drying-off practices on dairy farms in northern Germany (Bertulat et al. 2015), 79.6% of participating farms practiced blanket (whole herd) antimicrobial dry cow therapy. In the Czech Republic, data for the entire dairy sector indicates that 74% of the cows at drying off were treated in 2010; however, since this period antimicrobial dry-off practices have decreased to 65% in 2016 due to the promotion of the use of selective drying off (Pokludova L, unpublished data). Selective antimicrobial dry cow therapy is an alternative approach based on the presence of an intramammary infection using bacteriology, California Mastitis Test and individual somatic cell counts. In Denmark and the Netherlands, the preventive use of antimicrobials in dry cows is prohibited. MARAN (2015) found there has been a reduction in antimicrobial use in dairy cattle, including a shift away from critically important antimicrobials in dry cow therapy, with no negative udder health effects compared to that seen in previous studies where herds were smaller and before the restriction in antimicrobial use (Santman-Berends et al. 2016). Use of "on-farm" culturing systems can also help to minimize prophylactic use of antimicrobials either in cases of mastitis in dairy cattle or use as prophylaxis for metritis (cows, mares) (Lago et al. 2011; FVE 2016).

In 2014, 81% of the antimicrobials used in Canada on broiler farms were for prevention purposes, from which part administered in the feed was 84%. They were primarily intended to prevent necrotic enteritis caused by *Clostridium perfringens* and coccidiosis (CSCRA 2016). Possible alternatives used as a preventive tool to avoid or minimize the use of antimicrobials are probiotics, prebiotics, enzymes, organic acids, immunostimulants, bacteriocins, bacteriophages, phytogenic feed additives, phytocides, nanoparticles, and essential oils (Mehdi et al. 2018).

The antimicrobial growth promoter (mainly represented by zinc bacitracin) was also used in European countries with intensive rabbit farming (Maertens 2007). Those countries with high numbers of rabbit farms (e.g., Spain, Italy, France, to a smaller extent also the Czech Republic—EMA 2018), were after the ban of the antimicrobial growth promoters faced with an issue of different management of mainly gastrointestinal tract infections. Ever since, VMPs containing bacitracin (either in the water soluble, or premix pharmaceutical forms) that are available on the European market with indications of control and reduction of mortality rate of clostridial epizootic enterocolitis are broadly used for mass medication of rabbits. Despite many efforts, there is still lack of adequate alternatives, including vaccination possibilities to minimize the use of antimicrobials in rabbits.

1.1 Regulatory Considerations

Routine antimicrobial prophylaxis/metaphylaxis can be seen as uniquely veterinary concepts since equivalent examples in human medicine are very rare. This also demonstrates the major difference in common uses of antimicrobials between human versus veterinary medicine. With such distinct differences then it becomes confusing to discuss concepts of prudent/rational antimicrobial use or an antimicrobial classification system that specifies "safer" classes for veterinary use, when antimicrobial treatment is not the common use.

Within the European Union, the new veterinary regulations restrict antibiotic prophylactic use to individual animals, and only when justified by a veterinarian. VMP regulations further stipulate that antimicrobials should not be used routinely to compensate for poor hygiene or inadequate animal husbandry practices such as poor farm management, including improper care of animals. Metaphylaxis should only be used where the risk of spreading a contagious bacterial disease is high and no other appropriate alternatives are available. These rules for use of VMPs are also reflected in the regulation on medicated feed (see also Table 3).

Regulatory approval for antimicrobial mass medication indications is common and does not always stem from clinical trials. Instead, it tends to be based on the type of VMP formulation and a belief that if efficacy has been demonstrated for treatment indications then it is sufficient for prophylaxis/metaphylaxis. However, the "at-risk" intended-to-treat population for prophylaxis/metaphylaxis has distinctly different characteristics from the diseased intended-to-treat population for treatment that greatly influences the efficacy of antimicrobial mass medications (Table 4).

1.2 Drivers for Antimicrobial Mass Medications

Drivers for routine antimicrobial mass medications include traditional beliefs and common animal husbandry practices. Examples include the antiquated structure of both the feedlot and the veal calf industries (Ives and Richeson 2015). A major obstacle to the successful management and control of bovine respiratory disease (BRD) in these cattle populations is associated with the segmented infrastructure of the feedlot (beef) and veal-calf industries. Calves progress through the production phase, changing ownership at any and all points, resulting in transporting, mixing of cattle and other stressors, which provides ample opportunity for opportunistic pathogens associated with bovine respiratory disease to colonize the lower respiratory tract (Ives and Richeson 2015). Also, super-sized all-in/all-out, open shed designs or free-range for chickens, large open tanks for farmed fish, high stocking density for feedlot/veal calves and fattening pigs are drivers for antimicrobial mass medications. While some of these management systems are seen as an improvement in animal welfare, it creates an impractical/impossible scenario to separate sick individuals in cases of an infectious disease outbreak, thus leading to routine mass medication for each production cycle. Currently, there is no epiphany about the design of large-scale animal production systems that allow both for adequate animal welfare and the ease of treating sick individuals. For specific production systems (e.g., feedlot/veal calf industries) radical changes are necessary since it is well known that the infrastructure, including extended transport of animals, comingling, poor biosecurity, and other stressors are the main drivers for regular infectious disease outbreaks. This persistent infrastructure also impedes alternatives to antimicrobial mass medications, since the major predisposition factors are present at the time of arrival to the facilities. Furthermore, no other antimicrobial VMP formulation has been designed that allows for reduced labor costs and low stress of the treatment of sick animals within large groups.

Currently, the need for antimicrobial prophylaxis/metaphylaxis in food animals is based on beliefs of nonstandardized risk/s that a group/s of animals will contract a major bacterial disease (i.e., morbidity) and/or die from the disease (i.e., mortality), and typically without knowledge of the type/s of pathogens involved or antimicrobial susceptibilities. Efficacy is not guaranteed, but varies considerably according to the antimicrobial class used, antimicrobial resistance, placement weight, location, season, and animal health risk/s (e.g., transport, comingling, weaning, vaccination status, and other risks). At the time the decision is made for prophylaxis then all animals are visibly healthy, but "at risk." This lack of

Table 4 Comparison of relevant concepts for approved VMP indications for treatment versus prophylaxis/metaphylaxis

Treatment group	Prophylaxis//Metaphylaxis group
Clinical signs are present	No clinical signs present. Thus, animals could be noninfected or in the incubation phase of the contagious bacterial disease. Unable to distinguish between noninfected and incubation phase animals
Immune system is compromised and/or not able to resolve the infection	Healthy immune system (resilience and tolerance) under the same management "stressors" as other in-contact diseased animals. Immune system is not compromised and also participates in the "prevention" of the disease
Antimicrobial treatment has been proven through clinical trials to assist the immune response to resolve the infection	Antimicrobial influence on the immune system is unknown, but may not necessarily assist the immune system, since some antimicrobial classes "modulate" the immune response and could increase the risk of infection
Defined target site/s based on known disease pathogenesis	Target site is neither defined nor known. Target site/s could either be to prevent colonization of pathogen on the mucosal surface or kill pathogens in the incubation phase
Defined goals for the product (e.g., alleviate clinical signs, prevent mortality, kill the pathogen)	Goals are different in that it is both to reduce/prevent morbidity and mortality. A goal also could be to prevent relapses
Bacterial target/s can be defined through culture/PCR and antimicrobial susceptibility	Bacterial target/s assumed to be the same clones as diseased individuals. However, many examples of food animal diseases/complexes as opportunistic infections from the individuals' microflora secondary to the same management "stressors" and/or viral disease
Treatment efficacy can be explained further with known PK/PD characteristics (e.g., AUC, C_{max}, T > MIC, etc.). Target site tissue/s is inflamed and may assist in antimicrobial penetration	PK/PD characteristics are unknown. Efficacy is more related to coverage of mucosal surfaces to prevent colonization or penetration into a noninflamed target site tissue/s that will imply different PK/PD characteristics
Efficacy assessments may be different for acute versus chronically diseased animals	Efficacy is unknown, but maybe more related to pathogens in long incubation phases versus short incubation phases
Treatment goals are defined based on clinical signs and pathogenesis with relatively few confounding factors	Many confounding factors (e.g., management, time, stressors, nutrition, immune responses, self-cure) that strongly influence efficacy
Treatment regimens also designed to have a minimal impact on the rate of development of antimicrobial-resistant bacteria	Impact on the rate of development of antimicrobial-resistant bacteria is *unknown* but expected to be high since healthy animals will be exposed, especially with oral medications
Dose can be established based on guidelines and defined criteria	Dose is unknown and criteria have not been established. Different environmental risk assessments may be needed
Dose based on minimal efficacious mg/kg for the clinical effect and a minimal number of days of treatment	Minimal efficacious concept is unknown for prophylaxis/metaphylaxis and should not be assumed to be the same as treatment dosing regimens. For example, prophylaxis/metaphylaxis efficacy dosing may be more related to the incubation time of the pathogen rather than concepts used to determine doses for treatment purposes

foresight will thus attract typically broad-spectrum antimicrobials, and possibly critically important antimicrobial classes. In Europe, metaphylaxis is initiated typically when animals within a group display some kind of clinical manifestation. The decision can also be based on different criteria and without knowledge of the type/s of pathogens involved or antimicrobial susceptibilities. An improvement in metaphylaxis would involve an appropriate definition that includes time to establish a diagnosed contagious bacterial pathogen/s with antimicrobial susceptibility results. Also, there can be a consideration for better diagnostic tools, like evidence-based cattle–clinical scoring systems, together with early detection of disease systems (e.g., BRD and reticulo–rumen temperature boluses). The space for new technologies—smart farming applications for mobile phones can be the sound for the near future.

If healthy young animals are administered antimicrobials then there is a measurable growth promotional effect, as reflected by average daily weight gain (ADW). This growth promotional effect should not be underestimated as a motivation for routine prophylaxis/metaphylaxis in food animals. For example, with the size of the US beef feedlot industry, a recent economic impact assessment, based primarily on ADW as well as mortality, reported that removal of routine prophylaxis/metaphylaxis could result in a net loss of annual revenue between $532.18 and 679.56 million US from reduced growth and mortality (Dennis et al. 2018). This is comparable to other studies that estimated short-term economic impacts of complete bans on antimicrobials in feed and water, used primarily for prophylaxis/ metaphylaxis, at $280.55 million US for beef producers (Mathews 2002), $45.36 million to $291.24 million for pork producers (Wade and Barkley 1992; Brorsen et al. 2002; Sneeringer et al. 2015), and $189.00 million for poultry producers (Sneeringer et al. 2015).

1.3 Infectious Disease Dynamics

The true purpose of antimicrobial prophylaxis/ metaphylaxis should be to prevent/control a bacterial disease "epidemic." A basic infectious disease model typically incorporates variables that describe the probability of transmission per animal contact, the number of contacts with the infectious animal per unit time and the duration animal/s are infectious. For example, the basic reproduction number (R_o) (the expected number of secondary infections resulting from infected individual/s in a population) describes that for any given infectious disease in a group than three outcomes are possible, including that the disease could die out ($R_o < 1$), become endemic ($R_o = 1$), or progress to an epidemic ($R_o > 1$). Eventually, saturation (the resulting decline in the number of susceptible individuals to infection) occurs over time with more stable population dynamics (Grassly and Fraser 2008). The "infectiousness" (characteristics of infected individuals that determines the rate of spread to the susceptible population that can be broken down into biological, behavioral and environmental components) and "susceptibility" (biological, behavioral, and environmental) of cohorts of food animals contribute to the likelihood of each of three possible outcome population scenarios (endemic, epidemic, disease die out) and further dependent on other factors, including:

1. Herd immunity—when a significant proportion of the population has immunity (e.g., vaccines, natural acquired, or colostral immunity). Thus, more difficult for diseases to spread between individuals if a proportion is already immune, breaking the "chain of infection."
2. Animal stress factors that promote immunosuppression (e.g., weaning, castration, and dehorning).
3. Animal husbandry practices that promote contagious diseases (e.g., stocking density, transport of animals, comingling animals from different sources, and poor biosecurity).
4. Characteristics of the bacterial clone involved in the disease (e.g., virulence factors, antigenicity, and previous exposure to the population) (Baptiste and Kyvsgaard 2017).

For example, at low attack rates (e.g., morbidity $< 15\%$), then R_o is typically low (<1) based on a small "offspring" distribution (the number of secondary infections as a function of infectiousness over time) that does not economically justify prior antimicrobial mass medications. Under these circumstances, the disease could die out, become endemic, or the low number of sick individuals could be treated, but less likely to progress to an epidemic (Fig. 1). Although a disproportionate amount of disease transmission results from a small fraction of infected individuals, the random effects among

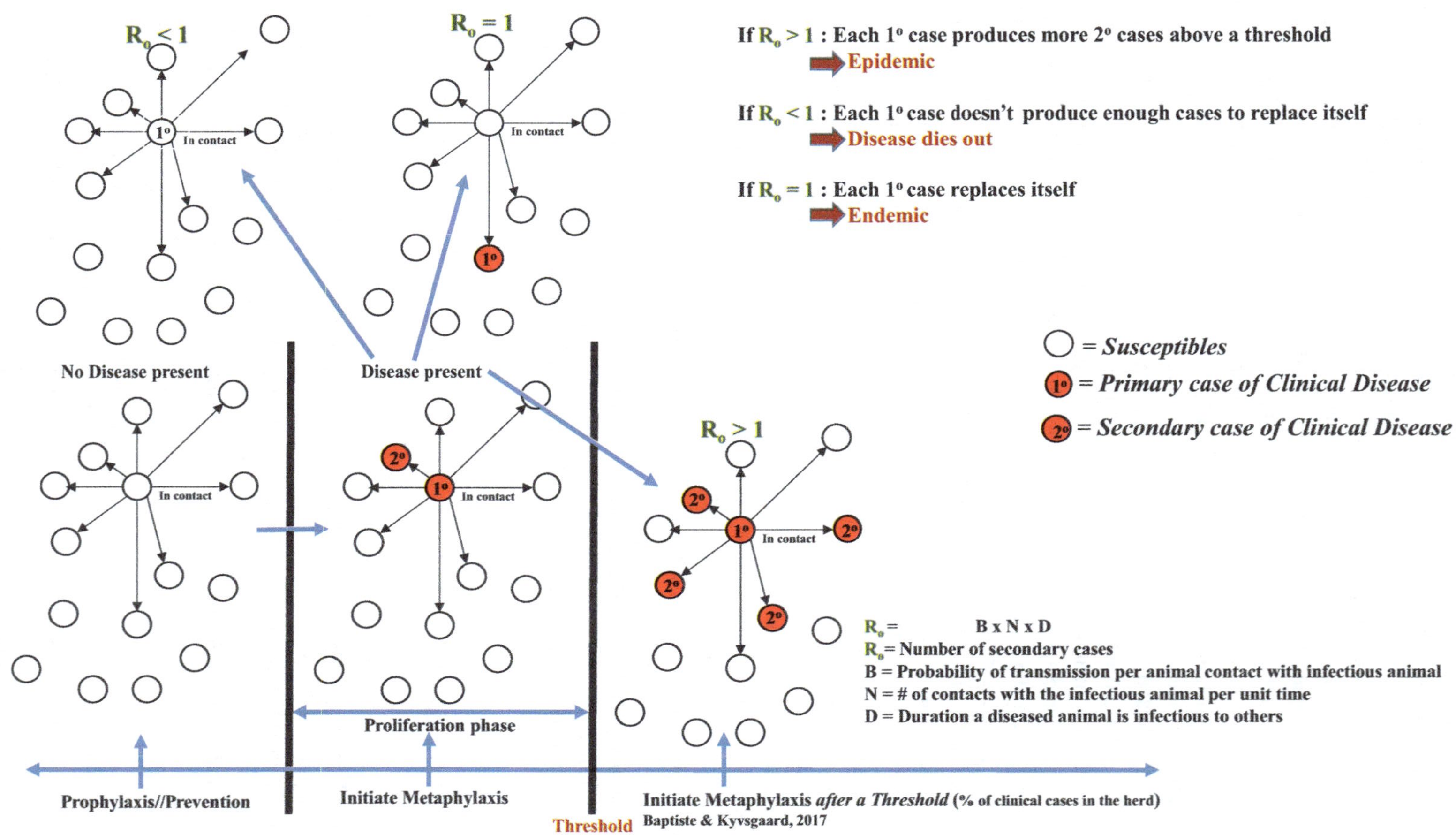

Fig. 1 Model of the three possible outcomes of infectious disease (resulting from one infected individual) considering development of reproduction number (Ro) (the expected number of secondary infections resulting from infected individual/s in a population) : disease could die out (Ro < 1), become endemic (Ro = 1), or progress to an epidemic (Ro > 1) and two scenarios of the initiation of metaphylaxis: initiation at (early) proliferation phase vs initiation after a threshold (certain % of clinical cases in the herd)

individuals tend to cancel each other out as the number of infected individuals increases, resulting in a more predictable progression to epidemic dynamics (Grassly and Fraser 2008). Therefore, a morbidity "threshold" should be part of a metaphylaxis definition that best incorporates the concepts of an infectious disease model by representing the best likelihood, in any given infectious disease scenario, that a given infectious disease will progress to an epidemic past a morbidity threshold (Fig. 1). This morbidity threshold definition can be further demonstrated using the concept of number-needed-to-treat (NNT—an epidemiological measure of an intervention on a population scale) as demonstrated on the example (Fig. 2). NNT is known to change according to the prevalence of disease (morbidity) where pharmacoeconomic benefits do not occur in populations with a low prevalence of the disease, compared to higher disease prevalence.

In this context, antimicrobial prophylaxis/metaphylaxis represents "barriers of defense" against an infectious disease outbreak, alongside other animal husbandry methods that should be employed before antimicrobial mass medications. The dilemma of antimicrobial prophylaxis is that it represents mass medication as a primary (first) "barrier of defense" to prevent disease, without employing other measures. In the veterinary scientific literature, there is no specific guidance or critical evaluation as to which types of bacterial diseases would justify antimicrobial prophylaxis versus metaphylaxis. With modern preventative animal husbandry measures then routine antimicrobial prophylaxis is no longer justifiable. Antimicrobial prophylaxis could be acceptable under exceptional circumstances, including:

- Contagious bacterial diseases where it is known that it will rapidly progress (e.g., 24 hours) to an epidemic and where mortality is a major outcome.
- Where there is not an effective vaccine available or other means to establish herd immunity, and there are no other recognized effective herd-health control measures.

There has been a lack of critical evaluations of antimicrobial mass medication practices. Baptiste and Kyvsgaard (2017) performed a meta-analysis of randomized clinical trials (RCTs) investigating antimicrobial prophylaxis/metaphylaxis for naturally occurring bovine respiratory disease. Bovine respiratory disease (BRD) represents the major indication for cattle antimicrobials worldwide. In total, 58 publications met the inclusion criteria summarizing 169 individual RCTs, spanning 50 years (1966–2016). Antimicrobial prophylaxis and metaphylaxis demonstrated moderate, yet highly variable efficacy in terms of relative risk reductions in BRD morbidity. These were dependent on the antimicrobial classes used, BRD attack rates (i.e., morbidity) and duration of the RCTs. Of the three different metaphylaxis definitions encountered (Group medication of cattle with pyrexia and no other symptoms; Group medication of cattle in contact with clinical BRD cattle; and Group medication of cattle when the BRD morbidity within the group $\geq$10%), the definition of BRD morbidity threshold $\geq$10% outperformed other definitions, in terms of relative risk reduction. Best relative risk reductions were from broad-spectrum critically important antimicrobials or combinations. BRD prophylaxis/metaphylaxis resulted in major antimicrobial consumption for highly variable short-term efficacy in terms of absolute risk reduction of morbidity/mortality (Baptiste and Kyvsgaard 2017). Antimicrobial mass medications could also be associated with negligible improvements or worsened BRD morbidity/mortality. Metaphylaxis had a similar impact as prophylaxis, in terms of relative and absolute risk reduction, but the potential for lower antimicrobial consumption with an appropriate morbidity threshold definition that eliminates the least efficient (highest NNT) possibilities and prevents an epidemic (Baptiste and Kyvsgaard 2017). Although RCTs results for BRD mortality were confounded by previous treatment of BRD cattle, the majority of RCTs reported zero mortality in control groups based on a "treatment-only" strategy of visual BRD cases, with no prior mass medication, as just an effective method of preventing mortality (Baptiste and Kyvsgaard

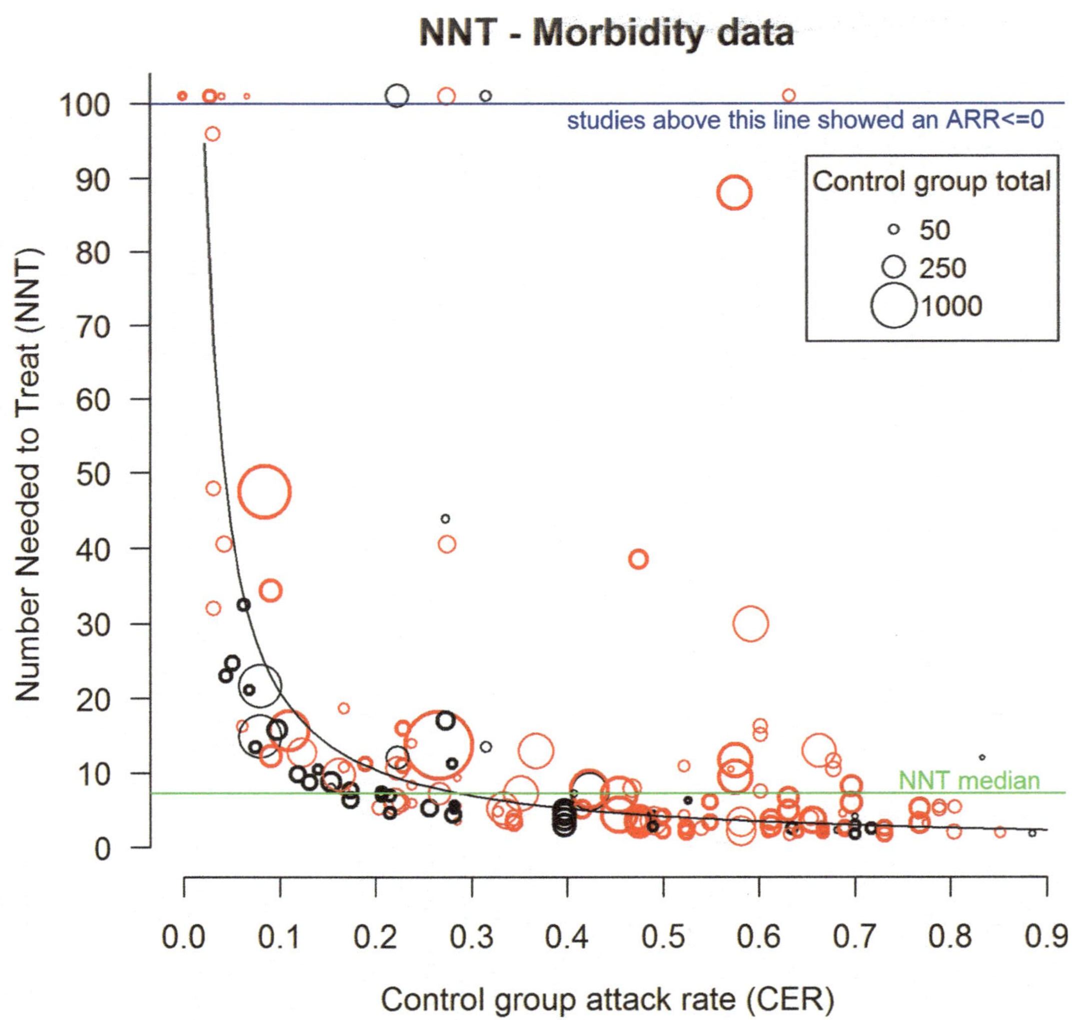

Fig. 2 NNT plot of bovine respiratory disease morbidity data from RCTs involving either antimicrobial prophylaxis (red circles) or metaphylaxis (black circles). Size of the circles reflects the sample size of the RCT (see the legend). Bolded circles are blinded RCTs. Greenline represents the overall NNT median value = 7.27. The curve represents the expected NNT as a function of CER assuming a uniform RR (=0.52) across all values of CER. CER is the morbidity identified from the control group in RCTs. NNT, number needed to treat; ARR, absolute risk reduction; RCT, randomized clinical trial. Plot reprinted with permission from Baptiste and Kyvsgaard (2017)

2017). Abell et al. (2017) also performed a meta-analysis on metaphylaxis for BRD, sponsored by the pharmaceutical industry. They found similar results that efficacies were dependent on the type of antimicrobial used as well as the expected BRD morbidity.

The WHO has published a guideline on the use of medically important antimicrobials (antimicrobial classes listed by the WHO as critically important antimicrobials, including subcategories of "important," "highly important," or "critically important" for human medicine) in food animals (WHO 2017). With recommendations supported by either systematic reviews and/or meta-analysis, the WHO does not recommend the use of medically important antimicrobial classes for routine prophylaxis/prevention, and does not recommend the use of critically important antimicrobial classes for metaphylaxis/control for food animals. Global action plans are designed to preserve such antimicrobial classes for the treatment of clinically important infections in humans. This is not a sustainable future goal if some of these antimicrobial classes are given routinely to healthy food animals for prophylaxis/metaphylaxis, and contributing to the widespread dissemination of antimicrobial resistance to both animal and human populations.

1.4 Individual Prophylaxis: Perioperative Prophylaxis

Surgical procedures in animals are another common reason for antimicrobial prophylaxis. The relative risk for surgical site infections is often assumed to be higher in farm animals than in human or companion animal surgery, because of the unsanitary operating environment in the field, depressed patient immune function in the periparturient period and the high probability of postoperative wound contamination (Dumas et al. 2016). Thus, the necessity for antimicrobial prophylaxis tends to be based on the type of surgery, field conditions for surgery, and the severity of potential postoperative infections. Furthermore, the development of surgical site infections is a complex interaction between the nature and the extent of microbial contamination, the virulence of microorganisms, the integrity of host innate and adaptive defense mechanisms, and factors that relate to the surgery itself.

Unrestricted antimicrobial prophylaxis can result in an increased risk of multiresistant nosocomial infections, general selection of resistant microorganisms, increased cost of hospitalization, and increased incidence of adverse effects for the host. Several studies in companion animals, horses, and cattle demonstrate that antimicrobial prophylaxis provides no benefit for clean surgical procedures. For example, a single preoperative dose of penicillin prior to rumenotomy in cattle is as effective in preventing postsurgical complications as a 7-day course of the same antibiotic (Haven et al. 1992). By definition, dirty surgical procedures require therapeutic rather than prophylactic antimicrobials.

2 Off Label and Cascade Use

Veterinarians are often faced with cases for which approved drugs are not available for the complete range of animal species and disease conditions encountered, or where "off-label use" may be more effective or appropriate. In EU legislation, it is considered implicit that veterinarians should follow summary of product characteristics (SPC) instructions for veterinary medicinal products (VMPs) (Article 106 of the Regulation (EU) 2019/6 on VMPs: "Veterinary medicinal products shall be used in accordance with the terms of the marketing authorization"). Acknowledging the fact that approved indications for VMPs might not cover all clinical needs, especially for minor species and rare diseases as well as also the need to avoid animal suffering, then regulations are in place to allow usage outside approved indications, for exceptional circumstances only. Such "off-label" use was defined in the European Directive 2001/82/EC as amended by Directive 2004/28/EC: "The use of a veterinary medicinal product that is not in accordance with the summary of the product characteristics (SPC), including the misuse and serious abuse of the product." Within the

Regulation (EU) 2019/6 on VMPs, as recently approved, there are defined conditions for the "Use of medicinal products outside the terms of the marketing authorization" that covers cascade use of medicinal products (both veterinary and human) and "magistraliter" in nonfood-producing animal species as well as food-producing terrestrial and aquatic species. The off-label use is not described explicitly here apart from the use outside the terms of marketing authorization describing just cascade use and namely with respect to indication and target species. Other references abroad as, e.g., CSCRA (2014) use the terms "off-label" and "extra-label drug use" (ELDU) interchangeably. In North America, the term "extra-label drug use" is defined in the Animal Medicinal Drug Use Clarification Act (AMDUCA) regulations from 1996. ELDU refers specifically to the use of an approved drug in a manner that is not in accordance with the approved label directions (SPC). This includes, but is not limited to, use in species not listed on the labeling, use for indications (disease or other conditions) not listed on the labeling, use at different dosage levels, frequencies, or routes of administration other than those stated in the labeling, as well as deviation from the labeled withdrawal time based on these different uses. Furthermore, ELDU is allowed for therapeutic purposes only. Also, in some countries where producers can purchase over-the-counter (OTC) antimicrobials, then if a producer/owner administers the drug not according to label directions, then this is known as "Non-Label Drug Use" and against the law. Thus, if drug residues are found in food-producing animals from "non-label drug use" then the producer is responsible and may be charged by the authorities for misuse of the drug. In the EU, there are no OTC antimicrobials, although there are examples of farmers in possession of antimicrobials without prescription/s (Rees et al. 2018).

The EU definition as set in the Directive 2001/82/EC as amended by Directive 2004/28/EC differs from others in that "off-label" also includes "misuse" and "serious abuse" of a VMP. Misuse in its broad sense refers to those of ineffective and inappropriate drug use, time, dose, and duration (OIE 2015; O'Neill 2015). Antimicrobial resistance can be accelerated by misuse of antimicrobials including both situations where antimicrobials are overused, especially for virus infections or poorly defined diseases, as well as overdose or low dose situations (e.g., growth promoters) (Ziping 2018). However, antimicrobial resistance is not simply the result of the antimicrobial use, particularly its misuse and overuse, but also the acceleration of a vicious cycle of high usage, leading to high antimicrobial resistance incidents, which in turn leads to further high antimicrobial usage (Goossens et al. 2005). In a veterinary prescription-based system, both responsible and prudent use (by farmers) is not sufficient for the reduction of antimicrobial misuse in food animal production, since in several countries veterinarians are allowed to obtain profits (markup) on the prescription and sales of veterinary medicines. Furthermore, risks and uncertainties that farmers perceive in disease prevention can also be a reason for the overuse of antimicrobials in food animal production (Ziping 2018). This can furthermore lead to misuse if off-label antimicrobials are used as a substitute for good farm management, biosecurity, optimal hospital hygiene, or lack of a precise diagnosis.

"Off-label" antimicrobial use is a complex subject. For example, within the EU the approval of drugs for sale follows the requirements of EU Directives, whereas the practice of veterinary medicine has fallen traditionally within each EU Member States' (MSs) jurisdictions. Also, legislation and regulations governing the practice of veterinary medicine vary between EU MSs. Thus, the off-label use of VMPs is perceived differently in EU member states. Furthermore, the availability of authorized veterinary antimicrobial products is variable in different countries for various reasons and leads to a lack of authorized products for all indications and especially for minor species. Thus, there are clinical situations in which off-label drug use is necessary. Within the EU, this scenario is recognized in Regulation (EU) 2019/6 on VMPs which allows the use of other than approved medicinal products. From

Articles 112, 113, and 114, a concept known as "the cascade principle" was developed. The principle of the *cascade* is that if no suitable veterinary medicine is authorized in the member state to treat an indication, the veterinary surgeon responsible for the animal may, "in particular to avoid causing unacceptable suffering exceptionally treat the animal/s concerned" in accordance with the following sequence in descending order of priority:

- A veterinary medicine authorized in the relevant Member State or another Member State for use in the same or other food-producing animal species for the same indication, or for another indication.
- If there is no such product, then either a veterinary medicine authorized for use in nonfood-producing animals and same indication.
- If there are not any products defined above, then a medicinal product for human use in accordance with Directive 2001/83/EC or Regulation (EC) No 726/2004.
- If there is no product referred to above, a VMP prepared extemporaneously in accordance with the terms of a veterinary prescription.

It should be noted that previously the Directive 2001/82/EC stipulated not only "indication" but more broadly "condition" as a reason for using the cascade principles.

Common reasons (conditions) related to off-label and/or cascade uses in veterinary practice include:

- *Lack of authorized medicines for certain (minor) species and indications*
- *Unintentional off-label* (i.e., due to lack of harmonized SPCs)
- *Unmet medical need*
- *Alternative routes of administration*
- *Individual patient characteristics* (e.g., neonates, geriatrics, aggressive animals)
- *Complex conditions* (e.g., dysbacteriosis)
- *Practical considerations* (e.g., preference for orally administered VMPs, preference for shorter withdrawal period)
- *Alternative posologies*
- *Combination treatments*
- *Nonantibacterial purposes* (e.g., immunomodulatory effects)

In the United States and Canada, there is another valid approach to off-label drug usage in food-producing animals. In these countries, it is the responsibility of the prescribing veterinarian to calculate an appropriately prolonged withdrawal time based on the scientific information that is available for that particular drug, or engage in direct consultation with the Food Animal Residue Avoidance Databank (FARAD). The exceptions are those substances specifically banned by law for certain species of food animals (e.g., cephalosporins in cattle and fluoroquinolones in poultry), following a specific risk identification and negative assessment. Furthermore, the FDA has established a list of substances that are prohibited for extra-label uses in all food-producing animals, including chloramphenicol, furazolidone, other nitrofurans, fluoroquinolones, glycopeptide, and sulfonamide drugs in lactating dairy cattle, with the exception of approved use of sulfadimethoxine, sulfabromomethazine, and sulfaethoxypyridazine (e.g., extra-label animal drug use; fluoroquinolones and glycopeptides; order of prohibition (US FDA 1997).

The European *cascade* principle as defined previously by the Directive 2001/82/EC gives preference to VMPs authorized for different conditions within the same species. The Regulation (EU) 2019/6 defines use under the cascade in the same or in another food-producing animal species for the same indication, or for another indication. This preference is based on the assumption that interspecies pharmacokinetic differences are less predictable than pharmacokinetic differences as a result of physiologic or pathologic changes. If a product is used outside the terms of authorization (e.g., at a higher dosage) or the product is only authorized for a different species, minimum statutory EU withdrawal periods are stipulated by law as listed in Table 5.

This approach illustrates the application of the "precautionary principle" through the

Table 5 Minimum statutory EU withdrawal periods are stipulated by Regulation 6/2019/EC

Food commodity	Withdrawal period (WP) should *not be less than*
Meat and offals	The longest WP provided in the SPC of the product used for meat and offal *multiplied by 1.5*
	If the medicinal product is not authorized for food-producing animals: *28 days*
	If the medicinal product has a zero WP and is used in a different taxonomic family than the target species authorized: *1 day*
Milk	The longest WP for milk provided in the SPC of the product used for any animal species *multiplied by 1.5*
	If the medicinal product is not authorized for animals producing milk for human consumption: *7 days*
	If the medicinal product has a zero WP: *1 day*
Eggs	The longest WP for eggs provided in the SPC of the product used for any animal species *multiplied by 1.5*
	If the product is not authorized for animals producing eggs for human consumption: *10 days*
Aquatic species	The longest WP for any of the aquatic species indicated in the SPC of product used *multiplied by 1.5 and expressed as degree-days*
	If the medicinal product is authorized for food-producing terrestrial animal species, the longest WP for any of the food-producing animal species indicated in the SPC *multiplied by a factor of 50 and expressed as degree-days, but not exceeding 500 degree-days*
	If the medicinal product is not authorized for food-producing animal species: *500 degree-days*
	If the highest withdrawal period for any animal species is zero: *25 degree-days*

specification of relatively conservative withdrawal times. The purpose of the current EU legislation is to allow for off-label usage when needed for important medical reasons, specifically to avoid animal suffering. In this context, it is assumed that the legislation is used on rare occasions only; however, in veterinary practice there are gaps within approved veterinary medicines for many species and indications. This does result in off-label use of antimicrobials as common use and outside the intent of the legislation.

Both the European and the United States' approaches limit off-label use in food-producing species. The European approach primarily does this by limiting off-label use to those drugs for which an MRL has been established by the regulatory authorities for any food-producing animals/commodities, whereas the North American approach can include several drugs (apart from those specifically banned by law), if the user can reliably prevent violations of drug residues. The latter approach potentially increases the number of drugs that can be used off-label by including those drugs that are approved in other countries as well as those drugs for which information related to food safety has been published in the scientific literature (Gehring et al. 2006). In Europe, it should be noted that although the use of substances without a MRL in food-producing animals is not in-line with EU legislation (i.e., prohibited), the establishment of MRLs is not linked to the possible off-label use of the substance and that the risk assessment of MRLs does not take into account any possible off-label use. Recently, legal provisions stipulating the consideration of MRLs have been set for national and European surveillance authorities that establish criteria for control in animal tissues and products derived from animals treated by VMPs used off-label. For ensured consumer protection, specific rules were prepared to specify which MRLs apply in each particular case (basically, the use of the lowest MRLs established for target tissues of related or not related species needs to be considered). Off-label use of VMPs is directly under the responsibility of prescribing veterinarian, and therefore the veterinarian must estimate an appropriate withdrawal period with consideration of rules in legislation (as in detail described above) and pharmacological properties of substance(s) within the formulation of the chosen product. One should not forget that the reasons for off-label use of VMPs is mainly to avoid unacceptable suffering of animals (for more details, please refer to the chapter "Status Quo in International Context").

There are no official data or monitoring on off-label use of antimicrobials. Within the EU the only official registration related to off-label drug usage is through the European pharmacovigilance network, by reporting adverse events.

Preconceived beliefs that all off-label antimicrobial usage is bad practice does represent misunderstandings, as there are examples of its necessity in veterinary medicine. However, there are instances of misuses of off-label antimicrobials. Thus, there are examples of "appropriate" and "inappropriate" or misuse of antimicrobials. Some examples of "appropriate" off-label antimicrobial include those with good documentation of a clinically relevant infection where either other approved antimicrobials will not suffice or other factors (PK/PD, dose, route-of-administration) will lead to an off-label antimicrobial use providing more optimal evidence-based patient therapy. Certain target sites (e.g., joints, tendons, eyes, brain/spinal cord, pleura, peritoneum, and reproductive organs) can lead to crippling or fatal diseases and uncommon for pharmaceutical companies to seek regulatory approval. In some cases, off-label antimicrobial usage can represent therapeutic gaps, and if scientifically valid then it could form the basis for a pharmaceutical company to seek regulatory approval and expand the SPC.

Views on appropriate/inappropriate off-label use are constantly evolving. For example, regular use of WHO CIAs not approved for veterinary use is raising major public health concerns. Further guidance in this area may be needed for veterinarians until "One Health" considerations become embedded. Treatment based on risk (e.g., mixing animals and transport of animals) or nonspecific clinical symptoms (e.g., fever, depression and undifferentiated diarrhea) is no longer considered part of prudent antimicrobial use. Combination antimicrobial treatments are also becoming questionable practices, especially if a multiresistant bacteria or mixed infections have not been identified. Most of the combination antimicrobial treatments are used empirically to cover either a broader spectrum of pathogens and/or different tissues/tracts infections (typically gastrointestinal/respiratory). Even if some combinations can be justifiable by in vitro tests (e.g., synergy effect), the pharmacokinetic interaction of the active substances may not act synergistically at the site of infection.

3 Treatment, National and Local Treatment Guidelines, Practice-based Protocols, Antimicrobial Stewardship in Veterinary Medicine

Prudent use of antimicrobials is the judicious practice of medical principles, as "the cost-effective use of antimicrobials which maximises clinical therapeutic effect while minimizing both drug-related toxicity and the development of antimicrobial resistance." (WHO 2001). The EU has published veterinary antimicrobial prudent use guidelines (European Commission 2015) as well as the OIE (OIE 2018b). This includes an accurate diagnosis, short-term effective first antimicrobial professional prescriptions based on microbial sensitivity or proven efficacy (RCTs, safety, PK/PD, spectrum of activity) and low impact on selecting antimicrobial-resistant bacteria.

Furthermore, prudent use can be particularly challenging in animals. For example, not only is there the target pathogen/s to consider affecting animal health, including a low impact on antimicrobial resistance, but also the influence of antimicrobial use on nontarget zoonotic bacteria. Best examples include the use of critically important antimicrobials in poultry production. Campylobacterosis is a leading cause of human bacterial enteritis (bloody diarrhea, fever, abdominal cramps, and vomiting lasting for approximately 5–7 days) in Europe (Spina et al. 2015), as well as one of the most costly foodborne diseases in Europe and worldwide (Skarp et al. 2016). Studies employing multilocus sequence typing and mathematical modelling have revealed that chickens are the most common reservoir/source of human *Campylobacter* spp. infections, with attrition rates varying from 38 to 77%, whereas cattle are regarded as the second most common source, with attrition rates varying between 16 and 54% (Skarp et al. 2016). Furthermore, ciprofloxacin-resistant *Campylobacter* spp. from livestock sources has become very common. Studies conducted in accordance with the licensed dosage (10 mg/kg – 50 ppm) revealed that fluoroquinolone resistance and clonal expansion developed rapidly and persisted in *Campylobacter jejuni*, when enrofloxacin was administered via drinking water to poultry (McDermott et al. 2002; Luo et al. 2003; Van Boven et al. 2003; Griggs et al. 2005; Humphrey et al. 2005). Furthermore, Takahashi et al. (2005) concluded that regardless of the enrofloxacin dosage used (15 or 50 ppm for 4 days—Dutch studies; 40 ppm for 5 days—US studies; 25 or 50 ppm for 5 days—US studies, 50 ppm for 10 days—Dutch studies; 10 mg per kg body weight per bird for 5 days—UK studies; 50 ppm for 3 days—Japanese study), rapid emergence of ciprofloxacin-resistant *C. jejuni* occurs. Only substantially higher doses than authorized can prevent/reduce persistence of fluoroquinolone-resistant *Campylobacter* spp. and *Salmonella* spp. in poultry (Stapleton et al. 2010; Li et al. 2017). Also, the poultry industry has a unique "pyramid" structure whereby farms/countries do not produce all stages of poultry production, and thus routinely import certain stages of the poultry production. Thus, antimicrobial treatment of parent flocks can lead to the dissemination of resistant bacteria and genes throughout the production pyramid, including subsequent generations on numerous farms in different countries. The best example of this occurred with the use of ceftiofur in ovo and in 1-day-old chicks that resulted in the widespread dissemination of extended-spectrum beta-lactamase (ESBL)—and/or AmpC-producing bacteria (Baron et al. 2014).

There are further challenges to prudent use of antimicrobials in veterinary medicine. For example, there are several common food animal bacterial pathogens that are not amenable to rapid routine culture and susceptibility methods (e.g., *Lawsonia* spp., *Brachyspira* spp., *Mycoplasma* spp., *Ornithobacterium rhinotracheale*, and *Dichelobacter nodosus*). Also, clinical breakpoints are not established for all veterinary pathogens and antimicrobials. In human medicine, methodologies as well as interpretative criteria based on both qualitative routine laboratory testing and more precise quantitative (MICs) testing are well established (e.g., CLSI and EUCAST). In veterinary medicine, VETCAST has been recently established and firstly introducing standardized methodology and clinical breakpoints, which is particularly challenging due to differences in PK/PD in different animal species.

Even approved indications for VMPs do not always imply prudent use. For example, several common diseases in animals as described in VMP indications are disease complexes or syndromes (e.g., bovine respiratory disease, swine respiratory disease, piglet postweaning diarrhea, and canine kennel cough), often used without precise definitions. These disease complexes are the amalgamation of various viruses, parasites, and other stressors that predispose animals to a variety of opportunistic bacterial pathogens existing commonly as commensals, or as biofilms within the upper respiratory tract and tonsils. In other words, several common disease complexes in food animals describe a collection of visual clinical symptoms that could be caused by virus/s, parasite/s, bacteria, or combinations thereof, but several antimicrobial VMPs have approved treatment indications for these disease complexes without further guidance.

Antimicrobial treatment doses that have been defined several decades ago are not typically based on well-established PK/PD data—especially data considering population PK modeling. There is also a major issue of the prophylaxis/metaphylaxis flock/herd medication regarding the appropriate dosing of all animals (animal flock/herd hierarchy, animals at the different phases prior/at the beginning of disease with different demands as for (medicated) water or feed). Also, some antimicrobial VMP treatment indications can provide benefits to groups of animals, but not in a traditional understanding of prudent use. There are examples of diseases that are systemic infections (e.g., chicken colibacillosis and Salmonellosis in calves) but there are approved oral antimicrobial VMPs that are not appreciably absorbed from the gastrointestinal tract (e.g., colistin, spectinomycin, neomycin, and apramycin). In other words, these oral antimicrobial VMPs will not achieve therapeutic minimum inhibitory concentrations (MIC) at the target site of infection. Instead, it is believed that these VMPs can provide a benefit by decolonizing the gastrointestinal tract of the pathogens that are known to cause these systemic infections, either through subsequent dust inhalation (e.g., chicken colibacillosis) or penetration from the gastrointestinal tract (e.g., salmonellosis). However, there are concerns about using WHO CIAs for decolonizing the gastrointestinal tract in food animals.

The cheapest, quickest method to decide on antimicrobial administration is with visual clinical signs. However, this comes with substantial inaccuracies for disease complexes or syndromes, involving multiple possible pathogens, such as BRD, as well as the fact that common clinical signs (e.g., depression, anorexia, and fever) are not pathognomonic. Comparing BRD field diagnostics to pulmonary lesions evident at slaughter reveals several cattle (e.g., >50%) with lung lesions, not previously identified and treated for BRD (Thompson et al. 2006; Tennant et al. 2014). Hierarchical Bayesian latent class meta-analysis comparing BRD clinical signs to slaughter lung lesions revealed an estimated predicted diagnostic sensitivity and specificity of 0.27 (95% CI = 0.01–0.96) and 0.92 (0.14–1.00), respectively (Timsit et al. 2016). These limitations contribute largely to the justification for routine antimicrobial prophylaxis/metaphylaxis in place of identifying, diagnosing with appropriate culture methods, and treating sick animals.

Some time ago an increasing trend in veterinary medicine has been the publication of treatment guidelines from species-specific veterinary associations, or veterinary specialist societies. These are intended typically for national use and dedicated to a specific animal species. Over the last 10 years, several animal species-specific treatment guidelines have been produced in EU member states (e.g. AT, BE, FI, SW, DK, NL, FR, and UK), including booklets, tables, and mobile apps. Some guidelines are driven and endorsed by national competent authorities and/or the national veterinary organizations. These guidelines typically take the format of listing common relevant diseases for an animal species with recommendations for first choice, second choice treatment options for each disease. Recent updates of such guidelines have started to be more closed so-called antimicrobial stewardship, as they also include alternative/preventive options (e.g., vaccination, preventive measures as for biosecurity as well as alternatives to antimicrobials usually naturally based—prebio-, probio-, synbiotics, enzymes, and phyto-additives).

Pioneering in setting of antimicrobial stewardship principles was human sector. In general, principles are very similar if not same and can be applied also in the veterinary sector. Five R strategy of antimicrobial stewardship considers: Responsibility, Reduction, Refinement, Replacement, and Review (FAAST 2019). In Table 6, main principles applicable in veterinary medicine of food-producing animals are summarized (and also more detailed in activities involved in each principle) considering current approaches and roles of stakeholders to be involved (Monnier et al. 2018, AVMA 2018). Please refer for more info to chapter "EU Policies and Regulatory Surroundings".

There is a wide variety of animal treatment guidelines, where there is no standardized approach. The publication of final treatment guidelines typically does not include the methods used for the basis of treatment recommendations. As such, evidence-based treatment guidelines are not straightforward, but involve an interplay between registered VMP approved indications, traditional veterinary practices, veterinary legislation, pharmacokinetic/pharmacodynamic considerations (e.g., target site concentrations), as well as MIC considerations for the various treatment options and pathogens involved. Past this, it is further unclear regarding public health aspects in treatment guidelines in terms of using national surveillance data on antimicrobial resistance for veterinary pathogens, the basis of CIA recommendations and recommendations for broad-spectrum and combination treatments.

Table 6 Antimicrobial stewardship principles (modified using Monnier et al. 2018; AVMA 2018)

Principle	Way to perform
Undertake the stewardship principles	Tailored stewardship plan development (plan to contain: Preventive measures/tools, if treatment needed—responsible selection/prescription of antimicrobial, antimicrobial susceptibility testing whenever possible, optimal dose/dosing schedule, proper administration, feedback on treatment success/failure, analysis of outcomes) Prioritization/focusing on the conditions where antimicrobials most frequently (over) used to concentrate relevant activities on appropriate targets Leader/s of stewardship and roles of individual stakeholders to be involved All relevant stakeholders engagement
Consider alternative tools for prevention, control, treatment using nonantibiotic tools preferably	Identify the most important preventive measures/tools relevant for exact production system as well as barriers that can block their use Involve relevant stakeholders to: • Identify tailored strategies (husbandry management, biosecurity) • Remove the blocks and set incentives to include in practice all relevant preventive tools • Build sustainable system of transferring the above strategies in practice
Identify an issue, and if medical, diagnose properly the disease	Identify available tools necessary for proper diagnosis including (rapid) diagnostics to be used Whenever possible use simple disease/troubles signal systems that can be performed by farmers, technicians taking everyday care on animals Do not forget new technologies (smart farming) if useful for certain disease (early) detection Final medical diagnosis should be performed by veterinarian on the spot (preferably by attending veterinarian with knowledge of herd/flock anamnesis)
Select antimicrobial/s judiciously	Identify obstacles for appropriate antimicrobial prescription/use Perform antimicrobial susceptibility testing, whenever relevant—Use evidence-based approach Follow principles of rational use as well as local guidelines on responsible use including advice on first, second, and third choice antimicrobials (criteria should also include consideration of risks of AMR/public health concerns consideration) Check for outcomes of treatment, in the case of issues identified, perform assessment (if necessary pharmacovigilance announcement)
Perform evaluation of extent and correctness of antimicrobial use	Monitor the use of antimicrobials, including benchmarking (use, e.g., sector-specific nationally set systems) Perform a feedback to veterinarians (ensure confidentiality) Analyze the results, anonymously publish the results of analysis highlighting main troubles, main mistakes, proposal for improvement, and proposal of best practices
Educate, establish evidence-based expertise, change behavioral models	Explain the antimicrobial stewardship, including all its principles (preventive measures, alternative tools, conditions when antimicrobial drugs are not needed and conditions of use when needed) Educate stakeholders adequately to different level of their skills/roles how to implement appropriate existing clinical guidelines for antimicrobial use (e.g., veterinarians proper choice/dosing schedule prescription, farmers/technical staff proper

(continued)

Table 6 (continued)

Principle	Way to perform
	administration, including specific practical trainings, e.g., proper water supply to ensure that diseased animals can have an equal approach to medicated water with correctly calculated/ medicated concentration of antimicrobial to ensure proper dose to be ingested by the diseased animals) Support research related to prevention and alternative tools, antimicrobial drug use, and resistance (including measures to minimize its transmission and spread)
Maintain the system sustainable and working	Keep up-to-date on strategies for disease prevention, update the portfolio of alternatives, and update antimicrobials of choice

Within veterinary practices, treatment guidelines tend not to be made in preference for standard operating procedures (SOPs). The larger the veterinary practice then the greater the need for SOPs. Many of these SOPs include the routine use of antimicrobials (e.g., for surgical procedures). Furthermore, to remain competitive more-and-more veterinary practices are seeking an international type of accreditation. For example, the ISO 9001 Quality Management Systems accreditation has become increasingly popular that further support developing SOPs. The pro of SOPs is a better standard level of practice on common procedures done at a given veterinary practice. However, this comes at the cost of promoting empirical use of antimicrobials, prophylaxis, and preventative use.

Also by their nature, treatment guidelines include off-label recommendations (e.g., different indications, doses, routes-of-administration), which may be based on veterinary specialist advice, local knowledge, peer-reviewed publications, or changes in bacterial resistance patterns from the original approval of various veterinary medicinal products. In this context, these treatment guidelines are defining "appropriate" off-label antimicrobial use outside of nationally approved products and the legislation. The dilemma occurs as to the basis of off-label antimicrobial recommendations in treatment guidelines and whether this can be used as an information source to define "appropriate" off-label antimicrobial use. For example, the priorities could be solely for animal species considerations (e.g., conservative broad-spectrum antimicrobials for individual companion animal medicine) without considerations of "one-health" perspectives for antimicrobial resistance. Also, such recommendations may not always be "in-concert" with national or EU surveillance programs that may show trends of public health aspects of antimicrobial resistance as well as target veterinary pathogens profiles of susceptibility/ resistance.

Well researched treatment guidelines have a role to assist veterinarians with recommendations, if they take into account modern research results (e.g., systematic reviews) as well as results of national or regional surveillance of antimicrobial resistance. For example, trends in antimicrobial resistance in animals can lead to situations where approved products are no longer efficacious for the treatment of certain diseases. This situation can occur for a number of years since regulatory procedures tend to lag behind trends in antimicrobial resistance. There is also the concern that treatment guidelines might encourage the use of empiric therapy without the use of culture and susceptibility testing.

Many commercial veterinary microbiology laboratories do not include all approved veterinary antimicrobials in their susceptibility panels, or the results for some specific pathogens are technically challenging for in vitro culture and antimicrobial susceptibility and not readily available; thus, treatment guidelines can serve a role for these gaps of unreported antimicrobial susceptibilities for that region/country. For example, commercially available MIC panels that are more relevant for human medicine are typically cheaper for laboratories than equivalent veterinary MIC panels. Some programes are operating throughout the European Member States based on samples from diseased animals and monitoring the susceptibility by broth microdilution method (i.e., minimum inhibitory concentrations) available (Schrijver et al. 2017). Routine susceptibility testing, especially for anaerobes, *Mycoplasma* spp., *Lawsonia*, etc. is not performed and therefore results are not commonly available. This increases the importance of evidence-based treatment guidelines of the data available, considering also their relevance and attempting to tailor the guiding principles to the conditions of the exact Member State to avoid overuse or misuse of antimicrobials. Such guidelines, as well as more complex approach known as antimicrobial stewardship, especially if followed by all stakeholders, represents significant progress in responsible use of antimicrobials.

References

Abell KM, Theurer ME, Larson RL, White BJ, Apley M (2017) A mixed treatment comparison meta-analysis of metaphylaxis treatments for bovine respiratory disease in beef cattle. J Anim Sci 95:626–635

AVMA (2018) Antimicrobial stewardship definition and Core principles. https://wwwavmaorg/KB/Policies/Documents/AntimicrobStewardshipDef_CorePrinciplesFlyer_052318pdf. Accessed 15 June 2019

Baptiste KE, Kyvsgaard NC (2017) Do antimicrobial mass medications work? A systematic review and meta-analysis of randomized clinical trials investigating antimicrobial prophylaxis or metaphylaxis against naturally occurring bovine respiratory disease. Pathog Dis (FEMS) 75(7): 1–12

Baron S, Jouy E, Larvor E, Eono F, Bougeard S, Kempf I (2014) Impact of third-generation-cephalosporin administration in hatcheries on fecal *Escherichia coli* antimicrobial resistance in broilers and layers. Antimicrob Agents Chemother 58(9):5428–5434

Bertulat S, Fischer-Tenhagen C, Heuwieser W (2015) A survey of drying-off practices on commercial dairy farms in northern Germany and a comparison to science-based recommendations. Vet Rec Open 2(1): e000068

Brorsen BW, Lehenbauer T, Ji D, Connor J (2002) Economic impacts of banning subtherapeutic use of antibiotics in swine production. J Agric Appl Econ 34 (3):489–500

Callens B, Persoons D, Maes D, Laanen M, Postma M, Boyen F, Haesebrouck F, Butaye P, Catry B, Dewulf J (2012) Prophylactic and metaphylactic antimicrobial use in Belgian fattening pig herds. Prev Vet Med 106:53–62

CSCRA (2014) Gouvernement du Canada: Extra-label drug use (ELDU) in animals. https://www.canada.ca/en/health-canada/services/drugs-health-products/veterinary-drugs/extra-label-drug-use.html. Accessed 4 June 2019

CSCRA (2016) Gouvernement du Canada: Canadian antimicrobial resistance surveillance system report. https://www.canada.ca/en/public-health/services/publications/drugs-health-products/canadian-antimicrobial-resistance-surveillance-system-report-2016.html. Accessed 4 June 2019

Dennis EJ, Schroeder TC, Renter DG, Pendell DL (2018) Value of arrival Metaphylaxis in U.S. cattle industry. J Agric Resour Econ 43(2):233–250

Dumas SE, French HM, Lavergne SN, Ramirez CR, Brown LJ, Bromfield CR, Garrett EF, French DD, Aldridge BM (2016) Judicious use of prophylactic antimicrobials to reduce abdominal surgical site infections in periparturient cows: part 1 – a risk factor review. Vet Rec 178:654–660

ECDC, European Centre for Disease Prevention and Control, EFSA, European Food Safety Authority and EMA European Medicines Agency (2015) ECDC/EFSA/EMA first joint report on the integrated analysis of the consumption of antimicrobial agents and occurrence of antimicrobial resistance in bacteria from humans and food-producing animals. https://www.ema.europa.eu/en/documents/report/ecdc/efsa/ema-first-joint-report-integrated-analysis-consumption-antimicrobial-agents-occurrence-antimicrobial_en.pdf. Accessed 4 June 2019

Edwards TA (2010) Control methods for bovine respiratory disease for feedlot cattle. Vet Clin North Am Food Anim Pract 26:273–284

EMA (2016) European medicines agency: revised guideline for the demonstration of efficacy for veterinary medicinal products containing antimicrobial substances (EMA/CVMP/627/01-Rev.1). https://www.ema.europa.eu/en/documents/scientific-guideline/final-guideline-demonstration-efficacy-veterinary-medicinal-products-containing-antimicrobial_en.pdf. Accessed 4 June 2019

EMA (2018) European Medicines Agency, European Surveillance of Veterinary Antimicrobial Consumption, 2018. Sales of veterinary antimicrobial agents in 30 European countries in 2016 (EMA/275982/2018). https://www.ema.europa.eu/en/documents/report/sales-veterinary-antimicrobial-agents-30-european-countries-2016-trends-2010-2016-eighth-esvac_en.pdf. Accessed 27 May 2020

European Commission (2015) Commission notice guidelines for the prudent use of antimicrobials in veterinary medicine (2015/C 299/04). Off J Eur Union, C 299:7–26

FAAST (2019) Farmed animal antimicrobial stewardship initiative. https://www.amstewardship.ca/faast-reviews/. Accessed 15 June 2019

FDA (2016) Food and drug administration summary: report on antimicrobials sold or distributed for use in food-producing animals. Table 11b fda.gov/downloads/ForIndustry/UserFees/AnimalDrugUserFeeActADUFA/UCM534243.pdf. Accessed 4 June 2019

FVE (2016) Federations of veterinarian of Europe: antimicrobial use in food-producing animals: replies to EFSA/EMA questions on the use of antimicrobials in food-producing animals and possible measures to reduce antimicrobial use. https://ema.europa.eu/documents/report/annex-replies-efsa/ema-questions-use-antimicrobials-food-producing-animals-eu-possible-measures-reduce-antimicrobial_en.pdf. Accessed 4 June 2019

Gehring R, Baynes RE, Riviere JE (2006) Application of risk assessment and management principles to the extralabel use of drugs in food-producing animals. J Vet Pharmacol Therap 29:5–14

Goossens H, Ferech M, Vander Stichele R, Elseviers M (2005) Outpatient antibiotic use in Europe and association with resistance: a crossnational database study. Lancet 365(9459):579–587

Grassly NC, Fraser C (2008) Mathematical models of infectious disease transmission. Nat Rev Microbiol 6:477–487

Griggs DJ, Johnson MM, Frost JA, Humphrey T, Jørgensen F, Piddock LVJ (2005) Incidence and mechanism of ciprofloxacin resistance in *Campylobacter* spp. isolated from commercial poultry flocks in the United Kingdom before, during, and after fluoroquinolone treatment. Antimicrob Agents Chemother 49:699–707

Haven ML, Wichtel JJ, Bristol DG, Fetrow JF, Spears JW (1992) Effects of antibiotic prophylaxis on postoperative complications after rumenotomy in cattle. J Am Vet Med Assoc 200(9):1332–1335. https://pubmed.ncbi.nlm.nih.gov/1601715/

Humphrey TJ, Jørgensen F, Frost JA, Wadda H, Domingue G, Elviss NC, Griggs JD, Piddock LVJ (2005) Prevalence and subtypes of ciprofloxacin-resistant *Campylobacter* spp. in commercial poultry flocks before, during, and after treatment with fluoroquinolones. Antimicrob Agents Chemother 49:690–698

Ives SE, Richeson JT (2015) Use of antimicrobial metaphylaxis for the control of bovine respiratory disease in high-risk cattle. Vet Clin N Am Food A 31:341–350

Klare I, Heier H, Claus H, Bohme G, Marin S, Seltmann G, Hakenbeck R, Antanassova V, Witte W (1995a) *Enterococcus faecium* strains with vanA-mediated high-level glycopeptide resistance isolated from animal foodstuffs and fecal samples of humans in the community. Microb Drug Resist 1:265–272

Klare I, Heier H, Claus H, Reissbrodt R, Witte W (1995b) VanA-mediated high-level glycopeptide resistance in *Enterococcus faecium* from animal husbandry. FEMS Microbiol Lett 125:165–171

Lago A, Godden SM, Bey R, Ruegg PL, Leslie K (2011) The selective treatment of clinical mastitis based on on-farm culture results: I. Effects on antibiotic use, milk withholding time, and short-term clinical and bacteriological outcomes. J Dairy Sci 94:4441–4456

Li J, Hao H, Cheng G, Wang X, Ahmed S, Shabbir MAB, Liu Z, Dai M, Yuan Z (2017) The effects of different enrofloxacin dosages on clinical efficacy and resistance development in chickens experimentally infected with *Salmonella* Typhimurium. Sci Rep 7:11676

Luo N, Sahin O, Lin J, Michel LO, Zhang Q (2003) *In vivo* selection of campylobacter isolates with high levels of fluoroquinolone resistance associated with gyrA mutations and the function of the CmeABC efflux pump. Antimicrob Agents Chemother 47:390–394

Maertens L (2007) Strategies for the reduction of antibiotic utilization during rearing, Giornate di Coniglicoltura ASIC. http://www.asic-wrsa.it/documenti/giornate2009/01_Maertens.pdf. Accessed 4 June 2019

MARAN (2015) Monitoring of antimicrobial resistance and antibiotic usage in animals in the Netherlands in 2014. http://www.wageningenur.nl/upload_mm/2/2/2/0ab4b3f5-1cf0-42e7-a460-d67136870ae5_NethmapMaran2015.pdf. Accessed 4 June 2019

Mathews KH (2002) Economic effects of a ban against antimicrobial drugs used in U.S. beef production. J Agric Appl Econ 34:513–530

McDermott PF, Bodeis SM, English LL, White DG, Walker RD, Zhao S, Simjee S, Wagner DD (2002) Ciprofloxacin resistance in *Campylobacter jejuni* evolves rapidly in chickens treated with fluoroquinolones. J Infect Dis 185:837–840

Mehdi Y, Létourneau-Montminy MP, Gaucher ML, Chorfi Y, Suresh G, Rouissi T, Brar SK, Côté C, Ramirez AA, Godbout S (2018) Use of antibiotics in broiler production: global impacts and alternatives. Anim Nutr 4:170–178

Monnier AA, Eisenstein B, Hulscher ME, Gyssens IC, DRIVE-AB WP1 Group (2018 Jun 1) Towards a global definition of responsible antibiotic use: results of an international multidisciplinary consensus procedure. J Antimicrob Chemother 73(Suppl_6):vi3–vi16

Moreno MA (2012) Survey of quantitative antimicrobial consumption in two different pig finishing systems. Vet Rec 171(13):325

Moreno MA (2014) Survey of quantitative antimicrobial consumption per production stage in farrow-to-finish pig farms in Spain. Vet Rec Open 1:e000002

Nilsson O (2012) Vancomycin resistant enterococci in farm animals – occurrence and importance. Infect Ecol Epidemiol 2:16959

O'Neill J (2015) Antimicrobials in agriculture and the environment: reducing unnecessary use and waste the review on antimicrobial resistence. https://amr-review.org/sites/default/files/Antimicrobials%20in%

20agriculture%20and%20the%20environment%20-%20Reducing%20unnecessary%20use%20and%20waste.pdf. Accessed 4 June 2019

OIE (2015) World organisation for animal health: fact sheets: antimicrobial resistance. Geneva: World Animal Health Organisation. http://www.oie.int/fileadmin/Home/eng/Media_Center/docs/pdf/Fact_sheets/ANTIBIO_EN.pdf. Accessed 4 June 2019

OIE (2018a) World organisation for animal health: OIE annual report on use of antimicrobial agents intended for use in animals.3rd report. https://www.oie.int/fileadmin/Home/eng/Our_scientific_expertise/docs/pdf/AMR/A_Third_Annual_Report_AMR.pdf. Accessed 27 May 2020

OIE (2018b) World organisation for animal health: responsible and pudent use of antimicrobial agents in veterinary medicine. http://www.oie.int/fileadmin/Home/eng/Health_standards/tahc/current/chapitre_antibio_use.pdf. Accessed 4 July 2019

Pardon B, Catry B, Dewulf J, Persoons D, Hostens M, De Bleecker K, Deprez P (2012) Prospective study on quantitative and qualitative antimicrobial and anti-inflammatory drug use in white veal calves. J Antimicrob Chemother 67:1027–1038

Rees GM, Barrett DC, Buller H, Mills HL, Reyher KK (2018) Storage of prescription veterinary medicines on UK dairy farms: a crosssectional study. Vet Rec 184 (5):153

Santman-Berends I, Swinkels J, Lam T, Keurentjes J, van Schaik G (2016) Evaluation of udder health parameters and risk factors for clinical mastitis in Dutch dairy herds in the context of a restricted antimicrobial usage policy. J Dairy Sci 99:2930–2939

Saraiva MMS, Moreira Filho ALB, Freitas Neto OC, Silva NMV, Givisiez PEN, Gebreyes WA, Oliveira CJB (2018) Off-label use of ceftiofur in one-day chicks triggers a short-term increase of ESBL-producing *E. coli* in the gut. PLoS ONE 13(9):e0203158

Schrijver R, Stijntjes M, Rodríguez-Baño J, Tacconelli E, Babu Rajendran N, Voss A (2017) Review of antimicrobial resistance surveillance programmes in livestock and their meat in Europe, with a focus on antimicrobial resistance patterns in humans. Clin Microbiol Infect 24(6):1–14

Seal BS, Lillehoj HS, Donovan DM, Gay CG (2013) Alternatives to antibiotics: a symposium on the challenges and solutions for animal production. Anim Health Res Rev 14(1):78–87

Skarp CPA, Hänninen ML, Rautelin HIK (2016) Campylobacteriosis: the role of poultry meat. Clin Microbiol Infect 22:103–109

Smith RA, Stokka GL, Radostits OM, Griffin DD (2001) Health and production management in beef feedlots. Herd Health: Food Anim Prod 14:581–633

Sneeringer S, MacDonald JM, Key N, McBride WD, Mathews K (2015) Economics of antibiotic use in U.S. livestock production. https://www.ers.usda.gov/publications/pub-details/?pubid=45488. Accessed 4 June 2019

Spina A, Kerr KG, Cormican M, Barbut F, Eigentler A, Zerva L, Tassios P, Popescu GA, Rafila A, Eerola E, Batista J, Maass M, Aschbacher R, Olsen KE, Allerberger F (2015) Spectrum of enteropathogens detected by the FilmArray GI panel in a multicentre study of community-acquired gastroenteritis. Clin Microbiol Infect 21:719–728

Stapleton K, Cawthraw SA, Cooles SW, Coldham NG, La Ragione RM, Newell DG, Ridley AM (2010) Selecting for development of fluoroquinolone resistance in a campylobacter jejuni strain 81116 in chickens using various enrofloxacin treatment protocols. J Appl Microbiol 109:132–1138

Swann MM, Baxter KL, Field HI (1969) Report of the joint committee on the use of antibiotics in animal husbandry and veterinary medicine. HMSO, London

Takahashi T, Ishihara K, Kojima A, Asai T, Harada K, Tamura Y, (2005) Emergence of fluoroquinolone resistance in *Campylobacter jejuni* in chickens exposed to enrofloxacin treatment at the inherent dosage licensed in Japan. J Veterinary Med Ser B 52(10):460–464

Tennant TC, Ives SE, Harper LB, Renter DG, Lawrence TE (2014) Comparison of tulathromycin and tilmicosin on the prevalence and severity of bovine respiratory disease in feedlot cattle in association with feedlot performance, carcass characteristics, and economic factors. J Anim Sci 92:5203–5213

Thompson PN, Stone A, Schultheiss WA (2006) Use of treatment records and lung lesion scoring to estimate the effect of respiratory disease on growth during early and late finishing periods in south African feedlot cattle. J Anim Sci 84:488–498

Timmerman T, Dewulf J, Catry B, Feyen B, Opsomer G, de Kruif A, Maes D (2006) Quantification and evaluation of antimicrobial drug use in group treatments for fattening pigs in Belgium. Prev Vet Med 74:251–263

Timsit E, Dendukuri N, Schiller I, Buczinski S (2016) Diagnostic accuracy of clinical illness for bovine respiratory disease (BRD) diagnosis in beef cattle placed in feedlots: a systematic literature review and hierarchical Bayesian latent-class meta-analysis. Prev Vet Med 135:67–73

Urban-Chmiel R, Grooms DL (2012) Prevention and control of bovine respiratory disease. J Livest Sci 3:27–36

US FDA (1997) U.S. Food and drug administration: extralabel animal drug use; fluoroquinolones and glycopeptides; order of prohibition. Fed Regist 62:27944–27947

USDA (2011) United States Department of Agriculture: Part IV: Health and and health management on U.S. feedlots with a capacity of 1000 or more head. https://www.aphis.usda.gov/animal_health/nahms/feedlot/downloads/feedlot2011/Feed11_dr_PartIV.pdf. Accessed 4 June 2019

Van Boven M, Veldman KT, de Jong MC, Mevius DJ (2003) Rapid selection of quinolone resistance in *Campylobacter jejuni* but not in *Escherichia coli* in individually housed broilers. J Antimicrob Chemother 52:719–723

Wade MA, Barkley AP (1992) The economic impacts of a ban on subtherapeutic antibiotics in swine production. Agribusiness 8:93–107

WHO (2001) World health organization: WHO global health strategy for containment of antimicrobial resistance. http://www.who.int/drugresistance/WHO_Global_Strategy_English.pdf. Accessed 4 June 2019

WHO (2017) World health organization: WHO guidelines on use of medically important antimicrobials in food-producing animals https://www.ncbi.nlm.nih.gov/books/NBK493702/. Accessed 4 June 2019

Ziping WU (2018) Antimicrobial use in food animal production: situation analysis and contributing factors. Front Agr Sci Eng 5(3):301–311

Laboratory Investigations and Result Interpretation

Kateřina Nedbalcová and Lucie Pokludová

Abstract

Proper clinical diagnosis, knowledge of the animal/s history, clinical experience as well as results of laboratory tests, support evidence-based, and correct decision on any use of antimicrobials. In the case of diagnosed disease with bacterium as a suspected etiological agent, in which resistance or even multiresistance can be expected, results of laboratory susceptibility testing are one of the essential tools for proper choice of an appropriate antimicrobial to be used for treatment. Each laboratory process starts with proper sampling. Chosen recommendations regarding bacteriological sampling are mentioned as well as sampling for yeast or microscopic fungi detection, which is briefly touched. Brief summary describing process of investigation and its limitations together with certain guidance for or links to the interpretation of laboratory results are given. To catch current progress and development of new methods of rapid antimicrobial susceptibility testing, a brief overview is given with view that some of those methods originally designed for human medicine could be, in the near future, also used in veterinary medicine. Laboratory data can significantly support the clinical and empirical decision for therapy using the proper antimicrobial considering the results of susceptibility testing. As a further benefit, data on susceptibility and resistance profiles allows us to follow the trends in susceptibility of the pathogens of concern. Proper samples taken by valid techniques and laboratory results on the efficacy of certain antimicrobials in vitro can contribute to success of treatment and minimizing of resistance development and spread.

Keywords

Sampling · Routine laboratory techniques · Laboratory diagnostics · Susceptibility testing · Resistance · Results interpretation

1 Sampling

Infectious diseases caused by various species of microorganisms are very common in herds of livestock animals. They occur either in chronic or in acute form, and cause great economic losses due to the suffering of animals from chronic stress. The acute course of some infections is accompanied by serious clinical signs with increased morbidity and mortality of diseased animals, resulting in increased treatment costs.

K. Nedbalcová (✉)
Veterinary Research Institute, Brno, Czech Republic
e-mail: nedbalcova@vri.cz

L. Pokludová
Institute for State Control of Veterinary Biologicals and Medicines, Brno, Czech Republic
e-mail: pokludova@uskvbl.cz

L. Pokludová (ed.), *Antimicrobials in Livestock 1: Regulation, Science, Practice*,
https://doi.org/10.1007/978-3-030-46721-0_8

Antimicrobials play an irreplaceable role in medicine for the treatment of bacterial infections and, currently, there are no alternative drugs to treat bacterial infections with appropriate antibacterial efficacy. In recent years, there has been a significant decline in the discovery and introduction of new antimicrobial molecules into clinical practice (Livermore 2012). However, the effectiveness of antimicrobials is seriously threatened by the increasing and rapidly spreading resistance of bacterial populations to them and this dangerous trend occurs globally. The cause of the increase in resistance is frequent overuse and misuse of antimicrobial agents as well as shortcomings in the field of prevention and control of infections, facilitating the spread of resistant microbes in intensive farming of livestock, which is concentrated in large quantities in a small space (Aarestrup et al. 2004). One of the most serious tasks that veterinary and human medicine faces are the effort to reduce excessive, and often useless, use of antibiotics, particularly antimicrobials with broad-spectrum of effectiveness and antimicrobials of higher generations. One way to support the above-mentioned effort in veterinary medicine is to test the susceptibility of specific pathogenic microorganisms occurring in certain herds or populations of animals and, depending on their current susceptibility, to select an effective antibiotic for therapy and determine the best regimen of treatment. According to general requirement for prudent use of antimicrobials, it is also part of various antimicrobial resistance control programmes (Official Journal of the European Union 2015/C 299/04).

The correct implementation of antimicrobial susceptibility testing (AST) is guided by several principles. Proper isolation and identification of the disease-causing bacterial agent from relevant samples taken by proper techniques are crucial for any further analysis. Before sampling, the assessment of a veterinarian based on differential diagnostics and on all possible causes of the disease is essential. As a necessary prerequisite, there should be the decision that the disease is of microbial etiology and the consideration which body sites of animals are affected (respiratory tract, intestines, joints, and others) and which method of sampling for bacteriological laboratory examination based on the site of infection will be the best one. Also, sampling tools/kits, as well as transport conditions, should be thoroughly chosen. A key role in the diagnostic process is performed by the attending veterinarian, who knows the anamnesis of herd/flock and individual animals and who decides on the treatment of the disease based on previous experiences and, above all, on clinical diagnostic procedures. However, in the case of suspicion of bacterial infection, it is advisable and of great importance to make all necessary steps to ensure that vet can rely on the results of laboratory examinations (Pokludova et al. 2018).

Each sample that is transported to the laboratory has to be properly described—what sample it is and when it was sampled. The sampling method also differs if sampling is carried out in vivo or postmortem. The transport of samples to the laboratory is another important point of sampling. The general principle is that the specimens corresponding to the predicted disease should be taken sterilely using sterile sampling tools or kits and should be placed in sterile and impermeable packings or containers suitable for transporting the collected sample types and be transferred to the environment at the recommended temperature. Special care should be paid to samples with suspected strictly anaerobic bacteria (e.g., *Bacteroides fragilis*, *Clostridium* spp.) and bacteria with specific growth requirements (e.g., *Actinobacillus pleuropneumoniae*, *Haemophilus* spp., *Campylobacter* spp.). The samples stored in this way should be transported as soon as possible to the laboratory, if possible, within 24–48 h. The transport time that ensures viability of the microorganism in the collected sample, also depends on the type of specimen and the temperature during transport. During the sampling, the prevention of contamination of the sample from the external environment is very important. The recommendations for sampling for bacteriology investigation in the laboratory are summarized in Table 1.

Table 1 The recommendations for sampling for bacteriology investigation

Animal species or category	Body site of sampling	Time of sampling	Type of sample	Storage of sample	Temperature	Recommended time of transport to the laboratory
Pigs	Respiratory tract	In vivo	Nasal or nasal secret swab	In transport medium	Room temperature	During 24 h
			Tonsillar swab			
		Post mortem	Swab from lung	In transport medium	Room temperature	During 24 h
			Sample of lung tissue with lesions	Sterile container or other sterile and impermeable packings	Refrigerate	As soon as possible
					Freeze	More than 24 h
Cattle			Bronchoalveolar lavage	Sterile containers	Refrigerate	As soon as possible
Horses					Freeze	More than 24 h
Poultry			Cadaver (piglets, chickens)	Sterile containers	Refrigerate	As soon as possible
					Freeze	More than 24 h
Pigs	Gastrointestinal tract	In vivo	Rectal swab	In transport medium	Room temperature	During 24 h
Cattle						
Horses			Sample of feces	Sterile containers or other sterile and impermeable packings	Refrigerate	As soon as possible
Poultry					Freeze	More than 24 h
		Post mortem	A part of the ligature of the intestine or colon	Sterile container or other sterile and impermeable packings	Refrigerate	As soon as possible
					Freeze	More than 24 h
			Samples of tissue of other organs of the gastrointestinal tract	Sterile containers	Refrigerate	As soon as possible
					Freeze	More than 24 h
			Cadaver (piglets, chickens)	Sterile containers	Refrigerate	As soon as possible
					Freeze	More than 24 h
Pigs	Urogenital tract	In vivo	Samples of urine	Sterile containers or tubes	Refrigerate	During 24 h
			Samples of sperm	Sterile containers or tubes	Refrigerate	As soon as possible
			Swab from urogenital tract	In transport medium	Room temperature	During 24 h
Cattle		*Post mortem*	Swab from urogenital tract	In transport medium	Room temperature	During 24 h
Horses			Samples of tissue or organs of urogenital tract	Sterile containers or other sterile and impermeable packings	Refrigerate	As soon as possible
Poultry					Freeze	More than 24 h
Pigs	Skin	In vivo or *post mortem*	Swab of injury	In transport medium	Room temperature	During 24 h
			Swab of dermatitis, exanthema	In transport medium	Room temperature	During 24 h
Cattle			Swab of abscess or pus	In transport medium	Room temperature	During 24 h
Horses			Scraping of dry skin lesions	Sterile containers or tubes with sterile physiological saline solution	Refrigerate	During 24 h
Poultry						
Dairy cows	Milk	In vivo	The milk from clean and dry teats after preparation	Sterile leak-proof containers	Refrigerate	As soon as possible
					Freeze	More than 24 h
			The milk from bulk tanks	Sterile leak-proof containers	Refrigerate	As soon as possible
					Freeze	More than 24 h

The sample should be sent to the laboratory together with a request form for laboratory investigation that should contain information about the sample origin and identification data about the veterinarian sending the sample. The main important data that should be mentioned in the request form are identification of the husbandry/place of the sample origin, name of herd or owner's name, species and age category of the animal from which the sample originates, date of sampling, if the sampling was performed in vivo or postmortem, exact sampling site (tissue sampled, swab collected), anamnesis of the herd and diseased animal with data of history of disease in herd, description of clinical signs and where appropriate *postmortem* finding at the necropsy, data about the previous course of therapy, if any, which antimicrobials were used for treatment at the level of herd and at the level of the sampled animal, and other important information in relation to the sample recognized by the veterinarian as appropriate.

After receiving samples, the first cultivation is usually performed on nonselective agar medium, such as blood agar. The conditions of incubation are chosen by the predicted possible pathogen, which is considered to be responsible for the outbreak of the disease. The relevant individual colonies of primo-culture are subsequently inoculated into selective growth media and, after incubation, this culture is suitable for further laboratory tests, such as further typing of bacteria and detection of virulence factors. For example, in the routine laboratory testing, the first cultivation on blood agar plates together with a streak of *Staphylococcus aureus* culture and sub-cultivation on chocolate agar is recommended for the isolation of *Actinobacillus pleuropneumoniae* or *Haemophilus parasuis* in samples of the respiratory tract of pigs with pneumonia. Where the sub-cultivation of pathogen is successful, the AST can be performed. Equally important is the current accurate identification of the pathogenic microorganism as a source of health problems, in particular for the selection of tested antimicrobial agents with regard to the efficacy of specific antimicrobials against individual pathogens.

2 What Should the Practitioner Know About Laboratory Investigation?

The laboratory investigation is a very useful tool for veterinary practitioners in the selection of effective therapy of bacterial infections of individual animals or the incidence of infections in livestock animals. The correctly performed laboratory investigation of correctly taken samples provides precise and complete identification of causative agents of infection, including some important properties of them that can influence the treatment strategy both at individual animal level and at herd/flock level. The results enable the targeted therapy of infections by selection of appropriate drugs and especially in the case of quantitative methods, e.g., test for Minimum Inhibitory Concentration (MIC), the dose, route of administration, and dosing schedule can be better adjusted. The laboratory investigation also plays a role in various control and prevention programs that have to prevent the occurrence of outbreaks and spread of infectious diseases.

Some basic knowledge of laboratory investigation accompanied by proper communication between the veterinary practitioners and laboratories is an important tool to gain as much information as possible with respect to the rational use of antimicrobials.

Veterinarians should know in general that part of the laboratory microbiological investigation is culturing microbes on basic/selective media and should be aware of the influence of wrong sampling techniques as well as negative influence of wrong transport conditions on the performance and outcomes of the laboratory tests. They should also consider that currently used routine culturing techniques take some time (also depending on the type of microorganism and its speed of multiplication/growth parameters). Veterinarians should also know other methods of confirmation of certain microbes (PCR), methods for laboratory investigation of virulence and toxin production. Laboratories should provide on their websites a list of provided tests together with brief explanation. The tick-field form accompanying each sample can also facilitate cooperation of the vet with lab staff and ensure proper tests to be performed by the laboratory.

The veterinarian should be aware of the limitations of disk diffusion susceptibility testing (i.e., almost qualitative results reported only) and benefits of MICs tests (quantitative results, possible impact on correct dose setting, from a long time perspective, the possibility of resistance trends evaluation). It is also important to know the role of certain antimicrobials involved in antibiogram (most of them being group/class representatives and representing directly results important for choice of antimicrobials, but others can be indicative of certain resistance profiles/ emerging resistance). Considerations in the assignment of antimicrobial agents to specific test/report groups include clinical efficacy, prevalence of resistance, minimizing emergence of resistance, cost, regulatory agency–approved clinical indications for use, and current consensus recommendations for first choice and alternative agents. Tests of selected agents may be useful for infection control and/or monitoring purposes.

According to the data on microbe identification and its (quantitative) susceptibility, further consideration can be made by veterinarians, such as switching to another suitable antimicrobial or combination of antimicrobials in the treatment failure or decision which antimicrobial is the most proper one not only based on results in vitro but based on MIC and considering pharmacokinetic characteristics (choice of the proper route of administration, dosing schedule, and dose). It should be highlighted that the final choice made by the veterinarian is not based solely on results of AST, but also on thorough consideration of the health status of the animal/ s and also possible impacts on residue depletion (withdrawal period), adverse reactions, interactions, contraindications, and possible use in age/production categories, etc.

The interpretation of the results of AST, the knowledge of the categorization of individual antimicrobials to pharmacological groups and cross-resistance within these groups is essential for the veterinarian. For example, clindamycin is an agent of the lincosamides group of antimicrobials and, therefore, the results of clindamycin susceptibility testing will be valid for dairy cows for other representatives of lincosamides—lincomycin and pirlimycin. Another example is that if the laboratory determined the *Staphylococcus aureus* strain as methicillin resistant (MRSA), it is useless to administer beta-lactam antimicrobials (although there are exceptions from higher generations of cephalosporins or beta lactam/beta-lactamase inhibitors but these molecules are authorized for use in human medicine). More examples are given in the Sect. 4.3 Expert rules.

3 Routine Laboratory Techniques Used for Susceptibility/ Resistance Testing

The performance of AST is important for confirmation of susceptibility to chosen empirical antimicrobial agents and for the detection of resistance or types of resistance in individual bacterial isolates. The aim of AST is to assure susceptibility to drugs for particular bacterial infections (Jorgensen and Ferraro 2009). The AST has to be performed in accordance with internationally accepted procedures. The methodologies for AST of bacteria from animal sources are given and published, especially by the Clinical Laboratory Standards Institute (CLSI), but also other national institutions, for example, Comité de l'Antibiogramme de la Société Française de Microbiologie (CA-SFM) among others (Schwarz et al. 2010). The OIE Terrestrial manual—Guideline 2.1 Laboratory methodologies for bacterial antimicrobial susceptibility testing (2012) listed some guidelines that are currently available for AST published by British Society for Antimicrobial Chemotherapy (BSAC, UK), Clinical Laboratory and Standards Institute (CLSI, USA), Comité de l'Antibiogramme de la Société Française de Microbiologie (CA-SFM, France), Commissie Richtlijnen Gevoeligheidsbepalingen (CRG, The Netherlands), Deutsches Institut für Normung (DIN, Germany), European Committee on Antimicrobial Susceptibility Testing (EUCAST),

Japanese Society for Chemotherapy (JSC, Japan), and Swedish Reference Group for Antibiotics (SRGA, Sweden). The interpretation of the results should be performed, whenever possible, according to the standards based on which the tests were performed (special caution should be paid once the standards differ in some critical parameters of test conditions).

The methodologies of CLSI are currently considered as a major standard for veterinary AST. The documents of CLSI contain description of the exact method of AST implementation, recommendation for testing, criteria for quality control testing, the interpretative criteria for some veterinary pathogens and for veterinary drugs, valid for individual animal species (dogs, cats, horses, cattle, swine, and chicken), and other important principles and recommendations for routine testing by veterinary microbiology laboratories (CLSI 2008, 2013a, b, 2018a). All methodological documents are regularly updated and, since the methods and interpretive criteria can be changed over time, it is important to follow the latest edition. Documents from other institutions are primarily based on them.

Recently, in 2015, the Veterinary Committee on Antimicrobial Susceptibility Testing (VetCAST) was established as a subcommittee of the European Committee on Antimicrobial Susceptibility Testing (EUCAST). Its remit is to define clinical breakpoints for antimicrobial drugs used in veterinary medicine in Europe. The main VetCAST aims are to advise on all aspects of AST for bacterial pathogens of animal origin and animal bacteria with zoonotic potential and to permit the standardization of AST methodology to ensure reproducibility of data between laboratories for estimating the prevalence of resistance (Toutain et al. 2017).

All AST documents mentioned above allow AST of many different bacterial species. However, for some specific veterinary pathogens, such as *Haemophilus parasuis*, no approved internationally standardized methodology exists. It is possible to adopt a method approved for a phylogenetically closely related organism, but it must be stated clearly that the method used has not been approved for the tested species and the method of testing was modified from a method of AST testing for another member of a related species or the same genus and the chosen methodology must be validated according to recommendations, for example, in CLSI document M37-A3 (2008).

Usually, the phenotypic susceptibility testing is performed and when required, it is supplemented with molecular analysis for the presence of resistance genes. The internationally accepted methods for AST are agar disk diffusion test, dilution tests (agar dilution test, broth dilution test, and microdilution test), and antibacterial gradient diffusion test (Etest). The results obtained by disk diffusion method are qualitative and dividing of isolates into categories of susceptibility is made according to the size of the zones of growth inhibition around the disk with antimicrobials. The dilution and gradient diffusion methods allow quantitative testing of susceptibility or resistance of bacteria based on determination of minimal inhibitory concentration (MIC) of antimicrobials. The MIC is determined for individual bacterial isolates against each tested antimicrobial agent and it is the lowest concentration of an antimicrobial that inhibits bacterial growth in culture incubation under normal conditions. Based on the AST results, the isolates can be divided into three categories of susceptibility—susceptible, intermediate, or resistant (Jorgensen and Ferraro 2009). However, the zone diameters may not correspond precisely to the MIC breakpoints. Regression line analysis should not be used to extrapolate MIC values from measurements of zones of inhibition because, in many cases, the relationship, while mathematically correct, cannot be considered comparable to an MIC derived by actual dilution testing for a given isolate (CLSI 2008).

The exact definitions of the susceptibility categories and examples of breakpoints and the interpretation of AST results are given in the following text and in Table 2 according to CLSI document VET08, word for word (CLSI 2018a).

Susceptible (S) A category defined by a breakpoint that implies that isolates with an MIC at or below or zone diameters at or above the

Table 2 Example of breakpoints and interpretive categories (CLSI 2013a)

		Interpretative categories and zone diameter breakpoints, nearest whole mm			Interpretive categories and MIC breakpoints μg/ml		
Antimicrobial agent	Disk content	S	I	R	S	I	R
X	30 μg	≥20	15–19	≤14	≤4	8–16	≥32
Y	–	–	–	–	≤1	2	≥4
Z	10 μg	≥16	–	–	≤1	–	–

susceptible breakpoint are inhibited by the usually achievable concentrations of antimicrobial agent when the dosage recommended to treat the site of infection is used, resulting in likely clinical efficacy.

Intermediate (I) A category defined by a breakpoint that includes isolates with MICs or zone diameters within the intermediate range that approach usually attainable blood and tissue levels and for which response rates may be lower than for susceptible isolates; *NOTE*: The intermediate category implies clinical efficacy in body sites for which the drugs are physiologically concentrated or when a higher than normal dosage of a drug can be used. This category also includes a buffer zone, which should prevent small, uncontrolled, technical factors from causing major discrepancies in interpretations, especially for drugs with narrow pharmacotoxicity margins.

Resistant (R) A category defined by a breakpoint that implies that isolates with an MIC at or above or zone diameters at or below the resistant breakpoint are not inhibited by the usually achievable concentrations of the agent with normal dosage schedules and/or that demonstrate MICs or zone diameters that fall in the range in which specific microbial resistance mechanisms are likely, and clinical efficacy of the agent against isolates has not been reliably shown in isolates with similar phenotypes.

Non-susceptible (NS) A category used for isolates for which only a susceptible breakpoint is designated because of the absence or rare occurrence of resistant strains. Isolates for which the antimicrobial agent MICs are above or zone diameters are below the value indicated for the susceptible breakpoint should be reported as non-susceptible; *NOTE 1*: An isolate that is interpreted as non-susceptible does not necessarily mean that the isolate has a resistance mechanism. It is possible that isolates with MICs above the susceptible breakpoint that lack resistance mechanisms may be encountered within the wild-type distribution subsequent to the time the susceptible only breakpoint was set; *NOTE 2*: The term "non-susceptible" should not be used when describing an organism/drug category with intermediate and resistant interpretive categories. Isolates that are in the categories of "intermediate" or "resistant" could be called "not susceptible" rather than "non-susceptible."

For antimicrobial agent X with breakpoints in Table 2, the susceptible breakpoint is ≤4 μg/ml or ≥20 mm and the resistant breakpoint is ≥32 μg/ml or ≤14 mm. For some antimicrobial agents (e.g., antimicrobial agent Y), only MIC breakpoints may be available. For these agents, the disk diffusion zone diameters do not correlate with MIC values. Technical issues may also preclude the use of the disk diffusion method for some agents. For some antimicrobial agents (e.g., antimicrobial agent Z) only a "susceptible" category exists. For these agents, the absence or rare occurrence of resistant strains precludes defining any result categories other than "susceptible." For strains yielding results suggestive of a "nonsusceptible" category, organism identification and antimicrobial susceptibility test results should be confirmed. In examples Y and Z, a dash mark (–) indicates that a disk is not available or that breakpoints are not applicable.

Currently, the EUCAST (2019a) gives new definitions of susceptibility categories (S, I, and R) with the modified definition of the "I-category" but the abbreviation in reports is still "I":

- *S—Susceptible, standard dosing regimen*: A microorganism is categorized as "Susceptible, standard dosing regimen," when there is a high likelihood of therapeutic success using a standard dosing regimen of the agent.
- *I—Susceptible, increased exposure*∗: A microorganism is categorized as "Susceptible, Increased exposure∗" when there is a high likelihood of therapeutic success because exposure to the agent is increased by adjusting the dosing regimen or by its concentration at the site of infection.
- *R—Resistant*: A microorganism is categorized as "Resistant" when there is a high likelihood of therapeutic failure even when there is increased exposure.

∗Exposure is a function of how the mode of administration, dose, dosing interval, infusion time, as well as distribution and excretion of the antimicrobial agent will influence the infecting organism at the site of infection.

Furthermore, the term "Area of technical uncertainty (ATU)" was introduced in susceptibility testing where a warning is needed to alert the laboratory to the uncertainty of the AST result. The ATU is defined by one (or on occasion more) MIC value or by one (or a range of) inhibition zone diameter. It is not a susceptibility testing category like S, I, and R and it does not interfere with the interpretation of results. The warning affects the laboratory, not the clinician, and the laboratory needs a strategy to (1) ascertain the correctness (e.g., repeat the test or perform an alternative test) or (2) to report the uncertainty of the result (report results in the ATU as "uncertain" with a comment or report results in the ATU as "R," if there are several good alternatives in the AST report this may be the easiest and safest option). ATU warnings are listed in EUCAST Breakpoint Tables, version 9.0 (EUCAST 2019b). For further updates is advised to regularly follow information on web page of the EUCAST (https://eucast.org/).

The parallel performance of quality controls (QC) with reference strains is always necessary for AST of isolates. Lists of approved reference strains with acceptable zone diameter ranges or MIC are included in the AST documents. The reference strains must be relevant to the tested bacterial species (Schwarz et al. 2010). The reference strains with specific characteristics as, e.g., production of extended-spectrum beta-lactamases (ESBLs) are also indicated in the respective guidance documents.

3.1 Disk Diffusion Susceptibility Test

The agar disk diffusion tests are a commonly used method for the determination of susceptibility/resistance of bacterial pathogens in veterinary diagnostic laboratories. The test is recommended for rapidly growing bacterial pathogens. Susceptibility testing of anaerobic bacteria isolated from animals should be performed by a dilution method in accordance with procedures in CLSI document M 11. The use of disk diffusion tests is not recommended. The method is based solely on the presence and size or absence of a zone of inhibition (CLSI 2012, 2013a). The method is relatively cheap and flexible as for the set of antimicrobials to be tested comparing e.g. microdilution sets, where types of antimicrobials are fixed.

The following points are considered as critical for the correct test performance: (1) use disks with the correct antimicrobial concentrations (and storage of disks in appropriate conditions to protect them against humidity); (2) the correct density of active growing inoculum for inoculation of agar plates; (2) the correct selection of agar medium for tested bacterial species (including correct thickness of agar layer); (4) observe the required conditions and time of incubation for the tested species; and (5) assess the results according to available interpretative criteria in the relevant AST guidelines.

The disks with antimicrobial agents in required concentrations of various manufacturers are commercially available. The density of inoculum is always adjusted with sterile saline or broth to 0.5 McFarland standard, which corresponds to the density of bacterial culture 1–2 × 10^8 CFU/ml. The divalent cation adjusted Mueller–Hinton agar (CAMHA) for non-fastidious aerobe bacteria is

Table 3 Standard methods for AST of some fastidious and specific veterinary pathogens by disk diffusion method (CLSI 2013b)

Organism	Medium	Incubation
Histophilus somni and *Actinobacillus pleuropneumoniae*	Chocolate MHA	35 °C/24 h; 5 ± 2% CO_2
Campylobacter spp.	MHA with 5% defibrinated sheep blood	36–37 °C/48 h or 42 °C/24 h 10% CO_2; 5% O_2; 85% N_2 or microaerophilic environment
Streptococcus spp.	MHA + 5% sheep blood	35 °C/20–24 h; 5 ± 2% CO_2
Pasteurella multocida and *Mannheimia haemolytica*	MHA + 5% sheep blood	35 °C/18–24 h

MHA—Mueller–Hinton agar

used. The CAMHA plates with inoculated culture and disks with antimicrobials are incubated aerobically at 35 °C for 18–24 h. The conditions of standard procedures for AST by agar disk diffusion test of some fastidious and special problem veterinary pathogens are listed in Table 3. The categories of susceptibility have to be evaluated according to the size of the inhibition zone (CLSI 2013b).

3.2 Broth and Agar Dilution Tests

Broth and agar dilution techniques may be used to measure quantitatively the in vitro activity of an antimicrobial agent against bacterial cultures. The microdilution method involves the use of small volumes of broth in sterile microdilution trays with round or conical bottom wells, each containing 0.05–0.1 ml of broth (CLSI 2013a). General principles apply to the correct performance of dilution tests and similarly to disk diffusion test. They are (1) correct preparation of a series of tubes, wells, or agar plates with broth or agar medium for tested bacterial species with various concentrations of the antimicrobial agents; (2) the correct density of active growing inoculum for inoculation of tubes, wells, or agar plates; (3) meeting the required conditions and time of incubation for the tested species; and (4) assessment of results according to available interpretative criteria in the AST guidelines.

The results of dilution tests are determination of MICs providing information about the concentration of the antimicrobial agent needed at the infection site to inhibit the infectious organism. The MICs have been usually determined using prepared twofold dilutions of antimicrobials (e.g., 1, 2, 4, 8, 16, 32 mg/L) in a liquid growth medium in tubes (broth dilution method), wells in microdilution trays (microdilution method) or agar plates (agar dilution method). The MICs obtained by dilution tests do not represent an absolute value, because the real MIC is somewhere between the lowest test concentration that inhibits the growth of tested bacterial culture and the next lowest test concentration. Therefore, the true MIC value cannot be accurately determined, and this should be reported as equal to, or less than the lowest tested concentration (CLSI 2013a).

Broth Dilution and Microdilution Procedure The cation adjusted Mueller–Hinton broth (CAMHB) is recommended as the medium for AST of commonly and rapidly growing non-fastidious pathogens. The tubes with CAMHB and antimicrobials are inoculated by inoculum so that after inoculation each tube or well in microdilution trays contains approximately 5×10^5 CFU/ml. The results of the test are interpreted after incubation at 35 °C for 16–20 h. The conditions of standard procedures for AST by broth dilution and microdilution tests of some fastidious and special problem veterinary pathogens are listed in Table 4 (CLSI 2013b).

Agar Dilution Procedure The cation adjusted Mueller–Hinton agar (CAMHA) is recommended for routine AST of aerobic and facultative anaerobic bacteria. The agar plates with antimicrobials are rapidly inoculated by inoculum with a density

Table 4 Standard methods for AST of some fastidious and special problem veterinary pathogens by dilution methods (CLSI 2013b)

Organisms	Method	Medium	Incubation
Staphylococcus hyicus	Broth microdilution	CAMHB + thymidine phosphorylase (0.2 IU/ml); for sulphonamides and trimethoprim only	35 °C/18–24 h
Histophilus somni and *Actinobacillus pleuropneumoniae*	Agar dilution	Chocolate MHA	35 °C/24 h; 5 ± 2% CO_2
	Broth microdilution	VFM	
Campylobacter spp.	Agar dilution	MHA + 5% defibrinated sheep blood	36–37 °C/48 h or 42 °C/24 h 10% CO_2; 5% O_2; 85% N_2 or microaerophilic environment
	Broth microdilution	CAMHB + 2.5% to 5% lysed horse blood	
Streptococcus spp.	Agar dilution	MHA + 5% sheep blood	35 °C/20–24 h; 5 ± 2% CO_2
	Broth microdilution	CAMHB + lysed horse blood (2.5–5% v/v/)	35 °C/20–24 h
Listeria spp.	Broth microdilution	CAMHB + lysed horse blood (2.5–5% v/v/)	35 °C/20–24 h
Pasteurella multocida and *Mannheimia haemolytica*	Broth microdilution	CAMHB	35 °C/18–24 h

CAMHB cation-adjusted Mueller–Hinton broth
MHA Mueller–Hinton agar
VFM veterinary fastidious medium

corresponding to 0.5 McFarland standard; it is approximately 1–2 × 10^8 CFU/ml on the agar surface. After incubation at 35 °C for 16–20 h, the bacterial growth on agar plates is assessed. The modification of the test procedure for fastidious organisms is shown in Table 4 (CLSI 2013b).

3.3 Antimicrobial Gradient Test

Antimicrobial gradient method uses the principle of antimicrobial concentration gradient in an agar medium as a means of determining susceptibility. The tests based on this method are commercially available and they are often routinely used in veterinary laboratories. Usually, the tests contain strips with a dried antimicrobial concentration gradient and on the upper surface, there is a scale of concentrations. The MIC is determined by intersection of the lower part of the ellipse-shaped growth inhibition area with the test strip (Jorgensen and Ferraro 2009).

3.4 Specific Tests

The performing of routine antibiotic susceptibility tests (determination of inhibition zone or MIC) is insufficient in some cases, when it is necessary to determine the types of resistance, the mechanisms of its formation or to identify genes that are responsible for the emergence of resistance. For this purpose, some other specific or complementary tests are carried out; the following is a brief description of the principle of interpretation and interpretation of some specific tests. The following listed tests are based on current CLSI and EUCAST methodologies CLSI 2018a, b; EUCAST 2019c).

High-Level Aminoglycoside Resistance (HLAR) Screening HLAR is a significant acquired resistance factor for AST of *Enterococcus* spp. infections, where enterococci could be etiological agents occurring especially in poultry, in which also multiresistant strains used to be detected. All enterococci naturally have

Table 5 Interpretation of the HLAR tests

		Resistance	Inconclusive	Susceptible
Gentamicin	Disk diffusion test Disk 120 μg	6 mm	7–9 mm	≥10 mm
	Broth dilution test 500 μg/ml	Any growth	–	No growth
Streptomycin	Disk diffusion test Disk 300 μg	6 mm	7–9 mm	≥10 mm
	Broth dilution test 1000 μg/ml	Any growth	–	No growth

Table 6 Interpretation of the oxacillin/cefoxitin tests

			Susceptible	Resistant
Oxacillin	MIC tests	*Staphylococcus aureus*	≤2 μg/ml	≥4 μg/ml
		CoNS	≤0.25 μg/ml	≤0.5 μg/ml
Cefoxitin	MIC tests	*Staphylococcus aureus*	≤4 μg/ml	≥8 μg/ml
		CoNS	–	–
Cefoxitin	Disk diffusion test	*Staphylococcus aureus*	≥22 mm	≤21 mm
		CoNS	≥25 mm	≤24 mm

low-level resistance to aminoglycosides, which invalidates the use of the disk test with usual concentrations of antimicrobial agents. HLAR is only meaningful for a testing method. When an enterococcal strain has high-level resistance to the aminoglycoside, there is no synergism and combination therapy with a beta-lactam drug will not have the desired bactericidal effect. Therefore, it is important to detect the presence of high-level resistance in order to predict aminoglycoside synergy. HLAR screening is carried out by testing susceptibility of enterococci to gentamicin or streptomycin by performing disk diffusion test with a high content of antimicrobials or dilution tests using high tested concentrations of antimicrobials. The interpretation of the HLAR tests according to EUCAST and CLSI is shown in Table 5.

Methicillin-Resistant Staphylococci These strains have been isolated from different infections in food-producing animals and horses, especially mastitis of cattle, respiratory tract infection of horses, staphylococcal dermatitis of poultry, and among pig populations, also from asymptomatic carriers. Strains that are oxacillin and methicillin resistance, historically termed methicillin-resistant *S. aureus* (MRSA), are resistant to all beta-lactam agents, including cephalosporins and carbapenems, although they may be susceptible to the newest class of MRSA active cephalosporins (e.g., ceftaroline). Strains of MRSA causing healthcare-associated infections are often multiplied resistant to other commonly used antimicrobial agents, including erythromycin, clindamycin, fluoroquinolones, and tetracycline, while strains causing community-associated infections are often resistant only to beta-lactam agents and erythromycin, and may be resistant to fluoroquinolones. Moreover, MRSA strains with decreased susceptibility to vancomycin have been reported. Staphylococcal resistance to oxacillin/methicillin occurs when an isolate produces an altered penicillin-binding protein, PBP2a, which is encoded by the *mec*A gene. The variant penicillin-binding protein binds beta-lactams with lower avidity, which results in resistance to this class of antimicrobial agents. Other species of straphylococci can also be methicillin resistant (especially of importance in companion animals are *Staphylococcus pseudintermedius* strains). Simple oxacillin/cefoxitin tests (MIC tests and disk diffusion test) are recommended for the detection of methicillin-resistant *S. aureus* or coagulase-negative staphylococci. The interpretation of the oxacillin/cefoxitin tests according to EUCAST and CLSI is shown in Table 6.

Rapid Inducible Beta-Lactamase Screen Test For determination of beta-lactam antibiotic resistance in a target bacterial strain, the strain is grown in the presence of both a beta-lactamase-inducing antibiotic and a beta-lactam indicator antibiotic (which kills or inhibits the growth of bacteria) unable to hydrolyze beta-lactam antibiotics. Growth, indicative of drug resistance in the target strain, is monitored by detecting a fluorophore released by the enzymatic cleavage of a metabolizable fluorogenic compound. The sample is inoculated into growth media containing both an inducing and an indicator noninducing beta-lactam antibiotic. It is considered that adding growth media already inoculated with a sample to the inducing and indicator antibiotics either dried or resuspended in a compatible buffer is the equivalent of inoculating growth media containing the two antibiotics, so long as both drugs are present simultaneously during incubation. The inducing antibiotic is selected from the group comprising the beta-lactam antibiotics that have been shown empirically to cause induction of the beta-lactamase gene with concomitant high-level expression of the enzyme. Inducer antibiotics include cefoxitin, imipenem, sulbactam, and clavulanic acid; cefoxitin is preferred.

One of the resistance genes that codes for beta-lactamase is *Amp*C that can be detected in Enterobacteriales. It has to be induced by, e.g., clavulanic acid and cefoxitin as strong inducers (ampicillin is considered as moderate inducer). Enterobacteriales that are reported as resistant to cefoxitin or to amoxicillin/clavulanic acid must be reported as resistant to all penicillins and cephalosporins.

ESBL Tests (Screen + Confirmation) Extended-spectrum beta-lactamases (ESBL) are enzymes that mediate resistance to extended-spectrum (third generation) cephalosporins (e.g., ceftazidime, cefotaxime, and ceftriaxone) and monobactams (e.g., aztreonam) but do not affect cephamycins (e.g., cefoxitin and cefotetan) or carbapenems (e.g., meropenem and imipenem). The frequent producers of ESBL are Gram-negative bacteria such as *Pseudomonas* spp., the members of the family Enterobacteriales, e.g., *Escherichia coli*, *Klebsiella pneumoniae*, and others. The sensitivity of screening for ESBLs in enteric organisms can vary depending on which antimicrobial agents are tested. The use of more than one of the five antimicrobial agents suggested for screening will improve the sensitivity of detection. Cefpodoxime and ceftazidime show the highest sensitivity for ESBL detection. ESBL activity in certain bacteria can be detected by using standard disk diffusion susceptibility test methods incorporating specific cephalosporins (cefotaxime and ceftazidime) in combination with a beta-lactamase inhibitor (clavulanic acid) and measuring the resulting zones of inhibition. The interpretation of the ESBL tests according to EUCAST and CLSI is shown in Table 7.

Table 7 Interpretation of the ESBL tests

	Positive test	
	Disk diffusion (mm)	MICs (μg/ml)
Cefpodoxime	≤22	≥2
Ceftazidime	≤22	≥2
Aztreonam	≤27	≥2
Cefotaxime	≤27	≥2
Ceftriaxone	≤25	≥2

Screening for Carbapenemases Carbapenemases are emerging resistance determinants in Gram-negative pathogens, including Enterobacteriales, *Pseudomonas*, and *Acinetobacter*. These carbapenem-hydrolyzing enzymes may be the cause of resistance of a broad variety of beta-lactams and also may result in the spread of resistance to multiple classes of antimicrobials like fluoroquinolones, aminoglycosides, and some potentiated sulphonamides. The first cause of suspicion of carbapenemase production in a clinical isolate is an increase in carbapenem MIC or a decrease in the inhibition zone diameter. The carbapenem susceptibility ranges for Enterobacteriales, *Pseudomonas*, and *Acinetobacter* are shown in Table 8. However, detection of carbapenemase producers based only on MIC values or zone diameters may lack

Table 8 Breakpoints of carbapenem susceptibility according to European (EUCAST 2019c) and US (CLSI 2018a, b) guidelines

		EUCAST						CLSI					
		MIC breakpoint (mg/L)			Zone diameter breakpoint (mm) for 10 μg disk			MIC breakpoint (mg/L)			Zone diameter breakpoint (mm) for 10 μg disk		
Microorganism	Carbapenem	S	I	R	S	I	R	S	I	R	S	I	R
Enterobacteriales	Doripenem	1	2	4	24	21–23	20	1	2	4	23	20–22	19
	Ertapenem	0.5	1	2	25	22–24	21	0.5	1	2	22	19–21	18
	Imipenem	2	4–8	16	22	16–21	15	1	2	4	23	20–22	19
	Meropenem	2	4–8	16	22	16–21	15	1	2	4	23	20–22	19
Pseudomonas	Doripenem	1	2	4	25	22–24	21	2	4	8	19	16–18	15
	Ertapenem	–	–	–	–	–	–	–	–	–	–	–	–
	Imipenem	4	8	16	20	17–19	16	2	4	8	19	16–18	15
	Meropenem	2	4–8	16	24	18–23	17	2	4	8	19	16–18	15
Acinetobacter	Doripenem	1	2	4	24	21–23	20	2	4	8	18	15–17	14
	Ertapenem	–	–	–	–	–	–	–	–	–	–	–	–
	Imipenem	2	4–8	16	23	17–22	16	2	4	8	22	19–21	18
	Meropenem	2	4–8	16	21	15–20	14	2	4	8	18	15–17	14

sensitivity. Carbapenem MICs are expected to substantially rise only in the presence of an additional resistance mechanism, like permeability lesions due to outer membrane protein mutation, or simultaneous production of *Amp*C cephalosporinases or ESBLs. To avoid false-negative results, or to maximize detection sensitivity, it has been proposed to screen enterobacterial isolates for carbapenemase activity if they exhibit MICs of ertapenem greater than or equal to 0.5 mg/L or MICs of imipenem or meropenem greater than or equal to 1 mg/L, or to screen any enterobacterial isolate displaying a slight decrease in susceptibility to carbapenems compared with the wild-type organism. Given the variability of carbapenemase screening results by antibiotic susceptibility tests, detection of carbapenemase-producing organisms by phenotypic, culture-based techniques is an optional step that avoids delayed reporting of such strains to the clinic, in case genotypic tests are not readily available. They include the modified Hodge test, inhibitor-based tests, and the use of specific culture media (Hammoudi et al. 2014). The simplified scheme of detection of carbapenemases is shown in Fig. 1.

D-Test (Inducible AMR Clindamycin) Erythromycin (a macrolide) and clindamycin (a lincosamide) represent two distinct classes of antimicrobial agents that inhibit protein synthesis by binding to the 50S ribosomal subunits of bacterial cells. In staphylococci, resistance to both of these antimicrobial agents can occur through methylation of their ribosomal target site. Such resistance is typically mediated by *erm* genes. Clindamycin is an attractive agent for empirical therapy for suspected *S. aureus* infections because of its excellent pharmacokinetic (especially in anaerobic conditions) and pharmacodynamic properties. Clinical failures of clindamycin therapy for treatment of MRSA infections have been documented for strains that were clindamycin sensitive but erythromycin resistant (Siberry et al. 2003). Clindamycin resistance may be constitutive (when the rRNA methylase is always produced) or inducible (when rRNA methylase is produced only in the presence of an inducing agent and isolates are resistant to erythromycin but appear susceptible to clindamycin in routine testing). Routine antibiotic susceptibility tests cannot identify these strains. The D (inducible clindamycin resistance) test is employed to

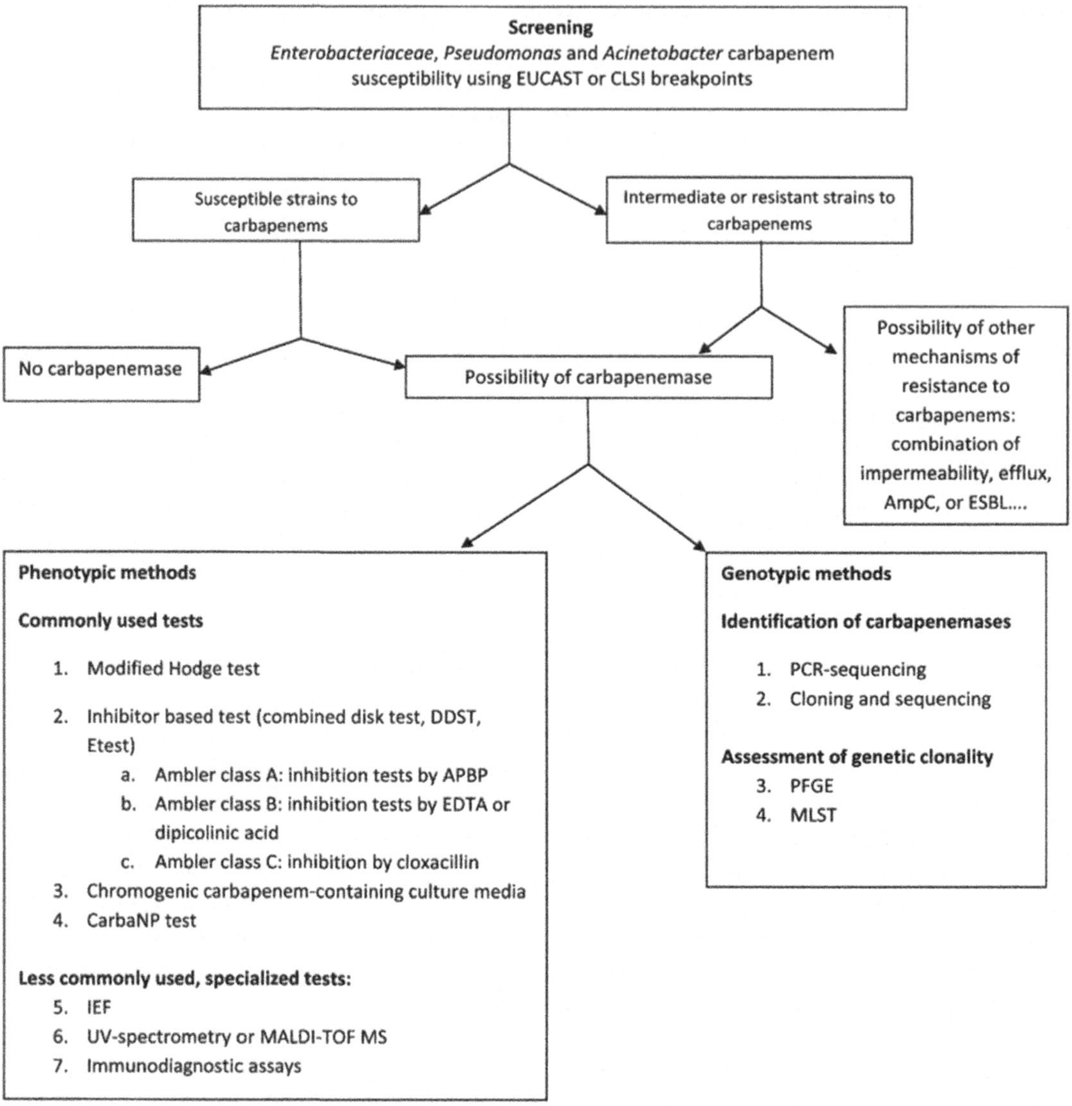

MIC, minimum inhibitory concentration; DDST, double disk synergy test; APBP; aminophenyl boronic acid; EDTA, ethylene diamine tetra-acetic acid; IEF, isoelectric focusing; MALDI-TOF MS, matrix-assisted laser desorption ionization-time of flight mass spectrometry; PCR, polymerase chain reaction; PFGE, pulsed field gel electrophoresis; MLST, multilocus sequence typing

Fig. 1 Simplified scheme for detection of carbapenemases

detect inducible clindamycin resistance. D-test is performed by disk diffusion method: the erythromycin disk (2 μg) and clindamycin disk (15 μg) are placed on an agar plate with *S. aureus* culture approximately 15 mm apart (measured edge to edge) and after incubation for 16–18 h at 37 °C the results are interpreted. A clear, D-shaped zone of inhibition around the clindamycin disk is designated as the D phenotype.

Colistin—Multiple AMR *Klebsiella* spp. and *Pseudomonas aeruginosa*

The emergence of multidrug-resistant Gram-negative bacteria (Enterobacteriales, *Pseudomonas*, *Acinetobacter*, *Klebsiella*, *Enterobacter*, and *Salmonella* species) is a threat to modern medicine and has been recognized worldwide. Colistin and polymyxin B remain part of the last line of antibiotics for multidrug-resistant Gram-negative

Table 9 The summary of known *mcr* genes

Gene	Year of identification	Source	Country	References	Note
mcr-1	2015	Multiple sources Enterobacteriales	China	Liu et al. (2016b)	Retrospective study of occurrence in China
mcr-2	2016	Porcine and bovine *Escherichia coli*	Belgium	Xavier et al. (2016)	
mcr-3	2017	Porcine *Escherichia coli*	China	Yin et al. (2017)	
mcr-4	2017	Porcine *Salmonella* and *Escherichia coli*	Italy, Spain, Belgium	Carattoli et al. (2017)	
mcr-5	2017	Poultry and food *Salmonella* Paratyphi B	Germany	Borowiak et al. ((2017)	
mcr-6	2017	Slaughtered pigs *Moraxella pluranimalium*	Great Britain	AbuOun et al. (2017)	Gene *mcr-6* was originally named *mcr-2.2*
mcr-7	2018	Chicken *Klebsiella pneumoniae*	China	Yang et al. (2018)	
mcr-8	2018	Pigs and human *Klebsiella pneumoniae*	China	Wang et al. (2018)	
mcr-9	2019	*Salmonella enterica* Serotype Typhimurium	United States	Carroll et al. (2019)	

bacteria, such as carbapenem-resistant Enterobacteriales. Current EUCAST and CLSI recommendations are for broth microdilution method for the determination of MIC of colistin. At present, quite a lot of work has been done on resistance testing of Gram-negative pathogens to colistin using other methods; the latest ones are, for example, the studies of Chew et al. (2017), Javed et al. (2018), or Carroll et al. (2019). It is also well known that the resistance of isolates to colistin is encoded by the *mcr* genes. Therefore, isolates designated as colistin resistant using routine AST are further tested for the presence of these genes (Caniaux et al. 2017). Currently known *mcr* genes (*mcr-1*, *mcr-2*, *mcr-3*, *mcr-4*, *mcr-5*, *mcr-6*, *mcr-7*, *mcr-8*, and *mcr-9*) are given in Table 9. The *mcr-1* gene has also recently been confirmed to confer cross-resistance with bacitracin, another antimicrobial from the polypeptide pharmacological group (Xu et al. 2018).

3.5 Selection of Drugs for Antimicrobial Susceptibility Testing

The drugs that are selected for performing AST should be most appropriate for the tested organism isolated and the site of infection must also be taken into account (Jorgensen 1993; Jorgensen and Ferraro 2009). Also, speaking about the food-producing animals there should be considered that despite the good susceptibility or broad antibacterial spectrum some of the antimicrobials are banned for use in food-producing animals, at least in the EU (e.g., chloramphenicol and metronidazole) due to the issue of safety for the consumer (listed in Table II of the Regulation EC/37/2010). Therefore, it is redundant to test them for therapeutic purposes, but they can be tested as class representatives (e.g., chloramphenicol interpreted for florfenicol). Selecting the most appropriate antimicrobial agents to test and report is a decision best made by each laboratory in consultation with veterinarians, infectious disease practitioners, and clinical pharmacologists, reflecting also valid legal provisions and in current days also local or national antibiotic policies The CLSI documents (currently CLSI 2018a) provide recommendations for selection of antimicrobials for AST and divide them into four groups (A–D). Each group includes a selection of tested antimicrobials separately for isolates originating from swine, cattle, bovine mastitis, poultry, horses, and dogs and cats. In general, routine testing should include only one representative of each group of related drugs (class) with an activity that is nearly identical against a spectrum of organisms, and for which interpretive results would be nearly always the same (CLSI 2013a).

The following text (definition of group A–D) is cited word for word from CLSI (2018a):

Group A includes antimicrobial agents with veterinary-specific breakpoints and interpretive categories that are considered appropriate for inclusion in a routine, primary testing panel for food and companion animals, as well as for routine reporting of results for specified organism groups. The recommended hierarchy for reporting is to first report group A agents over those using human medical breakpoints, because these compounds have demonstrated an acceptable level of correlation between in vitro susceptibility test results and clinical outcome.

Group B includes antimicrobial agents that use human medical breakpoints and interpretive categories and are next in the hierarchy to report. These agents may perform adequately, but outcome for many veterinary applications has not been demonstrated. The veterinary laboratory may use its discretion to decide whether to selectively report the results from testing these agents.

Group C includes antimicrobial agents that are regulatory agency approved for use in the specific animal species. Although QC data are available for these agents, they do not have veterinary- or human-specific CLSI-approved breakpoints and interpretive categories. These agents may be approved for use in other animal species and have veterinary-specific breakpoints in those animals. However, reporting interpretive categories determined by breakpoints set for a particular animal species is not recommended for application to other animal species because there are differences in dosages and pharmacokinetics between animals and people and between animal species. Thus, these agents should be reported selectively before extra-label use agents (group D) but after agents in group B.

Group D includes agents that are not approved but may be used in an extra-label (in EU is used term off label) manner per the Animal Medicinal Drug Use Clarification Act (AMDUCA) guidelines in the United States and per similar regulations in other countries for the listed animal. These supplemental agents may be selectively tested and selectively reported. Group D agents may be included in testing for monitoring antimicrobial resistance patterns or for surveillance programs (e.g., oxacillin, vancomycin, and carbapenems).

In addition, the CLSI documents provide a detailed overview of antimicrobial agents divided according to classes of antimicrobials with notices about selection of agents from some of the larger classes tested routinely and there are also recommendations based on spectrum of their efficacy and suitability for AST of individual species of bacteria, including information about appropriate supplementary tests, for example, beta-lactamase production testing (CLSI 2013a).

4 Data Obtained from Laboratory Investigation and Their Interpretation According to the Current Status of Knowledge

The interpretation of data obtained from laboratory investigation is a very important part of AST that has to follow the rules presented in the guidelines of internationally recognized institutions such as CLSI, EUCAST, and others mentioned in the previous subchapters. Correct performance of AST is necessary for the following successful antimicrobial therapy, and compliance with the rules of the correct interpretation of results will also allow comparison of the results among laboratories, which can serve as a basis for evaluating the state of susceptibility/resistance of individual pathogens at different levels in various resistance monitoring programs. Harmonization of the interpretation of results is of key importance also for the presentation of results in research publications. Schwarz et al. (2010) published an editorial article in the *Journal of Antimicrobial Chemotherapy*, summarizing the main recommendations for AST of bacteria obtained from animals. The main part of this article is focused on interpreting the results of AST. The recommendations in this article are considered by professionals as the gold standard, so the text in this chapter will be based on this publication.

4.1 Interpretation of AST Results in Individual Isolates

The main reason for implementing AST is to categorize individual bacterial isolates as susceptible, intermediate, or resistant to each tested antimicrobial based on the results obtained using standard tests—MIC (dilution and gradient methods) or diameter of inhibition zone (disk diffusion test) (Schwarz et al. 2010). Generally, two different types of interpretative criteria of AST are available: clinical breakpoints and epidemiological cut-off (ECOFF) values (Bywater et al. 2006). The ECOFF values well correspond to clinical breakpoints of susceptibility in many cases but, on the other hand, for some antimicrobials and pathogens, these values can be very different (see examples in Figs. 2 and 3 according to EUCAST 2019c). The ECOFF values of ampicillin and cefepime for isolates of *Escherichia coli* are also shown in the Figures. The cut-off values $\leq$8 mg/L for ampicillin correspond to the clinical breakpoint of susceptibility to ampicillin in *E. coli* isolates (susceptible: MIC $\leq$8 mg/L; intermediate: -; resistant: $\geq$16 mg/L). A quite different situation is with cefepime which has the ECOFF value $\leq$0.125 mg/L but the clinical breakpoint of susceptibility to cefepime in *E. coli* isolates according to MIC is much higher (susceptible: MIC $\leq$1 mg/L; intermediate: 2–4 mg/ml; resistant: $\geq$8 mg/L).

Epidemiological cut-off values should be used for MIC distributions of bacteria without clinical context and clinical breakpoints must be applied if data are intended to guide a therapeutic approach. The term "breakpoint" should be used exclusively for clinical breakpoints and "susceptible," "intermediate," and "resistant" or "non-susceptible" categories should also be reserved for classifications made in relation to the therapeutic administration of antimicrobial agents. When reporting data using epidemiological cut-off values, the term "resistant" is inappropriate; instead, bacteria should be reported as "wild-type" if the MIC or zone diameter falls within the wild-type range, or "non-wild-type" if the MIC is higher or the zone diameter is smaller than the wild-type range (Schwarz et al. 2010).

The interpretative criteria, as clinical breakpoints, are established by analysis of (1) microbiologic data according to comparison of MICs and zone sizes of a large number of bacterial strains with known mechanisms of resistance; (2) pharmacokinetics and pharmacodynamics data; and (3) results of clinical studies (Jorgensen and Ferraro 2009; Toutain et al. 2017).

Some AST guidelines, currently CLSI document Vet08 (CLSI 2018a) and veterinary recommendations of CA-SFM (2018) provide information and approved veterinary-specific breakpoint tables for bacteria of animal origin that are currently available. The breakpoints are presented for individual bacterial families or genera or species. In the CLSI document, the breakpoints are presented according to validity for individual species of animals or even for specific diseases. Unfortunately, the veterinary-specific clinical breakpoints are not defined for all antimicrobials and pathogens of animals, moreover, some of the breakpoints are defined with using certain dose of antimicrobial, dosing schedule, and route of administration, what is essential from pharmacokinetic perspective. The CLSI documents listed some approved breakpoints derived from human breakpoints for these cases. However, their true value for veterinary pathogens is unknown (Schwarz et al. 2010). Thus, these breakpoints should be used for interpretation of AST of veterinary pathogens very carefully and only for supporting information.

Despite the principles of interpretation of veterinary AST results mentioned above, we have

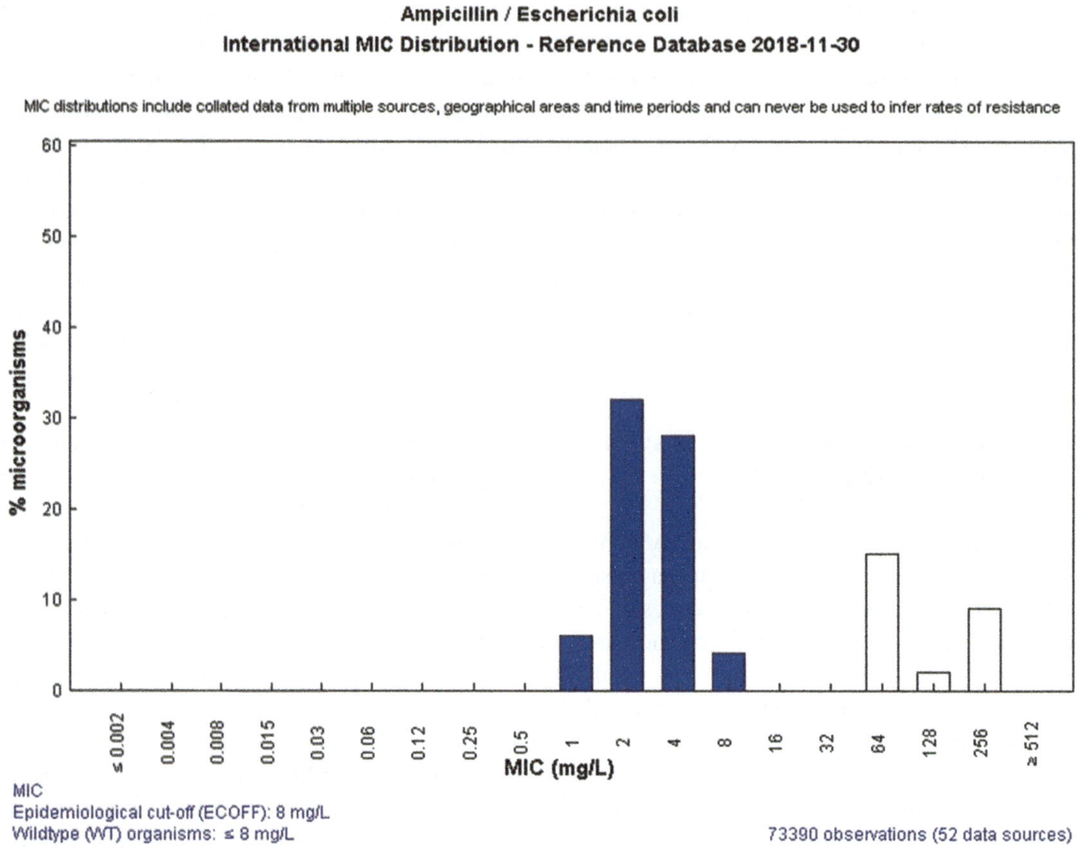

Fig. 2 MIC distribution and ECOFF for ampicillin in *E. coli* isolates (https://eucast.org/)

allowed ourselves to suggest AST interpretation criteria of important bacterial pathogens of some farm animal species—swine, cattle, poultry, and horses (Tables 10–13 with their sources of interpretative criteria are shown). However, even if the AST interpretation according to the tables proposed by us is used, it should be borne in mind that the general principles for the evaluation of results of AST in animal pathogens, as formulated by Schwarz et al. (2010) are still valid.

In Tables 10–13, the veterinary-specific breakpoints to categorize susceptible, intermediate, and resistant isolates are highlighted by green color. These interpretative criteria are taken from CLSI documents (2013b, 2016, 2018a) and CA-SFM (2018) and they are listed according to these documents, with respect to relevant species of animals, bacterial families, genera or species, and, where appropriate, also the type of disease. The breakpoints, which were derived from human breakpoints, are highlighted by yellow color. Their sources are interpretative criteria of EUCAST (2019b) and CLSI (2018b). Since we supposed that most readers of this publication will come from Europe, we preferred interpretations according to European standards (EUCAST), and only in cases where the interpretation of AST for the respective bacterial families, genera, or species was not available in EUCAST standards, we were looking for possible interpretative criteria in the latest documents of CLSI. For the cases, where interpretation of AST results is not possible for some antimicrobials or animal species, the cells in the tables are marked by red color.

The multiresistance (multidrug resistance) is another phenomenon, which is monitored within the AST of bacteria. There is no universally accepted definition of multiresistance. According to results of phenotypic susceptibility testing,

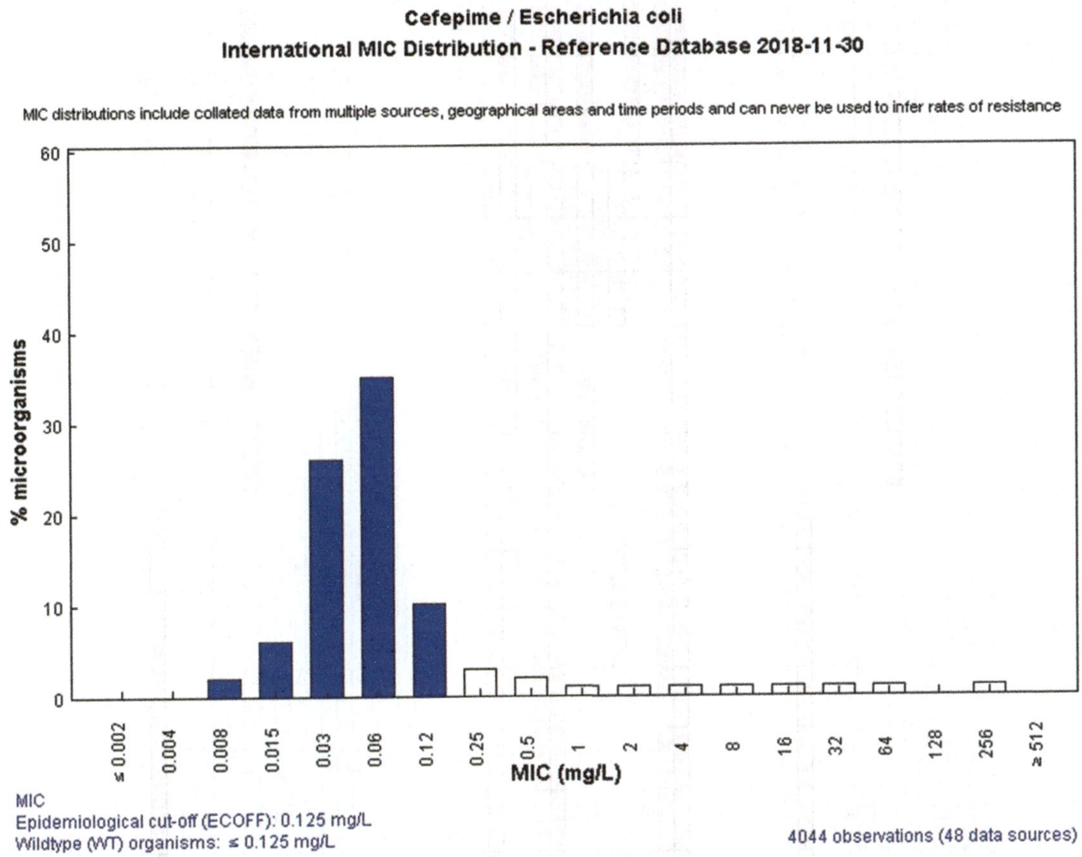

Fig. 3 MIC distribution and ECOFF for cefepime in *E. coli* isolates (https://eucast.org/)

resistance to three or more classes of antimicrobial agents can be referred to as multiresistance. Resistance to one agent from one antimicrobial class can be reasonably extrapolated to resistance or reduced susceptibility to other members of that class due to the same mechanisms of resistance mediation. However, single class representatives cannot always be validly defined, e.g., beta-lactams and aminoglycosides. When AST is supplemented with molecular analysis for the resistance genes, bacterial isolates exhibiting the presence of three or more resistance genes or mutations are considered to be resistant. An exception is cases where a single resistance gene or a gene complex associate resistance to more than one class of antimicrobial agents (Schwarz et al. 2010).

The growing number of multidrug-resistant strains leads to increasingly frequent use of different combinations of two and more antibiotics or antibiotics and other drugs in practice despite the relatively low knowledge of the effectiveness of this approach in terms of all possible drug combinations (Worthington and Melander 2013). Drug combination therapy can be a promising strategy to extend the life span of our antimicrobials. However, all drug combinations used in treatment must be carefully selected to minimize the evolution of resistance, either by carefully determining drug pairs that hinder the acquisition of resistance mechanisms, or by screening for combinations that inhibit growth and show reduced vulnerability to resistance (Hill and Cowen 2015). Combinations of antimicrobials are further employed to broaden the spectrum of bacteria to which the antimicrobial therapy is targeted (for treatment of polymicrobial infections or therapy of critically patients require empiric therapy before antimicrobial susceptibility can be determined) and

Table 10 Interpretative criteria for veterinary specific pathogens of swine, cattle, and poultry according to MIC determination (mg/L) and zone of inhibition (mm)

Antibiotic group	Antibiotic	Interpretative criteria																				
		Poultry			Cattle			Cattle			Cattle			Pigs			Pigs			Pigs		
		P. multocida			*P. multocida*			*M. haemolytica*			*H. somni*			*P. multocida*			*S. suis*			*A. pleuropneumoniae*		
		S	I	R	S	I	R	S	I	R	S	I	R	S	I	R	S	I	R	S	I	R
penicillins	penicillin	≤0.25	0.5	≥1	≤0.25	0.5	≥1	≤0.25	0.5	≥1	≤0.25	0.5	≥1	≤0.25	0.5	≥1	≤0.25	0.5	≥1	≤0.25	0.5	≥1
	ampicillin	≤0.03	0.06-0.12	≥0.25	≤0.03	0.06-0.12	≥0.25	≤0.03	0.06-0.12	≥0.25	≤0.03	0.06-0.12	≥0.25	≤0.5	1	≥2	≤0.5	1	≥2	≤0.5	1	≥2
cephalosporins	ceftiofur				≤2	4	≥8	≤2	4	≥8	≤2	4	≥8	≤2	4	≥8	≤2	4	≥8	≤2	4	≥8
macrolides	tildipirosin				≤8	16	≥32	≤4	8	≥16	≤8	16	≥32	≤4	-	-	≤4	-	-	≤16	-	-
	tilmikosin	≤16	-	≥32	≤8	16	≥32	≤8	16	≥32	≤8	16	≥32	≤16	-	≥32	≤16	-	≥32	≤16	-	≥32
	tulathromycin				≤16	32	≥64	≤16	32	≥64	≤16	32	≥64	≤16	32	≥64	≤16	32	≥64	≤64	-	-
	gamithromycin				≤4	8	≥16	≤4	8	≥16	≤4	8	≥16									
amphenicols	florfenicol	≤2	4	≥8	≤2	4	≥8	≤2	4	≥8	≤2	4	≥8	≤2	4	≥8	≤2	4	≥8	≤2	4	≥8
tetracyclines	tetracycline	≤0.5	1	≥2	≤2	4	≥8	≤2	4	≥8	≤2	4	≥8	≤0.5	1	≥2	≤0.5	1	≥2	≤0.5	1	≥2
	doxycycline	≤1	-	≥2	≤1	-	≥2	≤1	-	≥2	≤1	2	4	≤1	-	≥2	≤1	-	≥2	≤1	-	≥2
fluoroquinolones	enrofloxacin	≤0.25	0.5-1	≥2	≤0.25	0.5-1	≥2	≤0.25	0.5-1	≥2	≤0.25	0.5-1	≥2	≤0.25	0.5	≥1	≤0.5	1	≥2	≤0.25	0.5	≥1
aminocyclitols	spectinomycin	≤32	64	≥128	≤32	64	≥128	≤32	64	≥128	≤32	64	≥128	≤32	64	≥128	≤32	64	≥128	≤32	64	≥128
diterpens	tiamulin	≤4	8-16	≥32										≤4	8-16	≥32	≤4	8-16	≥32	≤16	-	≥32
potentiated sulphonamides	trimethoprim/sulfamethoxazole	≤0.25/4,75	-	≤0.5/9.5	≤0.25/4,75	-	≤0.5/9.5	≤0.25/4,75	-	≤0.5/9.5	≤0.5/9.5	1/19	≥2/38	≤0.25/4,75	-	≤0.5/9.5	≤0.25/4,75	-	≤0.5/9.5	≤0.5/9.5	1/19	≥2/38

Antibiotic group	Antibiotic disk	Interpretative criteria																				
		Poultry			Cattle			Cattle			Cattle			Pigs			Pigs			Pigs		
		P. multocida			*P. multocida*			*M. haemolytica*			*H. somni*			*P. multocida*			*S. suis*			*A. pleuropneumoniae*		
		S	I	R	S	I	R	S	I	R	S	I	R	S	I	R	S	I	R	S	I	R
penicillins	penicillin																					
	ampicillin																					
cephalosporins	ceftiofur 30 µg				≥21	18-20	≤17	≥21	18-20	≤17	≥21	18-20	≤17	≥21	18-20	≤17	≥21	18-20	≤17	≥21	18-20	≤17
macrolides	tildipirosin 60 µg				≥21	18-20	≤17	≥20	17-19	≤16	≥17	14-10	≤9	≥19	-	-						
	tilmikosin 15 µg							≥14	11-13	≤10				≥11	-	≤10				≥11	-	≤10
	tulathromycin 30 µg				≥18	15-17	≤14	≥18	15-17	≤14	≥18	15-17	≤14	≥18	15-17	≤14				≥10	-	-
	gamithromycin 15 µg				≥15	14-12	≤11	≥15	14-12	≤11	≥15	14-12	≤11									
amphenicols	florfenicol 30 µg				≥19	15-18	≤14	≥19	15-18	≤14	≥19	15-18	≤14	≥22	19-21	≤18				≥22	19-21	≤18
tetracyclines	tetracycline																					
	doxycycline 30 µg																≥23	19-22	≤18			
fluoroquinolones	enrofloxacin 5 µg				≥21	17-20	≤16	≥21	17-20	≤16	≥21	17-20	≤16	≥23	19-22	≤18				≥23	19-22	≤18
aminocyclitols	spectinomycin 100 µg				≥14	11-13	≤10	≥14	11-13	≤10												
diterpens	tiamulin 30 µg													≥9	-	≤8				≥9	-	≤8
potentiated sulphonamides	trimethoprim/sulfamethoxazole 1.25/23.75 µg	≥23	-	22	≥23	-	≤22	≥23	-	≤22	≥23	20-22	≤19	≥23	-	≤22	≥19	16-18	≤15	≥23	20-22	≤19

Antibiotic group	Antibiotic	Interpretative criteria						
		Poultry	Cattle	Cattle	Cattle	Pigs	Pigs	Pigs
		P. multocida	*P. multocida*	*M. haemolytica*	*H. somni*	*P. multocida*	*S. suis*	*A. pleuropneumoniae*
penicillins	penicillin	VET 08*	VET 08	VET 08	VET 08	VET 08	VET 08	VET 08 **
	ampicillin	VET 08*	VET 08	VET 08	VET 08	VET 08	VET 08	VET 08
cephalosporins	ceftiofur		VET 08	VET 08	VET 08	VET 08	VET 08 **	VET 08
macrolides	tildipirosin		VET 08	VET 08	VET 08	VET 08	VET 08 **	VET 08
	tilmikosin	VET 08*	VET 08 **	VET 08	VET 08 **	VET 08	VET 08 **	VET 08
	tulathromycin		VET 08	VET 08	VET 08	VET 08	VET 08 **	VET 08
	gamithromycin		VET 08	VET 08	VET 08			
amphenicols	florfenicol	VET 08*	VET 08	VET 08	VET 08	VET 08	VET 08	VET 08
tetracyclines	tetracycline	VET 08*	VET 08	VET 08	VET 08	VET 08	VET 08	VET 08
	doxycycline	EUCAST	EUCAST	EUCAST	EUCAST	EUCAST	M 100 28S	EUCAST
fluoroquinolones	enrofloxacin	VET 01 S	VET 08	VET 08	VET 08	VET 08	VET 08	VET 08
aminocyclitols	spectinomycin	VET 08*	VET 08	VET 08	VET 08	VET 08	VET 08*	VET 08*
diterpens	tiamulin	Jones et al. (2002)				Jones et al. (2002)	Jones et al. (2002)	VET 08
potentiated sulphonamides	trimethoprim/sulfamethoxazole	EUCAST	EUCAST	EUCAST	EUCAST	EUCAST	M 100 28S**	EUCAST

*breakpoints derived from isolates from different target animal species

**breakpoints extrapolated from one of the bacterial species

	missing
human isolates	EUCAST 2018 - European Committee on Antimicrobial Susceptibility Testing, URL: //www.eucast.org/
veterinary isolates	CLSI 2018. Vet08, Performance Standards for Antimicrobial Disk and Dilution Susceptibility Tests for Bacteria Isolated from Animals.
	CLSI 2015. VET01-S, Performance Standards for Antimicrobial Disk and Dilution Susceptibility Tests for Bacteria Isolated from Animals.
	Jones RN, et al. 2002. J Clin Microbiol. 40: 461-5.

Table 11 Interpretative criteria for pathogens of horses according to MIC determination (mg/L) and zone of inhibition (mm)

		Interpretative criteria																				
		Enterobacteriales			*Escherichia coli*			*Staphylococcus* spp.			*Staphylococcus aureus*			*Streptococcus equi* subsp. *zooepidemicus* and subsp. *equi*			*Streptococus* spp.			*Rhodococcus equi*		
Antibiotic group	**Antibiotic**	S	I	R	S	I	R	S	I	R	S	I	R	S	I	R	S	I	R	S	I	R
	penicillin							≤0.5	1	≥2	≤0.5	1	≥2	≤0.5	1	≥2	≤0.5	1	≥2	≤0.12	-	≥0.25
penicillins	ampicillin	≤8	-	≥16	≤8	-	≥16							≤0.25	-	-	≤0.5	1-2	≥4			
	cefazolin	≤2	4	≥8	≤2	4	≥8							≤2	4	≥8	≤2	4	≥8			
	ceftiofur	≤2	4	≥8	≤2	4	≥8	≤2	4	≥8	≤2	4	≥8	≤0.25	-	-	≤2	4	≥8			
cephalosporins	cefquinom	≤2	4	≥8	≤2	4	≥8															
	amikacin - foals	≤2	4	≥8	≤2	4	≥8	≤2	4	≥8	≤2	4	≥8	≤2	4	≥8	≤2	4	≥8			
	amikacin - adults	≤4	8	≥16	≤4	8	≥16	≤4	8	≥16	≤4	8	≥16	≤4	8	≥16	≤4	8	≥16			
aminoglycosides	gentamicin	≤2	4	≥8	≤2	4	≥8	≤1	-	≥2	≤1	-	≥2							≤1	-	≥2
fluoroquinolones	enrofloxacin	≤0.12	0.25	≥0.5	≤0.12	0.25	≥0.5	≤0.12	0.25	≥0.5	≤0.12	0.25	≥0.5	≤0.12	0.25	≥0.5	≤0.12	0.25	≥0.5			
	tetracycline	≤4	8	≥16	≤4	8	≥16	≤1	2	≥4	≤1	2	≥4	≤1	2	≥4	≤1	2	≥4	≤4	8	≥16
	doxycycline	≤0.12	0.25	≥0.5	≤0.12	0.25	≥0.5	≤0.12	0.25	≥0.5	≤0.12	0.25	≥0.5	≤0.12	0.25	≥0.5	≤0.12	0.25	≥0.5			
tetracyclines	minocycline	≤0.12	0.25	≥0.5	≤0.12	0.25	≥0.5	≤0.12	0.25	≥0.5	≤0.12	0.25	≥0.5	≤0.12	0.25	≥0.5	≤0.12	0.25	≥0.5			
potentiated sulphonamides	trimethoprim/sulfamethoxazole	≤2/38	4/76	≥8/152	≤2/38	4/76	≥8/152	≤2/38	4/76	≥8/152	≤2/38	4/76	≥8/152	≤1/19	2/38	≥4/76	≤1/19	2/38	≥4/76	≤2/38	-	≥4/76
ansamycins	rifampicin							≤0.06	0.12-0.5	≥1	≤0.06	0.12-0.5	≥1	≤0.06	0.12-0.5	≥1	≤0.06	0.12-0.5	≥1	≤1	2	≥4
	azithromycin							≤1	2	≥4	≤1	2	≥4	≤0.25	0.5	≥1	≤0.25	0.5	≥1	≤2	4	≥8
	clarithromycin							≤1	2	≥4	≤1	2	≥4	≤0.25	0.5	≥1	≤0.25	0.5	≥1	≤2	4	≥8
macrolides	erythromycin							≤1	2	≥4	≤1	2	≥4	≤0.25	0.5	≥1	≤0.25	0.5	≥1	≤0.5	1-4	≥8
amphenicols	chloramphenicol	≤8	-	≥16	≤8	-	≥16	≤8	-	≥16	≤8	-	≥16	≤8	-	≥16	≤8	-	≥16	≤8	16	≥32

		Interpretative criteria																				
		Enterobacteriales			*Escherichia coli*			Coagulase negative *Staphylococcus* spp.			*Staphylococcus aureus*			*Streptococcus equi* subsp. *zooepidemicus* and subsp. *equi*			*Streptococus* spp.			*Rhodococcus equi*		
Antibiotic group	**Antibiotic disc**	S	I	R	S	I	R	S	I	R	S	I	R	S	I	R	S	I	R	S	I	R
	penicillin 1 U																			≥29	-	≤28
penicillins	ampicillin 10 µg	≥14	-	≤13	≥14	-	≤13															
	cefazolin																					
	ceftiofur 30 µg	≥21	18-20	≤17	≥21	18-20	≤17	≥21	18-20	≤17	≥21	18-20	≤17	≥22	-	-						
cephalosporins	cefquinom 30 µg	≥21	17-22	≤16	≥21	17-22	≤16															
	amikacin - foals																					
	amikacin - adults 30 µg	≥16	13-15	≤12	≥16	13-15	≤12															
aminoglycosides	gentamicin 10 µg	≥16	13-15	≤12	≥16	13-15	≤12	≥22	-	≤21	≥18	-	≤17							≥23	-	≤22
fluoroquinolones	enrofloxacin																					
	tetracycline 30 µg	≥15	12-14	≤11	≥15	12-14	≤11	≥22	19-21	≤18	≥22	19-21	≤18	≥23	20-22	≤19	≥23	20-22	≤19			
	doxycycline																					
tetracyclines	minocycline																					
potentiated sulphonamides	trimethoprim/sulfamethoxazole 1.25/23.75 µg	≥14	11-13	≤10	≥14	11-13	≤10	≥17	14-16	≤13	≥17	14-16	≤13	≥18	15-17	≤14	≥18	15-17	≤14			
ansamycins	rifampicin 5 µg							≥26	23-25	≤22	≥26	23-25	≤22	≥21	15-20	≤14	≥21	15-20	≤14			
	azithromycin																					
	clarithromycin																					
macrolides	erythromycin 15 µg							≥21	18-20	≤17	≥21	18-20	≤17	≥22	19-21	≤18	≥22	19-21	≤18			
amphenicols	chloramphenicol 30 µg	≥17	-	≤16	≥17	-	≤16	≥18	-	≤17	≥18	-	≤17	≥19	-	≤18	≥19	-	≤18			

	only for *Streptococcus* spp. β-hemolytic group
	only for *Streptococcus equi* subsp. *zooedidemicus*
	only for streptococci viridans

Antibiotic group	Antibiotic	Enterobacteriales	Escherichia coli	Staphylococcus spp.	Staphylococcus aureus	Streptococcus equi subsp. zooepidemicus and subsp. equi	Streptococus spp.	Rhodococcus equi
	Interpretative criteria							
	penicillin			VET 08	VET 08	VET 08	VET 08	EUCAST
penicillins	ampicillin	EUCAST	EUCAST			VET 08	EUCAST	
	cefazolin	VET 08**	VET 08			VET 08	VET 08	
	ceftiofur	VET 08*	VET 08*	VET 08*	VET 08*	VET 08	VET 08**	
cephalosporins	cefquinom	CA-SFM	CA-SFM					
	amikacin - foals	VET 08**	VET 08	VET 08	VET 08	VET 08	VET 08**	
	amikacin - adults	VET 08**	VET 08	VET 08	VET 08	VET 08	VET 08**	
aminoglycosides	gentamicin	VET 08**	VET 08	EUCAST	EUCAST			EUCAST
fluoroquinolones	enrofloxacin	VET 08**	VET 08	VET 08**	VET 08	VET 08	VET 08**	
	tetracycline	M 100 28S	M 100 28 S	EUCAST	EUCAST	EUCAST	EUCAST	VET 06 E1
	doxycycline	VET 08	VET 08	VET 08**	VET 08	VET 08	VET 08**	
tetracyclines	minocycline	VET 08	VET 08	VET 08*	VET 08	VET 08	VET 08	
potentiated sulphonamides	trimethoprim/sulfamethoxazole	EUCAST	EUCAST	EUCAST	EUCAST	EUCAST	EUCAST	VET 06 E1
ansamycins	rifampicin			EUCAST	EUCAST	EUCAST	EUCAST	VET 06 E1
	azithromycin			EUCAST	EUCAST	EUCAST	EUCAST	VET 06 E1
	clarithromycin			EUCAST	EUCAST	EUCAST	EUCAST	VET 06 E1
macrolides	erythromycin			EUCAST	EUCAST	EUCAST	EUCAST	VET 06 E1
amphenicols	chloramphenicol	EUCAST	EUCAST	EUCAST	EUCAST	EUCAST	EUCAST	VET 06 E1

*breakpoints derived from isolates from different target animal species

****breakpoints extrapolated from one of the bacterial species ⊠**

***breakpoints derived from *Haemophilus influenzae*

human isolates	EUCAST 2018 - European Committee on Antimicrobial Susceptibility Testing, URL: //www.eucast.org/
veterinary isolates	CLSI 2018. Vet08, Performance Standards for Antimicrobial Disk and Dilution Susceptibility Tests for Bacteria Isolated from Animals.
	CA-SFM 2018. Comité de l'Antibiogramme de la Société Francaise de Microbiologie. Recommandations Vétérinaires 2018.
	missing

Table 12 Interpretative criteria for Gram-negative pathogens according to MIC determination (mg/L) and zone of inhibition (mm)

Antibiotic group	Antibiotic	Interpretative criteria			Source
		S	I	R	
penicillins	ampicillin	≤8	-	≥16	EUCAST
	amoxicillin/clavulanic acid	≤8/2	-	≥16/2	EUCAST
cephalosporins	cephalotin	≤8	16	≥32	VET 01 S*
	cefotaxime	≤1	2	≥4	EUCAST**
	ceftiofur	≤2	4	≥8	VET 08
	cefquinom	≤2	4	≥8	CA-SFM
amphenicols	florfenicol	≤4	8	≥16	VET 08
tetracyclines	tetracycline	≤4	8	≥16	M 100 28 S
	doxycycline	≤4	8	≥16	M 100 28 S
aminoglycosides	apramycin	≤16	-	≥32	CA-SFM
	gentamicin	≤2	4	≥8	VET 08
potentiated sulphonamides	trimethoprim/sulfamethoxazole	≤2/38	4/76	≥8/152	EUCAST
fluoroquinolones	enrofloxacin	≤0.25	0.5-1	≥2	VET 08
	marbofloxacin	≤1	2	≥4	VET 08***
polymyxins	colistin	≤2	-	≥4	EUCAST

Antibiotic group	Antibiotic disk	Interpretative criteria			Source
		S	I	R	
penicillins	ampicillin 10 µg	≥14	-	≤13	EUCAST
	amoxicillin/clavulanic acid 20/10 µg	≥19	-	≤18	EUCAST
cephalosporins	cephalotin 30 µg	≥22	21-19	≤18	VET 01 S*
	cefotaxime 5 µg	≥20	17-19	≤16	EUCAST**
	ceftiofur 30 µg	≥21	18-20	≤17	VET 08
	cefquinom 30 µg	≥21	17-22	≤16	CA-SFM
amphenicols	florfenicol				VET 08
tetracyclines	tetracycline 30 µg	≥15	12-14	≤11	M 100 28 S
	doxycycline 30 µg	≥14	11-13	≤10	M 100 28 S
aminoglycosides	apramycin 15 µg	≥15	14-12	≤11	CA-SFM
	gentamicin 10 µg	≥16	13-15	≤12	VET 08
potentiated sulphonamides	trimethoprim/sulfamethoxazole 1.25/23.75 µg	≥14	11-13	≤10	EUCAST
fluoroquinolones	enrofloxacin 5 µg	≥23	17-22	≤16	VET 08
	marbofloxacin 5 µg	≥20	15-19	≤14	VET 08***
polymyxins	colistin				

*human-derived interpretative criteria

**** tested to identify specific resistance mechanisms ⍰**

***for feline isolates

human isolates	EUCAST 2018 - European Committee on Antimicrobial Susceptibility Testing, URL: //www.eucast.org/
	CLSI 2018. M 100 S28, Performance Standards for Antimicrobial Susceptibility Testing.
veterinary isolates	CLSI 2015. VET 01 S, Performance Standards for Antimicrobial Disk and Dilution Susceptibility Tests for Bacteria Isolated from Animals.
	CLSI 2018. VET 08, Performance Standards for Antimicrobial Disk and Dilution Susceptibility Tests for Bacteria Isolated from Animals.
	CA-SFM 2018. Comité de l'Antibiogramme de la Société Francaise de Microbiologie. Recommandations Vétérinaires 2018.

treatment by combinations of drugs also should allow the use of lower concentrations of drugs in combination and thus diminish the incidence of dose-related antibiotic toxicity (Moellering 1983; Leekha et al. 2011). In the practice, two the most common variants of combinations of drugs in antimicrobial treatment are combinations of two or more antibiotics (e.g., often using combinations of trimethoprim and sulfamethoxazole or penicillin and novobiocin) and antibiotics/adjuvant combinations (classic examples are combinations of beta-lactam antibiotics with beta-lactamase inhibitors, e.g., amoxicillin with clavulanic acid) (Worthington and Melander 2013). Once the individual veterinary medicinal product contains two or more antimicrobials it is considered as so-called fixed combination, where synergic or additive effect was proved. The combinations of antimicrobials are used for systemic treatment (usually oral or parenteral

Table 13 Interpretative criteria for Gram-positive pathogens according to MIC determination (mg/L) and zone of inhibition (mm)

Antibiotic group	Antibiotic	β-haemolytic streptococci S	β-haemolytic streptococci I	β-haemolytic streptococci R	Viridans streptococci S	Viridans streptococci I	Viridans streptococci R	Staphylococci coagulase + S	Staphylococci coagulase + I	Staphylococci coagulase + R	*S. aureus, S. lugdunensis* S	*S. aureus, S. lugdunensis* I	*S. aureus, S. lugdunensis* R	Staphylococci coagulase - S	Staphylococci coagulase - I	Staphylococci coagulase - R	Enterococci S	Enterococci I	Enterococci R
penicillins	penicillin	≤0.25	-	≥0.5	≤0.25	0.5-2	≥4	≤0.12	-	≥0.25	≤0.12	-	≥0.25	≤0.12	-	≥0.25	≤8	-	≥16
	ampicillin	≤0.25	-	-	≤0.5	1-2	≥4										≤4	8	≥16
	oxacillin	≤2	-	≥4	≤2	-	≥4				≤2	-	≥4	≤0.25	-	≥0.5			
cephalosporins	cefoxitin										≤4	-	≥8						
	ceftiofur	≤2	4	≥8	≤2	4	≥8	≤2	4	≥8	≤2	4	≥8	≤2	4	≥8			
macrolides	erythromycin	≤0.25	0.5	≥1	≤0.25	0.5	≥1	≤1	2	≥4	≤1	2	≥4	≤1	2	≥4	≤0.5	1-4	≥8
lincosamides	clindamycin	≤0.5	-	≥1	≤0.5	-	≥1	≤0.25	0.5	≥1	≤0.25	0.5	≥1	≤0.25	0.5	≥1			
	pirlimycin*	≤2	-	≥4	≤2	-	≥4	≤2	-	≥4	≤2	-	≥4	≤2	-	≥4			
amphenicols	chloramphenicol	≤8	-	≥16	≤8	-	≥16	≤8	-	≥16	≤8	-	≥16	≤8	-	≥16	≤8	-	≥16
tetracyclines	tetracycline	≤1	2	≥4	≤2	4	≥8	≤1	2	≥4	≤1	2	≥4	≤1	2	≥4	≤4	8	≥16
aminoglycosides	gentamicin				≤128	-	≥256	≤1	-	≥2	≤1	-	≥2	≤1	-	≥2	≤128	-	≥256
potentiated sulphonamides	trimethoprim/sulfamethoxazole	≤1/19	2/38	≥4/76	≤1/19	2/38	≥4/76	≤2/38	4/76	≥8/152	≤2/38	4/76	≥8/152	≤2/38	4/76	≥8/152			
fluoroquinolones	enrofloxacin	≤0.5	1	≥2	≤0.5	1	≥2	≤0.5	1-2	≥4	≤0.5	1-2	≥4	≤0.5	1-2	≥4	≤0.25	0.5-1	≥2
ansamycins	rifampicin	≤0.06	0.12-0.5	≥1	≤0.06	0.12-0.5	≥1	≤0.06	0.12-0.5	≥1	≤0.06	0.12-0.5	≥1	≤0.06	0.12-0.5	≥1	≤1	2	≥4
combinations of antimicrobials	penicillin/novobiocin*	≤1/2	2/4	≥4/8	≤1/2	2/4	≥4/8	≤1/2	2/4	≥4/8	≤1/2	2/4	≥4/8	≤1/2	2/4	≥4/8			

Antibiotic group	Antibiotic disk	β-haemolytic streptococci S	β-haemolytic streptococci I	β-haemolytic streptococci R	Viridans streptococci S	Viridans streptococci I	Viridans streptococci R	Staphylococci coagulase + S	Staphylococci coagulase + I	Staphylococci coagulase + R	*S. aureus, S. lugdunensis* S	*S. aureus, S. lugdunensis* I	*S. aureus, S. lugdunensis* R	Staphylococci coagulase - S	Staphylococci coagulase - I	Staphylococci coagulase - R	Enterococci S	Enterococci I	Enterococci R
penicillins	penicillin 1 U	≥18	-	≤17	≥18	12-17	≤11	≥26	-	≤25	≥26	-	≤25	≥26	-	≤25	≥15	-	≤14
	ampicillin 2 µg**	≥24	-	-	≥21	15-20	≤14										≥10	8-9	≤7
	oxacillin 1 µg	≥21	-	≤20	≥21	-	≤20												
cephalosporins	cefoxitin 30 µg										≥22	-	≤21	≥22	-	≤21			
	ceftiofur 30 µg	≥21	18-20	≤17	≥21	18-20	≤17	≥21	18-20	≤17	≥21	18-20	≤17	≥21	18-20	≤17			
macrolides	erythromycin 15 µg	≥21	18-20	≤17	≥21	16-20	≤15	≥21	19-20	≤17	≥21	18-20	≤17	≥21	18-20	≤17	≥23	14-22	≤13
lincosamides	clindamycin 2 µg	≥17	-	≤16	≥19	-	≤18	≥22	19-21	≤18	≥22	19-21	≤18	≥22	19-21	≤18			
	pirlimycin* 2 µg	≥13	-	≤12	≥13	-	≤12				≥13	-	≤12						
amphenicols	chloramphenicol 30 µg	≥19	-	≤18	≥21	18-20	≤17	≥18	-	≤17	≥18	-	≤17	≥18	-	≤17	≥19	15-18	≤14
tetracyclines	tetracycline 30 µg	≥23	20-22	≤19	≥23	19-22	≤18	≥22	19-21	≤18	≥22	19-21	≤18	≥22	19-21	≤18	≥19	15-18	≤14
aminoglycosides	gentamicin 10 µg							≥18	-	≤17	≥18	-	≤17	≥22	-	≤21	≥18	13-17	≤12
potentiated sulphonamides	trimethoprim/sulfamethoxazole 1.25/23.75 µg	≥18	15-17	≤14	≥19	16-18	≤15	≥17	14-16	≤13	≥17	14-16	≤13	≥17	14-16	≤13			
fluoroquinolones	enrofloxacin 5 µg	≥22	17-21	≤16	≥22	17-21	≤16	≥23	17-22	≤16	≥23	17-22	≤16	≥23	17-22	≤16			
ansamycins	rifampicin 5 µg	≥21	15-20	≤14	≥19	17-18	≤16	≥26	23-25	≤22	≥26	23-25	≤22	≥26	23-25	≤22	≥20	17-19	≤16
combinations of antimicrobials	penicillin/novobiocin* 10 U/30 µg	≥18	15-17	≤14	≥18	15-17	≤14				≥18	15-17	≤14						

*mastitis

** ampicillin 10 µg

Antibiotic group	Antibiotic	β-haemolytic streptococci	Viridans streptococci	Staphylococci coagulase +	*S. aureus, S. lugdunensis*	Staphylococci coagulase -	Enterococci
penicillins	penicillin	EUCAST	EUCAST	EUCAST	EUCAST	EUCAST	M 100 28S
	ampicillin	VET 08	EUCAST				EUCAST
	oxacillin	CASFM	CASFM		EUCAST	EUCAST	
cephalosporins	cefoxitin				EUCAST		
	ceftiofur	VET 08	VET 08	VET 01 S	VET 01 S, VET 08*	VET 08	
macrolides	erythromycin	EUCAST	M 100 28S	EUCAST	EUCAST	EUCAST	M 100 28S
lincosamides	clindamycin	EUCAST	EUCAST	EUCAST	EUCAST	EUCAST	
	pirlimycin*	VET 08	VET 08	VET 08	VET 08	VET 08	
amphenicols	chloramphenicol	EUCAST	M 100 28S	EUCAST	EUCAST	EUCAST	M 100 28 S
tetracyclines	tetracycline	EUCAST	M 100 28S	EUCAST	EUCAST	EUCAST	M 100 28 S
aminoglycosides	gentamicin		EUCAST	EUCAST	EUCAST	EUCAST	M 100 28 S
potentiated sulphonamides	trimethoprim/sulfamethoxazole	EUCAST	M 100 28S***	EUCAST	EUCAST	EUCAST	
fluoroquinolones	enrofloxacin	CASFM	CASFM	VET 08**	VET 08**	VET 08**	VET 08
ansamycins	rifampicin	EUCAST	M 100 28S***	EUCAST	EUCAST	EUCAST	VET 08
combinations of antimicrobials	penicillin/novobiocin*	VET 08	VET 08	VET 08	VET 08	VET 08	

* *S. aureus* - mastitis

**for canine and feline isolates

***breakpoints extrapolated from one of the bacterial species

human isolates	EUCAST 2018 - European Committee on Antimicrobial Susceptibility Testing, URL: //www.eucast.org/
	CLSI 2018. M 100 S28, Performance Standards for Antimicrobial Susceptibility Testing.
veterinary isolates	CLSI 2015. VET 01 S, Performance Standards for Antimicrobial Disk and Dilution Susceptibility Tests for Bacteria Isolated from Animals.
	CLSI 2018. VET 08, Performance Standards for Antimicrobial Disk and Dilution Susceptibility Tests for Bacteria Isolated from Animals.
	CA-SFM 2018. Comité de l'Antibiogramme de la Société Francaise de Microbiologie. Recommandations Vétérinaires 2018.
	missing

administration), local treatment, for example, in the treatment of mastitis or intrauterine infections and for topical treatment (usually as dermatologic, otologic, or ophthalmologic veterinary medicinal products).

We have to take into account that all antimicrobial agents administered in combination with other antimicrobials or drugs are always affected by each other in terms of effectiveness. The combined effect that is smaller than the sum of the effects of each single drug present in the mixture is termed antagonism. On the contrary, the higher effect of drug combination than the effect of each single administrated drug is synergism (Jawetz 1975). For example, the combination of certain beta-lactams and aminoglycosides exhibits synergistic activity against a variety of Gram-positive and Gram-negative bacteria and is used in the treatment of serious infections, for which rapid killing is essential (e.g., treatment of endocarditis caused by *Enterococcus* species with a combination of penicillin and gentamicin). In this setting, the addition of gentamicin to penicillin has been shown to be bactericidal, whereas penicillin alone is only bacteriostatic and gentamicin alone has no significant activity. For certain streptococci, similar synergistic combinations that result in more rapid clearance of the infecting microorganism can also be used to shorten the course of antimicrobial therapy (e.g., for endocarditis due to viridans group streptococci, a combination of penicillin or ceftriaxone with gentamicin for 2 weeks can be as effective as penicillin or ceftriaxone alone for 4 weeks (Drusano 1990; Levison 2004; Leekha et al. 2011).

Methods for laboratory testing of combinations are not used routinely (with exception of so-called fixed combinations as amoxicillin/clavulanic acid or sulfamethoxazole/trimethoprim, where certain ratio of the actives in commercially available disks altogether with exact interpretative criteria are utilized routinely). One of the reasons is difficult interpretation as well as the uncertainty of clinical outcomes results when testing newly set combinations/concentrations. Two methods used for assesment of antimicrobials´combinations: checkerboard assay and time-kill assay, are briefly introduced below.

Checkerboard approach can be performed as a two-dimensional adaptation of the standard broth microdilution method—i.e., bacterial inoculum is added to wells in a microplate, where two different antimicrobials are used in serial twofold dilutions (one antimicrobial horizontally and another one vertically). Test can be used mainly to confirm synergy, indifference or antagonism. For the final evaluation is used so-called fractional inhibitory concentration (FIC index), which expressing fractional inhibitory concentration. The FIC index calculation formula for drugs X and Y is as follows:

$$\begin{aligned}\text{FIC index} &= \text{FIC}_\text{X} + \text{FIC}_\text{Y}\\ &= (\text{X})/(\text{MIC}_\text{X}) + (\text{Y})/(\text{MIC}_\text{Y})\end{aligned}$$

The (X) is the concentration of drug X in a given well along the growth-no-growth interface; (MIC_X) is the control MIC of the organism to drug X alone; FIC_X is the fractional inhibitory concentration of drug X; and (Y), (MIC_Y), and FIC_Y are defined in the same fashion for drug Y (Rohner et al. 1989; Hsieh et al. 1993). FIC index values were interpreted following the conventional models: According to Odds (2003), Synergistic effect is observed when FIC index value ≤ 0.5; an Indifferent effect when $0.5 <$ FIC index value ≤ 4 and an Antagonistic effect when FIC index value >4; or according to EUCAST (2000), a Synergistic effect is observed when FIC index value ≤ 0.5; an Additive effect when $0.5 <$ FIC index value ≤ 1; an Indifferent effect when $1 <$ FIC index value <2 and an Antagonistic effect when FIC index value ≥ 2. The various software are currently developed for calculation of the awaited synergistic or antagonistic effect of drugs based on the FIC index determination.

Another commonly used method is the time-kill assay. This method is sometimes used for confirmation of results of the checkerboard method. In time-kill assay, colony counts are obtained at serial time points from liquid cultures in which bacteria are exposed to each antibiotic individually and to the two drugs together. This method provides information on bacterial killing over time.

It should be highlighted that in the current time practitioners both in human and in veterinary medicine are still waiting for reliable and also sufficiently speed AST for combinations, which are often used in cases of multidrug-resistant bacteria or failure of sole antibacterial therapy.

4.2 Interpretation of AST in Multiple Isolates of One Bacterial Species

The important parameters for reporting results of AST in multiple isolates of one bacterial species are MIC_{50} (equivalent to the median MIC value) and MIC_{90} (the 90th percentile). The MIC_{50} (MIC_{90}) represents the MIC value at which at least 50% (90%) of the isolates in a test population are inhibited. The MIC_{50} and MIC_{90} values should always be presented as concentrations on the standard dilution series. It should be noted that if the MIC_{50} and MIC_{90} are calculated for a small test population of, e.g., 10–30 strains, a few strains with high MICs will have a disproportionately high influence on the MIC_{50} and MIC_{90} values. So, the significance of presentation of MIC_{50} and MIC_{90} increases with the number of tested strains (Schwarz et al. 2010).

For the veterinarians in practice is of importance, when MIC_{50} and MIC_{90} are available for the combination of antimicrobial/target pathogen/disease/species of animal to be treated. Not only sufficient number of isolates of certain bacterial species tested, but also the disease/tissue location of pathogen is of importance (e.g., *Escherichia coli* from diarrhea/intestine, from mastitis/udder, from urinary tract/e.g., bladder infections have generally different characteristics, including susceptibility patterns).

Also should be highlighted that due to time and geographical variability of isolates, preferred are recent results from the locations linked to animals to be treated (e.g., farm, flock, herd data; (grand) parents flocks; farm, from which animals were imported).

Of great importance is also that the clinical breakpoints are set considering the data including specific variables related to medicine (the antimicrobial, dose, route of administration, duration and frequency of treatment) and a disease (animal species, animal age/production category, disease, specific individual pathogens) and these clinical breakpoints primarily apply to this specific situation. Clinical breakpoints are however used in other applications also, but the confidence in the interpretation decays dramatically together with movement away from the specific situation for which the clinical breakpoints were set (Apley 2017).

4.3 Expert Rules

Expert rules are a tabulated collection of expert knowledge on intrinsic resistances, exceptional resistance phenotypes, and interpretive rules that may be applied to antimicrobial susceptibility testing in order to reduce errors and make appropriate recommendations for reporting particular resistances EUCAST 2019c CLSI 2018a, b). The expert rules have been developed to assist clinical microbiologists and describe actions to be taken in response to specific antimicrobial susceptibility test results. They (1) provide a way to evaluate the accuracy testing methods; (2) aid in the recognition of common phenotypes; (3) can assist with verification of cumulative AST data; and (4) include recommendations on reporting, such as inferring susceptibility to other agents from results with one, suppression of results that may be inappropriate, and editing of results from susceptible to intermediate or resistant or from intermediate to resistant on the basis of an inferred resistance mechanism. They are based on current clinical and/or microbiological evidence. The expert rules also include intrinsic resistance phenotypes and exceptional resistance phenotypes, which have not yet been reported or are very rare. The applicability of expert rules depends on the MIC breakpoints used to define the rules. Setting appropriate clinical breakpoints, based on treating patients and not on the detection of resistance mechanisms, may lead to modification of some expert rules in the future (Leclercq et al. 2013).

Intrinsic resistance is found within the genome of bacterial species and gives the bacteria an ability to resist the activity of a particular antimicrobial agent. It is independent of antibiotic selective pressure and horizontal gene transfer. Intrinsic resistance may be due to:

- A lack of affinity of the drug for the bacterial target
- Inaccessibility of the drug into the bacterial cell
- Extrusion of the drug by chromosomally encoded efflux pumps
- Presence of drug degrading enzymes

In practice, knowledge of the intrinsic resistance of a pathogen is essential to avoid inappropriate antimicrobial therapy and to decrease the risk of acquired resistance (Kostyanev and Can 2017).

Intrinsic (inherent) resistance, as opposed to acquired and/or mutational resistance, is a characteristic of all or almost all isolates of the bacterial species. The antimicrobial activity of the drug is clinically insufficient or antimicrobial resistance is innate, rendering it clinically useless. Antimicrobial susceptibility testing is therefore unnecessary, although it may be performed as part of panels of test agents. In these species, "susceptible" results should be viewed with caution, as they most likely indicate an error in identification or susceptibility testing. Even if a susceptible result is confirmed, the drug should preferably not be used or, when no alternative is available, should be used with caution. In some cases, intrinsic resistance to an agent may be expressed at a low level, with MIC values close to the susceptible breakpoint, although the agent is not considered to be clinically active. There are also situations where the agent appears to be fully active in vitro (MIC values cannot be separated from those of the wild type) but is inactive in vivo. These are generally not mentioned in the tables, as they are rather a matter of therapeutic recommendations.

Examples of intrinsic resistance are Enterobacteriales resistant to glycopeptides or linezolid, *Proteus mirabilis* resistant to nitrofurantoin and colistin, *Serratia marcescens* resistant to colistin, *Stenotrophomonas maltophilia* resistant to carbapenems, Gram-positive organisms resistant to aztreonam, and enterococci resistant to fusidic acid (Leclercq et al. 2013).

The current CLSI document for veterinary AST—VET08 (CLSI 2018a) listed a summary of intrinsic resistance of veterinary pathogens in Appendix B. There are the Tables with intrinsic resistance for individual bacterial groups (e.g., Enterobacteriales, non-Enterobacteriales, staphylococci, enterococci or other Gram-negative and Gram-positive bacteria of importance in veterinary medicine. The Tables are accompanied by warnings and notes that may be considered as expert rules, which are defined in EUCAST.

The EUCAST expert rules also define exceptional resistance phenotypes that are phenotypes of resistance of some bacterial species to particular antimicrobial agents that have not yet been reported or are very rare. Exceptional resistance phenotypes should be checked, as they may also indicate an error in identification or susceptibility testing. If they are confirmed locally, the isolate should be further studied to confirm the exceptional phenotype, and sent to a reference laboratory or other laboratory with expertise in resistance mechanisms for independent confirmation. Exceptional resistance phenotypes may change, as resistance may develop and increase over time. There may also be local, regional or national differences, and a very rare resistance phenotype in one hospital, area, or country may be more common in another. Examples of exceptional phenotypes are *Streptococcus pyogenes* resistant to penicillin, *Staphylococcus aureus* resistant to vancomycin, *Enterococcus faecium* susceptible to ampicillin, Enterobacteriales resistant to carbapenems (in veterinary medicine rare but increasing), and anaerobes resistant to metronidazole (Leclercq et al. 2013). Comparing the above-mentioned examples (viewed by the optics of 2013) can be also seen that the situation is changing. As an example can be considered resistance to carbapenems, with the more frequent use of carbapenems in human medicine during the last years, Enterobacteriales resistant to carbapenems occurs with higher frequency in some countries (ECDC 2018).

The expert rules also include interpretative rules for some groups of antimicrobials as a help for the clinical use of antimicrobials.

Interpretative Rules for Beta-Lactam Agents (Leclercq et al. 2013)

- All staphylococci resistant to methicillin, oxacillin, and/or cefoxitin, or with positive test results for *mecA* gene or PBP2a producing, should be considered to be resistant to all available beta-lactams, with the exception of those specifically licensed (until now human medicine) for the treatment of infections caused by methicillin-resistant staphylococci. Nevertheless, rare penicillinase hyperproduction may result in borderline resistance to oxacillin (but not cefoxitin) in vitro, owing to the lability of oxacillin, but there is no evidence that penicillinase hyperproduction is clinically relevant.

- All streptococci susceptible to penicillin can be reported as susceptible to aminopenicillins, cephalosporins, and carbapenems. If an isolate is resistant to penicillin, identification, and susceptibility should be checked. Conversely, resistance to beta-lactams in *Streptococcus pneumoniae* is common, owing to the production of mosaic PBPs that lead to various patterns of beta-lactam resistance.
- All enterococci are considered to be intrinsically resistant to cephalosporins but resistance to ampicillin mediated by alterations to PBP5 is increasingly found, particularly in *Enterococcus faecium*. These alterations lead to decreased affinity for beta-lactams, including all penicillins and carbapenems.
- Interpretive reading of the antibiogram is commonly based on beta-lactams and beta-lactamases in Gram-negative bacilli, particularly on extended-spectrum beta-lactamase (ESBLs) producing bacteria. ESBL-positive organisms are resistant to penicillins, cephalosporins, and aztreonam. Special attention should also be paid to reduced susceptibility to carbapenems that may be related to carbapenemases A, B, and D. New expert rules highlight the uncertain therapeutic outcome of treatment with penicillin in combination with a beta-lactamase inhibitor for Enterobacteriales isolates that are intermediate or resistant to any third- or fourth-generation cephalosporins in infections other than those affecting the urinary tract. Another rule recommends discouraging the use of cefotaxime, ceftriaxone, or ceftazidime in monotherapy or suppressing the susceptibility testing results for these agents, owing to the risk of selecting resistance in AmpC producers.
- *Haemophilus influenzae*—Resistance to ampicillin should also be considered as resistance to amoxicillin. *H. influenzae* isolates with altered PBPs and beta-lactamase production are phenotypically resistant to amoxicillin–clavulanate and ampicillin–sulbactam. They should also be considered to be resistant to piperacillin–tazobactam and to first-generation and second-generation cephalosporins.
- *Neisseria gonorrhoeae*, isolates that are beta-lactamase positive should be considered to be resistant to benzylpenicillin, ampicillin, and amoxicillin.

Interpretative Rules for Macrolides, Lincosamides, and Streptogramins (Leclercq et al. 2013)

Although the macrolides, lincosamides, and streptogramins have different chemical structures, they share similar mechanisms of action, and can be affected by the same resistance mechanisms. Erythromycin is considered to be the class representative for 14-membered (clarithromycin) and 15-membered (azithromycin) ring macrolides, with the exception of ketolides (telithromycin). There is cross-resistance between erythromycin and the other 14-membered and 15-membered ring macrolide antibiotics. This resistance can occur with or without cross-resistance to clindamycin and lincosamides. For staphylococci and streptococci, isolates resistant to erythromycin but susceptible to clindamycin should be tested for inducible MLS_B resistance. For staphylococcal isolates that are simultaneously resistant to erythromycin and clindamycin or lincomycin, a warning of reduced susceptibility to the combination quinupristin–dalfopristin and loss of bactericidal activity should be included in the susceptibility test report. For streptococci, less clinical evidence is available but, similarly, isolates that are resistant to erythromycin and susceptible to clindamycin should be tested for inducible MLSB resistance and reported as clindamycin susceptible if the result is positive but with a warning that resistance may develop on prolonged treatment.

Despite the facts summarized above, there should be commented on that within macrolides and related groups as triamilide, ketolide, azalide group is hard to find a class representative. As a main reason can be considered the differences in chemical structure (as, e.g., substituents in ring structures) that can have influence on their properties—antibacterial, interaction with immunodefence mechanisms of host as well as direct pharmacokinetic parameters (Watts et al. 2018).

Interpretative Rules for Aminoglycosides (Leclercq et al. 2013)

Several mechanisms that compromise the activity of aminoglycosides have been described: (1) decreased permeability and/or accumulation of the aminoglycoside agents because of mutations affecting passive diffusion or active transport, porin and/or lipopolysaccharide alteration (only in Gram-negative organisms), and efflux pump hyperexpression; (2) target (ribosomal) modifications caused by mutations in ribosomal proteins and as a result of the action of new methylases affecting 16S RNA; and (3) some aminoglycoside-modifying enzymes. Phenotype recognition of these resistance mechanisms is generally more complex than for those affecting beta-lactam compounds. Decreased permeability and/or resistance mechanisms involving efflux pumps usually confer a low-level resistance phenotype affecting nearly all aminoglycosides. With the exception of those described in *P. aeruginosa*, resistance mediated by efflux pumps is difficult to infer from phenotypic susceptibility, but cross-resistance to other antimicrobial classes, such as fluoroquinolone or tetracycline agents, might indicate their potential presence. Ribosomal mutations are extremely rare and do not always endow high-level resistance. Conversely, 16S RNA methylation confers high-level resistance, mainly affecting 4,6-disubstituted compounds (such as kanamycin, gentamicin, tobramycin, amikacin, and netilmicin), but not 4,5-disubstituted compounds (such as neomycin and paromomycin), streptomycin, and/or the aminocyclitol agent spectinomycin.

Single class representatives cannot be used for aminoglycosides as resistance is not a class effect, i.e., there are numerous resistance genes specifying a wide variety of resistance mechanisms, as the examples can be listed resistance to streptomycin and spectinomycin different from resistance to gentamicin, kanamycin, and/or tobramycin (Schwarz et al. 2010).

Attention should be also paid when considered results of AST for different groups of microorganisms (EMA 2018):

- *E. faecalis*, *E. faecium*, *E. gallinarum*, *E. casseliflavus*—Aminoglycosides may appear to be active in vitro, but are not effective clinically and should not be reported as susceptible (also compare above with comments to HLAR).
- *Salmonella* spp.—Due to the pharmacokinetic properties of aminoglycosides (difficulty to penetrate into eukaryotic cells), also in vitro testing of aminoglycosides in *Salmonella* spp. should be commented as of limited clinical correlation/therapeutic effect, especially due to the intracellular location of the pathogen and also low pH within vacuoles of the phagosome limiting antibacterial effect (CLSI 2013a, b, 2018b).
- Anaerobic bacteria (e.g., *Clostridium* spp. and *Bacteroides* spp. are intrinsically resistant (CLSI 2018b).
- *K. pneumoniae* resistant to carbapenems with false gentamicin susceptibilities observed using Vitek 2 in isolates carrying *armA* (Arena et al. 2015).

Interpretive Rules for Quinolones (Leclercq et al. 2013)

In general, older quinolones have lower activity than more recently developed agents. This is more obvious with Gram-negative organisms, and is particularly evident in Enterobacteriales. However, particularly with resistance caused by mutations in topoisomerases, decreased susceptibility to one fluoroquinolone is reflected in reduced susceptibility to other fluoroquinolones (class resistance). In staphylococci and viridans group of streptococci, resistance to the less active, but not to the more active, fluoroquinolones indicates that a first-step mutation may be present. In this case, a warning should be added to the susceptibility testing report, alerting clinicians to the potential for selection of a higher-level resistance mechanism involving different mutations.

Inference of specific fluoroquinolone resistance mechanisms can be difficult in multidrug-resistant organisms, as they may have superimposed mechanisms affecting these compounds (low- and/or high-level resistance).

Current draft of the EUCAST (2019c) referring to Cavaco et al. (2008) as for expert rules for interpreting: when using ciprofloxacin as class representative to report if resistant to ciprofloxacin then also to other fluoroquinolones, but if susceptible to ciprofloxacin then report other fluoroquinolones according to the tests result (with current breakpoints cannot be detected status of susceptibility to levofloxacin through ciprofloxacin).

4.4 The Future (Coming from Human Medicine?)

Bacterial infections in the human and animal populations, their diagnosis and following antibiotic treatment are still one of topical medicine issues worldwide. A major threat is the increasing and spreading of various types of resistances in bacterial populations to antimicrobials. Crucial clinical microbiology laboratory responsibilities associated with patient management and treatment include isolating and identifying the causative bacterium and performing ASTs, which are labor-intensive, complex, imprecise, and slow (taking days, depending on the growth rate of the pathogen). Considering the importance of rapid treatment and the increasing prevalence of antibiotic-resistant bacteria, rapid and automated new diagnostic tools of AST are needed (Syal et al. 2017). Paradigm-shifting AST technologies must overcome the current slow culturing steps. For the future, clinical samples without the need for selection and/or enrichment to be directly used, ideally low cost and easily be performed, having additional features, such as identification of bacterial strains before AST and the ability to perform AST of polymicrobial infections (Syal et al. 2017) and to help choose the most appropriate antimicrobial or combination of antimicrobials, especially to initiate the treatment at an early stage of the outbreak of infection process.

Newer AST technologies, which are currently and actively carried out by clinical translators, are considered to be new technologies for the purpose of this review. With the growing clinical demand for fast AST, various new AST techniques are based on optical imaging (Mohan et al. 2013; Choi et al. 2014; Metzger et al. 2014), microchannel resonators (Godin et al. 2010; Longo et al. 2013; Etayash et al. 2016), and other biosensors (Sinn et al. 2012; Hayden et al. 2016). For example, optical detection of bacterial growth through length and cell numbers (Mohan et al. 2013; Choi et al. 2014; Price et al. 2014), forward beam scattering (Hayden et al. 2016), and measurement of vibration amplitude changes of magnetic beads (Kinnunen et al. 2011; Sinn et al. 2012). Microchannel resonators have also been used to detect the oscillation of nanometers associated with bacterial growth. Quantitative molecular or biochemical markers such as 16SrRNA 28, ATP 29, and luciferase 30 in bacterial cells are also used for fast AST. These approaches can greatly improve current AST commercial technologies, but they still rely on cultivation that is not universally applicable to anaerobes, slow-growing bacteria, and microorganisms that are not cultivated. In addition, most of these new technologies still require substantial sample preparation and pre-treatment steps such as enrichment of bacteria from patient specimens and cell lysis for the extraction of biochemical markers (Syal et al. 2017).

The emerging technologies, being actively pursued by commercial entities discussed above, promise rapid AST within a few hours. Furthermore, some of the technologies can be directly applied to patient samples without any sample pretreatment. However, further shortening the test time and applying them to slowly growing organisms will require innovative approaches (Syal et al. 2017). Some of the promising emerging and future alternatives for the identification and for antimicrobials susceptibility testing are expected to revolutionize the field of clinical diagnostics (Maugeri et al. 2019) are listed and briefly characterized in Table 14. The table is not

Table 14 Emerging technologies coming from human medicine (Puttaswamy et al. 2018; Maugeri et al. 2019 modified)

System	Principle and technology used	Time taken for AST	Directly from sample	Caveats	Notes
Emerging					
BacterioScan FLLS[a] *(Forward Laser Light Scatter)*	Uses laser light source with scattered intensity measurements for accurate optical density (OD) readings in the presence of antibiotics. OD can be measured accurately up to 2 orders of magnitude lower than traditional methods.	6 h: Fast growing to 18 h: Slow growing	Yes (originally for urine)	Not distinguishing live/dead bacterial cells	
Smarticles technology[b]	Uses DNA probes inside non-replicating bacteriophages that can specifically bind to particular bacterial genus combined with synthetically designed plasmids. These plasmids containing luciferase gene that gets activated on contact with resistant bacteria. The increase in luciferase activity is directly related to the increase in the bacterial numbers in the sample.	<4 h	Yes (originally from blood culture samples)		MIC determination possible
Accelerate Pheno System[c]	Automated sample preparation step separating from other "material" bacterial and yeast cells, which are then released into culture media, introduced to multichannel cassettes (each with different antimicrobial/ different concentrations) and immobilized. Dark-field microscopy is used to image the cells every 10 min to record the growth of the cells vs. no growth or cell lysis over time. These time-lapse images are analyzed by software.	Approx. 7 h	Yes (originally blood, also other samples possible)	In its current form can only process one sample at a time and hence may slow down the clinical workflow.	Identification and MIC determination possible. FDA approved
Life Scan System[d]	Uses micro-cantilevers over which microbial cell suspension is passed through. As the cells pass through the resonator, the frequency of the cantilever vibration changes is detected by the sensor and correlated to the biomass of the bacteria at different concentrations of antimicrobial.	Approx. 3–4 h or longer	No (*positive* blood culture samples and direct urine samples where *concentrations of bacteria are*	Dependent on the multiplication time of bacteria tested. Fast-growing microbes with 20–30 min generation time: 3–4 h to results. Slow-growing bacteria longer	

(continued)

Table 14 (continued)

System	Principle and technology used	Time taken for AST	Directly from sample	Caveats	Notes
	At different time points, the variables are compared and MIC determined.		*above* 10^4 *CFU/ml)*		
Future					
AFM Cantilever[e] *(Amplitude Fluctuation Measurement)*	Characterizes the real-time physical activity of the bacteria utilizing low-frequency fluctuations of the cantilever. The bacteria to be tested are immobilized on the surface of the cantilever and their movement causes an increase in the amplitude of the cantilever fluctuations which is sensed by the sensing chamber. The resistant bacteria showed either an increase in the fluctuations of cantilever or initial drop in activity due to metabolic shock followed by return to normal cellular activity, while susceptible strains showed a decrease in their activity that can be measured by cantilever fluctuation.	<1 h	No	May need very pure isolates of bacteria or sample preprocessing as the presence of other nonbacterial cells in direct patient samples may affect bacterial immobilization or cantilever fluctuations.	MIC determination possible
MAC System[f] *(Microfluidic Agarose Channel)*	The bacteria are mixed with liquid agarose and then injected into the microfluidic channel, where are immobilized. Through realtime microscopy, the single bacterial time-lapse images are obtained are processed to determine the growth of bacteria in the presence of different concentrations of antimicrobials in agarose matrix.	4–10 h	Unsure (probably patient samples needs pre-processing to isolate pure microbial cultures)	MIC values for slower growing organisms may potentially take much longer than 4 h	Results exhibited good correlation between the MIC values obtained by MAC system to that of the CLSI standards.
SERS AST[g] *(Surface Enhanced Raman Spectra)*	SERS substrate based nanoparticles embedded in nano-channels of anodic aluminum oxide. SERS pattern decreases in amplitude over time in the case of presence of susceptible bacteria, while the resistant strains do not show any significant change in their spectral pattern.	2 h (yet performed for *S. aureus*-oxacillin *E. coli*-imipenem only)	No (pure bacterial cultures needed—delay due preprocessing)	Further studies with clinically relevant bacteria needed (to determine a new breakpoint/establish that the current breakpoint values are valid for all cases).	MIC determination possible, good correlation with standard broth dilution method with no major errors.

fASTest[h]	Uses microfluidic channels into which bacterial cells are trapped and monitored for growth with microscopic imaging uses microfluidic. Growth rate calculations are done for each individual cell traps in reference rows (which do not contain any antibiotics) and treatment rows (containing antibiotics at different concentrations). Comparison against the reference population.	<1 h (yet performed for *E. coli* only)	Yes (urine) (from other samples might be needed pure cultures—delay due preprocessing)	Dependent on the multiplication time of bacteria tested. Fast-growing microbes shorter detection time than slow growing.	
Isothermal microcalorimetry[i]	Measures the heat flowrate of a given bacterial sample in suspension to determine the bacterial growth or lack thereof. Since heat produced is proportional to the reaction rate in the suspension, the signals can reflect the effect of antibiotic on the bacteria and in turn determine MIC. Also able to identify different modes of action such as inhibitors of cell wall synthesis, DNA or protein synthesis—potentially useful in determining and classifying new antimicrobials/drug discovery. Including *Mycobacterium tuberculosis* (Howell et al. 2012), *E. coli*, and *S. aureus* (von Ah et al. 2009).	Approx. 24 h	Yes	However, the time taken to obtain the MIC results is long. Also, there may be delays in the onset of heat production (in low initial bacterial counts in the sample). This can cause misinterpretation of the mode of action or MIC determination.	Determining MIC. Method can also be used to distinguish between bacteriostatic vs. bactericidal effect of antimicrobial.
PIT[j] *(Plasmonic Imaging and Tracking)*	A plasmonic imaging and tracking technique has been used to track 3D motions of single bacterial cells associated with metabolic viability, the motion of individual bacterial cells correlates the phenotypic motion with bacterial metabolism and antibiotic action. The method is built on an inverted optical microscope, where light from a				Could potentially be used to spatially resolve and identify bacterial cells even in a complex matrix of urine, serum, and other body fluid samples.

(continued)

Table 14 (continued)

System	Principle and technology used	Time taken for AST	Directly from sample	Caveats	Notes
	luminescence diode is directed onto the sensor chip made of gold-coated glass film with immobilized bacterial cells.				
mEIS[k]	Uses microfluidic impedance measurements to determine capacitance changes in the presence of different concentrations of antibiotics. In the presence of an Alternating Current electric field, the bacteria present in their specially designed microfluidic channels can store charge across their membrane and hence act as capacitors. In the case of the death of bacteria (above MIC values of antibiotics), the bacterial membrane potential is lost, and as a result, the capacitance of the suspension containing bacteria-antibiotic will decrease. The concentration of the antimicrobial at which no change in the signal is seen over time or when a decrease in the signal is observed, is considered their MIC.	<4 h (the system does not need to wait to detect bacterial growth-which is limited by the generation time of the bacteria, but detect bacterial death in the presence of antibiotics).	Yes (no need for pure cultures)	Further studies with clinically relevant bacteria needed to validate the system and perform relevant quality control.	Determining MIC. Method can also be used to distinguish between bacteriostatic vs. bactericidal effect of antimicrobial.

[a]Hayden et al. (2016)
[b]Opens Helix (2015)
[c]Chantell (2015)
[d]Schneider et al. (2001)
[e]Longo et al. (2013)
[f]Choi et al. (2013)
[g]Liu et al. (2016a)
[h]Baltekin et al. (2017)
[i]Von Ah et al. (2009) and Howell et al. (2012)
[j]Syal et al. (2015a, b, 2017)
[k]Puttaswamy et al. (2018)

exhaustive and also other tools that especially help in confirmation of bacteria/certain species as a causative agent of infections are promising for the future. One of them so-called E-nose has been discovered for identification of *Mycobacterium tuberculosis* from patient breath through recognition of chemical fingerprint patterns based on an array of semi-selective sensors for volatile organic compounds (Maugeri et al. 2019). Despite the fact that some of them could be also finally used in veterinary medicine, especially in veterinary hospitals and clinics, there are several limitations among these probably not only the price, but also proper and speedy handling of the sample as well as difficulties to receive pure cultures what is more typical especially for blood samples collected from patients in hospitals. The methods that are designed for tests of urine could be with higher probability introduced in veterinary medicine, especially in small clinical practice, with pre-requisite of catheterized samples of urine.

References

Aarestrup F, Seyfarth A, Angen O (2004) Antimicrobial susceptibility of *Haemophilus parasuis* and *Histophilus somni* from pigs and cattle in Denmark. Vet Microbiol 101:143–146

AbuOun M, Stubberfield EJ, Duggett NA, Kirchner M, Dormer L, Nunez-Garcia J, Randall LP, Lemma F, Crook DW, Teale C, Smith RP, Anjum MF (2017) *mcr-1* and *mcr-2* variant genes identified in *Moraxella* species isolated from pigs in Great Britain from 2014 to 2015. J Antimicrob Chemother 72:2745–2749

Apley MD (2017) Susceptibility testing in veterinary medicine: what you can and can't conclude from antimicrobial susceptibility testing. Pharmacology II, Kansas State University, p 21. https://vetmed.illinois.edu/wp-content/uploads/sites/20/2017/04/Apley-Susceptibility-testing-Pharm-II-2017.pdf

Arena F, Tommaso F, Vaggeli G, Terenzi G, Pecile P, Rossolini GM (2015) Accuracy of different methods for susceptibility testing of gentamicin with KPC carbapenemase-producing *Klebsiella pneumoniae*. Diagn Microbiol Infect Dis 81:132–134

Baltekin O, Boucharin A, Tano E, Andersson DI, Elf J (2017) Antibiotic susceptibility testing in less than 30 min using direct single-cell imaging. Proc Nat Acad Sci USA 20:558

Borowiak M, Fischer J, Hammerl JA, Hendriksen RS, Szabo I, Malorny B (2017) Identification of a novel transposon-associated phosphoethanolamine transferase gene, *mcr-5*, conferring colistin resistance in d-tartrate fermenting *Salmonella enterica* subsp. *enterica* serovar Paratyphi B. J Antimicrob Chemother 72:3317–3324

Bywater R, Silley P, Simjee S (2006) Antimicrobial breakpoints—definitions and conflict requirements. Vet Microbiol 118:158–159

Caniaux I, van Belkum A, Zambardi G, Poirel L, Gros MF (2017) MCR: modern colistin resistance. Eu J Clini Microbiol Infect Dis 36:415–420

Carattoli A, Villa L, Feudi C, Curcio L, Orsini S, Luppi A, Pezzotti G, Magistrali CF (2017) Novel plasmid-mediated colistin resistance *mcr-4* gene in *Salmonella* and *Escherichia coli*, Italy 2013, Spain and Belgium, 2015 to 2016. Euro Surveill 22:30589

Carroll LM, Gaballa A, Guldimann C, Sullivan G, Henderson LO, Wiedmann M (2019) Identification of novel mobilized colistin resistance gene *mcr-9* in a multidrug-resistant, colistin-susceptible *Salmonella enterica* serotype Typhimurium isolate. mBio 7:1–6

CA-SFM (2018) Recommandations vétérinaires 2018. Comité de l'antibiogramme de la Société Francaise de Microbiologie. SFM, Paris, p 15

Cavaco LM, Frimodt-Møller N, Hasman H, Guardabassi L, Nielsen L, Aarestrup FM (2008) Prevalence of quinolone resistance mechanisms and associations to minimum inhibitory concentrations in quinolone-resistant *Escherichia coli* isolated from humans and swine in Denmark. Microbial Drug Resist 14:163–169

Chantell C (2015) Multiplexed automated digital microscopy for rapid identification and antimicrobial susceptibility testing of bacteria and yeast directly from clinical samples. Clin Microbiol News 37:161–167

Chew KL, La MV, Lin RTP, Teo JWP (2017) Colistin and polymyxin B susceptibility testing for carbapenem-resistant and *mcr*-positive *Enterobacteriaceae*: comparison of Sensititre, MicroScan, Vitek 2, and Etest with broth microdilution. J Clini Microbiol 55:2609–2616

Choi J, Jung YG, Kim J, Kim S, Jung Y, Na H, Kwon S (2013) Rapid antibiotic susceptibility testing by tracking single cell growth in a microfluidic agarose channel system. Lab Chip 13:280–287

Choi J, Yoo J, Lee M, Kim E, Lee JS, Lee S, Joo S, Song SH, Kim EC, Lee JC, Kim HC, Jung YG, Kwon S (2014) A rapid antimicrobial susceptibility test based on single-cell morphological analysis. Sci Transl Med 6:267–274

CLSI (2008) Development of in vitro susceptibility testing. Criteria and quality control parameters for veterinary antimicrobial agents. CLSI document M37-A3. Approved Guideline, 3rd edn. Clinical and Laboratory Standards Institute, Wayne, p 43

CLSI (2012) Methods for antimicrobial susceptibility testing of anaerobic bacteria. CLSI document M11-A8. Approved Standard, 8th edn. Clinical and Laboratory Standards Institute, Wayne, p 39

CLSI (2013a) Performance standards for antimicrobial disk and dilution susceptibility tests for bacteria isolated from animals. CLSI document Vet 01-A4. Approved Standard, 4th edn. Clinical and Laboratory Standards Institute, Wayne, p 80

CLSI (2013b) Performance standards for antimicrobial disk and dilution susceptibility tests for bacteria isolated from animals. CLSI document Vet 01-S2. 2nd Informational Supplement. Clinical and Laboratory Standards Institute, Wayne, p 70

CLSI (2016) Methods for antimicrobial susceptibility testing of infrequently isolated or fastidious bacteria isolated from animals. CLSI document VET 06, 1st edn. Clinical and Laboratory Standards Institute, Wayne, p 101

CLSI (2018a) Performance standards for antimicrobial disk and dilution susceptibility tests for bacteria isolated from animals. CLSI document Vet 08, 4th edn. Clinical and Laboratory Standards Institute, Wayne, p 170

CLSI (2018b) Performance standards for antimicrobial susceptibility testing. CLSI document M 100, 28th edn. Clinical and Laboratory Standards Institute, Wayne, p 258

Drusano GL (1990) Human pharmacodynamics of beta-lactams, aminoglycosides and their combinations. Scand J Infect Dis Suppl 74:235–248

ECDC (2018) Rapid risk assessment: carbapenem-resistant *Enterobacteriaceae*—first update 4 June 2018. European Centre for Disease Prevention and Control, Stockholm. https://vetmed.illinois.edu/wp-content/uploads/sites/20/2017/04/6.2-Apley-Susceptiblity-Testing-EVP-2017-1.pdf. Accessed 6 June 2019

EMA (2018) Reflection paper on use of aminoglycosides in animals in the European Union: development of resistance and impact on human and animal health (EMA/CVMP/AWP/721118/2014). https://www.ema.europa.eu/en/documents/scientific-guideline/reflection-paper-use-aminoglycosides-animals-european-union-development-resistance-impact-human_en.pdf. Assessed 6 June 2019

Etayash H, Khan MF, Kaur K, Thundat T (2016) Microfluidic cantilever detects bacteria and measures their susceptibility to antibiotics in small confined volumes. Nat Commun 7:12947

EUCAST (2000) Terminology relating to methods for the determination of susceptibility of bacteria to antimicrobial agents. Clin Microbiol Infect 6:503–508

EUCAST (2019a) New definitions of S, I and R. European Committee on Antimicrobial Susceptibility Testing. http://www.eucast.org/newsiandr/. Assessed 6 June 2019

EUCAST (2019b) Area of technical uncertainty (ATU) in antimicrobial susceptibility testing. European Committee on Antimicrobial Susceptibility Testing. http://www.eucast.org/fileadmin/src/media/PDFs/EUCAST_files/Breakpoint_tables/Area_of_Technical_Uncertainty_-_guidance_2019-1.pdf. Assessed 6 June 2019

EUCAST (2019c) European committee on antimicrobial susceptibility testing. http://www.eucast.org. Assessed 6 June 2019

Godin M, Delgado FF, Son S, Grover WH, Bryan AK, Tzur A, Jorgensen P, Payer K, Grossman AD, Kirschner MW, Manalis SR (2010) Using buoyant mass to measure the growth of single cells. Nat Methods 7:387–390

Hammoudi D, Moubareck CA, Sarkis DK (2014) How to detect carbapenemase producers? A literature review of phenotypic and molecular methods. J Microbiol Methods 107:106–118

Hayden RT, Clinton LK, Hewitt C, Koyamatsu T, Sun Y, Jamison GPerkins R, Tang L, Pounds S, Bankowski MJ (2016) Rapid antimicrobial susceptibility testing using forward laser light scatter technology. J Clin Microbiol 54:2701–2706

Hill JA, Cowen LE (2015) Using combination therapy to thwart drug resistance. Future Microbiol 10:1719–1726

Howell M, Wirz D, Daniels AU, Braissant O (2012) Application of a microcalorimetric method for determining drug susceptibility in *Mycobacterium* species. J Clin Microbiol 50:16–20

Hsieh MH, Yu CM, Yu VL, Chow JW (1993) Synergy assessed by checkerboard. A critical analysis. Diagn Microbiol Infect Dis 16:343–349

Javed M, Ueltzheffer V, Heinrich M, Siegrist HJ, Wildermuth R, Lorenz FR, Neher RA, Willmann M (2018) Colistin susceptibility test evaluation of multiple-resistance-level *Pseudomonas aeruginosa* isolates generated in a morbidostat device. J Antimicrob Chemother 73:3368–3374

Jawetz E (1975) Synergism and antagonism among antimicrobials drugs. A personal perspective. West J Med 123:87–91

Jorgensen JH (1993) Selection of antimicrobial agents for routine testing in clinical microbiology laboratory. Diagn Microbiol Infect Dis 16:245–249

Jorgensen JH, Ferraro MJ (2009) Antimicrobial susceptibility testing: a review of general principles or contemporary practises. Med Microbiol 49:1749–1755

Kinnunen P, Sinn I, McNaughton BH, Newton DW, Burns MA, Kopelman R (2011) Monitoring the growth and drug susceptibility of individual bacteria using asynchronous magnetic bead rotation sensors. Biosens Bioelectron 26:2751–2755

Kostyanev T, Can F (2017) The global crisis of antimicrobial resistance. In: Pulcini C, Ergonul O, Can F, Beović B (eds) Antimicrobial stewardship—developments in emerging and existing infectious diseases. Academic, Cambridge, pp 3–12

Leclercq R, Canton R, Brown DFJ, Giske CG, Heisig P, MacGowan AP, Mouton JW, Nordmann P, Rodloff AC, Rossolini GM, Soussy CJ, Steinbakk M, Winstanley TG, Kahlmeter G (2013) EUCAST expert rules in antimicrobial susceptibility testing. Clin Microbiol Infect 19:141–160

Leekha S, Terrell CL, Edson RS (2011) General principles of antimicrobial therapy. Mayo Clin Proc 86:156–167
Levison ME (2004) Pharmacodynamics of antimicrobial drugs. Infect Dis Clin N Am 18:451–465, vii
Liu CY, Han YY, Shih PH, Lian WN, Wang HH, Lin CH, Hsueh PR, Wang JK, Wang YL (2016a) Rapid bacterial antibiotic susceptibility test based on simple surfaceenhanced Raman spectroscopic biomarkers. Sci Rep 6:23375
Liu YY, Wang Y, Walsh TR, Yi LX, Zhang R, Spencer J, Doi Y, Tian G, Dong B, Huang X, Yu LF, Gu D, Ren H, Chen X, Lu L, He D, Zhou H, Liang Z, Liu JH, Shen J (2016b) Emergence of plasmid-mediated colistin resistance mechanism MCR-1 in animals and human beings in China: a microbiological and molecular biological study. Lancet Infect Dis 16:161–168
Livermore DM (2012) Current epidemiology and growing resistance of Gram-negative pathogens. Korean J Inter Med 27:128–142
Longo G, Alonso-Sarduy L, Rio LM, Bizzini A, Trampuz A, Notz J, Dietler G (2013) Rapid detection of bacterial resistance to antibiotics using AFM cantilevers as nanomechanical sensors. Nat Nanotechnol 8:522–526
Maugeri G, Lychko I, Sobral R, Roque ACA (2019) Identification and antibiotic-susceptibility profiling of infectious bacterial agents: a review of current and future trends. Biotechnol J 14:1700750
Metzger S, Frobel RA, Dunne WM (2014) Rapid simultaneous identification and quantitation of *Staphylococcus aureus* and *Pseudomonas aeruginosa* directly from bronchoalveolar lavage specimens using automated microscopy. Diagn Microbiol Infect Dis 79:160–165
Moellering RC (1983) Rationale for use of antimicrobial combinations. Am J Med 75:4–8
Mohan R, Mukherjee A, Sevgen SE, Sanpitakseree C, Lee J, Schroeder CM, Kenis PJ (2013) A multiplexed microfluidic platform for rapid antibiotic susceptibility testing. Biosens Bioelectron 49:118–125
Odds FC (2003) Synergy, antagonism, and what the chequerboard puts between them. J Antimicrob Chemother 52:1
Official Journal of the European Union, 2015/C 299/04 (2015) COMMISSION NOTICE—guidelines for the prudent use of antimicrobials in veterinary medicine. https://ec.europa.eu/health//sites/health/files/antimicrobial_resistance/docs/2015_prudent_use_guidelines_en.pdf. Accessed 6 June 2019
OIE (2012) Terrestrial manual. World Organisation for Animal Health [cit. 2018-11-30]. Guideline 2.1. Laboratory Methodologies for Bacterial Antimicrobial Susceptibility Testing. http://www.oie.int/fileadmin/Home/fr/Our_scientific_expertise/docs/pdf/GUIDE_2.1_ANTIMICROBIAL.pdf. Accessed 6 June 2019
Opens Helix I (2015) Roche gobbles Smarticles. New Yorker 26:186
Pokludova L, Pratova H, Kucharovicova I, Nedbalcova K, Bures J (2018) Recommendation for sample collection for microbiologic examination, interpretation of results of bacterial susceptibility testing in small animal practises. Veterinarni klinika 15:13–21. [publication in Czech]
Price CS, Kon SE, Metzger S (2014) Rapid antibiotic susceptibility phenotypic characterization of *Staphylococcus aureus* using automated microscopy of small numbers of cells. J Microbiol Methods 98:50–58
Puttaswamy S, Lee B, Amighi B, Chakraborty S, Sengupta S (2018) Novel electrical method for the rapid determination of minimum inhibitory concentration (MIC) and assay of bactericidal/bacteriostatic activity. J Biosens Bioelectron S 2:003
Rohner P, Herter C, Auckenthaler R, Pechere JC, Waldvogel FA, Lew DP (1989) Synergistic effect of quinolones and oxacillin on methicillin-resistant *Staphylococcus* species. Antimicrob Agents Chemother 33:2037–2041
Schneider CB, Harris K, Khera P, Strenn KS (2001) Rapid antimicrobial ausceptibility tests by mass measurement on a 96-well plate. Nature Nanotechnol 18:222
Schwarz S, Silley P, Simjee S, Woodford N, van Duijkeren E, Johnson AP, Gaastra W (2010) Assessing the antimicrobial susceptibility of bacteria obtained from animals. J Antimicrob Chemother 65:601–604
Siberry GK, Tekle T, Caroll K, Dick J (2003) Failure of clindamycin treatment of meticillin-resistant *Staphylococcus aureus* expressing inducible clindamycin resistance *in vitro*. Clin Infect Dis 37:1257–1260
Sinn I, Albertson T, Kinnunen P, Breslauer DN, McNaughton BH, Burns M, Kopelman R (2012) Asynchronous magnetic bead rotation microviscometer for rapid, sensitive, and label-free studies of bacterial growth and drug sensitivity. Anal Chem 84:5250–5256
Syal K, Wang W, Shan X, Wang S, Chen HY, Tao N (2015a) Plasmonic imaging of protein interactions with single bacterial cells. Biosens Bioelectron 63:131–137
Syal K, Iriya R, Yang Y, Yu H, Wang S, Haydel SE, Chen HY, Tao N (2015b) Antimicrobial susceptibility test with plasmonic imaging and tracking of single bacterial motions on nanometer scale. ACS Nano 10:845–852
Syal K, Mo M, Yu H, Iriya R, Jing W, Guodong S, Wang S, Grys TE, Haydel SE, Tao N (2017) Current and emerging techniques for antibiotic susceptibility tests. Theranostics 7:1795–1805
Toutain PL, Bousqued-Melou A, Damborg P, Ferran AA, Mevius D, Pelligard L, Veldman KT, Lees P (2017) En route towards European clinical breakpoints for veterinary antimicrobial susceptibility testing: a position paper explaining the VetCAST approach. Front Microbiol 8:1–13
Von Ah U, Wirz D, Daniels A (2009) Isothermal micro calorimetry—a new method for MIC determinations: results for 12 antibiotics and reference strains of *E. coli* and *S. aureus*. BMC Microbiol 9:106
Wang X, Wang Y, Zhou Y, Li J, Yin W, Wang S, Zhang S, Shen J, Shen Z, Wang Y (2018) Emergence of a novel mobile colistin resistance gene, *mcr-8*, in

NDM-producing *Klebsiella pneumoniae*. Emerg Microb Infect 7:122

Watts JL, Sweeney MT, Lubbers BV (2018) Antimicrobial susceptibility testing of bacteria of veterinary origin. Microbiol Spectr 6. https://doi.org/10.1128/microbiolspec.ARBA-0001-2017

Worthington RJ, Melander C (2013) Combination approaches to combat multi-drug resistant bacteria. Trends Biotechnol 31:177–184

Xavier BB, Lammens C, Ruhal R, Kumar-Singh S, Butaye P, Goossens H, Malhotra-Kumar S (2016) Identification of a novel plasmid-mediated colistin-resistance gene, *mcr-2*, in *Escherichia coli*, Belgium, June 2016. Euro Surveill 21:30280

Xu F, Zeng X, Hinenoya A, Lin J (2018) MCR-1 confers cross-resistance to bacitracin, a widely used in-feed antibiotic. mSphere 3:e00411–e00418

Yang YQ, Li YX, Lei CW, Zhang AY, Wang HN (2018) Novel plasmid-mediated colistin resistance gene *mcr-7.1* in *Klebsiella pneumoniae*. J Antimicrob Chemother 73:1791–1795

Yin W, Li H, Shen Y, Liu Z, Wang S, Shen Z, Zhang R, Walsh TR, Shen J, Wang Y (2017) Novel plasmid-mediated colistin resistance gene *mcr-3* in *Escherichia coli*. mBio 8:e00543–e00517

Wider Context of Antimicrobial Resistance, Including Molecular Biology Perspective and Implications for Clinical Practice

Lucie Pokludová and Hana Prátová

Abstract

Molecular methods provide the possibility of investigating the genetic background of the antimicrobial resistance in more detail. The majority of resistant bacteria carry resistance genes either on chromosomes or, what is from an epidemiological perspective more dangerous, on horizontally transferable mobile genetic elements. However, there are also described other "insusceptibility" mechanisms of bacteria, which can complicate treatment. Among those belong tolerance, persistence, dormancy, bacterial indifference, or heteroresistance. In light of the abovementioned mechanisms, genetic and phenotypic characteristics are many times questioned for the predictive value of the in vitro susceptibility results. However, there are other factors to be further considered when we think about clinical/therapeutic implications. Phenotypic testing results can be confirmed and outstanding issues can be answered using deep insight via the molecular methods that can provide proof of evidence of existing mechanisms of resistance, including distinguishing among intrinsic and acquired resistance, exact localization of resistance genes on chromosomes or mobile genetic elements (mainly plasmids), and colocalization with other genes of interest. Also, insights into mechanisms of resistance genes transfer, via conjugation, transformation, or transduction, can help from an epidemiological point of view to trace the origin of resistance and the routes of transfer. A short overview of the molecular methods as well as resistance mechanisms to different classes of antimicrobials is given. To outline the complexity of resistance directly to antimicrobials (primarily antibiotics), cross-resistance and co-resistance to biocides (generally), disinfectants, antiseptics, heavy metals, and further substances used in different parts of the agriculture and human medicine with a possible influence on antibiotic resistance are outlined within this chapter to give veterinarians in practice information on possible consequences of the use of different substances on selection/co-selection of antimicrobial resistance.

Keywords

Antimicrobial resistance mechanisms · Transfer of antimicrobial resistance · Acquired resistance · Cross-resistance · Co-resistance · Mobile genetic elements · Molecular methods · Biocides · Heavy metals · Tolerance · Persistence

L. Pokludová (✉) · H. Prátová
Institute for State Control of Veterinary Biologicals and Medicines, Brno, Czech Republic
e-mail: pokludova@uskvbl.cz

L. Pokludová (ed.), *Antimicrobials in Livestock 1: Regulation, Science, Practice*,
https://doi.org/10.1007/978-3-030-46721-0_9

1 Definitions of Antimicrobials and Antimicrobial Resistance

Despite the fact that antimicrobial resistance (AMR) is considered to be a very urgent issue during at least the last decade, there is still unclosed debate on its definition. It should be started with distinguishing the terms antimicrobial and antibiotic. Regulation EU 6/2019, on veterinary medicinal products (European Commission 2019), defines the term "antimicrobial" in relation to medicines, i.e., as any substance with a direct action on microorganisms used for treatment or prevention of infections or infectious diseases, including antibiotics, antivirals, antifungals, and anti-protozoals and it should be distinguished from the subcategory "antibiotic" which is in current meaning considered as acting against bacteria causing infectious diseases. In relation to public health, as well as animal health considering the broadness of information portfolio, in fact the majority of the activities, published studies, and concerns are targeted on antibiotics (i.e., antibacterials). As each rule has usually some exemptions, it should be highlighted that, e.g., sulfonamides, a group of antimicrobials (belonging among three mostly used groups of veterinary antimicrobials in Europe), are this exemption as they act on both bacteria and protozoa and are also used in practice to treat and prevent bacterial and protozoal diseases.

With regard to the definition of antimicrobial resistance, it is even more complicated. There could be some "official" definitions: according to Regulation EU 6/2019 "antimicrobial resistance" (AMR) is defined as "the ability of micro-organisms to *survive* or to grow in the presence of a concentration of an antimicrobial agent which is usually sufficient to inhibit or kill micro-organisms of the same species" (European Commission 2019). Even within this definition there is something missing as based on the current article by Balaban et al. (2019); the definition should be broadened—"survive *and replicate*" to distinguish better resistance from persistence and also clearly indicate the danger of spread by "replication" of bacteria. Also, the term "inhibit", i.e., to act against microbes and slow down their vital processes or at least keep them static, can be understood in a different way (see persistence and tolerance). The term "kill" is only undisputable and indicates microbicide effect of the antimicrobial/s, but not going into more details on AMR, it can be simplified as "the ability of bacteria to survive and replicate" and in such cases continue to either cause the infectious disease or to survive in certain environment (human and animal body, water, soil, solid surfaces, etc.) and further disseminate.

Different views on AMR and its definition considering not only theory but practical impacts should be further elaborated. Currently, being in "direct contact" with AMR the practical view on AMR can probably best describe experts from clinical diagnostic laboratories as well as clinicians (mostly those working in intensive care units of hospitals, but more and more frequently also those working in veterinary hospitals or as herd/flock veterinarians). They could say that AMR is not only about survival and/or replication of microorganisms under certain concentrations of antibiotics (what can clearly see people from clinical laboratory in their everyday practice), but especially clinicians would amend that clinical antimicrobial resistance causes further spread of the resistant bacteria (therefore support for broadening the definition with the word "replicate" of bacteria), causing damage to patients/animals, in serious cases causing even organ failure and can finally lead to total treatment failure and patient death (or even deaths of a huge number of animals in the flock/herd). Therefore, it is crucial to identify potential sources of resistant bacteria and ways of their transmission in the particular environment (e.g., animal husbandry) in order to handle properly the problem of antibiotic resistance of bacteria. Moreover, not only experts on AMR have started to be aware and afraid of multiresistance, i.e., microbes harbor various resistance genes to multiple structurally non-related antibiotics. Multiresistance of pathogenic bacteria has severe consequences in clinical practice. Multiresistant (MDR) bacteria can survive and/or replicate under the action of different antimicrobials from different classes or their combinations. When the antibiotic of choice is switched to another medicinal product containing different active substance/s (even from another pharmacological group), it does not help. MDR bacteria can even cause very severe issues or can be fatal once spread in the hospital (herd/flock) to the other patients (animals). Comparing to decades ago, such cases are encountered more and more by clinicians in companion and livestock practices.

Further possibility of AMR perception and definition considers phenotypic "insusceptibility" (persistence, tolerance, heteroresistance) *versus* inherited resistance. Resistance (see also further description in more detail) may arise via mutations (e.g., leading to modification of the antibiotic target in bacterial cell). Resistance can also be acquired by horizontal gene transfer (HGT) of genes encoding resistance (e.g., genes encoding antibiotic inactivating enzymes or efflux pumps) that confer resistance phenotype to the bacterial population. Once established it can be transmitted to subsequent generation, or even more, once located on mobile genetic element, it can be spread horizontally (to the same species and/or to other species, even those that are not closely related).

There can be also phenotypic "resistance" that was described in situations, where no specific resistance gene was found (Corona and Martinez 2013) and "resistance" was associated with specific processes such as stationary growth phase, persistence, tolerance (Balaban et al. 2019), or heteroresistance (Band and Weiss 2019). The practical impact is that these situations are not commonly considered in standard clinical laboratory susceptibility testing and can be one of the reasons of not full compliance of the laboratory results with clinical outcomes, which can cause treatment failure. Bacterial resistance, reduced susceptibility, or full susceptibility to antibiotics (not considering patient factors) is highly dependent on the bacterial metabolism, phases of growth of bacterial population, and also global metabolic regulators that can modulate bacterial phenotype (Corona and Martinez 2013).

Several mechanisms that in fact could mimic the resistance phenotype in the laboratory and/or in the clinical picture of the disease development *versus* treatment outcome are as follows:

Dormancy (stationary growth and metabolism phase) "reflects the state of a bacterium that does not grow and has decreased activity when compared with growing cells or even typical stationary phase cells. This term is often also used for single cells that are viable but do not grow despite environmental conditions that support growth. Dormant bacteria are often tolerant to many antibiotics because of their growth arrest or their decreased metabolism. Dormancy is not necessary pre-requisite of tolerance or persistence" (Balaban et al. 2019).

Bacterial indifference has been mostly described for specific germs in human medicine—as, e.g., *Mycobacterium tuberculosis, Staphylococcus aureus*, or *Streptococcus pyogenes*. Based on the results of the studies on these species, it has been suggested that the existence of drug indifferent cells with minimal metabolism, slow growth, or stopping/slowing down of cell division, can be the cause of relapse after antibiotic treatment (Fitoussi et al. 1997; Clement et al. 2005; Corona and Martinez 2013). Bacterial indifference can also be considered with respect to various antimicrobials, taking into account their mechanism of action (e.g. the efficacy of beta-lactams in the stage of the cell division in bacterial culture susceptible to these antimicrobials is well known and therefore indifference can be observed in stationary phase of growth of bacterial culture). On the other hand, current knowledge also counts with more deep information on interaction with host, i.e., location of these bacteria (caverns, abscesses, intracellular occurrence, joints, and cartilages), which further complicates clinical outcomes of treatment. Considering the specific case of veterinary importance—e.g., mastitis and *Staphylococcus aureus* invasion strategy, several factors should be taken into account which play a significant role in the efficacy of treatment by antimicrobials, e.g., the ability of *S. aureus* to reside inside the host cells by surviving the neutrophil arsenal upon phagocytosis or by invading mammary epithelial cells, formation of small colony variants or L-forms and induction of formation of (micro-)abscesses and fibrosis. Especially intracellular location of *S. aureus* is counted as important contributing factor to the issue of therapeutic failure (Rainard et al. 2018). Treatment failure can be caused also due to the fact that intracellular staphylococci are not in a "metabolic state of susceptibility" to the antibiotic, so they can be considered as under indifference stage (Craven and Anderson 1980).

Persistence as population level phenomenon is defined as the ability of a subpopulation to survive exposure to a bactericidal drug concentration (antibiotics with bactericidal action are the only considered, e.g., β-lactams and fluoroquinolones). Biphasic killing curves indicate the presence of two subpopulations with individual killing rates. The clonal culture consists of cells that are killed fast by the antibiotic and tolerant cells that may survive. Features distinguishing persistence from resistance were defined by Balaban et al. (2019) as follows (corresponding to the Fig. 1):

1. The biphasic killing curve (not all bacteria in a clonal culture are killed at the same rate).
2. In the conditions without antibiotics, persister cells regrow and their progeny give rise to a subpopulation whose rate of susceptibility to antibiotics is to the same extent as the parental population's susceptibility.
3. In the populations having concentrations far above MIC, the level of persistence, namely, the size of the persister subpopulation, will only weakly depend on the concentration of the antibiotic.
4. In contrast to resistant cells, persister bacteria cannot replicate in the presence of the antibiotic any better than the non-persister cells but are killed at a lower rate than the susceptible population from which they arose. This characteristic allows us to distinguish persistence from heteroresistance, a phenomenon in which a small subpopulation transiently displays a substantially (more than eightfold) higher MIC.

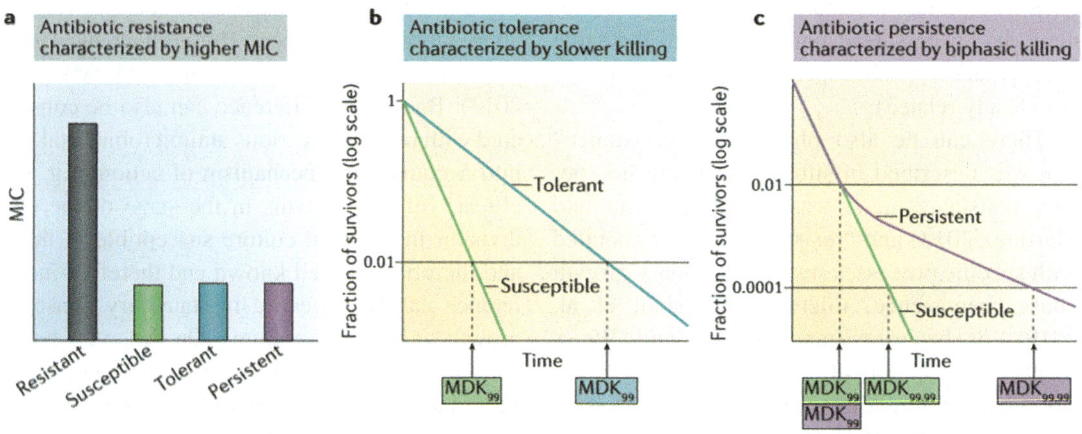

Fig. 1 Graphic characteristics of antimicrobial resistance, tolerance, and persistence (reprint, Balaban et al. 2019). Antibiotic resistance, tolerance, and persistence. Resistance, tolerance, and persistence are distinct responses to antibiotic treatment that lead to increased survival compared with susceptible cells. (**a**) To inhibit the growth of resistant bacteria, a substantially higher minimum inhibitory concentration (MIC) of the antibiotic is needed than for susceptible bacteria. Notably, persistence and tolerance do not lead to an increase in the MIC compared with susceptible bacteria. (**b**) By contrast, tolerance increases the minimum duration for killing [MDK; for example, for 99% of bacterial cells in the population (MDK_{99})] compared with susceptible bacteria. (**c**) Persistence leads to a similar MIC and a similar initial killing of the bacterial population compared with susceptible bacteria; however, the MDK for 99.99% of bacterial cells in the population ($MDK_{99.99}$) can be substantially higher owing to the survival of the persister cells. Note that pure exponential killing of the susceptible strain is rarely observed because most bacterial cultures have some level of persistence. The data shown are only illustrations and not actual measurements. Parts (**b**) and (**c**) were adapted by Balaban et al. 2019 from Brauner et al. 2016

It can be expected that several subpopulations (consisting of persister cells) can develop from parental population. Finally, not only bimodal but even multimodal killing curve may occur. Persister cells can originate from a subpopulation of tolerant bacteria; the term "heterotolerance" can be explained in such cases (see below) (Balaban et al. 2019).

Tolerance is within population perspective interchangeable with the term "persistence." Meylan et al. (2018) studied tolerance of four pathogens of Gram-negatives (*E. coli, P. aeruginosa*) and Gram-positives (*S. aureus* and *M. tuberculosis*). According to outcomes of this study, antibiotic tolerance can be considered as the capacity of genetically susceptible bacteria to survive the lethal effects of antibiotic treatment, which is expressed in another, more detailed wording by Balaban et al. (2019) as the ability of the population to survive the duration of a transient antibiotic treatment several times above the minimum inhibitory concentration (MIC) without a resistance mechanism. Longer treatments, for example, by having a lower killing rate, but without a change in the MIC, can be also survived by tolerant population (Balaban et al. 2019). Tolerant populations survive the period of antibiotic treatment better, with unchanged MIC, which means weak dependence on the antibiotic concentration. What characterizes their slower killing, even at high concentrations of the antibiotic, is the time required to kill a large fraction of the population. The variable "*minimum duration to kill 99% of the population (MDK 99)*" can be defined and assessed. Similarly to persistence—mechanisms such as dormancy, reduced metabolism, and reduced ATP levels have also been described (Balaban et al. 2019). For a better understanding, please refer to the graphical explanation in Fig. 1.

An *antibiotic-tolerant cell* is "a cell that survives treatment with an antibiotic, without showing a particular resistance mechanism and that can regrow after removal of the antibiotic" (Balaban et al. 2019). Often, tolerant cells are non-growing before antibiotic exposure, but not necessarily. While more susceptible bacteria are killed within a certain time span of the treatment, tolerant cells can survive (Balaban et al. 2019).

As proposed by El-Halfawy and Valvano (2015), *heteroresistance* is defined as "a population-wide variation of antibiotic resistance, where different subpopulations within an isolate exhibit various susceptibilities to a particular antimicrobial agent." From the methodological point of view, the authors define it even more precisely: the presence of subpopulation of cells with capacity to growth at concentrations of antibiotics at least eightfold higher than the highest concentration that does not affect the replication of the dominant population. Recent study by Nicoloff et al. (2019) shows how important can be this phenomenon from the laboratory interpretation and finally also influencing clinical perspective. According to the results of the study, where clinical isolates of pathogens such as *E. coli, Salmonella enterica, Klebsiella pneumoniae,* and *Acinetobacter baumannii* (in total, susceptibility of 41 isolates to 28 antibiotics were tested, which creates in total 766 combinations from which 27.4% were considered as cases of heteroresistance based on above-cited definition criteria). Using whole genome sequencing (WGS) and also quantitative PCR (qPCR) for confirmation of the level of DNA amplification and/or plasmid copy number, Nicoloff et al. (2019) proved that a majority of heteroresistance cases were unstable. Spontaneous tandem amplifications, typically including known resistance genes, were the reason for increased resistance of the subpopulations. As heteroresistance can be considered as one of the reasons for discrepancy of MIC results and treatment outcomes (potential treatment failure due to heteroresistance), both El-Halfawy and Valvano (2015) and Nicoloff et al. (2019) called for standardized definitions and protocols to identify heteroresistance in clinical pathogens. As is clear from the above-tested pathogens, they are of importance not only from the perspective of human medicine, but also on veterinary side we should take the phenomenon of heteroresistance into consideration.

As commented above, currently it has been recognized that, even in the absence of any antibiotic resistance, many bacterial infections are hard to treat and tend to relapse (such as tuberculosis, lung infections in people with cystic fibrosis, systemic infections with *Salmonella*, tonsillitis, and urinary tract infections as well as infections of animals, e.g., mastitis caused by *Staphylococcus aureus* and infections of dogs caused by *Staphylococcus pseudintermedius*). Non-growing bacteria and high-persister-forming mutants are selected over time in patients exposed to repeated doses of antibiotics. The guidelines for in vitro measurements of persistence have been recently released by Balaban et al. (2019) designing protocols adapted to the clinical evaluation of antibiotic persistence.

2 Predictive Value of the In Vitro Susceptibility Results and Factors to be Further Considered

Despite the new facts mentioned above, even in standard susceptibility testing, doctors as well as veterinarians are asking what is the predictive value of in vitro susceptibility testing and why infections associated with causative agents with good susceptibility confirmed in vitro are in some cases untreatable in practice? The underlying reasons are most likely multifactorial including factors associated with sampling and laboratory testing:

- *Factors of correct/proper sampling* (the first step) can highly influence the results of phenotypic testing. Several factors are of key importance: avoiding contamination by choosing and performing proper sampling techniques (e.g. whenever possible to gain real causative pathogen); choosing the right transportation system; sufficient speed of transport to laboratory. Quite challenging is also processing and analysis of samples, once polymicrobial infection (i.e. at least two causative pathogens) is expected:
  - *Example—pure culture collected postmortem from inner organ lesions in laboratory detecting* E. coli *vs. polymicrobial culture from, e.g., rectal swab transported to laboratory (with mixture of various* E. coli *strains, but with different virulence and resistance properties). It is clear how difficult, if at all possible, it is to distinguish the "real" causative/pathogenic* E. coli *strain for further culturing and susceptibility testing from rectal swab.*
- *Factors of laboratory selection of the pathogen/s of importance as causative agent for further testing* (identification, susceptibility testing):
  - *Example—polymicrobial culture that needs purification and gaining, e.g., two possible causative pathogens (broiler chicks* E. coli *and* Enterococcus faecium*); especially "tricky" step is choosing the "right colony" on Petri dish, which will be further cultured and investigated.*
- *Factors of the choice of clinical breakpoints (CBPs)* distinguishing susceptible or resistant bacteria, especially when the case from practice is not in accordance with the combination of variables used for setting of the CBP (these variables include antimicrobial substance, species of bacteria, animal species (target tissue), disease, route of administration, dosing regimen). Each deviation and extrapolation can bring uncertainty and affect the predictive value and probability of proper correlation.
  - *Example: CBP available for oxacillin—*Staphylococcus aureus*—dairy cow (udder)—mastitis—intramammary—xx mg oxacillin in single dose repeated (how many times in the case of* Staphylococcus*?)—but in practice, from correctly taken sample has been isolated coagulase-negative (COAN)* Staphylococcus. *The question here is still connected with* Staphylococcus *species, but these COANs have usually very different properties including antimicrobial resistance patterns.*
- *Factors of heteroresistance:*
  - *Example: heteroresistance has been recently described as the possible cause of the treatment failure* (Band and Weiss

2019), *and recently, also several papers dealing with the issue of mechanism of heteroresistance* (Andersson et al. 2019), *its genetic background* (Nicoloff et al. 2019), *as well as studies of antimicrobial combinations fighting against treatment failures in infections caused by heteroresistant bacteria* (Band et al. 2019) *are available.*

- Factors of the host:
  - *Location of infection (pharmacokinetic properties and concentration of antimicrobial that can be in real situation achieved at site of infection)*

    Example: abscess with limited/no accessibility of antimicrobial

    Example: big molecules (e.g., colistin sulfate) administered orally do not penetrate to system from the gastro-intestinal tract
  - *Whole immunostatus of the diseased animals (and the role that can play in the exact case of infection)*

    Example: antimicrobial can act as bacteriostatic, but immunodefense mechanisms of the animal can help to recover fully from the infection; in case of immunocompromised animal, infections are with more serious course and more difficult to treat.

    Example: Interactions of the antimicrobial with host immunodefense mechanisms (e.g., exposure of bacterial cells to clavulanic acid resulted in alteration of bacterial cell wall integrity and changes in expression of surface proteins of the bacteria. Alterations of surface charges and hydrophobicity can influence the rate of phagocytosis and the extent of intracellular killing of bacteria. An effect on the rates of phagocytosis and the intracellular killing functions of polymorphonuclear leukocytes was demonstrated in experimental studies)
  - *Stage of disease (acute—beginning stage, chronical, recurrent/treatment failure)*
- Factors related to the active substance/medicinal product
  - *Type of active substance and its pharmacokinetic properties (e.g., level of liposolubility)*

    Example: tetracycline-resistant bacteria can be still in vivo *doxycycline-susceptible as doxycycline crosses the cellular membranes more easily thanks to its lipophilic properties*
  - *Stability/instability of active substance (e.g., in certain environmental conditions/pH)*

    Example: Orally administered VMPs and their stability once distributed in drinking water of poor quality in metal distribution systems can lead to finally low amount of the active substance in treated animals that causes underdosing and therefore insufficient efficacy
  - *Route of administration*

    Example: intravenous administration of the antimicrobial (with bactericidal concentration-dependent mechanism of action)—more quick achievement of the maximum concentration in plasma and therefore recommended for (serious/progressive) systemic infections, where bactericidal effect is needed; MIC testing of susceptibility/resistance is beneficial to adjust precisely the dose (especially in cases where range of doses is authorized)

    Example: administration of the intramammary VMP directly in teat—concentration at the/near the site of infection
  - *Bacterial killing effect (time vs. concentration dependence)*

    Example: importance of following dosing intervals precisely in antimicrobials with high level of time dependence to keep levels at/above minimum inhibitory concentration of the target pathogen

- Factors related to bacterial population (in relation to success of action of antimicrobial)
  - Resistance, tolerance, persistence, heteroresistance (see above)
  - Ability of biofilm formation in target tissue in the case of mastitis:

    Example: mastitis causing Staphylococcus aureus *strains are able to create biofilms (within biofilms high increase of the MICs is known that complicate the treatment of infection)* (Amorena et al. 1999)

From the group of factors related to bacteria, except for the testing of resistance phenotype in the diagnostic laboratory, further detection and characterization of resistance genes by molecular methods can be also performed. These methods at present offer a powerful tool for the precise detection and description of resistance mechanisms in bacteria and provide detailed insights into the epidemiology of antibiotic resistance. However, molecular methods alone (not accompanied with conventional laboratory methods comprising cultivation, susceptibility testing, and microscopy) are not considered usually valid for choice of an antibiotic for treatment in clinical practice, because viable bacterial cells are not analysed by these methods.

3 Antimicrobial Resistance from the Microbiological and Molecular Point of View

The widespread use, or even abuse of antibiotics in treatment, prophylaxis, and metaphylaxis both in human medicine and agriculture (mainly in animal husbandry), or in other fields of human activity, has severe consequences on bacterial ecology and evolution. Concentrations of antimicrobials and other substances with potential for co-selection of antimicrobial resistance in the environment associated with human activities increase a relative abundance of resistant bacteria and resistance genes while population of susceptible bacteria is minimized in certain ecological niches. It has been acknowledged that selective pressure of antimicrobials increases rates of horizontal gene transfer (HGT) and other evolutional processes, like mutation or recombination (Gillings 2013). A high rate of spontaneous mutations and widely prevalent DNA transfer mechanisms in bacteria are critical contributors to the emergence of the phenomenon of antibiotic resistance (Peterson and Kaur 2018). Bacterial genomes are extremely flexible and able to adapt to different conditions, which allow them to survive; therefore due to the use of antimicrobial, there can be expected sooner or later the appearance of the resistant bacteria, with different potential to adapt also to other environmental/host specific conditions (e.g., virulence factors, biofilm formation ability) (Peterson and Kaur 2018).

From a microbiological point of view, inherited antibiotic resistance can be divided into two principal types: *intrinsic* (also called inherent, innate, non-acquired, or primary) or *acquired* (secondary) (Fernández and Hancock 2012). Intrinsic resistance is given by inherent properties of a particular microorganism that limit or prevent action of particular antibiotic (Fernández and Hancock 2012; Pulcini et al. 2017). Intrinsic resistance is reflected in wild-type antimicrobial patterns of all or almost all representatives of certain species/genus. Despite the fact that intrinsic resistance can be so common among the certain species/genus and susceptibility testing is unnecessary, it can be recommended and is considered beneficial to warn veterinarians that such resistance patterns exist (e.g., note that in certain antimicrobials, even not tested within exact panel of tests toward etiological agent of concern, intrinsic resistance occurs in this microbe).

Intrinsic resistance can occur due to any of the following mechanisms (Pulcini et al. 2017):

- Lack of affinity of the drug for its molecular target
- Decreased uptake of the drug
- Active efflux of the drug out of the cell
- Native production of enzymes that inactivate the drug

For example Gram-positive cocci are intrinsically resistant to colistin; in the case of *Enterococcus* spp. isolates, cephalosporins, aminoglycosides (except for high-level resistance testing), clindamycin, and trimethoprim-sulfamethoxazole may appear active in vitro but are not effective clinically and enterococci should not be reported as susceptible. Resistance of anaerobes to aminoglycosides can be also considered as intrinsic (CLSI 2018).

Intrinsic resistance can be mediated due to low cell wall permeability (one of the cell protective mechanism) and active efflux of any bacterial cell toxic/inhibitory molecules (e.g., antibiotics, or generally antimicrobials, biocides, and heavy metals). Such mechanisms are well described as small outer membrane porins and multidrug efflux transporters in *Pseudomonas aeruginosa*, respectively (Zgurskaya et al. 2015).

Acquired resistance can be considered the most important from an epidemiological point of view as it arises due to mutations of chromosomal genes or acquisition of new genetic material via horizontal transfer. Regardless of the mechanism, originally susceptible bacterium becomes resistant (Fernández and Hancock 2012). Acquired resistance in bacteria is the most threatening public health concern as its development and dissemination is directly connected to antibiotic use. Nowadays, it can be clearly seen that mobile genetic elements (i.e., insertion sequences, integrons, transposons, plasmids, or integrative conjugative elements) as carriers of antibiotic resistance genes play a pivotal role in the development and dissemination of antibiotic resistance in bacteria (Partridge 2011).

Information on the genetic background of resistance provided by molecular methods can be highly valuable for epidemiological and clinical purposes. Molecular methods enable precise detection of resistance genes, description of their localization (chromosome vs. mobile genetic elements), or their colocalization with other important genes including those encoding virulence, non-essential metabolic pathways, and resistance to other substances like disinfectants or heavy metals. Moreover, strain typing methods (e.g., core genome/whole genome multilocus sequence typing or single-nucleotide polymorphism (SNP)-based analyses following whole genome sequencing) associated with detection of antibiotic resistance genes can bring comprehensive epidemiological information and contribute to infection control.

The remainder of this chapter is targeted on acquired antimicrobial resistance including description of molecular mechanism to individual antibiotic groups, the most important mobile genetic elements associated with dissemination of antibiotic resistance, means of horizontal transfer, and molecular methods used for detection of antibiotic resistance genes.

4 Mechanisms of Resistance to Individual Groups of Antimicrobials

Antimicrobials used in veterinary medicine act against essential processes in bacterial cell including synthesis of bacterial cell wall, protein synthesis, replication of DNA, and metabolic pathways. Within each of these processes, antimicrobials affect specific chromosomally encoded molecular targets and limit their normal function. Bacteria can protect these targets via alteration of their primary structure due to mutation/s or via posttranslational modification of the targets. Despite these changes causing inefficient antibiotic biding, the molecular targets retain their normal activity. Moreover, bacteria can modulate uptake of antibiotics via porins and/or their removal from the cell via efflux pumps due to mutations in respective genes and/or in their regulatory regions and direct interaction between antibiotic and respective regulator. Furthermore, bacteria can acquire new genes coding for enzymes that hydrolyze or modify the antibiotic (Blair et al. 2015). A list of these resistance mechanisms is presented below. However, as novel genes and alleles are continuously identified, the list is not exhaustive.

It should be noted that resistance to the same antimicrobial agent can be mediated by different mechanisms that can be encoded by chromosome and/or plasmid/s. Mechanisms of action of individual antibiotic groups and corresponding mechanisms of resistance are summarized in Table 1. In some cases, the same resistance gene/mechanism is found in a wide variety of bacteria, whereas in other cases, resistance genes or mechanisms appear to be limited to certain bacterial species or genera. It should be also highlighted that a huge amount of available bibliography is targeted on the detection of mechanisms of resistance and resistance genes of culturable bacteria and that a lot of work is also concentrated on recognition of the whole pool of resistance genes, mechanisms of resistance, and means of transfer. However, real impact on the efficacy of known and therapeutically used antimicrobials, due to resistance genes, level of their expression, as well as co-location of genes for resistance and those determining virulence/ability of bacteria to invade host organism, is still waiting to be discovered via metagenomics and other molecular methods, utilizing knowledge and, where appropriate, still using phenotypic methods as well as clinical studies. Also, it should be highlighted that bacteria of human, animal, and environmental microbiomes can bear and transfer a wide variety of the resistance genes.

Up to this date, following essential groups of the resistance mechanism have been described (Fernandez and Hancock 2012; Blair et al. 2015; Munita and Arias 2016; van Duijkeren et al. 2018; Peterson and Kaur 2018):

1. **Resistance mechanisms connected with molecular target**
   causes resistance to β-lactams, glycopeptides, aminoglycosides, fluoroquinolones, MLS: macrolides, lincosamides, streptogramins, and others
   (a) **Target protection**
   Examples:
   • *TETRACYCLINES—ribosome protective proteins [e.g.,* tet*(M),* tet*(O),* tet*(S),* tet*(T)]*
   • *FLUOROQUINOLONES—quinolone resistance proteins (*qnrA, qnrB, qnrC, qnrD, qnrE, qnrS, qnrVC*)*
   (b) **Alteration of the primary structure of the target site via mutation**
   Encoded chromosomally.
   Examples:
   • *TETRACYCLINES, STREPTOMYCIN—mutation in the 16S rDNA*
   • *MACROLIDES—mutation in the 23S rDNA*
   • *FLUOROQUINOLONES—mutation of genes encoding gyrase and/or topoisomerase IV*
   • *TIAMULIN—mutation in the gene for the ribosomal protein L3*
   (c) **Modification of the target site via enzymatic alteration**
   Example:
   • *MACROLIDES, LINCOSAMIDES, STREPTOGRAMINS—rRNA methylase (*erm*)*
   (d) **Modification/replacement/alteration of the target site**
   Examples:
   • *PENICILLINS—synthesis of alternate low-affinity/altered specificity targets (PBPs) that reduce or completely block antibiotic (penicillins) from associating with the target (*mecA, mec *in methicillin-resistant* S. aureus, S. pseudintermedius*)*
   • *SULFONAMIDES—dihydropteroate synthase (e.g.,* sul1, sul2, sul3*)*
   • *TRIMETHOPRIM—dihydrofolate reductase (e.g.,* dfrA, dfrB, dfrD, dfrG, dfrK*)*
   • *COLISTIN—alteration of lipopolysaccharides (e.g.,* mcr-1-5*)*
   (e) **Target bypass**
   involves generation of additional antibiotic targets or subunits that are not susceptible to binding of the antibiotic.
   Example:
   • *SULFONAMIDES/TRIMETHOPRIM—"bypass" of the metabolic pathway. Bacteria inhibit the antibiotic action via target overproduction*

2. **Decreased permeability of the cell wall (Gram-negative bacteria)**
   Examples:
   • *AMINOGLYCOSIDES, β-LACTAMS, TETRACYCLINES, CHLORAMPHENICOLS, FLUOROQUINOLONES and influence of porins in bacterial cell wall on antibiotic*

resistance: size-selective defined channels allowing hydrophilic molecules to permeate across cell wall. Can be influenced also by charge of the amino acids present in the channel. Resistance to antibiotics can be caused by downregulation of porins or replacement of porins in cell wall with more selective channels.

3. **Efflux pumps (cell wall):** active transport of the antibiotic out of the cell using ATP or proton gradients. Efflux pumps are common structures of bacterial cell walls; however, when overexpressed, they can significantly contribute to resistance phenotype. Several efflux pumps transport only a limited set of substrates; however, many remove a wide range of distinct substrates including antibiotics out of the cell and are known as multidrug resistance efflux pumps.
   (a) **ABC family** (EfrAB, MsrC in enterococci)
   (b) **MFS family** (EfmA in *E. faecium*, QepA, QepA2 in *E. coli*, NorB, NorC in *S. aureus*)
   (c) **MATE family** (MepA in *S. aureus*)
   (d) **SMR family** (SsmE in *Serratia marcescens*)
   (e) **RND family** (AcrAB-TolC in *E. coli* and MexAB-OprM in *P. aeruginosa)*
   Examples:
   • *TETRACYCLINES—MFS family [*tet*(A)-(E)]*
   • *AMPHENICOLS—MFS family (*floR, fexA*)*
   • *MACROLIDES (14-, 15-membered)—MFS family (*mefA*)*
   • *PLEUROMUTILLINS, LINCOSAMIDES: ABC family [*vga*(A),* vga*(C),* vga*(E)]*
   • *MACROLIDES: ABC family (*msrA*)*
   • *CHLORAMPHENICOL, FLUOROQUINOLONES (*blt, norA*)*

4. **Enzymatic modification**
   Examples:
   • *AMINOGLYCOSIDES: addition of acetyl, phosphate, or adenyl groups to aminoglycosides by N-acetyltransferases (*aac*), O-phosphotransferases (*aph*), and O-adenyltransferases (*ant*), respectively*
   • *CHLORAMPHENICOL: acetyltransferases (*cat*)*
   • *MACROLIDES: phosphotransferases (*mph*)*
   • *LINCOSAMIDES: nucleotidyltransferases (*lnu*)*

5. **Enzymatic inactivation**

   Examples:
   • *β-LACTAMS: β-lactamases (*bla*) hydrolyze the antibiotic. Extended-spectrum beta-lactamases (*bla_{TEM}, bla_{SHV}, bla_{OXA}, $bla_{CTX\text{-}M}$*), AmpC beta-lactamases (*bla_{CMY}, bla_{DHA}, bla_{MOX}, bla_{FOX}, bla_{ACC}*), and carbapenemases (*bla_{KPC}, bla_{IMP}, bla_{VIM}, bla_{NDM}, bla_{OXA}*) are nowadays of special concern.*
   • *MACROLIDES: esterases (*ere*)*
   • *TETRACYCLINES: oxidoreductases [*tet*(X),* tet*(47)-(56)]*

Table 1 Mechanisms of antimicrobial action and resistance in classes of antimicrobials used in veterinary medicine (modified according CLSI 2018)

Antimicrobial class	Antimicrobial subclass	Antimicrobial agent	Mechanism of antimicrobial action	Resistance mechanism	Comments
Aminocyclitols		Apramycin Spectinomycin	Inhibition of protein synthesis (target 30S ribosomal subunit)	Enzymatic modification Efflux Target site (ribosome) mutation	
Aminoglycosides		Amikacin Dihydrostreptomycin Framycetin Gentamicin Kanamycin Neomycin Streptomycin	Inhibition of protein synthesis (target 30S ribosomal subunit)	Enzymatic modification Decreased cell wall permeability Target site (ribosome) modification/mutation Efflux	
Ansamycins		Rifaximin	Inhibition of mRNA synthesis	Target (RNA polymerase) mutation	
β-lactams + β-lactam inhibitor		Amoxicillin + clavulanic acid	Inhibition of cell wall synthesis, protection against β-lactamases	Enzymatic inactivation (β-lactamases)	
Cephems	Cephalosporins 1st	Cefacetrile Cefalonium Cefadroxil Cefazolin Cefalexin Cefalotin Cefapirin	Inhibition of cell wall synthesis	Enzymatic inactivation (β-lactamases) Decreased cell wall permeability Altered penicillin-binding proteins Efflux	
	Cephalosporins 2nd	Cefaclor Cefuroxime			Not used in VMPs in EU according to ESVAC 2017
	Cephalosporins 3rd	Cefoperazone Cefovecin Ceftiofur *Cefotaxime* *Ceftazidime*			Cefotaxime, ceftazidime used only in laboratory testing—detection of resistance
	Cephalosporins 4th	Cefepime Cefquinome			
Carbapenems		*Doripenem, Ertapenem Imipenem, Meropenem*			

Folate pathway antagonists	Sulfonamides	For example, Sulfadoxine, Sulfamethoxazole Sulfaclozine Sulfadimethoxine Sulfadiazine Sulfathiazole	Folate pathway antagonists	Decreased cell wall permeability Production of drug-insensitive enzymes Overexpression of sensitive enzymes Efflux	
	Diaminopyrimidines	Trimethoprim			
Glycopeptides	*Glycopeptides*	*Vancomycin, Avoparcin*	Inhibition of cell wall synthesis	Target site (cell wall) modification	
	Lipopeptides	*Teicoplanin*			
Lincosamides		Clindamycin Lincomycin Pirlimycin	Inhibition of protein synthesis (target 50S ribosomal subunit)	Target site (ribosome) modification/mutation Enzymatic inactivation Efflux Ribosomal protection	
Macrolides	14-Membered rings	Erythromycin Oleandomycin	Inhibition of protein synthesis (target 50S ribosomal subunit)	Target site (ribosome) modification/mutation Decreased cell wall permeability Enzymatic inactivation Efflux	
	15-Membered rings	Gamithromycin Tulathromycin			
	16-Membered rings	Spiramycin Tildipirosin Tilmicosin Tylosin Tylvalosin			
Nitroimidazoles		Metronidazole		Altered drug-activating enzymes	Banned for food-producing animals
Nitrofuran derivatives		Furazolidone			Banned for food-producing animals
Orthosomycins		Avilamycin		Decreased ribosomal binding	

(continued)

Table 1 (continued)

Antimicrobial class	Antimicrobial subclass	Antimicrobial agent	Mechanism of antimicrobial action	Resistance mechanism	Comments
Penicillins (penicillinase-labile)	Penicillins	Penicillins Penethamate Pheneticillin	Inhibition of cell wall synthesis	Reduced cell wall permeability Altered penicillin-binding proteins Enzymatic inactivation (β-lactamases) Efflux	
	Aminopenicillins	Amoxicillin Ampicillin Metampicillin			
Penicillins (penicillinase-stable)		*Methicillin* Nafcillin Oxacillin Cloxacillin Dicloxacillin			Methicillin used only in laboratory testing—detection of resistance
Phenicols		Chloramphenicol Florfenicol Thiamphenicol	Inhibition of protein synthesis (target 50S ribosomal subunit)	Target site (ribosome) modification/mutation Decreased cell wall permeability Enzymatic inactivation Efflux Ribosomal protection	Chloramphenicol banned for food-producing animals
Pleuromutilins		Tiamulin Valnemulin	Inhibition of protein synthesis (target 50S ribosomal subunit)	Target site (ribosome) modification/mutation Decreased cell wall permeability Efflux Ribosomal protection	
Polyether ionophore		Monensin Narasin			
Polypeptides		Bacitracin Polymyxin B Colistin	Inhibition of cell membrane synthesis	Lipopolysaccharide modification Efflux	

Quinolones	Quinolones	Nalidixic acid Oxolinic acid Flumequin Cinoxacin	Inhibition of DNA synthesis	Target (gyrase, topoisomerase IV) mutation Decreased cell wall permeability Efflux Target site protection Enzymatic inactivation	Ciprofloxacin used only in laboratory testing—detection of resistance
	Fluoroquinolones	*Ciprofloxacin* Danofloxacin Difloxacin Enrofloxacin Ibafloxacin Marbofloxacin Norfloxacin Orbifloxacin Pradofloxacin			
Streptogramins		Virginiamycin	Inhibition of protein synthesis (target 50S ribosomal subunit)	Target site (ribosome) modification/mutation Enzymatic inactivation Efflux Ribosomal protection	
Tetracyclines		Chlortetracycline Doxycycline Oxytetracycline Tetracycline	Inhibition of protein synthesis (target 30S ribosomal subunit)	Efflux Target site (ribosome) modification/mutation Enzymatic inactivation Ribosomal protection	
Others		Nitrofurantoin Novobiocin			

Note 1: Some of antimicrobials have no MRL established (cefadroxil, cefovecin, ibafloxacin, norfloxacin, cinoxacin, polymyxins B)—authorized (primarily intended) to be used in non-food-producing animals

Note 2: Mostly listed antimicrobials used in veterinary medicine according to ESVAC and CLSI, but also listed some antimicrobials primarily used for detection of specific resistance patterns

Note 3: Antimicrobials listed in italics have not been authorized yet for use in animals (in the European region)

5 Cross-Resistance and Co-resistance to Antibacterials and Other Substances

5.1 Cross-Resistance

Bacterial cross-resistance can be defined as resistance to multiple distinct antimicrobial agents conferred by a single molecular mechanism (Colclough et al. 2019). It occurs when antimicrobials share a route of access to the cytoplasm, bind to the same target, or are involved in the same pathway leading to the inhibition of growth or cell death (Baker-Austin et al. 2006). This phenomenon is best described in the context of cross-resistance among antibiotics within the same pharmacological class but also among antimicrobials from different pharmacological classes. Cross-resistance has been also many times described among antibiotics and disinfectants, biocides, or solvents and between antibiotics and heavy metals. An example of such cross-resistance is the efflux system AcrAB–TolC in *E. coli* which confers resistance to multiple pharmacological classes of antimicrobials but also to metals, dyes, and detergents (Anes et al. 2015).

Speaking about cross-resistance among antibiotics, several examples can be mentioned (Petinaki and Papagiannitsis 2017):

Cross-resistance within one pharmacological class/subclass of antibiotics:

- Aminoglycosides: modifying enzymes may confer resistance to several members of the aminoglycoside class.
- Tetracyclines (in veterinary medicine tetracycline-oxytetracycline-chlortetracycline): having the same mechanism of action/resistance.
- Macrolides and their subgroups (MS_B-phenotype, *msr(*A) coding for ABC transporter): associated with resistance only to 14- (clarithromycin, erythromycin$_{(vet)}$, roxithromycin) and 15-membered ring macrolides (azithromycin, gamithromycin$_{(vet)}$, tulathromycin$_{(vet)}$) and streptogramin B, while 16-membered ring macrolides (e.g., spiramycin$_{(vet)}$, tildipirosin$_{(vet)}$, tilmicosin$_{(vet)}$, tylosin$_{(vet)}$, tylvalosin$_{(vet)}$) and lincosamides (lincomycin$_{(vet)}$, clindamycin$_{(vet)}$) remain active.
- Macrolides (M-phenotype, *mph*(C) coding for macrolide phosphotransferase).
- PBP modification leading to cross-resistance to several members of β-lactam group.

Cross-resistance among different classes of antibiotics:

- Macrolides, lincosamides, streptogramins (MLS_B phenotype, *erm* encoding rRNA methylases)[1]
- Lincosamides, pleuromutilins, and streptogramins A, but macrolides and streptogramin B remain active (PLS_A-phenotype; *vga*(A), *vga*(C), *vga*(E), and *lsa* (E) coding for ABC transporters)

Review by Kampf (2018) evaluated MEDLINE search for 13 biocidal agents at sublethal concentrations used (with respect to antibiotic tolerance, antibiotic resistance, horizontal gene transfer, and efflux pumps) in Gram-negative species. In cells adapted to:

- Benzalkonium chloride, new resistance was most frequently found to be representative of aminopenicillins—ampicillin (eight species), representative of the third-generation cephalosporins—cefotaxime (six species), and a member of sulfonamide class—sulfamethoxazole (three species), some of them with relevance for *Enterobacter cloacae* or *Escherichia coli*.
- Chlorhexidine, new resistance was frequently observed to ceftazidime (third cephalosporins), sulfamethoxazole and imipenem (carbapenems)

[1] According to the rules of EUCAST, if a staphylococcal isolate with an inducible MLS_B phenotype is detected, it must be reported as resistant and considered adding this comment to the report "Clindamycin may still be used for short-term therapy of less serious skin and soft tissue infections as constitutive resistance is unlikely to develop during such therapy."

(eight species each), followed by cefotaxime and tetracycline (seven species each).

Cross-resistance to antibiotics and biocides was identified regarding triclosan, octenidine, sodium hypochlorite, and didecyldimethylammonium chloride. However, cross-resistance to antibiotics and biocides was not found after exposure to ethanol, propanol, peracetic acid, polyhexanide, povidone iodine, glutaraldehyde, and hydrogen peroxide at low levels.

Another part of similar review based on the same methodology by Kampf (2019) evaluated 13 biocidal agents and Gram-positive species that were able to exhibit a tolerance or even resistance to various antibiotics after exposure to sublethal concentrations of selected biocidal agents used for disinfection, especially benzalkonium chloride, chlorhexidine digluconate, and triclosan. In cells adapted to:

- Benzalkonium chloride, new resistance was often encountered to representative of aminopenicillins—ampicillin (seven species), representative of the third-generation cephalosporins—cefotaxime, a member of sulfonamide class—sulfamethoxazole (six species each), and ceftazidime (five species), some of them with relevance for *Enterococcus faecium* and *Enterococcus faecalis*.
- Chlorhexidine, new resistance was detected to member of carbapanem class—imipenem (ten species), and cephalosporins—cefotaxime, ceftazidime, and tetracycline (seven species each) as well.
- Triclosan and cephalosporin ceftazidime cross-resistance was identified in eight species.

Very uncommon cross-resistance was found for didecyldimethylammonium chloride or hydrogen peroxide. Using the data from MEDLINE search, Kampf (2019) concluded that cross-resistance to antibiotics and biocides was not found after sublethal exposure to substances such as glutaraldehyde, ethanol, propanol, peracetic acid, octenidine, povidone iodine, sodium hypochlorite, and polyhexanide.

Based on above-quoted reviews published in 2018 and 2019, and analyses of the data, Kampf concluded that it seems necessary to scrutinize the use of biocidal agents for disinfection. Based on a thorough evaluation, agents with lower or no potential to cause antimicrobial tolerance or resistance should be preferably recommended for use.

Generally, it is recommended to design for exact conditions tailored plans of rotation of disinfectants (based also on consideration which antimicrobials are used and which microbes are of concern for certain settings) and consider establishment of the adequate risk mitigation measures. The issue of cross-resistance to biocidal substances and antibiotics should be investigated in its complexity (see below the comments related to fitness, biofilms, and additive/synergistic effects).

5.2 Co-resistance

Co-selection of antimicrobial resistance among bacteria exposed to biocides used as disinfectants, antiseptics, and preservatives and to heavy metals (particularly copper and zinc) used as antiinfectives, growth promoters, and therapeutic agents for some livestock species is a matter of concern and discussion in both human and veterinary areas. Especially zinc oxide appeared as emerging issue several years ago, due to environmental risks, but also regarding co-selection of antibiotic resistance. Experimental/observational evidence showed that exposure to these non-antibiotic antiinfectives can induce or co-select for bacterial adaptations that result in decreased susceptibility or even resistance to one or more antibiotics (Wales and Davies 2015).

Acquired resistance is associated with changes in the organism caused by mutations including altering of genes expression, or by lateral transfer of mobile genetic elements (MGEs). Colocalization of genes encoding decreased susceptibility or resistance to different, non-related substances [antibiotics, biocides, heavy metals, herbicides (glyphosate)] on MGEs, such as plasmids and transposons, raises the possibility of transfer of co-resistance among bacteria. Class 1 integrons can be considered as obvious example (for further description, see below).

Some debates also arose as for cost–benefit of keeping/replication/expression and transfer of the (large) MGEs for bacterial cells. According to some studies, co-selection of decreased susceptibility/resistance genes to biocides and antibiotics does not have to be associated a priori with fitness cost. Moreover, prolonged exposure may enhance selection of adaptations leading to restoration of bacterial fitness (Russell 2003). However, the fitness costs can militate against survival of many adapted (including those resistant) strains with respect to shifting and reestablishing bacterial communities in different niches such as farm environments (Sheridan et al. 2013). Nonetheless, some adaptations, such as biofilm capability, may prove immediately an advantage for survival. Strains showing reduced susceptibility or even resistance can be in certain conditions maintained by using, except extensive antibiotic use, heavy metals (Wales and Davies 2015). However, in practice, the particular circumstances where an appropriately high selective pressure is applied at the right time to maintain/intensify a particular reduction in antibiotic susceptibility that has been generated by biocide use may not be especially common. Indeed, instead of co-selection, biocides and antibiotics, once properly used/administered, dosed, and chosen for specific situation, can have additive or synergistic antimicrobial effects when applied together (Zanini et al. 2014).

Paramount evidence that exposure of bacteria to members of one class of antimicrobial agents can affect susceptibilities to other classes either "true" antimicrobials or biocides comes from in vitro experimental studies. However, data from practice have provided conflicting evidence as to the likely clinical significance of biocide-induced co-selection of antibiotic-resistant pathogens. Health care-associated pathogens, biocides used in hospitals, or more generally human-care facilities are more frequently investigated and also more times quoted, than data for similar effects on the diverse microbiota on farms and along the food chain. Links exist for the important zoonotic pathogens but survey data suggest that, for now, counterselective processes largely limit the effects of co-selective pressures. Studies, from livestock animal and environment, providing evidence of existing interlinks between use of heavy metals and resistance to antimicrobials that identified genetic linkages with antibiotic resistance and heavy metals are acknowledged. Analysis of the data, mostly coming from the swine and poultry sectors, gave the evidence of increased horizontal gene transfer and co-selection of resistance to antimicrobials by heavy metals (Wales and Davies 2015).

Considering biocides, heavy metals, and antimicrobials, acquired and adaptive mechanisms associated to a different extent with co-resistance and/or cross-resistance include biofilm capability, multidrug efflux, altered cell wall and cellular membrane permeability, and target site mutation and overexpression (Wales and Davies 2015).

Pal et al. (2015) performed a review on co-occurrence patterns of resistance genes using publicly available, fully sequenced bacterial genomes ($n = 2522$) and plasmids ($n = 4582$). Biocide/metal resistance genes (BMRGs) were identified in 86% of bacterial genomes, and their colocalization with antibiotic resistance genes (ARGs) was found in 17% of the cases.

Results also showed that both BMRGs and ARGs were harbored on the same plasmids of *Escherichia*, *Staphylococcus*, *Salmonella*, and *Klebsiella* species. All studied external environments (<0.7%) showed colocalization of BMRGs and ARGs at low level, differently from humans and domestic animals (5% and 7%, respectively) where co-occurrence was more common. Authors of this study warned that these results could be biased by the fact that above-mentioned bacterial species as well as the animal husbandries and hospitals are more frequently investigated compared to other sources/bacterial species (Pal et al. 2015).

Pal et al. (2015) also documented that plasmids with both BMRGs and ARGs were more likely to be conjugative and carried toxin–antitoxin systems than plasmids without resistance genes. These additional characteristics can promote their longevity in bacterial populations even in the absence of selection pressure by antibiotics, biocides, or metals. Copper, silver, arsenic, antimony, cobalt, nickel, cadmium, iron, zinc, mercury, and QACs were identified by Pal et al. (2015), as being all potential co-selectors for strains resistant to, e.g., sulfonamides, β-lactams, amphenicols, tetracyclines, and aminoglycosides. As a main driver for the overrepresentation of co-occurrences between ARGs and BMRGs in these environments, use of antibiotics rather than exposure to biocides or metals was hypothesized. Interestingly, it was also indicated that over 70% of plasmids and 14% of all genomes lacked known resistance genes.

5.2.1 Resistance and Co-resistance to Biocides

Starting to think about biocides, one will be confronted with an essential difficulty. This is definition of resistance to biocides that almost differ from definitions of resistance to antimicrobials/antibiotics (Maillard 2018). In biocides, historically, but also at present, terms such as resistance, tolerance, decreased/reduced susceptibility, insusceptibility, and acquired reduced susceptibility are used across bibliography. Usually, for practical reasons, once guidance documents of either use or laboratory testing of MICs and MBCs of biocides are set, a certain range of concentrations as well as time of exposure are recommended to be used/tested.

As pointed out by Maillard (2018), from an academic point of view, definition of strain/cell resistant to biocides can be:

- "a bacterial strain that is not killed by a biocide concentration to which majority of bacterial species are susceptible."
- "a bacterial cell in a culture that survives biocide exposure that kills the majority of the bacterial population in that culture."

From the perspective of mechanisms of bacterial resistance to biocides (either reversible or irreversible), as outlined by Maillard (2018), following scenarios should be considered:

- Short exposure (disruption of active membrane transporting functions—decoupling of oxidative phosphorylation, inhibition of respiration, and catabolic/anabolic functions)
- Prolonged exposure [disruption of metabolic processes, disruption of replication, loss of membrane integrity leading to leakage of essential compounds (ions as well as sugars, nucleotides, nucleosides, amino acids, proteins), coagulation of intracellular material, cell lysis]

Mechanisms of resistance/decreased susceptibility:

(A) ***Decrease of biocide concentration in the bacterial cell***

Reduced penetration

Reduced penetration is well known in bacterial endospores, Gram-negatives and mycobacteria. Role of lipopolysaccharides, mycolic acid in mycobacteria, porins and membrane compound alterations are of importance for this mechanism.

Efflux pumps (Piddock 2006)

Well documented are efflux pumps particularly in *Staphylococcus aureus* (including MRSA), *Escherichia coli, Acinetobacter baumannii, Salmonella* spp., *Campylobacter* spp., and also non-fermenting rods as *Pseudomonas aeruginosa, Burkholderia cepacia, Stenotrophomonas maltophilia.*

In *Enterobacteriaceae* single-component TolC-independent multidrug resistance efflux pumps were described (Slipski et al. 2018): CDF and PACE families. These efflux systems are of growing concern as they are rapidly spread between members of *Enterobacteriaceae* on conjugative plasmids and other mobile genetic elements, emphasizing their importance to antimicrobial resistance.

Expression of the genes encoding efflux systems differs and can be induced by biocide alone or by different classes of antimicrobials where efflux system has been described as one of the resistance mechanisms (Blanco et al. 2016; Slipski et al. 2018).

Chromosomally encoded bacterial efflux pumps families, with respect to bacterial multidrug efflux (e.g., antimicrobials as well as biocides and in some cases also heavy metals), are as follows: **RND** and **ABC** (multidrug transporter, both antimicrobials and biocides), **MFS** (proven to transport, e.g., benzalkonium, cetrimide, chlorhexidine as well as, e.g., chloramphenicol, tetracyclines (TC, DOX), fluoroquinolones (CIP, NOR, OFL), aminoglycosides (KAN, NEO, GEN), trimethoprim), **SMR** (proven to transport, e.g., acriflavine, benzalkonium, cetrimide, as well as, e.g., chloramphenicol, tetracyclines (TC, DOX), fluoroquinolones (CIP, NOR), erythromycin, trimethoprim), **MATE** (extruding, e.g., aminoglycosides, fluoroquinolones, QAC, biguanides), **CDF** (proven to transport, e.g., chlorhexidine, confer also heavy metal resistance/tolerance, e.g., zinc), and **PACE** (proven to transport chlorhexidine, QACs as acriflavine and benzalkonium) families (Piddock 2006; Slipski et al. 2018). However, efflux pumps may be also encoded and horizontally transferred by plasmids (Li and Nikaido 2009)—e.g., **SMR** (proven to transport chlorhexidine, QACs as acriflavine and benzalkonium) and **MFS** and **CDF** families (Slipski et al. 2018).

(B) ***Enzymatic modification/degradation*** (Maillard 2018; SCENHIR 2009; Demple 1996)

Some bacterial enzymes can change chemical structure of biocides or degrade them so that they are less effective. Further data from field conditions are needed to confirm whether this mechanism is relevant for the high concentrations of biocides used in practice. Several bacterial species are able to produce degradative enzymes constitutively. The presence of catalase, superoxide dismutase, and alkyl hydroperoxidases decreases susceptibility of bacteria to oxidizing agents/peroxygens. Enzymatic reduction of the cation to the metal (affecting metallic ions presented in different biocides) was also described.

(C) ***Physiological and metabolic changes*** (Curiao et al. 2016; Hashemi et al. 2019)

Despite the fact that different physiological and metabolic pathways were considered to have an influence on the decreased susceptibility, tolerance, or resistance to biocides, there is lack of exact evidence of the extent as well as clinical practice impacts due to respective mechanisms. The (multiple) metabolic pathways alterations of *Salmonella enterica* to triclosan, chlorhexidine, and benzalkonium activity has been studied.

Also, growth in biofilm can usually significantly reduce susceptibility or can cause resistance of some bacteria in the biofilm, making more difficult to fight against them via disinfecting/biocide agents.

(D) ***Mutations***

Mutations, either random/spontaneous or those that develop due to the pressure of the extraneous substances (e.g., selection pressure of antibiotics), were described in *Escherichia coli* and *Pseudomonas aeruginosa*, as well as in *Acinetobacter baumannii* especially in relation to efflux pumps (Zhu et al. 2010; Chen et al. 2009). Moreover, there were also described mutations in *Salmonella enterica* serovar Typhimurium that resulted in de-repression of the multidrug efflux pump AcrAB-TolC in multidrug resistant mutants accompanied by a mutation in another gene contributing to resistance phenotype. Furthermore, *Salmonella enterica* sv. Typhimurium mutants highly resistant to both quinolones and the biocide triclosan were described (Webber et al. 2015).

Despite the fact that an essential mechanism of resistance has been described, it should be noted that a broad spectrum of biocidal substances/combination of substances/different biocidal formulations exists and relatively small amount of them were (thoroughly) tested for resistance. The issue of nonharmonized testing methods is also of importance. As was mentioned by Maillard (2018) and confirmed from several sources (SCENHIR 2009; ECHA database 2019), it has been recognized that in most studies, in vitro testing of resistance to compounds representing certain classes, e.g., triclosan (phenolics), chlorhexidine (cationic biocides), quarternary ammonium compounds (particularly cetylpyridinium chloride and benzalkonium chloride), isothiazolinones, glutaraldehyde (alkylating agents), and some of oxidizing agents and iodine-containing compounds were performed using different methods, different level of precision, and in different period (which can also limit the studies due to availability of methods for testing at certain period). Being more exact in commenting of the studies, one of the key limitations, comparing investigations and level of knowledge/pool of studies available for antibiotics (with clinical isolates to be tested), the majority of the studies with biocides were performed in laboratory conditions as in vitro tests and also a big pool of those studies were performed with type cultures, which can be on one hand considered as advantage to provide as a "model" of mechanism of action/first step for standardization of biocide-susceptibility testing and on the other hand hardly to be interpreted for clinical/field conditions. This is also in line with the study by Škaloud et al. (2003), in which the susceptibility of field isolates of STEC O157, O26 and reference strains from collections to different concentrations of sodium hypochlorite, sodium benzensulfochloramid (chloramine B), glutaraldehyde with glyoxal, peracetic acid, and lactic acid was tested via conductance assay. As the method allowed us to evaluate not only final effect of biocides, but also growth curves, it was also proven that differences between the tested toxin-producing field isolates and reference strains in evaluation of growth curves exist.

Recently, not only by scientific papers, but also by documents released by regulatory bodies - e.g. in EU Scientific Committee on consumer safety (SCCS, 2010) and in US by FDA (2016) are the concerns targeted on the situations, in which bacteria (especially those with importance for human and animal health) become clinically resistant to antibiotics following the exposure/ selection pressure of biocide (or certain route of biocide administration, its concentration, and type of biocidal product). The issue of co-resistance and cross-resistance seems to bring increasing attention to the use of biocides as co-selectors of AMR.

Maillard (2018) commented that there exist conflicting evidence and ongoing discussions on cross-resistance to biocides and antibiotics and also on inducers of this resistance. On the other hand, it is the opinion of the authors of this chapter that some mechanisms of resistance of bacteria, especially decreased uptake, which can be caused by efflux pumps as well as the changes in permeability of the bacterial cell wall or the bacterial membrane, can be influenced and cross-resistance can exist in both antimicrobials and biocides. Also, it should be mentioned that with whole genome sequencing techniques, an increasing number of genes/mutations identified that are responsible for resistance to biocides and for resistance to antibiotics can be expected.

Moreover, co-selection of resistance to biocides and antibiotics was documented in the paper by Webber et al. (2015) in which mutations within both *ramR* and *gyr*A appeared after biocide exposure in vitro. These mutations are commonly seen in clinical isolates of *Salmonella* and other *Enterobacteriaceae*, and therefore, their selection by biocide is a concern. The mutations identified after exposure to biocides from different classes with different mechanisms of action indicate that there are convergent pathways of the survival of antimicrobial stress present in *Enterobacteriaceae*. Yet the exact technique has not been determined, which can bring proof of evidence if antibiotic or biocide molecule is stronger selector and if identical mutations in the same genes can arise more frequently due to selective pressure of both antibiotics and biocides (Webber et al. 2015; Braoudaki and Hilton 2004).

Considering in vitro experiments with bacterial cultures again, mutants emerged readily after exposures to different biocides and were present after only two exposures (sub-cultures). Once selected, mutants were stable; there was no evidence of accumulation of multiple mutations after further biocide exposure. No great fitness costs of selected mutations when tested in competitive index experiments against the parental strain were determined. The authors confirmed for the *ramR, gyrA, and fabI* mutations in clinical and veterinary isolates that they carry no prohibitive fitness cost (Webber et al. 2015).

Further research especially considering clinical/field isolates and clinical/field conditions of biocide use (concentration, exposure, type of surface on which biocide is used, surrounding conditions of use—e.g., detritus, pus, blood, and mechanic dirty, as well as biofilm presence can significantly influence not only efficacy of biocide but as well as concentration available to act on microbes and create selection pressure on microbes) is needed. Also, field isolates from different species/genera of bacteria representing Gram-positive, Gram-negative, specific microorganisms such as *Mycoplasma* spp., *Mycobacterium* spp.; sporulating microbes, and other specific groups to better elucidate differences in mechanism of action as well possible co-selection of resistance should be considered. "Hot spots" for co-selection of resistance for both groups of substances (e.g., hospital wastewater, husbandries wastewater, and feces/slurry from stables) should also be considered. Real samples collected from farms where antimicrobials are extensively used altogether with biocides should be subjects of isolation, identification, and genetic testing in laboratory conditions, where under controlled conditions, simulating such real situation can contribute to elucidate this emerging issues.

5.2.2 Resistance and Co-resistance to Heavy Metals

Zinc and copper are essential trace elements and are many times used either as additives to comply with the nutritional requirements or therapeutically (zinc).

Especially zinc is used across Europe for therapeutic, prophylactic, or metaphylactic purposes predominantly in swine, where target category is weaning piglets. Zinc in currently authorized medicinal products is present in the form of zinc oxide, but also as zinc chloride or zinc sulfate as well as chelatized form. The reason for the administration is mainly to prevent or minimize post-weaning diarrhea and/or post-weaning scouring. As the European Commission banned the use of veterinary medicinal products containing zinc (coming into effect in 2022), currently farmers try to find alternatives and phase the zinc-containing VMPs out. The reason for this ban, as indicated with the referral procedure outcomes, was due to the environmental risks as well as risks for co-selection of resistance to antimicrobials.

Speaking more generally about resistance to heavy metals, the evidence-based studies should be thoroughly selected and considered that, for example, the pH of the test medium may affect the study results and outcomes. Also, addition of the metals tested into the culturing medium can influence the bacterial growth and hence interpretation of the results (Cavaco et al. 2010). Despite these technical difficulties, several proper studies exist that identified bacteria of animal origin in which resistance to copper and zinc in different forms was described (Aarestrup and Hasman 2004). Just for the short overview and as the examples can be mentioned, for both copper and zinc, resistance of some bacterial species that are most common in major food-producing species: *E. coli*, *Enterococcus faecium*, *Enterococcus faecalis*, *Salmonella* spp., *Staphylococcus aureus*, and *Staphylococcus hyicus* in chicken, cattle, and pigs (Aarestrup and Hasman 2004).

Genes that were detected to be linked to resistance (or might be in specific cases spoken about decreased susceptibility/tolerance) to:

Copper—were investigated in isolates from poultry, cattle, and pigs: *pcoA*, *pcoD* (Gram-negatives, studies mainly on *Salmonella* spp. and *E. coli*), *copB* (*Staphylococcus aureus* and other selected species of staphylococci), *cueO* and *cueC* (isolates of *E. coli* from poultry), and *mco* (swine *S. aureus*).

Zinc—were investigated in isolates from live animals or their meat (pigs, poultry, cattle)—*czrC* (*Staphylococcus aureus* resistant to methicillin, *S. hyicus,* and other staphylococci), *czcD* (swine *Salmonella*), and *zntA* (*E. coli* from poultry) (Rensing et al. 2018).

The genomic islands were determined to be involved in copper resistance in *Enterobacteriaceae* (incorporated into either chromosomes or plasmids, linked frequently with yersiniabactin virulence factors) as well as in *Enterococcus* spp. (plasmid encoded, transferred via conjugation, but only present in genus *Enterococcus*, not proven to be transferrable to other *Firmicutes* (Rensing et al. 2018).

In the case of zinc, mechanism of resistance linked to efflux pumps was described (yet at least four systems are known as P_{1B}-type ATPases (*czrC* located within the SCC staphylococcal chromosomal cassette), CDF transporters, 2-TM-GxN transporters, and CBA—RND efflux systems). All these efflux systems transport zinc divalent cations (Nies 2003; Knoop et al. 2005; Scherer and Nies 2009; Rensing et al. 2018).

Co-selection of resistance to zinc and antibiotics was documented in LA-MRSA (livestock-associated *Staphylococcus aureus* resistant to methicillin) clonal complex (CC) 398 that is usually multidrug resistant and persists in environment with zinc for a long time. According to study by Cavaco et al. (2010), 74% of MRSA CC398 isolates from pigs in Denmark were resistant to zinc. Staphylococcal cassette chromosome *mec* (SCC*mec*) type V harbors *mecA*, *tet*(K), *crzC* (zinc and cadmium resistance gene). Another co-selection associated with zinc was documented in multiresistant *E. coli* isolates from pigs (along with genes conferring resistance to sulfonamides, tetracyclines, and ampicillin (Bednorz et al. 2013).

5.2.3 Resistance and Co-selection of Resistance to Others Chemicals

Evidence that antibiotic resistance evolution is influenced by exposure of bacteria to a wide range of substances belonging to different chemical groups may require us to make changes in how we manage both antibiotics and other manufactured and widely distributed chemical products. As multiple factors of extrinsic environment induce adaptive changes, a complexity is frequently ignored in standard studies of resistance (Kurenbach et al. 2018).

Within the experiments in study on *E. coli* and *Salmonella enterica* sv. Typhimurium, Kurenbach et al. (2018) showed that complex effects of exposures to non-therapeutic chemicals may undermine strategies to preserve the effectiveness of antibiotics through altering just their use. *Escherichia coli* or *S. enterica* sv. Typhimurium were exposed to the herbicide formulation Roundup (containing glyphosate) which led to the increase of MIC of the fluoroquinolone antibiotic ciprofloxacin, as occurred also after exposure of *S. enterica* sv. Typhimurium to the herbicide formulation Kamba (containing 2,4-dichlorophenoxyacetic acid) (Kurenbach et al. 2015).

Within human medicine, a lot of studies have been performed yet in relation to antimicrobial resistance and also seems that further active substances administered in human medicinal products can cause some unexpected effects and events. Experiments linking frequently used drugs (e.g., antidepressants) with the possibility of causing/co-selecting antimicrobial resistance were done. Jin et al. (2018) published the results of the experiments using whole genome sequencing of the mutants. Result of the study revealed that ROS-mediated mutagenesis (e.g., deletion, insertion, and substitution) of DNA-binding transcriptional regulators (e.g., *marR*, *rob*, *sdiA*, *cytR*, and *crp*) upregulates the expression of efflux pumps and may further enhance the antibiotic efflux. Within the experiments, it was determined that exposure of *Escherichia coli* to antidepressant fluoxetine at 5–100 mg/L after repeated subculture for 30 days promoted the mutation frequency, resulting in increased resistance against several antibiotics, including two very frequently used both in human and veterinary medicine—amoxicillin and tetracycline, as well as rarely clinically used chloramphenicol.

A very recent study by Wang et al. (2019) also investigated another active pharmacological substance present in frequently prescribed human medicinal products—antiepileptic carbamazepine. Its properties as for increase of horizontal transfer of plasmid-borne resistance genes were proven. Being cumulative, carbamazepine is often detected as one of the highest pharmaceutical residues in aquatic environments (groundwater, surface water, wastewater, and drinking water). Wang et al. (2019) brought the data signalizing that carbamazepine could significantly enhance conjugation frequency. This property was observed regarding conjugation between the same and among distinct genera. Their findings were supported by phenotypic tests and through results of MinION plasmid sequencing, genome-wide RNA sequencing, and proteomic analysis. Carbamazepine induces a series of responses that included increased levels of reactive oxygen species (ROS), triggering the SOS response, increased cell membrane permeability, and the increased generation of pilus.

Despite the fact that there should be further more in depth studies performed, the findings of the studies by Kurenbach et al. (2015, 2018), Jin et al. (2018), and Wang et al. (2019) changed the perception related to the dissemination of antibiotic resistance and show how development of resistance is complex and can be enhanced by non-antibiotic pharmaceuticals/chemicals. It warns us and forces us to rethink the spread of antimicrobial resistance genes in the environment as well as to investigate the broad range of substances/factors that can promote resistance dissemination. These papers also showed how important it is to improve molecular methods and accurate analysis of the results coming from those methods as well as how important it is to create multidisciplinary teams involving specialists from different area of expertise.

6 Spread of Antibiotic Resistance: Mobile Genetic Elements and Mechanisms of Horizontal Transfer

Besides chromosomally encoded resistance, antibiotic resistance can arise also as a consequence of acquisition of new genes via horizontal transfer. The origin of antibiotic resistance genes commonly found in resistant clinical strains is likely from chromosomes of commensal (non-pathogenic) and environmental bacteria (e.g., origin of *qnrA*: *Shewanella algae*; $bla_{CTX\text{-}M}$: *Kluyvera* spp.) in which they apparently played other functions than being involved in antibiotic resistance (Martinez and Baquero 2014). In their original species, resistance genes are not intrinsically mobile and their spreading is enabled by mobile genetic elements (MGEs) (Partridge 2011).

MGEs can be divided into two groups: MGEs promoting DNA mobility within one bacterial cell (1) and MGEs accountable for transfer of DNA among individual bacterial cells (2). The first group of MGEs captures and mobilizes new resistance genes from chromosome of the original species and transfers them to other DNA molecules present in the bacterial cell. This group of MGEs comprises, e.g., insertion sequences, gene cassettes, integrons, and transposons. The second group of MGEs comprises plasmids or integrative conjugative elements (ICEs—which are ranked among genomic islands) that facilitate horizontal transfer of resistance genes via conjugation, transformation, or transduction (Partridge 2011; Partridge et al. 2018).

Actions of various MGEs and interactions between them significantly contribute to the high plasticity and rapid evolution of bacterial genomes that lead to substantial adaptive capacity of bacteria. In the context of antibiotic resistance, the main driver for development of antibiotic resistance regions and dissemination of antibiotic resistance genes via MGEs is selective pressure resulting from antibiotic use.

6.1 Intracellular MGEs

Insertion sequences (IS) and transposons are distinct segments of bacterial DNA capable of their own movement along with associated resistance genes from one location to another on the same or different DNA molecule in the cell (Partridge et al. 2018).

IS (<2.5 kb) are compact mobile elements that harbor one or two transposase genes (Fig. 2) (Siguier et al. 2015; Partridge et al. 2018). Insertion sequences can move to a new location by either conservative ("cut-and-paste") or replicative ("copy-and-paste" and "copy-out-paste-in") processes depending on the IS. Conservative mechanism consists in simple excision followed by insertion of the IS to the target sequence. During so-called copy-and-paste mechanism, a

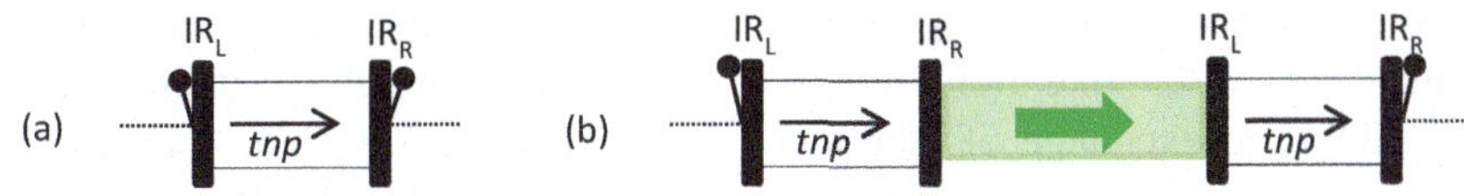

Fig. 2 Schematic illustration of insertion sequences and composite transposons adopted from Partridge (2011). (**a**) A typical insertion sequence, (**b**) a composite transposon. IR_L and IR_R, left and right inverted repeats (tall black bars); *tnp*, transposase (thin black arrows); IS are shown as open boxes; antibiotic resistance genes as green block arrow; direct repeats (created as a result of transposition) as lines with circle at the end

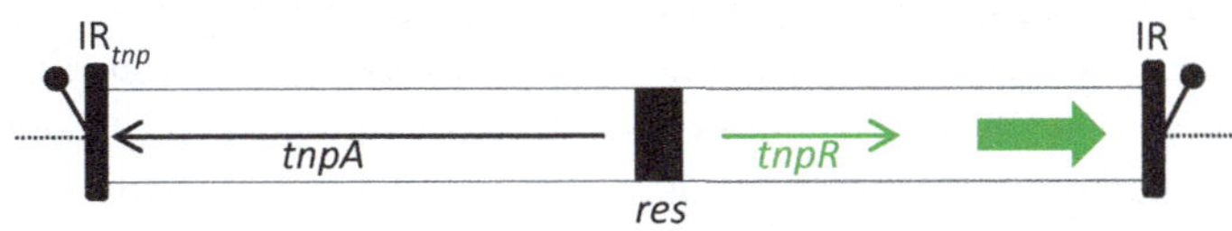

Fig. 3 Schematic illustration of Tn*3* family transposon adopted from Partridge (2011). IR and IR_{tnp}, inverted repeats (tall black bars); *tnpA*, transposase (thin black arrow); *res*, resolution site (small black box); *tnpR*, resolvase (thin green arrow); antibiotic resistance gene is shown as green block arrow; direct repeats (created as a result of transposition) as lines with circle at the end

cointegrate of a donor and target DNA linked by replicated IS split by resolvase leading to two molecules harboring a copy of the IS. During the second mechanism, "copy-out-paste-in," replication of the IS creates a double-stranded circular IS copy that integrates into the target site. Two identical or closely related IS can constitute a composite transposon that is able to move the DNA segment between these two IS as a unit (Fig. 2). Moreover, there are also IS-related elements (IS*Ecp1*-like and IS*CR*) that act as single copies when they mobilize adjacent genes. IS are associated with capture of many resistance genes; for example, IS*Ecp1* apparently mobilized $bla_{CTX\text{-}M}$ (ESBL) from *Kluyvera* chromosome. Another example is mobilization of *mcr-1* conferring resistance to colistin which is facilitated by a composite transposon-type structure of IS*Apl1* from *Moraxella* chromosome (Partridge et al. 2018). Resistance to tetracyclines can be caused by different mechanisms. Genes encoding efflux pumps accountable for tetracycline removal from the bacterial cell were found in association with insertion sequences and composite transposons, e.g., *tet*(C)/IS*26*, *tet*(K)/IS*257*, *tet*(B)/Tn*10*, *tet* (H)/Tn*5706*, *tet*(31)/IS*CR2* (Kehrenberg et al. 1998; Chalmers et al. 2000; Partridge 2011; Partridge et al. 2018). Ribosomal protection protein is encoded by the gene *tet*(M) associated with Tn*5385* (Flannagan et al. 1994).

Regarding antibiotic resistance, **unit transposons (Tn)** are represented by two important families—Tn*3* and Tn*7*-like. Tn*3* family transposons consist of *tnpA* (encoding transposase), *tnpR* (resolvase), resolution (*res)* site and may harbor also a "passenger" gene. Tn*3* family transposons are flanked by 38-bp inverted repeats (Fig. 3). The mechanism of transposition is replicative ("copy-and-paste"). Whereas Tn*7*-like transposons carry several genes associated with transposition whose mechanism varies among members of the family. Important member of Tn*3* family is, e.g., Tn*1546* that is associated with widespread dissemination of *vanA* gene cluster encoding resistance to vancomycin in enterococci due to its location on conjugative plasmids. Tn*21* and related transposons of Tn*3* family often harbor *mer* operon coding for mercury resistance and/or a class 1 integron (see below). Gene *tet* (A) encoding tetracycline efflux pump is also associated with Tn*1721* transposon of Tn*3* family. Tn*552*-like elements belonging to Tn*7* family harbor gene *blaZ* coding for penicillin resistance in *S. aureus* (Partridge et al. 2018).

Besides the role of IS and Tn in capture and transfer of resistance (and other) genes between DNA molecules, they can also affect expression of adjacent genes or disrupt coding sequences due to their insertion. Moreover, as these MGEs can be present in multiple copies within the genome, they can facilitate processes like homologous recombination (leading to exchange of identical or similar sequences), rearrangements, or deletions of DNA (Poirel et al. 2012; Darmon and Leach 2014; Siguier et al. 2015).

Gene cassettes are small MGEs (up to 1 kb) that contain usually a single gene and a recombination site *attC* and lack promoter sequences (Fig. 4). These MGEs can exist in a circular form; however, they are usually found within a variable region of integrons (Partridge et al. 2018). **Integrons (In)** consist of an integrase gene (*intI*), a recombination site (*attI*), and a promoter (Pc) (Fig. 4) (Domingues et al. 2012; Partridge et al. 2018). Integrase can insert gene cassettes into the variable region of integron creating a cassette array using site-specific recombination between *attI* and *attC* or two *attC* sites. When inserted into integron, gene cassettes can be expressed from the Pc promoter. However, as the distance from promoter sequence increases, expression of gene cassettes is decreased. Integrons are divided into classes according to integrase sequence with class 1 integrons being the most commonly encountered in resistant bacteria from clinical samples. The common type of class 1 integrons (so-called clinical or *sul1*-type) contains a truncated *qacEΔ1* gene and intact *sul1* gene encoding sulfonamide resistance in addition to its variable cassette array (Fig. 4) (Partridge et al. 2018). Integrons do not code any proteins that could facilitate their own movement. However, as can be seen on the example of class 1 integrons, the segment *intI1*/*attI1*/Pc was likely captured from chromosome by a Tn*5053* family transposon. Therefore, the class 1 integrons can be transferred to different sites via transposition thanks to their association with a transposon (Fig. 4) (Partridge 2011; Partridge et al. 2018).

In Gram-negative bacteria, the role of class 1 integrons in dissemination of antibiotic resistance is better documented than in clinically relevant Gram-positive genera like *Staphylococcus* and *Enterococcus*. In Gram-negatives, genes encoding resistance to β-lactams (e.g., bla_{VIM}, bla_{IMP}, bla_{GES}, and some bla_{OXA}), aminoglycosides (e.g., *aacA4*/*aac(6′)-Ib*), streptomycin and spectinomycin (*aadA*), and trimethoprim (*dfr*) were found on gene cassettes (Partridge et al. 2018).

Integrons play a significant role in dissemination of resistance genes, as they can harbor multiple antibiotic resistance genes within its variable region. Integrons, typically transferred by plasmids or enclosed in transposons, performing the task of resistance gene dissemination, play an important role in the revealing of multidrug-resistant "Super Bugs" (Xu et al. 2011).

6.2 Intercellular MGEs

PLASMIDS are double-stranded autonomously replicating extrachromosomal DNA molecules harboring non-essential genes and are considered

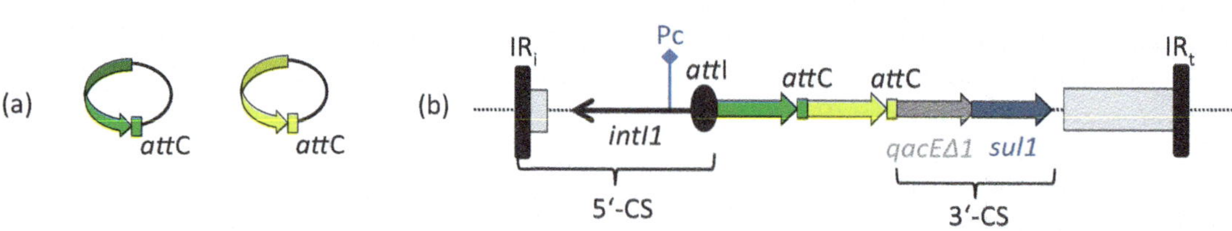

Fig. 4 Schematic illustration of gene cassettes and class 1 integron associated with a transposon adopted from Partridge (2011). (**a**) Circular form of gene cassettes; (**b**) general structure of a class 1 integron associated with a transposon. Parts of a truncated transposon are shown as shaded boxes. 5′-CS, 5′ conserved segment comprises IR_i, *intI1* and *attI*; 3′-CS, 3′ conserved segment contains *qacEΔ1* and *sul1*; IR_i and IR_t, inverted repeat adjacent to *intI1* and transposon region, respectively (tall black bars); *intI1*, integrase gene (thin black arrow); Pc, promoter sequence (blue thin line with rhombus at the end); *attI*, recombination site of the integron (black oval); *attC*, recombination site of gene cassettes (small green and yellow boxes); antibiotic resistance genes along with *qacEΔ1* are shown as colored block arrows

as the most responsible for global spread of resistance (Couturier et al. 1988; del Solar and Espinosa 2000; Carattoli 2013).

Plasmids possess a modular, mosaic structure that can be divided into so-called backbone and variable region. Plasmid backbone contains discrete regions of genes ensuring plasmid replication and copy control, stable maintenance, and propagation via conjugation or mobilization (Norman et al. 2009). Variable region harbors "accessory elements" coding for traits that are beneficial for the host in a particular environment (Fig. 5). Within these variable regions, a wide range of antibiotic and heavy metal resistance, virulence genes, or genes encoding non-essential metabolic pathways associated with intracellular MGEs can be found (Norman et al. 2009; Partridge 2011; Partridge et al. 2018).

Replication module is involved in vertical transmission of plasmids. Plasmids predominantly replicate via two mechanisms, either theta-replication or rolling-circle replication. This module also plays a role in plasmid copy control that mostly occurs at the level of replication initiation in order to decrease fitness burden for their hosts. Plasmids also code for systems ensuring their stable maintenance within division

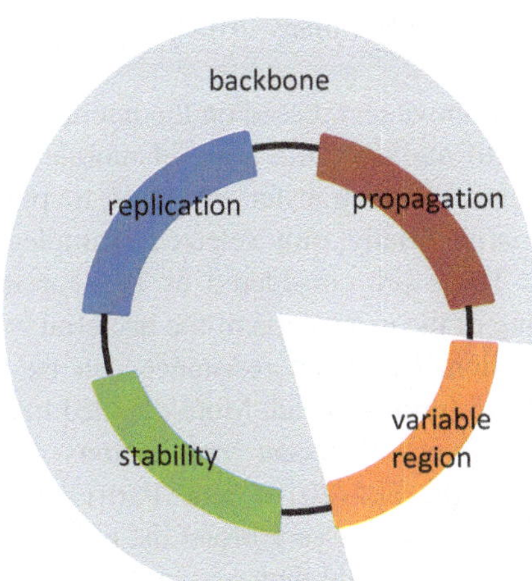

Fig. 5 Schematic illustration of a conjugative plasmid organization adopted from Norman et al. (2009). The plasmid contains four modules responsible for stability (green), replication (blue), propagation (red), and a variable region (orange)

of their hosts, e.g., active partitioning or post-segregational killing. Based on the ability of self-transfer, plasmids can be divided into conjugative and mobilizable. Conjugative plasmids are self-transferrable and code for all structures involved in their own transfer. On the contrary, mobilizable plasmids can be transferred horizontally only in the presence of a conjugative plasmid in the same cell as they usually do not harbor all genes necessary for conjugation (Norman et al. 2009).

Plasmids can be categorized according to their relatedness into incompatibility (Inc) groups. Two plasmids sharing the same elements involved in replication control cannot be stably propagated in one cell line and thus are related and belong to the same incompatibility group (Couturier et al. 1988; Carattoli 2013). Host spectrum of plasmids in Gram-negative bacteria is inferred from the mechanism of plasmid replication. Several groups of plasmids, e.g., IncF, IncI, IncX, or IncN, show a narrow host spectrum as they are usually found in enterobacteria. However, other groups like IncP plasmids show broad host range (Norman et al. 2009).

In *Enterobacteriaceae*, resistance plasmids (attributing resistance to commonly used antibiotics) can be large (up to at least 200 kb) and usually able to self-transfer via conjugation or small that can be mobilized. In clinical isolates of staphylococci, one or more plasmids responsible for resistance to antibiotics, heavy metals, and other substances were often encountered. These plasmids can be divided into three groups: small plasmids (1–10 kb) performing rolling-circle replication (1); multiresistance plasmids (>15 kb) (2); and larger, conjugative multiresistance plasmids. In enterococci, antibiotic resistance genes are frequently found on theta-replicating plasmids (Partridge et al. 2018).

Example of (large) plasmids conferring multiresistance as well as co-resistance that have clinical implications:

- *In* Klebsiella pneumoniae *as well as in* E. coli, *IncN plasmid designated pKP33 was found. pKP33 carries 11 antibiotic resistance genes accountable for resistance phenotype to* *multiple antibiotic classes covering β-lactams (*bla$_{CTX-M-15}$, bla$_{TEM-1}$, bla$_{OXA-1}$*), aminoglycosides (*aac(6′)-III, aac(6′)-lb, *and* strA, strB*), quinolones (*qnrB*), sulfonamides (*sul2*), and dihydrofolate reductase inhibitors (*folA*)* (Porse et al. 2016). *In order to ensure stable maintenance and propagation, pKP33 harbors the* stb *operon coding for factors involved in active segregation and regulation of conjugation* (Guynet et al. 2011). *Moreover, pKP33 encodes a putative toxin–antitoxin stability system (*ecoRII-dcm *restriction–antirestriction system) that may play a role in the inhibition of growth of plasmid-free cells after cell division* (Mruk and Kobayashi 2014).

- *Heavy metals (e.g., copper, silver, arsenic, antimony, cobalt, nickel, cadmium, iron, zinc, mercury) and biocides/disinfectants (as quarternary ammonium compounds) are all potential co-selectors for strains resistant to antimicrobials (e.g., sulfonamides, β-lactams, amphenicols, tetracyclines, and aminoglycosides).*

GENOMIC ISLANDS are defined as distinct regions of bacterial chromosome that were acquired horizontally. This broad group comprises a wide variety of elements, e.g., integrative and conjugative elements described in greater detail below, or elements like staphylococcal cassette chromosome (SCC*mec*) conferring resistance to methicillin in *S. aureus* that can be excised from bacterial chromosome and transferred horizontally via mechanisms mediated by a bacteriophage (Partridge et al. 2018).

INTEGRATIVE AND CONJUGATIVE ELEMENTS (ICEs, a.k.a. conjugative transposons) are diverse mobile elements found in both Gram-positive and Gram-negative bacteria. ICEs are self-transferable by conjugation, and in contrast to plasmids, ICEs integrate into the host chromosome and replicate as its part. They possess three modules in their backbone: (1) phage-like integration and excision module; (2) plasmid-like conjugation module; and (3) regulation module. These modules contain different array of genes that code for proteins operating by distinct mechanisms. Site-specific integration and excision of the element, frequently into a unique site on the chromosome of the host organism, are promoted by the gene encoding an integrase (Int) (Boyd et al. 2009). Some integrative and conjugative elements bear maintenance modules such as toxin–antitoxin systems (Wozniak and Waldor 2009) and additional partition systems that guarantee thriving vertical inheritance of these elements. In contrast to plasmids, ICEs are not found in extrachromosomal state, because they lack autonomous replication (Sultan et al. 2018). In addition to the self-transferability, ICEs have also been reported to be capable of mobilization of other genetic elements, such as the chromosome-borne integrative and mobilizable elements (IMEs) and cis-mobilizable elements (CIMEs) functioning as important transferable vehicles for virulence and resistance genes (Boyd et al. 2009).

Examples of genes transferred by integrative and conjugative elements that encode:

- *Heavy metals resistance, virulence, and biofilm formation* (Sultan et al. 2018)*;*
- *Resistance to tetracycline (Tn*916 *family)* (Roberts and Mullany 2011).

In the above text, based on the intention of the authors to introduce the issue of mobile genetic elements, role in transfer of AMR to practical veterinarian briefly, only selected examples were listed. These are considered as key for understanding of the exact parts of the bacterial genetic information, by which resistance can be transferred and further spread. More detailed information is given in recent publications by van Duijkeren et al. (2018) and Partridge et al. (2018) containing a very broad list of the mobile genetic elements and their association with resistance mechanisms, resistance genes, resistance phenotypes, and bacteria involved.

7 Means of Resistance Genes Transfer and Possible Implications

Mobile genetic elements can be transferred by different means of horizontal gene transfer (HGT)—transformation with extracellular DNA, transduction by bacteriophages, or conjugation involving plasmids or ICEs. All three HGT mechanisms are widely spread in nature, although certain species of bacteria tend to employ one mechanism more heavily over the others (Barlow 2009). For example, streptococci can become naturally competent and thus participate effectively in transformation, whereas enterobacteria commonly use conjugative plasmids for transfer of genetic information.

But veterinarians and clinicians would ask what level of contribution of either transformation or transduction in transferring resistance genes can be in clinical practice (hospitals/stables). Because of high density of bacteria, phages, and plasmids, several environments are considered to be very suitable for gene transfer, e.g., sewage and wastewater treatment plants, hospital effluents, aquaculture, agricultural and slaughterhouse waste, as well as any surfaces covered with biofilms. Human and animal gut microbiome is also an ideal hot spot for HGT. Regarding conjugation, settings with high density of bacteria, such as the human or animal gut, biofilms, hospitals, stables (herd/flocks), aquacultures, and co-infection conditions, facilitate HGT. All three HGT mechanisms are subject to limitations imposed by the host range of the incoming plasmid or the phage, the restriction modification systems of the host, ability to form cell-to-cell effective connections, fitness cost of acquiring a new genetic element, as well as the ability of the donor DNA to recombine with the recipient DNA (Thomas and Nielsen 2005; Domingues et al. 2012). Except horizontal gene transfer, vertical transfer dependent on the ability of a mobile genetic element to replicate autonomously helps establish new genetic properties in a bacterial lineage. The most evolutionary successful conjugative plasmids, such as the incompatibility group IncP, have a broad host range (Davies and Davies 2010), which facilitates their transfer to and maintenance in distantly related phyla (Klümper et al. 2015). The ability of DNA (or mobile genetic elements) to persist in the environment also affects success of HGT. Cell-to-cell effective connection essential in conjugation provides better protection to DNA (as in other conditions, naked DNA is prone to being degraded quickly). DNA encapsulated in a phage particle is more protected than naked DNA, but its further transfer can be limited by the narrow host range of a phage (von Wintersdorff et al. 2016).

7.1 Conjugation

Horizontal transfer of resistance plasmids via conjugation is considered to be the most prevalent in disseminating resistance genes in nature. Transfer process is performed between two viable bacterial cells by direct cell-to-cell contact or by a bridge-like connection between two cells. Considering the example of conjugation in *E. coli*, the donor cell harbors a conjugative plasmid that can be transferred through a conjugation pilus to the (plasmid-free) recipient cell. As a result of conjugation, both donor and recipient cells harbor the conjugative plasmid (Norman et al. 2009; Huddleston 2014). In Gram-positive bacteria, besides plasmids, integrative conjugative elements can be also transferred by conjugation (Partridge et al. 2018).

Gene transfer via plasmids goes through different species, genera, and even kingdoms depending on the host range of the plasmid. Various plasmids spreading carbapenemase, ESBL, and quinolone resistance genes among Gram-negative bacteria over very large geographical distances were found in different sources (Carattoli 2013).

Among Gram-positive bacteria, *S. aureus* is of the great clinical importance. Interestingly, according to the work published by Ramsay et al. (2016) only some part of the population of *S. aureus* is able to transfer resistance genes via conjugation (the assumption is around 5–6%, but this number can be heavily biased due sequencing data being available for selected pathogenic strains). Despite this assumption of low conjugative potential, it appears that the majority of non-conjugative plasmids, including most large multiresistance plasmids, are potentially mobilizable.

7.2 Transformation

Within transformation, recipient cells in the state of competence directly take up free DNA from the surrounding environment. Translocated DNA is either integrated into the recipient genome or can be recircularized in case of plasmids being transferred. The state of competence is transient and can be induced by certain conditions including occurrence of peptides or autoinducers, nutritional status, and other stressors including antibiotics. Therefore, antibiotic use was shown to facilitate transformation of many bacterial species (von Wintersdorff et al. 2016).

Although the physiological role of transformation is still debated, its main purpose is believed to be DNA repair or genetic diversification to enhance adaptability (Johnston et al. 2014). Indeed, transformation seems to have played an important role in evolution of antibiotic-resistant strains of the genus *Streptococcus*. Mosaic variants of penicillin-binding protein genes causing reduced affinity to β-lactams have also been reported in several *Streptococcus* species, implying the role of transformation in incorporating segments of foreign DNA (von Wintersdorff et al. 2016).

7.3 Transduction

Transduction is a process by which DNA is transferred from one bacterium to another by a bacterial virus—bacteriophage. Transduction does not require physical contact between the donor and the recipient cell in contrast to conjugation, and in comparison with transformation, it is DNAase resistant (Huddleston 2014).

Transduction is believed to play a major role in dissemination of antibiotic resistance in *Staphylococcus aureus*, although it has been shown to occur in many other bacteria at a rather low frequency (Peterson and Kaur 2018). Diverse strains of *S. aureus* can carry multiple accessory elements including phages, plasmids, transposons, genomic islands, as well as staphylococcal cassette chromosome SCC*mec* (most of which carry resistance genes) in their genomes (Haaber et al. 2017). Genes coding for penicillinase, metallo-β-lactamase, and tetracycline resistance transferred by transducing phages were reported in *S. aureus* (Varga et al. 2016). So-called phage-related chromosomal islands containing resistance and virulence genes in *S. aureus* chromosome are known to be transduced by bacteriophages at remarkably high frequencies. These islands harbor many antibiotic resistance genes, suggesting that transduction may contribute significantly to variability and evolution of resistance in *S. aureus* (Novick et al. 2010; Penadés and Christie 2015). Interestingly, interspecies and intergeneric transfer of elements of *S. aureus* pathogenicity islands was proven to occur between *S. aureus*, *S. epidermidis*, and even *Listeria monocytogenes*, showing a broader host range of staphylococcal phages (Maiques et al. 2007).

8 The Clinical and Epidemiological Relevance of Acquired AMR

Acquired resistance to antimicrobials is of high clinical and epidemiological importance. There are hundreds or even more scientific papers published documenting a huge amount of bacterial isolates carrying resistance genes carried by mobile genetic elements. There is also an increasing number of studies using new molecular methods bringing the evidence not only on exact location of genes in exact genetic context of individual strains of bacteria, but also investigating and tracing the routes of transmission as well as epidemiological consequences. Acquired resistance and especially (intra- and interspecies) horizontal transfer of antimicrobial resistance genes among bacteria in different environments can be considered as a big threat from the perspective of successful treatment of infectious diseases. On the other hand, well-designed, accurately performed, and correctly analyzed studies are of great need to bring the robust evidence to broaden our knowledge, e.g., the level of contribution of different antibiotic resistance genes reservoirs to the spread of antibiotic resistance (e.g., link animal to human), hot spots of AMR development and dissemination, as well as epidemiological links.

In the next part of the text, two examples of complex epidemiology of antibiotic resistance studied by molecular methods in two clinically important species of Gram-positive and Gram-negative bacteria are given. The first example deals with *Staphylococcus aureus*, representative of Gram-positive bacteria with pathogenic and zoonotic potential. The second example is *Escherichia coli*, representative of Gram-negatives with varying pathogenic potential covering commensal, opportunistically pathogenic, and pathogenic strains that might also act as zoonotic pathogens with varying resistance patterns. Both these examples underline the need of comprehensive molecular analyses of antibiotic-resistant bacteria in order to unravel complex epidemiology of antibiotic resistance.

According to the report provided by EFSA-ECDC (2018), dealing with the data from 2016 (this time yet voluntary monitoring data from some EU countries on MRSA), livestock associated-methicillin-resistant *Staphylococcus aureus* (LA-MRSA) is evidently widespread geographically and present in a variety of animal host species. The findings have underlined the need of appropriate molecular characterization of MRSA isolates. Detection of LA-MRSA, hospital-acquired MRSA (HA-MRSA), and community-acquired MRSA (CA-MRSA) from companion animals showed the importance of monitoring AMR not only in livestock (MRSA founded in all livestock species—live animals), but also in food commodities (e.g., milk), as well as in companion animals. Isolation of linezolid-resistant strains harboring *cfr* gene from pigs highlighted that the situation is constantly evolving (linezolid is not authorized for use in veterinary medicine in the EU). Detection of such a gene in swine isolates points to ongoing evolution of antibiotic resistance mediated by MGEs and might be associated with co-selection of resistance genes. Furthermore, investigation of presence or absence of certain virulence genes is also of great importance when assessing the significance of MRSA isolates. Figure 6 brings important information—e.g., HA-MRSA was found in samples from a pig and a cat, as well as CA-MRSA was identified in samples from a dog and a cat.

The second example led to serious consideration that well-designed studies, together with phenotype and genotype testing of not only AMR but also epidemiological relatedness, virulence, and other traits that help to track the epidemiological links, are still of great need. The study performed by Ludden et al. (2019), based on next-generation sequencing and analysis of mobile genetic elements, does not support the commonly accepted claim that, e.g., *E. coli* strains causing invasive disease or their resistance genes are commonly acquired from livestock. In the study, genomes of *E. coli* from a total of 1517 patients with bloodstream infection were compared with genomes of a total of 431 *E. coli* isolates from livestock farms and meat from the same area of the UK. Livestock and bloodstream isolates were genetically distinct populations based on core genome and accessory genome analyses. Despite the fact that identical antimicrobial resistance genes were found in livestock and human isolates, overlap in the mobile elements carrying these genes was not proven (Fig. 7).

It should be also mentioned that papers dealing with mathematical modeling of AMR and factors determining the spread of AMR as well as factors that could help set AMR mitigation measures/ antimicrobial stewardships and assessment of the success of these measures have started to be published recently (Cravo Oliveira Hashiguchi et al. 2019). A vast majority of them are focused on modeling AMR in relation to humans, either directly by modeling hospital or communities'

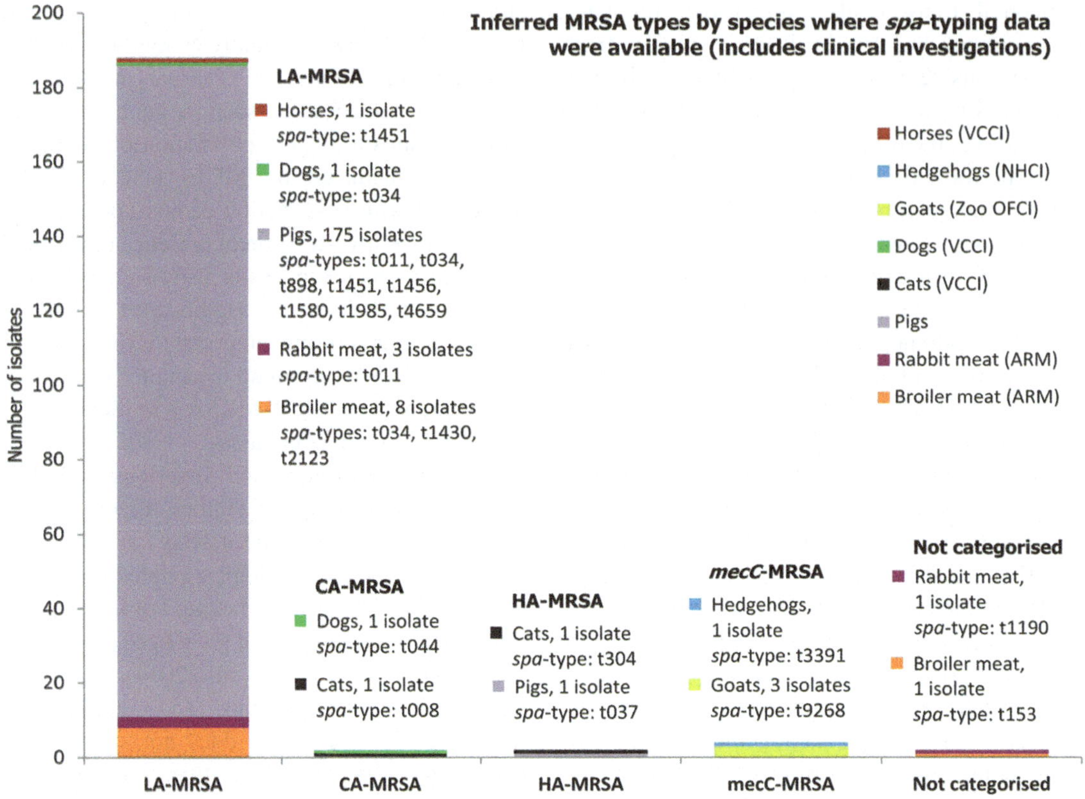

Fig. 6 Overview of MRSA types by animal species reported in 2016, including healthy animals and clinical investigations (EFSA-ECDC 2018)

human population or in bacteria related directly to human health. Only four models relating to animals were identified throughout studies available fulfilling the criteria of analysis by Birkegård et al. (2018). Animals might constitute a reservoir of antimicrobial resistance that can be spread to humans via food commodities of animal origin (e.g., meat, milk, eggs), the environment (feces, slurry used as fertilizers), or direct contact; thus, Birkegård et al. (2018) concluded that more attention should be paid to improving our understanding of AMR dynamics within livestock production systems and the environment. It can be expected that, together with gaining more information from molecular methods, also different mathematic modeling will help to fill some of the knowledge gaps in understanding of interlinks and routes of spread of acquired AMR among human (community and hospital)—animals (livestock, pet and wild)—environment.

9 Overview of Molecular Methods for Detection of AMR

In the routine clinical laboratory, mostly phenotypic detection methods are used for antimicrobial susceptibility testing, which brings in most times either qualitative or quantitative results signalizing the susceptibility or resistance of the concerned isolate. Despite this fact, molecular methods can serve as a powerful tool for identification and/or confirmation of particular resistance gene(s) in those isolates identified by phenotypic methods as suspect resistant. Moreover, molecular methods can show more detailed information about the pathogen than phenotypic methods.

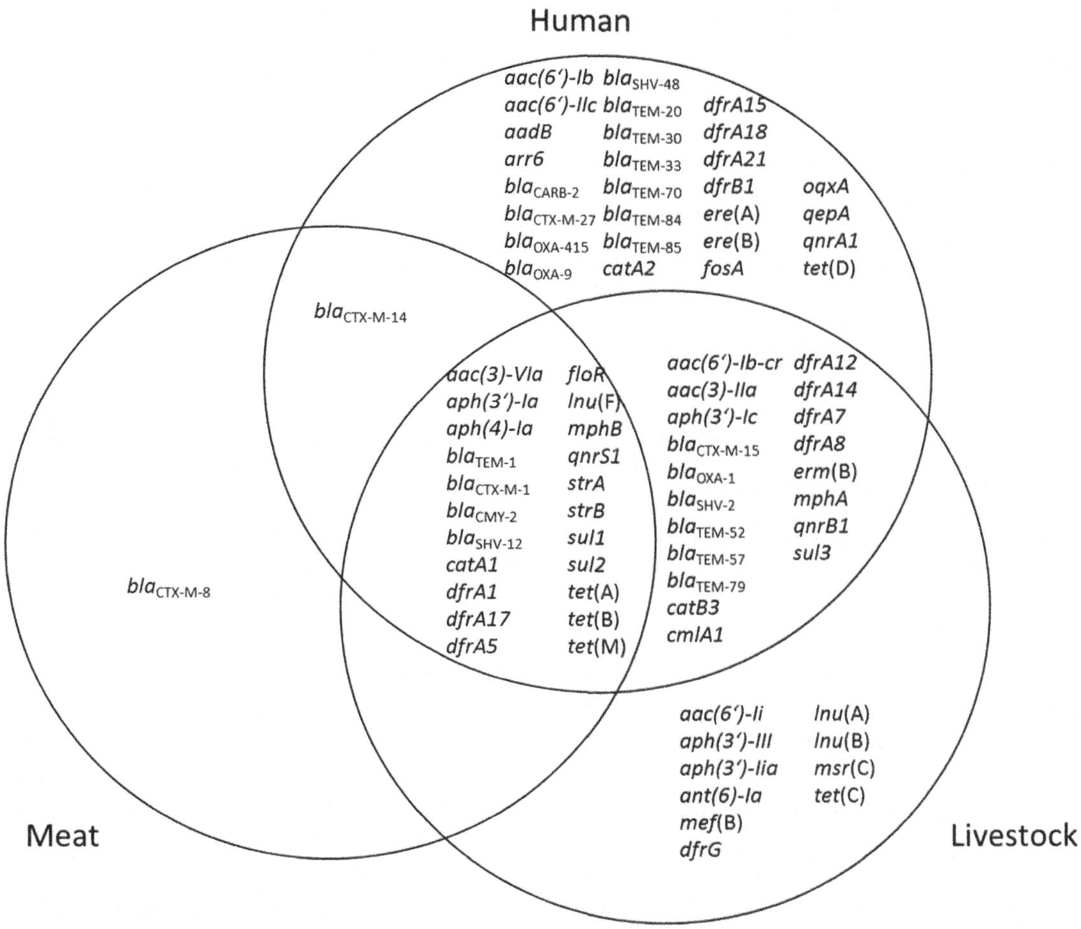

Fig. 7 Examples of documented resistance genes in *Escherichia coli* isolates from human (bloodstream), livestock (different species—cattle, pig, chicken, turkey), and meat (from previously mentioned animal species) samples (adopted from the study by Ludden et al. (2019)]

There could be also other reasons, than confirmatory, why molecular methods are used at present including those of importance from the perspectives of clinical use: results can be obtained relatively rapidly in comparison with phenotypic methods and they are relatively easy to perform. Almost routinely used are PCR detection of *mecA* gene encoding methicillin resistance in staphylococci, ESBL genes (encoding extended-spectrum β-lactamases), and PMQR genes (encoding plasmid-mediated quinolone resistance) detection, e.g., in enterobacteria. From the epidemiology perspective, the ability of methods like whole genome sequencing (WGS) to find and trace up the possible links and routes of spread of certain resistance pattern (e.g., in the last period *mcr*-harboring plasmids responsible for resistance to colistin) in association with clonal lineages of bacteria is of highest value. During the last decade, molecular methods have started to be more common even in (bigger) routine microbiological laboratories, but have become necessary standard of each research microbiology laboratory.

Before going into more details it should be highlighted that the standard plate cultivation is still of high importance and that especially techniques that ensure HIGH PURITY cultures (when working with isolates) are essential prior to the start of any molecular biology testing methods. Without pure cultures, many of molecular/biology analysis can provide misleading

Table 2 Examples of positives and negatives of phenotypic and molecular methods

Phenotype methods	Molecular methods
Pros • Lower costs • Easy to perform • Working with viable microbes • Interpretative criteria well established (especially in human medicine, in veterinary medicine for many clinical pathogens still missing) • Lower specificity—enable to discover a resistant phenotype encoded by a wide variety of genes, or not yet known resistance mechanisms • Mostly used in practice, routinely performed	Pros • Rapid techniques • In some methods, there is no need to culture bacteria (the methods allow to detect AMR genes also in bacterial species hardly culturable/non-culturable) • High specificity—possibility to detect exact gene(s)/ mechanism(s) of resistance
Cons • Time demands • Phenotypic results can be considered as indicative only in the case(s) of selected mechanisms of resistance and therefore performance of confirmation by molecular methods is required • Some resistance mechanisms (e.g., those conferring only decreased susceptibility to bacteria) are difficult to detect	Cons • False-positive results – Detection of deleted/non-functioning genes (isolate susceptible) • False-negative results – Only known mechanisms of resistance detected • Higher initial investment and operating costs • More demanding for qualification of the staff in order to perform proper analysis of the results

results or at least can make interpretation more difficult. Pros and cons of both approaches should be available for consideration of limitations (see Table 2).

To date, a plethora of molecular methods have been developed in order to detect antibiotic resistance in bacteria. However, the following section brings only a brief, non-exhaustive overview of molecular methods that belong to the most frequently used methods for antibiotic resistance gene detection in current laboratory practice.

9.1 Polymerase Chain Reaction

Polymerase chain reaction (PCR) is a method of cyclic amplification of DNA template in vitro and nowadays belongs to the most widely used techniques in molecular laboratories. Many modifications of PCR technique have been developed for special purposes and PCR products can be further analyzed using numerous molecular applications (e.g., restriction analysis to identify mutations in the amplified sequence, or to determine the amplicon sequence using Sanger sequencing).

9.1.1 Conventional PCR

Using PCR, DNA template is exponentially amplified using two short sequence-specific oligonucleotides (forward and reverse primer), heat-stable DNA polymerase, and deoxyribonucleotides (dNTP). Within PCR, three steps are cyclically repeated: double-stranded DNA denaturation occurring at 95 °C (1), primer annealing at 50–60 °C (2), and DNA extension at 72 °C (3). The amplicon can be visualized using ethidium bromide or other fluorescent dyes intercalating DNA in agarose gels after electrophoresis (Anjum et al. 2017). *Multiplex PCR*, a modification of conventional simplex PCR, consists of simultaneous amplification of multiple DNA targets (Anjum et al. 2017). Multiplex PCR is advantageous for detection of multiple resistance genes encoding a particular resistance phenotype (e.g., ESBL or carbapenemase production) (Lupo et al. 2013). Conventional PCR provides information about the presence or absence of target sequences (resistance genes); however, for detection of point mutation within the target genes, Sanger sequencing should be performed (Anjum et al. 2017). Positives and negatives of PCR and selected modifications are summarized in Table 3.

Table 3 Conventional PCR pros and cons (Lupo et al. 2013; Anjum et al. 2017)

Pros	Cons
• Rapid achievement of results: up to 5 h (24 h in the case of follow-up tests) • Easy to perform • Relatively low operating costs • Possibility of optimization of single PCRs into multiplex reactions	• Need for relatively high concentration of the target sequence—previous isolation, cultivation, and DNA extraction needed (BUT possibility to perform so-called colony PCR) • Presence of PCR inhibitors can decrease sensitivity of the method in clinical samples (hemoglobin/heparin in blood); salts of uric acids in urine (purine nucleotides break down the product), polysaccharides in feces, etc.) • Carcinogenicity of ethidium bromide used in electrophoresis

9.1.2 Real-Time PCR (Quantitative PCR, qPCR)

In comparison with conventional PCR, real-time PCR (a.k.a. quantitative PCR, qPCR) allows the detection and quantification of amplicon as the reaction proceeds via detection of the fluorescent signal (Lupo et al. 2013; Anjum et al. 2017). In order to quantify the initial concentration of the DNA target, the PCR cycle, in which exponential amplification starts, should be determined. In this so-called quantitative cycle (Cq) or threshold cycle (CT), the intensity of fluorescent signal exceeds a given threshold. In the exponential phase of the reaction, intensity of the fluorescence signal proportionally corresponds to input amount of the target DNA in the sample. Quantification of the DNA target can be either absolute or relative. Regarding absolute quantification, Cqs of the analyzed DNA target and serially diluted DNA standard are compared. This quantification method is suitable for determination of the copy number of the target sequence in genome or comparison of target DNA concentrations in two samples. Relative quantification is based on comparison of expression of DNA target and constantly expressed (housekeeping) gene. Within this quantification method, mRNA serves as a template that is transcribed into complementary DNA (cDNA) using reverse transcriptase. The results are expressed as fold change (increase or decrease) (Lupo et al. 2013).

Amplicon in real-time PCR can be detected using either fluorescent dyes intercalating any double-stranded DNA (e.g., SYBR Green) or fluorescently labeled probes complementary to the DNA target (e.g., TagMan probes). TagMan probes carry a "reporter" (a fluorophore) at the 5′ end and a "quencher" (prevents emission of fluorescence) located at the 3′ end. After hybridization of the probe to the target sequence, DNA polymerase cleaves off the fluorophore from the probe and therefore fluorescence can be emitted (Lupo et al. 2013).

Possible pros and cons of this method from the pragmatic point of view are outlined in Table 4.

Examples of use of real-time PCR in practice (Lupo et al. 2013)*:*

Single real-time PCR:

- *Distinguishing of alleles:*
  - *enterobacteria:* bla$_{SHV}$/bla$_{TEM}$ *coding for beta-lactamases with narrow spectrum as well as extended spectrum of substrates*
  - *enterobacteria:* bla$_{CTX\text{-}M}$*—ESBL (TaqMan probes)*
- *Determination of resistance gene copy number*
- *Quantification of the gene expression*
  - *porins OmpK36 in KPC-producing* Klebsiella pneumoniae *associated with carbapenem susceptibility*

Multiplex real-time PCRs (to distinguish different amplicons, the melting temperature curve analysis can be performed):

- *plasmid-mediated AmpC beta-lactamase, selected carbapenemase, plasmid-mediated quinolone resistance genes*

Table 4 Real-time PCR pros and cons (Bar et al. 2012; Lupo et al. 2013)

Pros	Cons
• Rapid technique (no need for electrophoresis or sequencing in some cases) • High sensitivity (especially regarding probes)—allow us to distinguish target sequence/different target sequences in multiplex reaction) • Lower limit of detection in comparison with culture methods (detection of genes directly from the clinical samples)	• Non-specific fluorescent dyes can intercalate into dimers of primers—in diagnostic, probes detecting the target sequence are used in the vast majority of cases • "In-house" qPCR: • Optimization needed—melting temperatures of primers and probes, and reaction kinetics should be taken into consideration • To ensure sufficiently good reproducibility, standardization of the DNA preparation step is essential, as well as accurate and thorough interpretation of results

9.2 DNA Microarray/DNA Chip

The method is based on DNA–DNA hybridization that enables simultaneous identification and partial characterization of a wide range of genes. DNA sample to be analyzed is (in most types) labeled with a fluorescent dye and hybridized with specific DNA probes spotted on the solid surface of the chip. The method shows whether the analyzed isolate harbors particular genes/alleles included in the array or not (Lupo et al. 2013; Anjum et al. 2017). Parallel identification of the isolate and detection of its AMR profile is possible in well-designed microarray.

Originally, glass slides and fluorescent dyes were used, which made the DNA microarrays expensive and also relatively time-consuming (Anjum et al. 2017). Different commercial microarrays are nowadays available, but also "in-house" microarrays concentrated, e.g., on β-lactamase screening were developed (Lupo et al. 2013). Pros and cons of DNA microarrays are summarized in Table 5.

Examples:

- *CapitalBio DNA microarray to identify* Mycobacterium tuberculosis *and detect its resistance profile: oligonucleotide probes designed to detect 16S rRNA gene of* Mycobacterium *species and detect mutations in genes* rpoB *(resistance to rifampicin) and* inhA *and* katG *(resistance to izoniazide)* (Zhang et al. 2012).
- *Alere Technologies adapted their microarrays to a simpler platform (DNA probes are bound at the bottom of a test tube or a 96-well plate) using the horseradish peroxidase for detection of successful hybridization, simplified protocols in which numerous DNA samples can be tested and the need of dual hybridization can be avoided. However, the total amount of DNA probes included in these microarrays is lower than on glass slides and this method allows detection of presence or absence of the tested genes rather than detection of gene expression* (Anjum et al. 2017).
- *Recent publication* (Torres Fink et al. 2019) *brought results where "AMR Direct Flow Chip Kit" was used for 210 bacterial isolates (and 30 control strains) harboring either one or more antimicrobial resistance genes including plasmid-encoded extended-spectrum β-lactamases (SHV, CTX-M) and carbapenemases (GES, SME, KPC, NMC/IMI, SIM, GIM, SPM, NDM, VIM, IMP, and OXA),* mecA, vanA, *and* vanB.

Table 5 DNA microarrays pros and cons (Lupo et al. 2013; Anjum et al. 2017)

Pros	Cons
• High analytical capacity • Detection of important single-nucleotide polymorphisms (SNPs)—allele detection • Screening of a large amount of isolates • Relatively easy to perform	• Relatively lower speed (6–8 h including DNA preparation) • Cross-reactions possible in specific cases • Detection of only known genes and those involved in the microarray platform • High costs of some commercial systems

9.3 Sanger Sequencing

Sanger sequencing is based on DNA synthesis catalyzed by DNA polymerase using dNTPs and fluorescently labeled dideoxyribonucleotides (ddNTPs; each is marked by a specific fluorophore) that terminate the synthesis process, as they lack 3′-OH group. As a result, fluorescently labeled amplicons of different length are produced and then separated using capillary electrophoresis. The sequence is deduced on the grounds of the detection of the fluorescent label specific for each base (Lupo et al. 2013).

The Sanger sequencing method was still widely used, for smaller-scale projects (e.g., detection of mutations in target genes), and for validation/confirmation of new-generation sequencing results. Furthermore, Sanger sequencing in comparison with short-read sequencing technologies (like Illumina) produces longer DNA sequence reads (>500 nucleotides) and sequencing errors are less frequent (Anjum et al. 2017).

Microfluidic Sanger sequencing is a technology based on Sanger sequencing method that generates long and accurate sequence data. The whole process of sequencing takes place on a chip using only small volumes of analyzed samples (nanoliters) (Paegel et al. 2003).

Examples of applications of microfluidic sequencing:

- *Single-nucleotide polymorphism (SNP) detection in target genes (e.g., ESBL detection)*
- *Single-strand conformation polymorphism heteroduplex analysis*

9.4 Whole Genome Sequencing and Whole Metagenome Sequencing

WGS and whole metagenome sequencing (WMS) data are nowadays usually generated using methods belonging to so-called second/new-generation sequencing. New-generation sequencing (NGS) comprises very sophisticated high-throughput sequencing methods that generate large amounts of sequence data of numerous DNA samples within a single sequence run (Anjum et al. 2017). General workflow of NGS includes several steps: library preparation (DNA or RNA fragments are ligated to adapters and barcodes specific for each sample), clonal amplification of the library, normalization, and sequencing (Deurenberg et al. 2017). Nowadays, sequencing platforms developed by Illumina or Ion Torrent belong to the most widely used (Anjum et al. 2017). Both these platforms are based on sequencing by synthesis. To infer the DNA sequence, fluorescence or hydrogen ions are detected in Illumina and Ion Torrent sequencers, respectively (Lupo et al. 2013). Sequences of appropriate quality and quantity that do not contain contaminant DNA should be used for follow-up analyses using bioinformatic approaches (Anjum et al. 2017).

In comparison with other molecular methods, WGS can identify and subtype multiple target genes simultaneously (Anjum et al. 2017). Therefore, all relevant information from the clinical perspective can be obtained relatively rapidly from the WGS data, covering genotype, serotype, multilocus sequence typing (MLST) profile, virulence, and antibiotic resistance gene profiles as well as the phylogenetic background of the tested strains with a considerably high discrimination (Deurenberg et al. 2017). In addition, WGS is more flexible compared to microarrays, as WGS enables us to include new target sequences in the analysis database and to reanalyze already sequenced DNA samples rapidly as a new target gene of interest emerges (Anjum et al. 2017).

However, sequencing data analyses represent undoubtedly the major challenge of NGS. Firstly, AMR genes and single-point mutations conferring antibiotic resistance phenotype are nowadays identified in sequences using numerous databases comprising relevant DNA or protein targets (Anjum et al. 2017). For the illustrative list of examples of the bioinformatic tools for detection/prediction of acquired resistance genes and specific genes (encoding antimicrobial targets or drug transport systems) affected by mutations

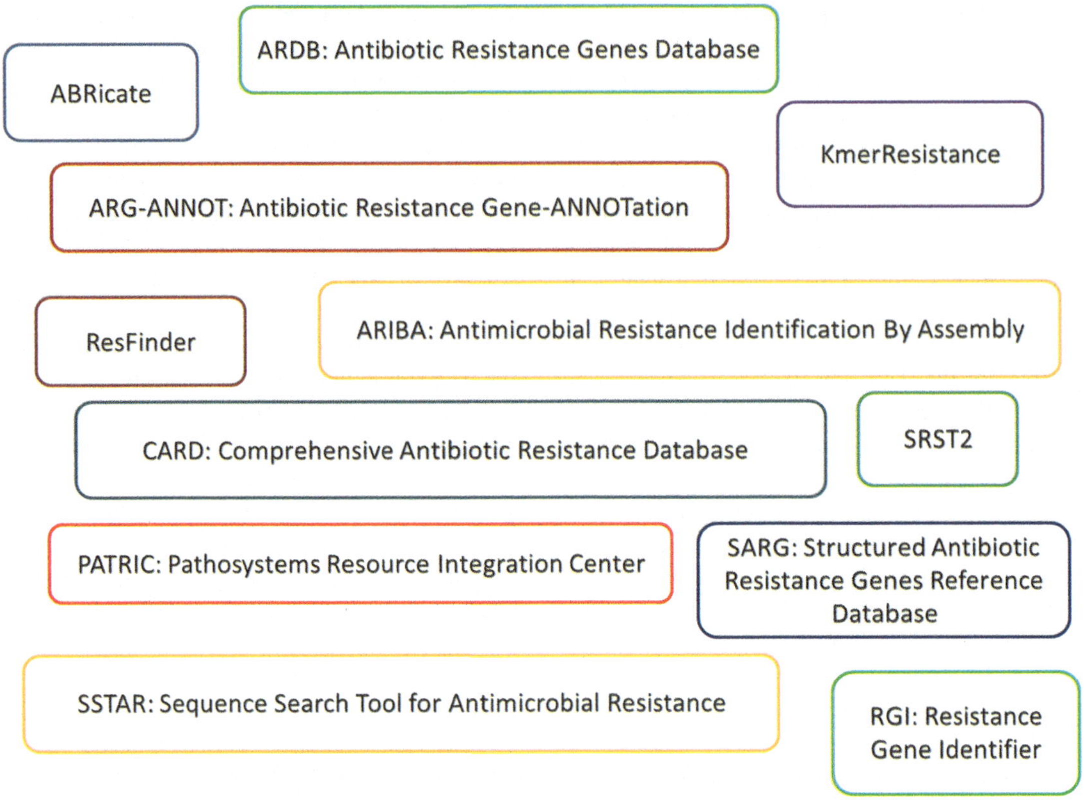

Fig. 8 Examples of the bioinformatic tools for detection of resistance genes in WGS data (Jia et al. 2017; Anjum et al. 2017; Oniciuc et al. 2017; Collineau et al. 2019; Su et al. 2019)

known to date, please refer to Fig. 8. AMR phenotype can be predicted based on sequence data analysed using these databases (Collineau et al. 2019). However, to cover all possible mechanisms conferring particular resistance phenotype including those not very well-characterized and/or species-specific in a single comprehensive database is very tricky. Moreover, the approach of using databases of well-characterized genes involved in resistance is not appropriate for the identification of new resistance genes or mutations (Anjum et al. 2017). Nevertheless, for the future it will be optimal and beneficial to create (and regularly update and also strictly check for correctness) the global, single, public database of all known AMR genes and mutations (DTU Food 2018). Also, it is of importance to mention that identification of relevant target genes from WGS data is commonly performed using two bioinformatic approaches—"mapping analysis of raw sequencing reads" and "BLAST-based analysis of (de novo assembled) draft genome contigs." However, both these approaches have their specific disadvantages and limitations (Anjum et al. 2017).

Compared to WGS, characterization of the whole microbiome originating from different sources, e.g., food, water, feces, soil, or environment, directly without preceding cultivation is enabled by metagenomics approaches that can be performed using NGS via either amplicon sequencing or whole metagenome sequencing (WMS) (Schloss and Handelsman 2005; Walsh et al. 2017). *Amplicon sequencing* is used for characterization of taxonomic composition (up to genus level) of the analyzed sample (e.g., food) with marker genes, such as 16S rRNA or 18S rRNA genes being amplified using PCR and sequences being aligned against a reference database (Oniciuc et al. 2017). Within *whole metagenome sequencing,* total genomic DNA extracted from a particular sample is studied (Franzosa et al. 2015). As a result, more information compared to amplicon sequencing can be obtained by this technique including more precise identification (up to species- or strain-level) and metabolic, virulence, or resistance profiles (Oniciuc et al. 2017).

Sequence data obtained through WGS and WMS can be exploited in many ways, e.g., in order to study the occurrence and distribution of AMR genes in bacteria from a wide range of sources (e.g., foods, food-related environments, and human and veterinary clinical specimens). Moreover, WGS data and metadata from food or clinical samples integrated with further knowledge and information can be included in the implementation of quantitative risk assessment frameworks of modeling resistance determinants occurrence and distribution (Oniciuc et al. 2017). Concerning phylogenetic analyses of bacterial isolates, WGS enables higher resolution than traditional typing methods such as multilocus variable-number tandem repeat analysis (MLVA), pulsed-field gel electrophoresis (PFGE), random amplified polymorphic DNA (RAPD), multiple-locus variable-number tandem repeat (VNTR), or multilocus sequence typing (MLST) (Oyarzabal and Kathariou 2014; Oniciuc et al. 2017). Regarding outbreak investigation, WGS can also provide useful information for design of a screening test specific for particular outbreak (Deurenberg et al. 2017). Moreover, confirmation of whether AMR genes are located on plasmids or integrated into the chromosome is key information from the perspective of AMR spread. If WGS analysis reveals that two or more AMR genes are located on the same genetic element, then co-selection of distinct AMR genes might occur (as it is sufficient to use one of the two or more antimicrobials only to select multiple AMR genes (Collineau et al. 2019).

New-generation sequencing has been already implemented in some clinical laboratories (yet more human centers are involved) and can be used for outbreak management (epidemiological tracing, selection of positive patients and their quarantine, check for carriers), molecular case finding, characterization and surveillance of pathogens (Oniciuc et al. 2017), targeted NGS using molecular inversion probes for detection of, e.g., ciprofloxacin resistance (Stefan et al. 2016), identification of bacterial species using the 16S-23S rRNA sequences (amplicon sequencing) (Sabat et al. 2017), taxonomy, metagenomics approaches on clinical samples (Willmann et al. 2015), and the determination of the transmission of zoonotic bacteria from animals to humans (Deurenberg et al. 2017). It seems according to individual studies for some species of interest that NGS methods could be also very promising for susceptibility testing of the slow-growing or hardly culturable microbes causing infections in both human [e.g., *Mycobacterium tuberculosis* (Ko et al. 2019)] or animals (*Brachyspira hyodysenteriae* (Card et al. 2018) or, e.g., *Mycoplasma* spp. (Görföl-Sulyok 2017), *Lawsonia intracellularis*, or in some anaerobes of veterinary importance, e.g., causing bovine digital dermatitis (Zinicola et al. 2015)—providing reliable results of the presence of AMR genes fast. Regarding association between resistance genes identified using WGS and phenotypic susceptibility, authors of a very recent paper by Hendriksen et al. (2019) reviewed references including those focused on foodborne pathogens (*Salmonella*, campylobacters, enterococci, *Staphylococcus aureus*, and *Enterobacteriaceae*) which have shown a high concordance (>96%) between the presence of known AMR genes or mutations and minimum inhibitory concentration of several antimicrobials at or above the epidemiological cutoff value or clinical breakpoint for resistance.

Infections located in gastrointestinal tract are very frequent, especially in young animals (De Briyne et al. 2014). Also, the gut is a known reservoir for antibiotic resistance genes, and especially the use of systemically acting antimicrobials has an impact on the intestinal resistome. Gut is considered as hot spot for horizontal gene transfer and the selection of resistant bacteria (Liu et al. 2012). NGS methods, or more recently, e.g., third-generation sequencing platforms such as single-molecule real-time (SMRT) and nanopore sequencing can be used to study the gut microbiota of poultry (Shang et al. 2018). During the course of antimicrobial treatment (fluoroquinolone, e.g., orally administered enrofloxacin frequently used in poultry) or during longer term (for chlortetracycline), using properly designed metagenomic studies can significantly help elucidate changes in the whole poultry microbiota and in resistome (Xiong et al. 2018). The above-cited innovative methods for analyzing of the factors determining antibiotic selection pressure can be, except other purposes, used to compare therapeutic regimens and their effect on the intestinal resistome. As for marketing authorizations of veterinary medicinal products containing new antimicrobials, outcomes of such studies can bring important information. For example, selection of treatment regimen with a low selective antibiotic pressure on the bacteria in the animal's gut that could result in a limited dissemination of antibiotic resistant bacteria will be of great benefit. Similar study with ciprofloxacin has already been performed in a human (Willmann et al. 2015).

Despite abovementioned information on the advances in technologies and approaches in genome sequencing, the document considering the role of WGS in (routine) antimicrobial susceptibility testing of bacteria that was published by EUCAST in 2017 (based on the available published data in 2015/2016) takes more cautious opinion. It concluded that available published evidence does not currently support the use of WGS-inferred susceptibility to guide clinical decision making. Experts from EUCAST pointed out that an absence of a resistance gene or mutation is not necessarily always associated with susceptible phenotype of the isolate. Therefore, robust evidence will be needed to show that the potential of WGS for very major errors does not adversely affect treatment outcomes. It seems likely these new methods could be considered preferably for slow-growing/unculturable microbes, where the speed of WGS-generated results is advantageous over traditional phenotypic susceptibility methods. In some microbes, WGS could be the only reliable method detecting resistance genes, even with very limited use—mostly in research laboratories rather than in routine laboratory practice. Yet, for most bacteria of clinical relevance and in most countries the current cost/need for expert knowledge including bioinformatics and speed of inferring antibiotic susceptibility from WGS data remain prohibitive to wide adoption in routine clinical laboratories (in comparison with AST using disk diffusion method, for example) (Ellington et al. 2017).

Fig. 9 brings SWOT analysis for consideration of factors playing a role once the NGS methods are considered to be involved in the laboratory testing being a part of, e.g., National Action Plans to fight against AMR. This analysis allows us to think also about aspects in which the NGS methods can bring benefits and threats.

9.5 MALDI-TOF MS

Matrix-assisted laser desorption ionization-time of flight mass spectroscopy (MALDI-TOF) is primarily used in clinical laboratories for bacterial species identification. However, it can be also used for detection of some resistance mechanisms. Nowadays, the method is not widely used for AMR detection due to its several limitations (Lupo et al. 2013; Anjum et al. 2017). Within MALDI-TOF MS, analyzed molecules (e.g., DNA, proteins, or peptides) are ionized using a laser. Molecule ions are separated according to their mass/charge ratio and their time of flight is measured resulting in a unique mass spectrum profile that is compared against a database of reference mass spectra (Anjum et al. 2017). Regarding resistance testing, MALDI-TOF MS allows us to detect resistance proteins (e.g., ESBL, carbapenemases) or antibiotics and their degradation products (e.g., carbapenem resistance). The main disadvantage of this method is associated with considerable costs of purchase and maintaining the instrument besides others. On the contrary, sample processing is rather inexpensive, and this high-throughput method provides relatively reliable results fast (Anjum et al. 2017).

Strengths

- High specificity and sensitivity
- High resolution capacity (clonal resolution)
- High extent parallel analysis of isolates
- High extent genotyping (complete antimicrobial resistance genes profile, virulence profile, identification of polymorfisms etc.)
- Relatively short time to gain results
- Availability of on -line databases and tools for the data analysis
- Around EU, laboratories equipped and experienced exist to perform such analysis in veterinary samples

Weakness

- Need for improvement of bioinformatic processing of the primary data, building of new/improvement of existing bioinformatic platforms
- High initial investment and also operational costs comparing to the traditional phenotype/older generation methods
- More complex laboratory methods/procedures
- Necessity of the optimisation and validation of methods
- Necessity for education/continual training and gaining experience of the laboratory staff as well as expert staff (need of bioinformatics of „new generation" => well experienced experts with certain NGS method
- False positive results can be also gained – of importance proper performance of all steps:
  - To avoid any contamination (microbiological phase);
  - To work precisely to avoid contamination within the process with imprecise DNA purification or at running of the DNA samples on the sequencer;
  - To evaluate gene expression properly (it is not possible to perform such evaluation based ONLY on NGS results)
  - In the case of the direct detection of the pathogen in the sample without culturing not possible to distinguish live or dead cells (viability of pathogen issue)

Opportunities

- Comparing to older technologies, gaining of the better overview of the current epidemiological situation in relation to AMR (e.g. „mining" of the information/data from isolates from different monitorings (zoonotic, indicator, target pathogen, soil/water/envi AMR profiling)
- Identifying of the sources and routes of spread of bacteria with „high risk/emerging" mechanism of resistance once the data from different laboratories across Europe/worldwide will be published (on-line databases), identification of epidemiological links/cross borders threat
- Help to satisfy growing demands on better quality/effectivity of antimicrobial therapy impacting also health care and welfare of animals, possibly having finally also impact on better quality of food of animal origin
- Raising of the qualification and building of the expert capacity in the area of AMR and related spheres
- Development/improvement of the NGS methods, finally can lead also to lowering the investigation costs
- Growing demands to biotech companies to made cheaper/easily available/higher quality of commercial solutions of different steps: extraction/purification of nucleic acids

Threats

- Lack of harmonisation of analysis of nucleic acids among laboratories (frequently own know-how based on own experience of the experts of individual laboratories)
- Risk of the improper treatment if based on NGS analysis of DNA (see also comments above related the need of phenotype/culturing methods)
- Demanding validation (QC) of the whole preparatory process and validation of NGS in the cases of use for the purpose of clinical (practice) outcomes to be produced

Fig. 9 SWOT analysis—use of NGS in a clinical laboratory in antimicrobial susceptibility testing (adopted from Costa et al. 2014; Pratova, unpublished data)

10 Conclusion

Despite the great advances in new molecular methods including NGS or third-generation sequencing that enable comprehensive analysis of antibiotic resistance, there are still open questions and a lot of work waiting to be done. These methods have started to be used in research conditions, but for application in routine laboratories, except the cost decrease, validation and standardization of these methods as well as education in proper data analysis using bioinformatic approaches are needed. As recently indicated by Hendriksen et al. (2019), there is also an undisputable need for standardization of pipelines and databases as well as phenotypic predictions based on the genomic data.

Advancing molecular methods can be of assistance to minimize the knowledge gaps and there can be identified examples of the exact use—e.g., further investigation of the rate of AMR gene expression. Also, association of resistance genes with mobile genetic elements and other important determinants, e.g., specific virulence factors characteristic for invasive isolates causing, e.g., human and animal infections, can be elucidated using NGS methods (Ludden et al. 2019). Further analysis might be also concentrated on investigation of co-selection potential (but not only directly in laboratory conditions, but utilize to a larger extent isolates recently gained from field/hospital conditions). Other thoughts and investigations could be targeted on fitness cost analysis of resistant and invasive bacteria harboring specific mobile genetic elements. New molecular methods might help us gain more understanding of all interlinks and find the way leading toward keeping in balance all ecosystems (including those microbial). These above examples cannot be considered as exhaustive and for sure need amendment, because we are still waiting for the answers to questions like "How it is possible that in such amount of resistance mechanisms, gene transfers, and pressure we still find susceptible strains and that human beings as well as animals can survive in this environment full of resistance"?

Mitigation strategies focused on limiting selective pressure should be investigated and used where reasonable in each area of human activities, especially those introduced by any means of artificially synthetized substances, for example by reducing unnecessary usage of antibiotics/biocide/pesticides and insecticides in agriculture that was proven to select/co-select antimicrobial resistance. It seems to be essential to make a "step back" and start using again less synthetic/artificial chemicals and to be closer to the nature.

References

Aarestrup FM, Hasman H (2004) Susceptibility of different bacterial species isolated from food animals to copper sulphate, zinc chloride and antimicrobial substances used for disinfection. Vet Microbiol 100 (1):83–89

Amorena B, Gracia E, Monzón M, Leiva J, Oteiza C, Pérez M, Alabart JL, Hernández-Yago J (1999) Antibiotic susceptibility assay for *Staphylococcus aureus* in biofilms developed in vitro. J Antimicrob Chemother 44(1):43–55

Andersson DI, Nicoloff H, Hjort K (2019) Mechanisms and clinical relevance of bacterial heteroresistance. Nat Rev Microbiol 17(8):479–496

Anes J, McCusker MP, Fanning S, Martins M (2015) The ins and outs of RND efflux pumps in *Escherichia coli*. Front Microbiol 6:587

Anjum MF, Zankari E, Hasman H (2017) Molecular methods for detection of antimicrobial resistance. Microbiol Spectr 5. https://doi.org/10.1128/microbiolspec.ARBA-0011-2017

Baker-Austin C, Wright MS, Stepanauskas R, McArthur JV (2006) Co-selection of antibiotic and metal resistance. Trends Microbiol 14(4):176–182

Balaban NQ, Helaine S, Lewis K, Ackermann M, Aldridge B, Andersson DI, Brynildsen MP, Bumann D, Camilli A, Collins JJ, Dehio C, Fortune S, Ghigo JM, Hardt WD, Harms A, Heinemann M, Hung DT, Jenal U, Levin BR, Michiels J, Storz G, Tan MW, Tenson T, Van Melderen L, Zinkernagel A (2019) Definitions and guidelines for research on antibiotic persistence. Nat Rev Microbiol 17(7):441–448

Band VI, Weiss DS (2019) Heteroresistance: a cause of unexplained antibiotic treatment failure? PLoS Pathog 15(6):e1007726

Band VI, Hufnagel DA, Jaggavarapu S, Sherman EX, Wozniak JE, Satola SW, Farley MM, Jacob JT, Burd EM, Weiss DS (2019) Antibiotic combinations that exploit heteroresistance to multiple drugs effectively control infection. Nat Microbiol 4(10):1627–1635

Bar T, Kubista M, Tichopad A (2012) Validation of kinetics similarity in qPCR. Nucleic Acids Res 40 (4):1395–1406

Barlow M (2009) What antimicrobial resistance has taught us about horizontal gene transfer. Methods Mol Biol 532:397–411

Bednorz C, Oelgeschlager K, Kinnemann B, Hartmann S, Neumann K, Pieper R, Bethe A, Semmler T, Tedin K, Schierack P, Wieler LH, Guenther S (2013) The broader context of antibiotic resistance: zinc feed supplementation of piglets increases the proportion of multi-resistant *Escherichia coli* in vivo. Int J Med Microbiol 303(6–7):396–403

Birkegård AC, Halasa T, Toft N, Folkesson A, Græsbøll K (2018) Send more data: a systematic review of mathematical models of antimicrobial resistance. Antimicrob Resist Infect Control 7:117

Blair JMA, Webber MA, Baylay AJ, Ogbolu DO, Piddock LJV (2015) Molecular mechanisms of antibiotic resistance. Nat Rev Microbiol 13:42–51

Blanco P, Hernando-Amado S, Reales-Calderon JA, Corona F, Lira F, Alcalde-Rico M, Bernardini A, Sanchez MB, Martinez JL (2016) Bacterial multidrug efflux pumps: much more than antibiotic resistance determinants. Microorganisms 4(1):14

Boyd EF, Almagro-Moreno S, Parent MA (2009) Genomic islands are dynamic, ancient integrative elements in bacterial evolution. Trends Microbiol 17:47–53

Braoudaki M, Hilton AC (2004) Low level of cross-resistance between triclosan and antibiotics in *Escherichia coli* K-12 and *E. coli* O5 compared to *E. coli* O157. FEMS Microbiol Lett 235:305–309

Brauner A, Fridman O, Gefen O, Balaban NQ (2016) Distinguishing between resistance, tolerance and persistence to antibiotic treatment. Nat Rev Microbiol 14 (5):320–330

Carattoli A (2013) Plasmids and the spread of resistance. Int J Med Microbiol 303(6–7):298–304

Card RM, Stubberfield E, Rogers J, Nunez-Garcia J, Ellis RJ, AbuOun M, Strugnell B, Teale C, Williamson S, Anjum MF (2018) Identification of a new antimicrobial resistance gene provides fresh insights into pleuromutilin resistance in *Brachyspira hyodysenteriae*, aetiological agent of swine dysentery. Front Microbiol 9:1183

Cavaco LM, Hasman H, Stegger M, Andersen PS, Skov R, Fluit AC, Ito T, Aarestrup FM (2010) Cloning and occurrence of *czrC*, a gene conferring cadmium and zinc resistance in methicillin-resistant *Staphylococcus aureus* CC398 isolates. Antimicrob Agents Chemother 54(9):3605–3608

Chalmers R, Sewitz S, Lipkow K, Crellin P (2000) Complete nucleotide sequence of Tn*10*. J Bacteriol 182:2970–2972

Chen Y, Pi B, Zhou H, Yu Y, Li L (2009) Triclosan resistance in clinical isolates of *Acinetobacter baumannii*. J Med Microbiol 58:1086–1091

Clement S, Vaudaux P, Francois P, Schrenzel J, Huggler E, Kampf S, Chaponnier C, Lew D, Lacroix JS (2005) Evidence of an intracellular reservoir in the nasal mucosa of patients with recurrent *Staphylococcus aureus* rhinosinusitis. J Infect Dis 2005 (192):1023–1028

CLSI (2018) Performance standards for antimicrobial disk and dilution susceptibility tests for bacteria isolated from animals, 4th edn. CLSI document Vet 08. Clinical and Laboratory Standards Institute, Wayne, p 170

Colclough A, Corander J, Sheppard SK, Bayliss SC, Vos M (2019) Patterns of cross-resistance and collateral sensitivity between clinical antibiotics and natural antimicrobials. Evol Appl 12(5):878–887

Collineau L, Boerlin P, Carson CA, Chapman B, Fazil A, Hetman B, McEwen SA, Parmley EJ, Reid-Smith RJ, Taboada EN, Smith BA (2019) Integrating whole-genome sequencing data into quantitative risk assessment of foodborne antimicrobial resistance: a review of opportunities and challenges. Front Microbiol 10:1107

Corona F, Martinez JL (2013) Phenotypic resistance to antibiotics. Antibiotics (Basel) 2(2):237–255

Costa P, Botelho A, Couto I, Viveiros M, Inacio J (2014) Standing of nucleic acid testing strategies in veterinary diagnosis laboratories to uncover *Mycobacterium tuberculosis* complex members. Front Mol Biosci 1:16

Couturier M, Bex F, Bergquist PL, Maas WK (1988) Identification and classification of bacterial plasmids. Microbiol Rev 52:375–395

Craven N, Anderson JC (1980) The effects of cloxacillin on staphylococci phagocytosed by bovine neutrophils. Res Vet Sci 29(1):57–62

Cravo Oliveira Hashiguchi T, Ait Ouakrim D, Padget M, Cassini A, Cecchini M (2019) Resistance proportions for eight priority antibiotic-bacterium combinations in OECD, EU/EEA and G20 countries 2000 to 2030: a modelling study. Euro Surveill 24(20):1800445

Curiao T, Marchi E, Grandgirard D, León-Sampedro R, Viti C (2016) Multiple adaptive routes of *Salmonella enterica* Typhimurium to biocide and antibiotic exposure. BMC Genomics 17:491

Darmon E, Leach DR (2014) Bacterial genome instability. Microbiol Mol Biol Rev 78(1):1–39

Davies D, Davies J (2010) Origins and evolution of antibiotic resistance. Microbiol Mol Biol Rev 74 (3):417–433

De Briyne N, Atkinson J, Pokludová L, Borriello SP (2014) Antibiotics used most commonly to treat animals in Europe. The Veterinary record 175(13):325

del Solar G, Espinosa M (2000) Plasmid copy number control: an ever-growing story. Mol Microbiol 37 (3):492–500

Demple B (1996) Redox signalising and gene control in the *Escherichia coli soxRS* oxidative stress regulon: a review. Gene 179:53–57

Deurenberg RH, Bathoorn E, Chlebowicz MA, Couto N, Ferdous M, García-Cobos S, Kooistra-Smid AMD, Raangs EC, Rosema S, Veloo ACM, Zhou K, Friedrich AW, Rossen JWA (2017) Application of next generation sequencing in clinical microbiology and infection prevention. J Biotechnol 243:16–24

Domingues S, da Silva GJ, Nielsen KM (2012) Integrons: vehicles and pathways for horizontal dissemination in bacteria. Mob Genet Elem 2(5):211–223

DTU Food (National Food Institute) (2018) How to use WGS for monitoring of AMR in bacteria – version August 2018. https://www.eurl-ar.eu/CustomerData/Files/Folders/25-resourcer/433_wgsforamrmonitoring-inclapp-aug2018-final.pdf. Accessed 8 Aug 2019

ECHA database (2019). https://echa.europa.eu/cs/information-on-chemicals/biocidal-active-substances. Accessed 21 Sept 2019

EFSA-ECDC (2018) European Food Safety Authority and European Centre for Disease Prevention and Control: The European Union summary report on antimicrobial resistance in zoonotic and indicator bacteria from humans, animals and food in 2016. EFSA J 16 (2):5182, 270 pp. https://doi.org/10.2903/j.efsa.2018.5182. Accessed 8 Aug 2019

El-Halfawy OM, Valvano MA (2015) Antimicrobial heteroresistance: an emerging field in need of clarity. Clin Microbiol Rev 28:191–207

Ellington MJ, Ekelund O, Aarestrup FM, Canton R, Doumith M, Giske C, Grundman H, Hasman H, Holden MTG, Hopkins KL, Iredell J, Kahlmeter G, Köser CU, Macgowan A, Mevius D, Mulvey M, Naas T, Peto T, Rolain JM, Samuelsen Ø, Woodford N (2017) The role of whole genome sequencing in antimicrobial susceptibility testing of bacteria: report from the EUCAST Subcommittee. Clin Microbiol Infect 23:2–22

European Commission (2019) Regulation (EU) 2019/6 of the European Parliament and of the Council of 11 December 2018 on veterinary medicinal products and repealing Directive 2001/82/EC. OJ L20, Volume 62

Fernández L, Hancock RE (2012) Adaptive and mutational resistance: role of porins and efflux pumps in drug resistance. Clin Microbiol Rev 25(4):661–681

Fitoussi F, Cohen R, Brami G, Doit C, Brahimi N, de la Rocque F, Bingen E (1997) Molecular DNA analysis for differentiation of persistence or relapse from recurrence in treatment failure of *Streptococcus pyogenes* pharyngitis. Eur J Clin Microbiol Infect Dis 16:233–237

Flannagan SE, Zitzow LA, Su YA, Clewell DB (1994) Nucleotide sequence of the 18-kb conjugative transposon Tn*916* from *Enterococcus faecalis*. Plasmid 32:350–354

Franzosa EA, Hsu T, Sirota-Madi A, Shafquat A, Abu-Ali G, Morgan XC, Huttenhower C (2015) Sequencing and beyond: integrating molecular 'omics' for microbial community profiling. Nat Rev Microbiol 13(6):360–372

Gillings MR (2013) Evolutionary consequences of antibiotic use for the resistome, mobilome, and microbial pangenome. Front Microbiol 4:4

Görföl-Sulyok KM (2017) Genetic diversity and antibiotic resistance of *Mycoplasma bovis*. PhD Thesis, University of Veterinary Medicine Doctoral School of Veterinary Science, Hungary. https://univet.hu/files/subpages/59/files/Görföl-SulyokK_értekezés.pdf. Accessed 11 Sept 2019

Guynet C, Cuevas A, Moncalián G, de la Cruz F (2011) The *stb* operon balances the requirements for vegetative stability and conjugative transfer of plasmid R388. PLoS Genet 7(5):e1002073

Haaber J, Penadés JR, Ingmer H (2017) Transfer of antibiotic resistance in *Staphylococcus aureus*. Trends Microbiol 25(11):893–905

Hashemi MM, Holden BS, Coburn J, Taylor MF, Weber S, Hilton B, Zaugg AL, McEwan C, Carson R, Andersen JL, Price JC, Deng S, Savage PB (2019) Proteomic analysis of resistance of Gram-negative bacteria to chlorhexidine and impacts on susceptibility to colistin, antimicrobial peptides, and ceragenins. Front Microbiol 10:210

Hendriksen RS, Bortolaia V, Tate H, Tyson GH, Aarestrup FM, McDermott PF (2019) Using genomics to track global antimicrobial resistance. Front Public Health 7:242

Huddleston JR (2014) Horizontal gene transfer in the human gastrointestinal tract: potential spread of antibiotic resistance genes. Infect Drug Resist 7:167–176

Jia B, Raphenya AR, Alcock B, Waglechner N, Guo P, Tsang KK, Lago BA, Dave BM, Pereira S, Sharma AN, Doshi S, Courtot M, Lo R, Williams LE, Frye JG, Elsayegh T, Sardar D, Westman EL, Pawlowski AC, Johnson TA, Brinkman FS, Wright GD, McArthur AG (2017) CARD 2017: expansion and model-centric curation of the comprehensive antibiotic resistance database. Nucleic Acids Res 45(D1):D566–D573

Jin M, Lu J, Chen Z, Nguyen SH, Mao L, Li J, Yuan Z, Guo J (2018) Antidepressant fluoxetine induces multiple antibiotics resistance in *Escherichia coli* via ROS-mediated mutagenesis. Environ Int 120:421–430

Johnston C, Martin B, Fichant G, Polard P, Claverys JP (2014) Bacterial transformation: distribution, shared mechanisms and divergent control. Nat Rev Microbiol 12(3):181–196

Kampf G (2018) Biocidal agents used for disinfection can enhance antibiotic resistance in Gram-negative species. Antibiotics (Basel) 7(4):110

Kampf G (2019) Antibiotic resistance can be enhanced in Gram-positive species by some biocidal agents used for disinfection. Antibiotics (Basel) 8(1):13

Kehrenberg C, Werckenthin C, Schwarz S (1998) Tn*5706*, a transposon-like element from *Pasteurella multocida* mediating tetracycline resistance. Antimicrob Agents Chemother 42:2116–2118

Klümper U, Riber L, Dechesne A, Sannazzarro A, Hansen LH, Sørensen SJ, Smets BF (2015) Broad host range plasmids can invade an unexpectedly diverse fraction of a soil bacterial community. ISME J 9(4):934–945

Knoop G-MM, Gebert M, Eifler K, Weyand K (2005) Transport of magnesium and other divalent cations: evolution of the 2-TM-GxN proteins in the MIT superfamily. Mol Genet Genomics 274:205–216

Ko DH, Lee EJ, Lee SK, Kim HS, Shin SY, Hyun J, Kim JS, Song W, Kim HS (2019) Application of next-generation sequencing to detect variants of drug-resistant *Mycobacterium tuberculosis*: genotype-phenotype correlation. Ann Clin Microbiol Antimicrob 18(1):2. https://doi.org/10.1186/s12941-018-0300-y

Kurenbach B, Marjoshi D, Amabile-Cuevas CF, Ferguson GC, Godsoe W, Gibson P, Heinemann JA (2015) Sublethal exposure to commercial formulations of the herbicides dicamba, 2,4-dichlorophenoxyacetic acid, and glyphosate cause changes in antibiotic susceptibility in *Escherichia coli* and *Salmonella enterica* serovar Typhimurium. mBio 6:e00009-15

Kurenbach B, Hill AM, Godsoe W, van Hamelsveld S, Heinemann JA (2018) Agrichemicals and antibiotics in combination increase antibiotic resistance evolution. PeerJ 6:e5801

Li X-Z, Nikaido H (2009) Efflux-mediated drug resistance in bacteria an update. Drugs 69(12):1555–1623

Liu L, Chen X, Skogerbø G, Zhang P, Chen R, He S (2012) The human microbiome: a hot spot of microbial horizontal gene transfer. Genomics 100:265–270. https://doi.org/10.1016/j.ygeno.2012.07.012

Ludden C, Raven KE, Jamrozy D, Gouliouris T, Blane B, Coll F, de Goffau M, Naydenova P, Horner C, Hernandez-Garcia J, Wood P, Hadjirin N, Radakovic M, Brown NM, Holmes M, Parkhill J, Peacock SJ (2019) One health genomic surveillance of *Escherichia coli* demonstrates distinct lineages and mobile genetic elements in isolates from humans versus livestock. mBio 10:e02693-18

Lupo A, Papp-Wallace KM, Sendi P, Bonomo RA, Endimiani A (2013) Non-phenotypic tests to detect and characterize antibiotic resistance mechanisms in *Enterobacteriaceae*. Diagn Microbiol Infect Dis 77 (3):179–194

Maillard JY (2018) Resistance of bacteria to biocides. Microbiol Spectr 6(2). https://doi.org/10.1128/microbiolspec.ARBA-0006-2017

Maiques E, Ubeda C, Tormo MA, Ferrer MD, Lasa I, Novick RP, Penadés JR (2007) Role of staphylococcal phage and SaPI integrase in intra- and interspecies SaPI transfer. J Bacteriol 189(15):5608–5616

Martinez JL, Baquero F (2014) Emergence and spread of antibiotic resistance: setting a parameter space. Ups J Med Sci 119(2):68–77

Meylan S, Andrews IW, Collins JJ (2018) Targeting antibiotic tolerance, pathogen by pathogen. Cell 172 (6):1228–1238

Mruk I, Kobayashi I (2014) To be or not to be: regulation of restriction–modification systems and other toxin–antitoxin systems. Nucleic Acids Res 42(1):70–86

Munita JM, Arias CA (2016) Mechanisms of antibiotic resistance. Microbiol Spectr 4(2). https://doi.org/10.1128/microbiolspec.VMBF-0016-2015

Nicoloff H, Hjort K, Levin BR, Andersson DI (2019) The high prevalence of antibiotic heteroresistance in pathogenic bacteria is mainly caused by gene amplification. Nat Microbiol 4(3):504–514

Nies DH (2003) Efflux-mediated heavy metal resistance in prokaryotes. FEMS Microbiol Rev 27:313–339

Norman A, Hansen LH, Sorensen SJ (2009) Conjugative plasmids: vessels of the communal gene pool. Philos Trans R Soc Lond Ser B Biol Sci 364 (1527):2275–2289

Novick RP, Christie GE, Penadés JR (2010) The phage-related chromosomal islands of Gram-positive bacteria. Nat Rev Microbiol 8(8):541–551

Oniciuc EA, Likotrafiti E, Alvarez-Molina A, Prieto M, Santos JA, Alvarez-Ordóñez A (2017) The present and future of whole genome sequencing (WGS) and whole metagenome sequencing (WMS) for surveillance of antimicrobial resistant microorganisms and antimicrobial resistance genes across the food chain. Genes (Basel) 9(5):268

Oyarzabal OA, Kathariou S (2014) DNA methods in food safety. Wiley, Hoboken, NJ

Paegel BM, Blazej RG, Mathies RA (2003) Microfluidic devices for DNA sequencing: sample preparation and electrophoretic analysis. Curr Opin Biotechnol 14 (1):42–50

Pal C, Bengtsson-Palme J, Kristiansson E, Larsson DG (2015) Co-occurrence of resistance genes to antibiotics, biocides and metals reveals novel insights into their co-selection potential. BMC Genomics 16:964

Partridge SR (2011) Analysis of antibiotic resistance regions in Gram-negative bacteria. FEMS Microbiol Rev 35(5):820–855

Partridge SR, Kwong SM, Firth N, Jensen SO (2018) Mobile genetic elements associated with antimicrobial resistance. Clin Microbiol Rev 31(4):e00088–e00017

Penadés JR, Christie GE (2015) The phage-inducible chromosomal islands: a family of highly evolved molecular parasites. Annu Rev Virol 2(1):181–201

Peterson E, Kaur P (2018) Antibiotic resistance mechanisms in bacteria: relationships between resistance determinants of antibiotic producers, environmental bacteria, and clinical pathogens. Front Microbiol 9:2928

Petinaki E, Papagiannitsis C (2017) Resistance of staphylococci to macrolides-lincosamides-streptogramins b (MLS_B): epidemiology and mechanisms of resistance, epidemiology. https://doi.org/10.5772/intechopen.75192. Accessed 21 Sept 2019

Piddock LJ (2006) Clinically relevant chromosomally encoded multidrug resistance efflux pumps in bacteria. Clin Microbiol Rev 19(2):382–402

Poirel L, Bonnin RA, Nordmann P (2012) Genetic support and diversity of acquired extended-spectrum beta-lactamases in Gram-negative rods. Infect Genet Evol 12(5):883–893

Porse A, Schønning K, Munck C, Sommer MO (2016) Survival and evolution of a large multidrug resistance plasmid in new clinical bacterial hosts. Mol Biol Evol 33(11):2860–2873

Pulcini C, Ergonul O, Can F, Beović B (2017) Antimicrobial stewardship. Elsevier Science

Rainard P, Foucras G, Fitzgerald JR, Watts JL, Koop G, Middleton JR (2018) Knowledge gaps and research priorities in *Staphylococcus aureus* mastitis control. Transbound Emerg Dis 65(Suppl 1):149–165

Ramsay JP, Kwong SM, Murphy RJ, Yui Eto K, Price KJ, Nguyen QT, O'Brien FG, Grubb WB, Coombs GW, Firth N (2016) An updated view of plasmid conjugation and mobilization in *Staphylococcus*. Mob Genet Elem 6(4):e1208317

Rensing C, Moodley A, Cavaco LM, McDevitt SF (2018) Resistance to metals used in agricultural production. Microbiol Spectr 6. https://doi.org/10.1128/microbiolspec.ARBA-0025-201

Roberts AP, Mullany P (2011) Tn*916*-like genetic elements: a diverse group of modular mobile elements conferring antibiotic resistance. FEMS Microbiol Rev 35(5):856–871

Russell AD (2003) Biocide use and antibiotic resistance: the relevance of laboratory findings to clinical and environmental situations. Lancet Infect Dis 3 (12):794–803

Sabat AJ, van Zanten E, Akkerboom V, Wisselink G, van Slochteren K, de Boer RF, Hendrix R, Friedrich AW, Rossen JWA, Kooistra-Smid AMDM (2017) Targeted next-generation sequencing of the 16S-23S rRNA region for culture-independent bacterial identification – increased discrimination of closely related species. Sci Rep 7(1):3434

SCCS (2010) Scientific Committee on Consumer Safety: opinion on triclosan antimicrobial resistance. https://ec.europa.eu/health/scientific_committees/consumer_safety/docs/sccs_o_054.pdf. Accessed 8 Aug 2019

SCENHIR (2009) Scientific Committee on Emerging and Newly Identified Health Risks, Assessment of the antibiotic resistance effects of biocides; 19th January 2009. https://ec.europa.eu/health/ph_risk/committees/04_scenihr/docs/scenihr_o_021.pdf. Accessed 8 Aug 2019

Scherer J, Nies DH (2009) CzcP is a novel efflux system contributing to transition metal resistance in *Cupriavidus metallidurans* CH34. Mol Microbiol 73:601–621

Schloss PD, Handelsman J (2005) Metagenomics for studying unculturable microorganisms: cutting the Gordian knot. Genome Biol 6(8):229

Shang Y, Kumar S, Oakley B, Kim WK (2018) Chicken gut microbiota: importance and detection technology. Front Vet Sci 5:254

Sheridan A, Lenahan M, Condell O, Bonilla-Santiago R, Sergeant K, Renaut J, Duffy G, Fanning S, Nally JE, Burgess CM (2013) Proteomic and phenotypic analysis of triclosan tolerant verocytotoxigenic *Escherichia coli* O157:H19. J Proteome 80:78–90

Siguier P, Gourbeyre E, Varani A, Bao TH, Chandler M (2015) Everyman's guide to bacterial insertion sequences. Microbiol Spectr 3(2). https://doi.org/10.1128/microbiolspec.MDNA3-0030-2014

Škaloud J, Pokludová L, Novotná R, Čížek A (2003) Evaluation by conductance assay of Shiga toxin-producing *Escherichia coli* (STEC) O157 and O26 and their sensitivity to selected disinfectants. Acta Vet Brno 72:101

Slipski CJ, Zhanel GG, Bay DC (2018) Biocide selective TolC-independent efflux pumps in *Enterobacteriaceae*. J Membr Biol 251(1):15–33

Stefan CP, Koehler JW, Minogue TD (2016) Targeted next-generation sequencing for the detection of ciprofloxacin resistance markers using molecular inversion probes. Sci Rep 6:25904

Su M, Satola SW, Read TD (2019) Genome-based prediction of bacterial antibiotic resistance. J Clin Microbiol 57:e01405-18

Sultan I, Rahman S, Jan AT, Siddiqui MT, Mondal AH, Haq QMR (2018) Antibiotics, resistome and resistance mechanisms: a bacterial perspective. Front Microbiol 9:2066

Thomas CM, Nielsen KM (2005) Mechanisms of, and barriers to, horizontal gene transfer between bacteria. Nat Rev Microbiol 3(9):711–721

Torres Fink I, Tormo Palop N, Borrás Salvador R, Buesa Gómez J, Gimeno Cardona C, Navarro Ortega D (2019) Evaluation of the DNA microarray "AMR Direct Flow Chip Kit" for detection of antimicrobial resistance genes from Gram-positive and Gram-negative bacterial isolated colonies. Enferm Infecc Microbiol Clin 37(7):454–457

U.S. FDA (2016) Safety and effectiveness of consumer antiseptic rub products; Topical antimicrobial drug products for over-the-counter human use; PROPOSED AMENDMENT OF THE TENTATIVE FINAL MONOGRAPH. Docket No. FDA-2016-N-0124. https://www.fda.gov/media/98943/download. Accessed 8 Aug 2019

van Duijkeren E, Schink AK, Roberts MC, Wang Y, Schwarz S (2018) Mechanisms of bacterial resistance to antimicrobial agents. Microbiol Spectr 6(1). Epub 2018/01/13. https://doi.org/10.1128/microbiolspec.ARBA-0019-2017

Varga M, Pantůček R, Růžičková V, Doškař J (2016) Molecular characterization of a new efficiently transducing bacteriophage identified in meticillin-resistant *Staphylococcus aureus*. J Gen Virol 97(1):258–268

von Wintersdorff CJ, Penders J, Van Niekerk JM, Mills ND, Majumder S, Van Alphen LB, Savelkoul PH, Wolffs PF (2016) Dissemination of antimicrobial resistance in microbial ecosystems through horizontal gene transfer. Front Microbiol 7:173

Wales AD, Davies RH (2015) Co-selection of resistance to antibiotics, biocides and heavy metals, and its relevance to foodborne pathogens. Antibiotics (Basel) 4 (4):567–604

Walsh AM, Crispie F, Claesson MJ, Cotter PD (2017) Translating omics to food microbiology. Annu Rev Food Sci Technol 8:113–134

Wang Y, Lu J, Mao L, Li J, Yuan Z, Bond PL, Guo J (2019) Antiepileptic drug carbamazepine promotes horizontal transfer of plasmid-borne multi-antibiotic resistance genes within and across bacterial genera. ISME J 13:509–522

Webber MA, Whitehead RN, Mount M, Loman NJ, Pallen MJ, Piddock LJ (2015) Parallel evolutionary pathways to antibiotic resistance selected by biocide exposure. J Antimicrob Chemother 70(8):2241–2248

Willmann M, El-Hadidi M, Huson DH, Schütz M, Weidenmaier C, Autenrieth IB, Peter S (2015)

Antibiotic selection pressure determination through sequence-based metagenomics. Antimicrob Agents Chemother 59(12):7335–7345

Wozniak RA, Waldor MK (2009) A toxin-antitoxin system promotes the maintenance of an integrative conjugative element. PLoS Genet 5(3):e1000439

Xiong W, Wang Y, Sun Y, Ma L, Zeng Q, Jiang X, Li A, Zeng Z, Zhang T (2018) Antibiotic-mediated changes in the fecal microbiome of broiler chickens define the incidence of antibiotic resistance genes. Microbiome 6:34

Xu Z, Li L, Shirtliff ME, Peters BM, Li B, Peng Y, Alam MJ, Yamasaki S, Shi L (2011) Resistance class 1 integron in clinical methicillin-resistant *Staphylococcus aureus* strains in Southern China, 2001–2006. Clin Microbiol Infect 17(5):714–718

Zanini SF, Silva-Angulo AB, Rosenthal A, Aliaga DR, Martínez A (2014) Influence of the treatment of *Listeria monocytogenes* and *Salmonella enterica* serovar Typhimurium with citral on the efficacy of various antibiotics. Foodborne Pathog Dis 11 (4):265–271

Zgurskaya HI, Löpez CA, Gnanakaran S (2015) Permeability barrier of Gram-negative cell envelopes and approaches to bypass it. ACS Infect Dis 1(11):512–522

Zhang Z, Li L, Luo F, Cheng P, Wu F, Wu Z, Hou T, Zhong M, Xu J (2012) Rapid and accurate detection of RMP- and INH-resistant *Mycobacterium tuberculosis* in spinal tuberculosis specimens by CapitalBio™ DNA microarray: a prospective validation study. BMC Infect Dis 12:303–303

Zhu L, Lin J, Ma J, Cronan JE, Wang H (2010) Triclosan resistance of *Pseudomonas aeruginosa* PAO1 is due to FabV, a triclosan-resistant enoyl-acyl carrier protein reductase. Antimicrob Agents Chemother 54(2):689–698

Zinicola M, Higgins H, Lima S, Machado V, Guard C, Bicalh R (2015) Shotgun metagenomic sequencing reveals functional genes and microbiome associated with bovine digital dermatitis. PLoS One 10(7): e0133674

Molecular Biology Perspective of Susceptibility and Resistance in Main Target Pathogens in the Respective Species and Antimicrobials of Concern

Lucie Pokludová

Abstract

Laboratory antimicrobial susceptibility testing providing phenotypic characterisation of bacteria is routinely used in everyday laboratory and clinical practice. During the last decade, the importance of molecular biology methods, including whole-genome sequencing, draws attention to the genotype background of resistance and can contribute to the surveillance of the emergence and spread of antibiotic resistance. This chapter aims is to provide a summary of the specific considerations for selected main target pathogens of major food-producing species—pigs, cattle and poultry from the perspective of molecular biology and genetics with respect to resistance to different pharmacological groups of antimicrobials. The genes for virulence factors are also mentioned where available. On example of the mobile genetic elements of importance is shown the genesis and spread of resistance. The insight to multidrug resistance and co-selection of resistance via an overview of recently published results of studies can help practitioners to better understand the complexity of the issue of resistance development and spread.

L. Pokludová (✉)
Institute for State Control of Veterinary Biologicals and Medicines, Brno, Czech Republic
e-mail: pokludova@uskvbl.cz

Keywords

Resistance genes Virulence genes · Genomics · Whole-genome sequencing · Mobile genetic elements · Plasmids · Integrons · Transposons · Resistance of swine pathogens · Resistance of cattle pathogens · Resistance of poultry pathogens

From the perspective of the everyday practice in the clinical diagnostic veterinary laboratory with specialisation on the bacteria associated with the infectious diseases of food-producing animals, there are usually groups of pathogens for which standard procedures are set in relation to bacterial identification and antimicrobial susceptibility testing. Once thinking on such grouping of bacteria, according to the, e.g. growth demands, there can be provided the list of bacterial species (or at least genera), for which the laboratory identification and susceptibility testing can help in deciding on the correct antimicrobial therapy of diseases primarily caused and/or associated with bacteria (e.g. multifactorial diseases, where primary infection or immunostatus allow to bacteria as a secondary pathogen to participate on a disease). Lists of such bacteria with importance/mostly tested in routine laboratories for pigs, poultry and cattle (specifically for cattle mastitis) are mentioned in Figs. 1, 2, 3 and 4 below in the parts of this chapter related to the respective animal species.

These elementary phenotypic procedures, if well performed, standardised, and validated can

L. Pokludová (ed.), *Antimicrobials in Livestock 1: Regulation, Science, Practice*,
https://doi.org/10.1007/978-3-030-46721-0_10

provide in routine practice results for everyday reporting to practitioners (please refer also chapter "Considerations Reflecting Possible Risks from Use of Antimicrobials"). As the importance of molecular biology methods, including whole-genome sequencing, draw the attention to the genotype background of resistance and can contribute to the surveillance of the emergence and spread of antibiotic resistance, following subchapters will provide a summary of the specific considerations for selected main target pathogens of pigs, cattle and poultry from the perspective of molecular biology and genetics with respect to resistance to different pharmacological groups of antimicrobials. Where appropriate also virulence factors or other factors promoting spread and or pathogenicity, either isolated or co-located with resistance genes are commented.

Over the past decade, advances in metagenomics have elucidated the richness and diversity of bacterial taxa and also was able to cover certain gaps with regard to investigation of resistance mechanisms and resistance spread. Results of such metagenomics studies, if properly analysed and communicated could provide us with the tool for better tailoring of the measures preventing or decreasing spread of resistance. Huge efforts have been invested in decrypting genomes of bacteria to better understand the forces that have shaped bacterial evolution. Horizontal gene transfer (HGT) as a key player of microbial diversification as well as further mechanisms (recombination, gene duplication, gene loss, etc.) influencing directly bacterial genome as well as combinations of factors influencing gene expression challenging our traditional view of bacterial clonality and species boundaries (Daubin and Szöllősi 2016). The role of mobile genetic elements (MGE) is considered as the key one, but recently, an increasing number of reports suggests that the transfer of these might represent a fragment of complex mosaic only (Gray et al. 2013; Dordet-Frisoni et al. 2014; Boritsch et al. 2016; Blesa et al. 2017; Husain et al. 2017). Metagenomic and other advancing molecular methods allow to study also the bacterial species, where previously not so much studies indicate possibilities of gene transfer (*Mycoplasma* spp., *Bacteroides* spp., *Mycobacterium* spp., etc.) and bring the information on the conjugative transfer of large chromosomal fragments across genomes and their subsequent recombination. New studies show that such mechanisms might be more prominent and complex than first envisaged, with several new emerging mechanisms (Dordet-Frisoni et al. 2014; Boritsch et al. 2016; Husain et al. 2017; Faucher et al. 2019) that differ from the canonical Hfr- (or *oriT*-based) transfers (Dordet-Frisoni et al. 2019). These latter ones were initially described in Hfr (High frequency of recombination) strains of *Escherichia coli* (Wollman et al. 1956) and are initiated from an origin of transfer (*oriT*) integrated into the donor chromosome. Transfers, considered as *oriT*-based, are characterised by a gradient, with genes closer to the *oriT* being more reliably and more frequently transferred, mainly because of physical constraints applying on large molecules during transfer. But very recently a novel form of chromosomal conjugative transfer occurring in *Mycoplasma* spp. that did not fit to classic *oriT*-based models was discovered (Dordet-Frisoni et al. 2019; Faucher et al. 2019).

Despite the fact that the most recent bibliography was searched and completed to provide the essential information on the bacteria frequently causing the diseases in swine, poultry and cattle above summary coming from them as well as conclusions indicating further research completion and continuation, therefore data listed below in this chapter cannot be exhaustive. The selection of the bacterial species was made predominantly according to the assortment of those tested in different EU member states in their national monitorings of antimicrobial resistance (in target pathogens) as published by Schrijver et al. (2018), as it seems, that those are of the most importance from different reasons.

1 General Considerations for Pathogens Occurring in Pig, Poultry, and Cattle

Thanks to molecular biology and also utilising genetic methods, we can describe a lot of characteristics including antimicrobial resistance and virulence patterns. Below are listed (alphabetical order chosen, due to different level of importance in different species) some pathogens of importance across pig, poultry and cattle sectors, and their brief characteristics using the most recent bibliography and reviews.

1.1 *Acinetobacter* spp.

Acinetobacter spp. is aerobic, rod-shaped, Gram-negative bacteria belonging to the Moraxellaceae family of the class Gammaproteobacteria and is considered a ubiquitous organism. Among them, *Acinetobacter baumannii* is the most clinically significant species with an extraordinary ability to accumulate antimicrobial resistance. Isolates are mainly described from human medicine and associated with nosocomial infections and were a little bit neglected in veterinary area (Wareth et al. 2019); recent bibliography, however, indicates *A. baumannii* as a veterinary nosocomial pathogen, e.g. in horse clinic (Walther et al. 2018) and describe also isolation across different animal species (van der Kolk et al. 2018).

Although *Acinetobacter* spp. can be identified to species level via MALDI-TOF/MS (utilising recent databases), molecular techniques are still of essential importance for genotyping and determination of clonal lineages. It appears that the majority of infections due to *A. baumannii* in veterinary medicine are nosocomial (van der Kolk et al. 2018). Despite that isolates yet described and published are usually associated with several types of infections, mainly in nonfood-producing animals, such as canine pyoderma, feline necrotising fasciitis, urinary tract infections in dogs and cats, equine thrombophlebitis and lower respiratory tract infection, foal sepsis and pneumonia in minks, several studies describe Acinetobacters also in food-producing animals. Therefore, *Acinetobacter*-associated infections have been already described in pigs, cattle, and poultry with yet not well-described etiology of exact diseases. Acinetobacters are considered dangerous, especially from the perspective of antimicrobial resistance and horizontal gene transfer, but according to the current state of knowledge—commensal clones are genetically unrelated to those nosocomial (van der Kolk et al. 2018). In food-producing animals, it has been shown that *Acinetobacter* spp. isolates were not Multi-Drug Resistant (MDR) and lacked significant antimicrobial resistance features such as Resistance Islands (RIs), class 1 integrons, and ISAba1 (if not counting some rare exceptions). Therefore, it can be suggested that MDR *A. baumannii* found in hospitals may not have directly evolved from animals and from food products made thereof (Hamouda et al. 2011). A few studies reported the presence of acquired carbapenemase genes in *A. baumannii* from food-producing animals, such as *bla* $_{\text{OXA-23}}$ in a cow and a pig in Lebanon and *bla* $_{\text{NDM-1}}$ in a pig in China (Al Bayssari et al. 2015; Zhang et al. 2013b). The above studies brought the evidence that further attention has to be paid to Acinetobacters as a potential reservoir of AMR genes and that there is a need for further genetic studies (van der Kolk et al. 2018).

A. baumannii is naturally (or intrinsically) resistant to the following antimicrobials: ampicillin; amoxicillin/clavulanic acid; cefazolin; cefotaxime; ceftriaxone; ertapenem; trimethoprim; and fosfomycin (Leclercq et al. 2013). Therefore, the list of antimicrobials that are usually active against wild-type *A. baumannii* infections is already short, consisting of carbapenems (doripenem, ertapenem, imipenem, and meropenem), combinations with carbapenems and beta-lactamase inhibitors (imipenem–relebactam, meropenem–varbobactam), polymyxins (colistin and polymyxin B), trimethoprim/sulfamethoxazole, tigecycline, selected fluoroquinolones and aminoglycosides (Michalopoulos and Falagas 2010; EUCAST 2020), but all from the above list should be subject of antimicrobial susceptibility testing to confirm the susceptibility at least in vitro.

Considering resistance, overexpressing intrinsic beta-lactamases is typical property of *A. baumanii*. Typical is also multidrug resistance based on different mechanisms: efflux genes, presence of aminoglycoside-modifying enzymes, chromosomal mutations in the quinolone resistance determining region (QRDR), enzymes targeting tetracycline-specific efflux, glycylcyclines as well as ribosomal methylation (Peleg et al. 2008; Doi et al. 2015; Clark et al. 2016; Wareth et al. 2019). Beta-lactamases, located chromosomally, that are overexpressed, can be represented by AmpC cephalosporinases and the OXA-51-like oxacillinases. The OXA-like oxacillinases have been associated with insertion sequence elements as, e.g. ISAba1 and ISAba3 (Doi et al. 2015; Hujer et al. 2005; Poirel and Nordmann 2006). In human, *A. baumannii* isolates have been described acquiring the carbapenemases—among them, the OXA-type carbapenemases (OXA-23, OXA-24/40, OXA-58 group, OXA-143 group, and OXA-235 group) as well as the KPC and OXA-48, and metallo-beta-lactamases as VIM and NDM. In animals, as summarised from different studies working with animal isolates from Germany (cat, companion animals, chicken geese, horses (Ewers et al. 2016, 2017; Wilharm et al. 2017; Walther et al. 2018), the following beta-lactamases has been detected and reported: OXA-23; OXA-69; OXA-68; OXA-385; OXA-314; OXA-71; and OXA-95. Carbapenem resistance has been identified in different *Acinetobacter* spp. including *A. baumannii* isolated from clinical infection cases in animals and being located on different mobile genetic elements including plasmids and transposons (very frequently $bla_{\text{OXA-23}}$) as well as on chromosome.

Efflux pumps belonging to the resistance nodulation–cell division (RND) family having a particular effect on resistance generation due to formation of a tripartite complex together with the periplasmic proteins belonging to the membrane fusion protein (MFP) family (AdeL and Adej transporters) and the outer membrane protein (OMP) channels. Due to these efflux mechanisms, drugs are pumped out directly to the external medium. Such pumps lead to the resistance to beta-lactams, aminoglycosides, fluoroquinolones, and structurally unrelated compounds (Rajamohan et al. 2010). Another RND family of exporters discovered was the AdeABC system, which is known to pump out mostly aminoglycosides, tetracyclines (also tigecycline), macrolides, fluoroquinolones as well as beta-lactams and chloramphenicol (Doi et al. 2015; Magnet et al. 2001; Longo et al. 2014).

Acquired resistance genes described in *Acinetobacter* spp. can also be responsible for mechanisms of resistance based on aminoglycoside-modifying enzymes, tetracycline efflux, sulphonamide resistance dihydropteroate synthase, and carbapenemases (Guardabassi et al. 2000; Bonnin et al. 2013; Lin and Lan 2014; Doi et al. 2015; Lee et al. 2017).

As for aminoglycoside resistance, there were reported aminoglycoside-modifying enzymes (e.g. AAC(6′)-I that is cryptic in several *Acinetobacter* spp. and confers, in the gene expression, resistance to netilmicin, tobramycin, gentamicin, and amikacin. Also, 16S rRNA methylase has been described conferring high-level resistance to amikacin, gentamicin, netilmicin, tobramycin, and kanamycin (Liou et al. 2006; Périchon et al. 2007). Recent German data indicate also presence of genes *aad*A1; *aph*(3′); and *aac*(3)-Ia from *A. baumannii* cat isolate also resistant to carbapenems (OXA-69) (Ewers et al. 2016).

In fluoroquinolones resistance, mainly point mutations of the topoisomerase and gyrase, with particular importance of GyrA Ser83Leu together with ParC Ser80Leu and Glu84 Lys amino acids substitutions have been described yet (Lupo et al. 2018).

The emergence of colistin-resistant *A. baumannii* is a serious public health concern as colistin—is considered last resort in human life-threatening infection. Colistin resistance has been attributed to the loss of lipopolysaccharide (LPS) and to mutations into the PmrAB operon that lead to the addition of phosphoethanolamine to the lipid A region of LPS through activation of the phosphoethanolamine transferase PmrC (Moffatt et al. 2010; Beceiro et al. 2011). The critical issue is that Acinetobacters, having already multiresistance, or even XDR (extensive drug resistance)—e.g. resistance to carbapenems, tigecycline, aminoglycosides, and fluoroquinolones can develop a high-level resistance to colistin and rifampicin under treatment, as a result of mutations in genes *pmr*B and *rpo*B, what has been already documented (Potron et al. 2019). Once outbreaks with such clones appear it is practically impossible to find effective treatment options.

There were also documented resistance to tetracycline encoded by *tet*A gene (Ewers et al. 2016), as well as tigecycline due to overexpression of the AdeABC multidrug efflux pump in *A. baumannii–calcoaceticus* complex (Ruzin et al. 2007; Magnet et al. 2001) as well as overexpression of AdeFGH (Coyne et al. 2011) and AdeIJK (Damier-Piolle and Magnet 2008). These efflux pumps are non-specific and therefore also impact other antimicrobials such as beta-lactams, chloramphenicol, tetracyclines, fluoroquinolones, macrolides, and aminoglycosides. Also, other efflux-pump regulation mechanisms have been linked with tigecycline resistance in *Acinetobacter* species (Sun et al. 2012a; Singh et al. 2013).

Genes encoding resistance to sulphonamides *sul*1 and chloramphenicol *cat*A1 have been already found in animal isolate (Ewers et al. 2016).

There is a challenge to use modern genetic methods effectively that can assist to identify virulence factors that help Acinetobacters to survive and to warn us on possible combinations of the mechanism of resistance as well as give us the chance to try to find new antimicrobials to defend. Newly virulence factors including porins, surface structures such as capsular polysaccharides and LPS, phospholipases; iron acquisition systems and outer membrane vesicles were described. Of importance are also differently acting proteins—regulatory proteins, biofilm-associated proteins, protein secretion systems, as well as several different types of binding proteins. Metabolism and the ability to survive in different conditions can be influenced by utilising peptide nitrogen sources more efficiently and the thickness of biofilms formed, respectively (Cerqueira and Peleg 2011; Peleg et al. 2016; Lee et al. 2017). Also, further investigation is needed to distinguish strains with pathogenic potential in animals, pathologies caused by *Acinetobacter* spp. strains and virulence factors that are present in such strains.

1.2 *Clostridium* spp.

Clostridia are anaerobic, heat-resistant endospores forming, Gram-positive rods. They are members of phylum Firmicutes, family Clostridiaceae. Most species are ubiquitous, but some pathogenic causing diseases as well as having toxic (entero, neuro, and Histotoxic) potential are described in animals (Zaragoza et al. 2019; Carter et al. 2014):

1. Enterotoxic, enteric diseases causing (*C. perfringens*, *C. difficile*, *C. spiriforme*)
2. Neurotoxic (*C. botulinum*, *C. tetani*)

3. Histotoxic (*C. chauvoei*, *C. novyi*, *C. septicum*, *C. sordellii*, *C. haemolyticum*, *C. perfringens*, and *C. colinum*)

1.3 *Clostridium perfringens*

Clostridium perfringens (A–G, based on types of toxins, of which, e.g. type A can cause gangrene, type C causing mainly neonatal hemorrhagic and necrotising enteritis, type D causative agent of enterotoxaemia mainly in small ruminants, type E causing, e.g. bovine hemorrhagic enteritis, enterotoxaemia in rabbits, some F types are associated with food poisoning and antibiotic-associated diarrhoea, type G can cause necrotic enteritis in chickens (with significant role NetB pore-forming toxin (Keyburn et al. 2010; Rood et al. 2018).

Genome-wide search on relevant toxin genes using sequence similarity search program, e.g. BLAST for toxinotyping on *C. perfringens* genomes is used (Kiu et al. 2017) and also Multiplex PCR approach is commonly used to amplify key toxins genes to classify *C. perfringens* into 7 (A–G) different toxinotypes according to the toxin genes combination (van Asten et al. 2009).

A recent large-scale genomic study on 56 strains *C. perfringens* strains, revealed a diverse pangenome (a repertoire of genes in a defined number of genomes), with only 12.6% core genes (= genes that are commonly present in each genome). The *C. perfringens* pangenome substantial genetic divergence suggests that there may be additional novel virulence-related genes encoded within the "accessory genome" in addition to the plasmid-borne toxins, known to be primarily responsible for specific disease pathologies. Plasmid-encoded genes/toxins parallel to chromosomally encoded genes in the accessory genome should be therefore analysed to gain more complete genomic picture of *C. perfringens* (Kiu et al. 2017).

Clostridium perfringens seems to be the biggest problem in turkeys and chickens, where resistances were reported to tetracyclines, sulphonamides, macrolides, and lincosamides and also to substances used in many parts of the world as growth promoters—bacitracin and virginiamycin. Most studies report susceptibility to beta-lactams (penicillin, amoxicillin, and ampicillin), amphenicols, fluoroquinolones; some also to glycopeptides (avoparcin and vancomycin) and avilamycin. Also, most frequently used ionophores show susceptibility (narasin, salinomycin, lasalocid, and monensin). But the profiles can vary in time and geographical areas across the world and more and more need to test resistance and have recent results for the region or farm seems vital.

Interestingly, studies have demonstrated that the biofilm formed by *C. perfringens* could protect the cells from exposure to atmospheric oxygen as well as to high concentrations of penicillin (Varga et al. 2008; Charlebois et al. 2014). Antimicrobial tolerance has been described also in previous work of Charlebois et al. (2012), where is described that the strains are able to survive/tolerate bacitracin, penicillin, lincomycin, tylosin, virginiamycin, and also ionophores (salinomycin, monensin, and narasin).

Park and Rafii (2014) shown in their comparative transcriptomic analysis that *C. perfringens* strains exposed to fluoroquinolones, except being resistant appeared also to had affected virulence (toxin production). Test for fitness cost in fluoroquinolone-resistant strains shown that fluoroquinolone resistance selection resulted in changes in various metabolic activities in different strains of *C. perfringens*. Within the experiments, there was proven that bacterial genotype, as well as exact structure of the fluoroquinolone, affected colonisation efficiency of the strains.

Genetic basis of the antimicrobial resistance in *Clostridium perfringens* known yet is limited to a certain group of antimicrobials and mechanisms (Li et al. 2017; Archambault and Rubin 2018):

- Tetracycline resistance mechanisms include oxidoreductase [resistance protein, *tet*A(P), is an inner-membrane protein that mediates the active efflux of tetracycline from the bacterial cell, also *tet*(B, K, L) are associated with efflux, ribosomal protection proteins are responsible for resistance, which is encoded, e.g. by *tet*(M,O,Q,W, 32 or B(P)].

- Bacitracin resistance due to the existence of ABC transporter and an overproduced undecaprenol kinase, encoded by gene *bcr*RABD; bacitracin resistance genetic determinant is located on integrative conjugative element typified by ICECp1.
- MLS_B: Caused by methylation of the target site and encoded by genes *erm* (B, C, F, G, Q); as macrolide and lincosamide resistance (mainly erythromycin and lincomycin) appears widespread therefore is considered ineffective in treating *C. perfringens* infections (by erythromycin, tylosin, and lincomycin).
- Lincosamides, where genes *lnu*(A, B, P) encode nucleotidyltransferase, which is responsible for resistance to lincosamides, genes are located on transposon-like insertion sequences.
- Chloramphenicol, florfenicol, erythromycin, and linezolid resistance caused by mutation in the gene *rpl*D, encoding protein L4 of the 50 ribosomal subunit of *C. perfringens* cells.
- Chloramphenicol as sole substance *cat*(P,Q) acetyltransferases, *cat*P has been located on chloramphenicol resistance integrative mobilisable elements typified by Tn*4451* and Tn*4452*.
- The genes, *mpr*F and *rpo*B (rifampin-resistant) have also been reported to be encoded (Li et al. 2017).
- Virginiamycin and sulphonamide resistances are mentioned by some studies (without specification of certain genes).

Recently, conjugative plasmids have been described, which belong to a large plasmid family that has a key role in the distribution of antibiotic resistance genes in *C. perfringens*. Genetic elements responsible for resistance to lincosamides (tISCpe8), bacitracin (ICECp1), chloramphenicol (*Tn 4451*), and tetracycline (plasmid pCW3) are found on such conjugative plasmids, what can be considered as dangerous from the perspective of MDR spread (Adams et al. 2018).

Anti-defensin gene *mpr*F (possibly involved in multidrug-resistant, including resistance against gentamicin) was recently reported in a large-scale genomic study of *C. perfringens* ($n = 56$ strains) to be present in 100% of the genomes (Kiu et al. 2017). Recent study brought information on *tet*A(P) in 75% of the 56 strains and higher prevalence than *tet*B(P)(42% of samples investigated). An aminoglycoside resistance gene *ant*(6)-Ib was determined in *C. perfringens* toxinotype C strain. Although mainly anaerobic bacteria like *C. perfringens* may have reduced transport of aminoglycosides intracellularly, strains sensitive to aminoglycosides (like gentamicin) at higher concentration indicates that *C. perfringens* might also have another acquired resistance to aminoglycosides (Udhayavel et al. 2017).

The number of characterised virulence factors is constantly increasing, with more than 20 toxins and hydrolytic enzymes identified to date in *C. perfringens* (Kiu and Hall 2018). A single strain cannot produce all these virulence factors (Freedman et al. 2016). The virulence factors of *C. perfringens* can be classified functionally as membrane-damaging enzymes, pore-forming toxins, intracellular toxins, and hydrolytic enzymes (Revitt-Mills et al. 2015).

1.4 *Clostridium difficile*

The causative agent of necrotic enterocolitis in animals and predominantly in humans as pseudomembranous colitis (many times associated with the use of antimicrobials). Animal species with large or expanded bowels are mostly exposed to *C. difficile* disease—horses, swine, rabbits, and guinea pigs (Archambault and Rubin 2018), recent studies, however, describe this germ is present in calves, foals, piglets, dogs, but also in poultry, with statement that neonatal animals are much more likely to be affected than adult animals (Brown and Wilson 2018). Especially in piglets, the pathophysiology of *C. difficile* is well-described—diarrhoea, dehydration, weight loss, and enteritis histologically similar to human lesions, and with high mortality (Knight and Riley 2019). In a 2009 Spanish study (Alvarez-Perez et al. 2009), 26% (140/541) of newborn piglets were found to have *C. difficile* in rectal swabs and 94% (132/140) were toxigenic strains (TcdA+, tcdB+). In a Belgian study of *C. difficile* prevalence in beef cattle farms, there was a higher colonisation rate of calves less than 6 months of age versus older, >11 months old, calves (Rodriguez et al. 2017). Knight and Riley (2019) investigated and summarised data from 86 studies, 23 worldwide countries and different sources (including animals, meat, vegetables, households, natural environment): the authors of the study indicated the prevalence in domestic pigs and piglets averages around 43%, ranging from 0% (Belgium, Switzerland) to 50% (USA, Slovenia) and 100% (Spain and The Netherlands). From the same comprehensive summarisation, made by Knight and Riley (2019), coming results for cattle and calves, where *C. difficile* prevalence averages accounted for around 14%, ranging from 0.5% (Switzerland) to 20% (Italy, Belgium, and the United States) to 50% (Australia and Canada). In ovine hosts (sheep and lambs) average prevalence calculated was 6% and in poultry (hens, broiler chickens) varying considerably (0.3% in the United States, to 29.0% in Zimbabwe and 62% in Slovenia) with mean 19% for poultry species. Other nonhuman animal reservoirs of *C. difficile* include cats and dogs (prevalence 0–100%), horses, and foals (3–33%).

Interestingly, *C. difficile* lineage identified in many of these animal studies is multilocus sequence type (MLST, ST) 11, predominated by RT078 and its close relatives RTs 033, 045, 066, 126, 127, and 288—all binary toxin positive, toxinotype V and cause *C. difficile* infections in human (Knight and Riley 2019).

Two toxins are essential for mediation of virulence belonging to the large clostridial cytotoxin family—toxin A (TcdA), an enterotoxin, and toxin B (TcdB) a cytotoxin, which both are encoded on a chromosome (Knight and Riley 2019).

In some animal species the association with the use of antimicrobials and occurrence of *C. difficile* has been described. The disease occurrence in horses was described to be associated with the use of erythromycin, trimethoprim/sulphonamides, beta-lactams, clindamycin, rifampicin, and gentamicin (Diab et al. 2013).

Whole-genome analysis of *C. difficile* RT078 strains in the Netherlands from 2002 to 2011 found that identical strains were shared between pigs and pig farmers, indicating transmission between the two groups (Knetsch et al. 2014). There was zero SNVs difference in their core genome. Strains also contain streptomycin (*Tn6235*, aphA1+) and tetracycline (*Tn6190*, tetM+) genes settled on identical mobile genetic elements. Other non-clonal strains suggested alternative reservoirs for the community spread of ribotype RT078, including wild animals and environmental sources. Knight et al. (2015) utilising genomic analysis shown that RT078 and RT027 porcine strains are similar to strains isolated from human *C. difficile* infections bringing a piece of confirmation of possible interspecies transmission. Brown and Wilson (2018), using different references (also WGS analysis from Australia, Knight et al. 2016) concluded that *C. difficile* genome analysis and differing demographic patterns between community/hospital *C. difficile* suggest a zoonotic origin in Australian community strains, particularly porcine-derived RT014/020. These studies considered of importance the interlink with antimicrobial use in the agricultural industry and possible transmission of resistant *C. difficile* strains. Findings with the investigation of microevolution in the core genome were extended by Knetsch et al. (2018) by data for 248 strains of *C. difficile* RT078 sourced from humans and animals in 22 countries. This study provided the first estimate of the population structure of global RT078 ribotype and yielded new insights into zoonosis from the perspective of potential and extent of spread. *Clostridium difficile* RT078 clonal population frequent movement (likely over a long time period) between animal and human hosts, with no geographical constraints, were highly supported by above-summarised studies. Also very recent data—comprehensive work of Knight and Riley (2019) showed that non-RT078 ST11 strains such as RTs 126, 127, 033, and 288 display a high zoonotic potential with possible importance from One Health perspective.

Acquired resistance in *C. difficile* isolates of animal origin to a range of antimicrobials has been described, including chloramphenicol, rifampicin, metronidazole, tetracyclines, and erythromycin as cited in Archambault and Rubin (2018). An investigation of global population of RT078 was proven containment of a broad spectra of genes encoding resistance to aminoglycosides and streptothricin (*aph*3′-III, *ant*6′-Ib, and *Sat*4A), erythromycin (*erm*B+), and tetracycline (*tet*M, *tet*O, *tet*32, *tet*40, and *tet*44). The gene *cde*A encoding a multidrug efflux transporter was found in all isolates (Knetsch et al. 2018).

In the ST11 study (Knight et al. 2019), half of all strains showed phenotypic resistance to one or more of tetracycline, moxifloxacin, erythromycin, and clindamycin, of which a quarter, predominantly RTs 126/078, were resistant to $\geq$3 of these agents. Chromosomal mutations in gyrA/B (fluoroquinolone resistance responsible); mobile genetic elements with genes encoding resistance to macrolides and lincosamides (*Tn6194*; ermB+) as well as genes conferring resistance to tetracycline (*Tn6190*; tetM+ and Tn6164 (as reported in animals by Corver et al. 2012) and tet44+) were described in RT078 on the specific genomic island. Alarming is the spread of clones around the world. *Clostridium difficile* (ribotype 078) detected in piglets in the Czech Republic belonging to the same cluster as same PCR ribotypes *Clostridium difficile* 078 isolates from Germany, Japan, and Taiwan (Krutova et al. 2018).

Another issue is origin of some genes from another bacterial species (e.g. phenotypically silent vanB2 transposon likely be transferred to *C. difficile* RT033 from *Enterococcus faecalis*)—evidenced in veal calf, isolated at a slaughterhouse in Australia. Also, *Erysipelothrix rhusiopathiae* seems to be the origin of the numerous aminoglycoside resistance gene clusters present in all ST11 sub-lineages of Clostridia (Knight et al. 2019).

1.5 *Escherichia coli*

Well known and easily culturable, Gram-negative, facultative anaerobic, rod-shaped, coliform bacterium, commonly found as the commensal in the gastrointestinal tract both animals and human (analysis of the human gut microbiota signalise that *E. coli* constitute about 0.1% of the bacteria there (Eckburg et al. 2005). Some strains with virulence and pathogenic factors present can cause serious infections of humans and animals (illness, especially reported in intensively reared young animals). Work made by Raimondi et al. (2019) also bring the evidence that even *E. coli* strains isolated from healthy human volunteers' intestine, they contain virulence factors (No. of strains 51, mostly have at least 6 virulence genes from 34 genes tested). Therefore, the border of "potential pathogenicity" seems to be thin, especially considering the cases where intestinal *E. coli* can be transferred to extraintestinal body parts, in human/animal under stress conditions (trauma, surgery, immunocompromised patients, husbandry stress conditions, etc.). Coming to the issue of resistance to antimicrobials, it is very hardly distinguished in many of (older) studies, if they include *E. coli* that was real causative agent of the disease or if the *E. coli* isolated was the commensal strain, also due to the above-mentioned facts. Further investigations are of need to prove how highly virulent/pathogenic strains are equipped with resistance genes. As the transfer of resistance can be both via virulent and via commensal *E. coli*, there exist currently a plethora of studies/projects dealing with both clinical isolates (e.g. systems of monitoring of pathogenic *E. coli* in different European countries—CZ, DE, DK, IT, NL, FI, NO, and SE according to EU-JAMRAI (2019) as well as European harmonised monitoring of commensal, indicator and zoonotic bacteria under EFSA umbrella (EFSA et al. 2019).

Several potential adherence/virulence factors in *E. coli* as flagella, fimbriae, the lipopolysaccharide (LPS) cell wall, capsula, outer membrane proteins (OMPs), adhesins, hemolysins, cytolysins, and siderophores were described (Kaper et al. 2004). These adherence/virulence factors, toxins, and effectors can differ in individual pathogenic strains and there will be of help to find some rapid diagnostic tool to distinguish commensal from pathogenic strains. According to pathological processes and toxin production, *E. coli* can be classified as diffusely adherent *E. coli* (DAEC), enteroaggregative *E. coli* (EAEC), enterohaemorrhagic *E. coli* (EHEC), enteroinvasive *E. coli* (EIEC), enteropathogenic *E. coli* (EPEC), and enterotoxigenic *E. coli* (ETEC), and extraintestinally pathogenic—uropathogenic *E. coli* (UPEC), which causes urinary tract infections and also meningitis-associated *E. coli* (MNEC) exists (Reygaert 2017).

In *Escherichia coli*, multidrug resistance, as well as resistance to the last resort antimicrobials, has been identified as a threat in both human and veterinary medicine due to growing evidence reported. Despite the intrinsic susceptibility of *E. coli* to almost all clinically relevant "anti-Gram-negative" antimicrobial agents (e.g. Raimondi et al. 2019–51 strains form healthy volunteers, 46 with full susceptibility), there exist a great capacity of acquisition and accumulation of resistance genes, mostly through horizontal gene transfer impacting greatly the portfolio of treatment options in clinical outbreaks caused by different *E. coli* strains. Mobile genetic elements, such as plasmids, transposons, integrons, insertion sequences, and genetic islands, contribute to the plasticity and great diversity of the *E. coli* genome, resulting in an extremely large pangenome of more than 16,000 genes (Kaas et al. 2012). The question of co-selection of resistance is also of importance. According to Poirel et al. (2018), tetracyclines or sulphonamides (two of the most used groups of antimicrobials in veterinary medicine) can also be responsible for co-selection of resistance to those antimicrobials, considered as critically important for human medicine, as long as all those determinants are located on the same (mobile) genetic elements. Different mechanisms can be involved in co-selection of multidrug resistance as harbouring and functional linkage of different resistance genes on the same plasmid, as well as upregulation of the AcrAB-TolC efflux pump and downregulation of porins and with regard to fluoroquinolones, e.g. introduction of the A87G mutation on GyrA (Li et al. 2019a). Extended-spectrum beta-lactamases encoding genes (conferring resistance to broad-spectrum cephalosporins and the monobactam aztreonam) together with carbapenemase encoding genes, plasmid-mediated quinolone resistance (PMQR) genes (conferring resistance to fluoroquinolones and quinolones), and *mcr* genes (conferring resistance to polymyxins) as well as 16S rRNA methylases (conferring pan-resistance to aminoglycosides), belong to the ensemble of genes present in XDR *E. coli.* Getting ill by such XDR strain, therefore, can be fatal.

There might be of importance to mention, the transfer of the resistance among the members (species) of the family Enterobacteriaceae. According to Rozwandowicz et al. (2018), there have been described 28 known plasmid types in Enterobacteriaceae distinguished by PCR-based replicon typing. Frequently reported plasmids: IncF, IncI, IncA/C, IncL (previously designated IncL/M), IncN, and IncH are of great epidemiological importance, as they include the greatest variety of resistance genes. Also should be considered the possibility of transfer of plasmids among bacteria of different origin (animal, human, and environmental) that is of importance from the perspective of both AMR spread as well as the persistence of AMR in different conditions (Fig. 1).

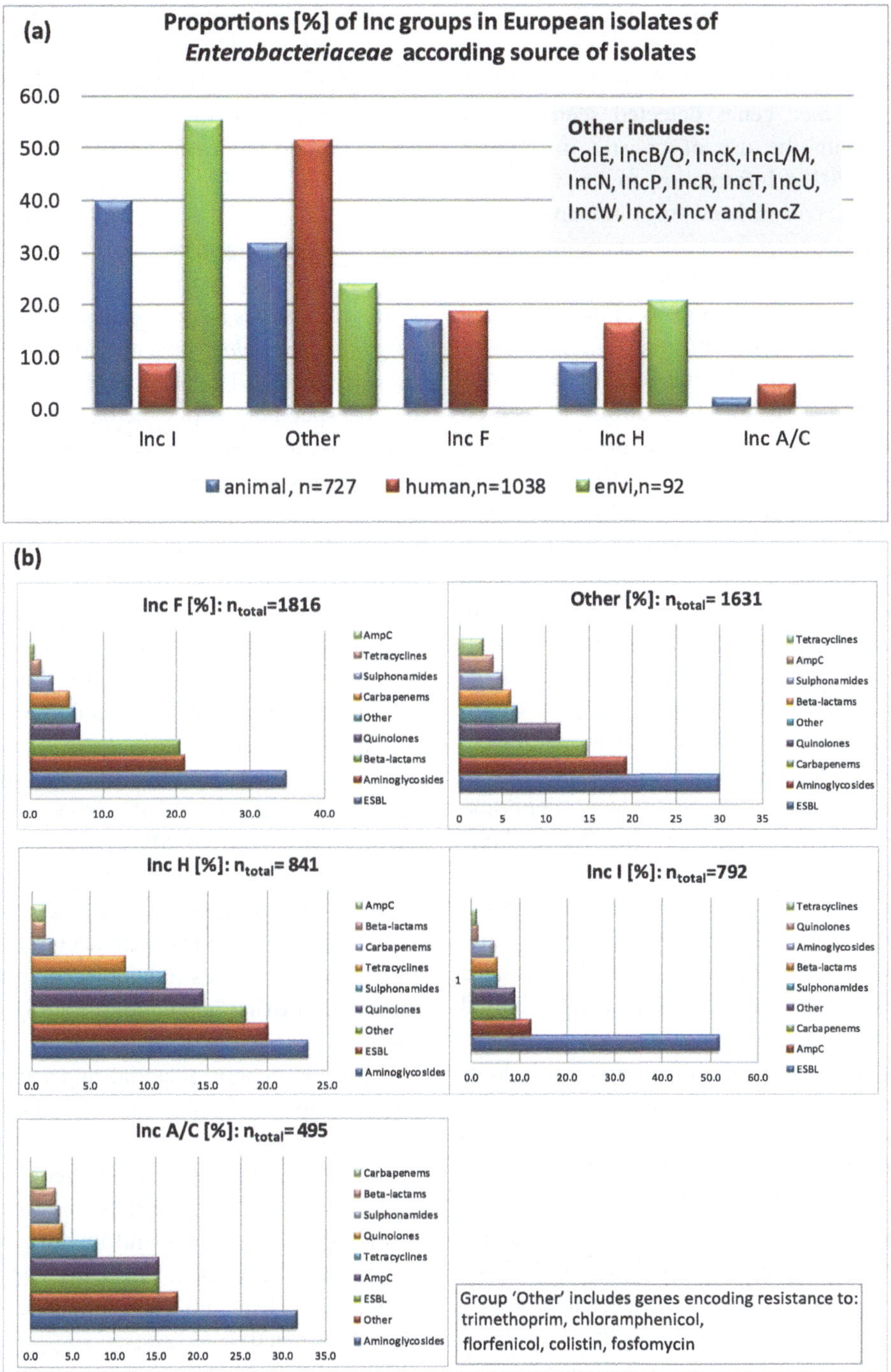

Fig. 1 Plasmids of different Inc types (modified according Rozwandowicz et al. 2018). (**a**) Relative proportions (%) of presence of Inc groups in European isolates of Enterobacteriaceae according source of isolates (Animals, $n=727$; Human, $n=1038$; Environment, $n=92$). (**b**) Distribution of genes encoding resistance to different antimicrobial classes carried by different plasmid Inc types (all sources of isolates (animals, human, environment) and areas (Europe, Asia, America) together

In the human medicine worldwide, and yet scarcely been detected in animals are carbapenemases. In contrary *E. coli* resistant to colistin with *mcr* genes detected seems rather associated with the use of colistin in animals (Poirel et al. 2018). Information on the prevalence of AMR in *E. coli* isolated from human, animal, food and environmental specimens coming from metanalysis and the comprehensive review of studies (2000–2018) made by Pormohammad et al. (2019) has been published. The authors of this study indicated with considering the results from 20 studies using disc diffusion method (DDM) and 19 MICs testing that the lowest prevalence of resistance was for colistin, with 0.8% (95% CI 0.2%–3.8%) and the highest for amoxicillin, with 70.5% (95% CI 57.5%–81%) in human *E. coli* isolates tested (DDM; in colistin DDM is not recomended as the test method, microdilution well validated tests should be used instead). Subgroup analysis from 2000 to 2018 (from both DDM and MICs testing) showed a significant increase in ciprofloxacin resistance. Despite the interesting results and outcomes of this meta-nalysis there are some important limitations as also pointed out by authors, e.g. the lack of comprehensive studies in different world regions as well as limited number of high quality designed/performed/documented and comparable studies reporting drug resistance from different sources, therefore results should be considered with caution.

As for resistance mechanism and example of genes encoding such resistance please refer to Table 1 (modified according to data from Poirel et al. 2018; Reygaert 2017; EUCAST 2020).

Importance of co-resistance, e.g. with tetracyclines, as antimicrobials, which are generally the most frequently used in majority of the European countries (EMA 2019) is supported by different findings as, e.g. Inc plasmids (e.g. IncHI2/IncP harbouring genes encoding for resistance to beta-lactams, sulphonamides, trimethoprim as well as aminoglycosides; IncI1 plasmids carry resistance genes conferring resistance to beta-lactams, aminoglycosides, and also, e.g. gene *qacG* for quaternary ammonium compounds resistance (disinfecting agents). It is hard to say, which substance is the main driver for co-selection as tetracyclines, beta-lactams, sulphonamides are the three "top" groups used mostly in food-producing animals (EMA 2019).

1.6 Mycoplasma

About 125 species have been described yet within the genus *Mycoplasma*, with parasitic characteristic, being host-dependent and many times tissue/host specific. Despite the fact that vaccination is possible, it is still not covering the whole spectrum and possible scenarios, so antimicrobials are used for prevention and/or treatment (Gautier-Bouchardon 2018). *Mycoplasma*, as currently classified, belong to the phylum Tenericutes, specifically to the class Mollicutes (characterised by the absence of the cell wall). The phylogeny of Mollicutes is interesting as regards the evolutionary degeneration of the genome (multiple reductions in genome size have occurred, the usual genetic code has been altered, but recently this paradigm is changing as proposed by Naito and Pawlowska (2016): Mycoplasma sexual competence is crucial for obligate intracellular mycoplasmas to defy the effects of Muller's ratchet, a process in which accumulation and fixation of deleterious mutations drive to extinction small, asexual populations. Pathogenic intracellular or extracellular mycoplasma populations with limited genetic material and lack in DNA repair components make them particularly vulnerable to the deleterious effect of Muller's ratchet. The capacity to horizontally exchange any part of their genome may contribute to the maintenance of genomic information in mycoplasma subpopulations, providing them with a means to rescue injured genomes, restoring deleted, or inactivated genes.

Mycoplasma belongs to the smallest replicable prokaryotic cells (0.2–0.3 μm), with exceptionally small genome (only 500–1000 genes, cca 0.2–0.5 Mb), with small amount of the G + C content, and specific stop codon—UGA (Faucher et al. 2019). Due to the small genome, a small

Table 1 Essential resistance mechanism listed according to pharmacological groups of antimicrobials and examples of genes encoding such resistance, which has been proven in *E. coli* animal source isolates

Antimicrobial agents	Mechanisms of resistance	Genetic basis
Beta-lactams Penicillins Cephalosporins Monobactams Carbapenems	Enzymatic inactivation (beta-lactamases, including, e.g. ESBLs) Active efflux Porin loss	AmpC-type enzymes (CMY-, DHA-, ACC-) encoded by different genes *amp*C *bla* genes (mobile genetic elements harboured) (TEM, SHV, CTX-M, NDM, IMP, VIM, OXA, KPC) *acrAB(tolC)*, *acrAD(tolC)*
Inhibitor of beta-lactamase (e.g. clavulanic acid)	Enzymatic inactivation (ESBLs)	Class A (Ambler classification) and group 2be (Bush–Jacoby classification) Spread via plasmids (Inc), Insertion sequences (e.g. IS*Ecp1*, IS*CR1*, IS*26*, IS*10*), transposones (Tn2), and integrones.
Aminoglycosides (e.g. amikacin, gentamicin, streptomycin)	Aminoglycoside modifying enzymes	*aac*, *ant*, and *aph*
	Modifying target: 16S rRNA or S5 and S12 ribosomal proteins	*amrA*, *rmtB* (*rmtD* and *rmtE*)
	Active efflux	*mdtEF(tolC)*
Tetracyclines Tetracycline	Limited uptake	*ompF* *acrAB(tolC)*
	Active efflux	*tet*(A), *tet*(B), *tet*(C), *tet*(D), *tet*(E), *tet*(G), *tet*(L), *tet* (M) plasmids, frequently with other genes, co-selection!
Phenicols	Enzymatic inactivation	*cat*
	Target site methylation (rRNA methylase)	*cfr*
	Limited uptake	*ompF*
	Active efflux	
	• Non-fluorinated—chloramphenicol • Fluorinated—florfenicol	*cml* *flo*R
FluoroquinolonesEnrofloxacin	Modified target—gyrase	
	Modified target—topoisomerase IV	*gyrA*
	Protect DNA from quinolone binding	*parC*
	Modification of fluoroquinolone (e.g. enrofloxacin)	Qnr-like proteins (*qnr* genes) AAC(6′)-Ib-cr acetyltransferase (gene *aac(6′)Ib-cr*)
	Limited uptake	*ompF*
	Active efflux	*acrAB(tolC)*, *acrEF(tolC)*, and *mdtABC(tolC)* PMQR as frequent mechanism of resistance transfer
Metabolic pathway inhibitors Trimethoprim/ Sulfamethoxazole	Target enzyme modification	TMP: *dfr*A, *dfr*B SUl: *sul*1, *sul*2, and *sul*3
Polymyxins Colistin	LPS-modifying enzymes	*pmr*CAB (mutations)*pmr* genes and *mgr*B, *pho*P and Q also identified
	MCR phosphoethanolamine transferase	*mcr* genes (up to 2020 *mcr*-10 last discovered, not all in *E. coli*)

number of cellular proteins are expressed, what leads to minimalisation of the enzymatic activity/metabolic pathways. Culturing of mycoplasma in vitro is therefore very demanding, as they require "additional host factors", which need to be provided in axenic, complex media (e.g. serum components as a source of fatty acids and cholesterol, metabolisable carbohydrates (glucose, arginine, and urea) as a source of energy. Due to this fact routine diagnostics is limited many times to serological methods (currently mostly PCR, but also enzyme-linked immunosorbent assay or rapid plate agglutination). Therefore, limited number of laboratories only performs phenotypical antimicrobial susceptibility testing (AST). Despite the fact that CLSI released standards for AST for certain *Mycoplasma* isolated in humans (CLSI 2011), for the AST of the animal isolates there is no harmonised standard—and no quality control strain/s for MICs testing (Felde et al. 2018; Gautier-Bouchardon 2018). Mostly broth dilution or agar dilution methods are used in different modified media (e.g. Hayflick's, Friis, Frey's, PPLO, media with arginine, NAD, turkey serum, as listed per different species in Gautier-Bouchardon (2018). For some (unstable) antimicrobials in vitro culturing of slow-growing *Mycoplasma* does not provide reliable (in vivo relevant) results due to degradation of antimicrobial in laboratory conditions. Even in strains detected as susceptible/resistant in vitro, host-linked factors can contribute to treatment failure or success—namely ability of antimicrobial to penetrate intracellularly in sufficiently static/cidal concentration; pH and cation balance. Some antimicrobials are mycoplasmacidal (e.g. fluoroquinolones), some mycoplasmastatic (e.g. tetracyclines) (Gautier-Bouchardon 2018). Multilocus Sequence Type of *Mycoplasma bovis* isolates and a method developed for rapid detection of point mutations involved in decreased susceptibility to macrolides, lincosamides, tetracyclines, and spectinomycin seems to be promising as a possible way of antimicrobial susceptibility testing in *Mycoplasma*, as demonstrated in the very recent study (Hata et al. 2019).

Intrinsic resistance of mycoplasma is typical for antimicrobials targeting the bacterial cell wall (or it synthesis), as, e.g. beta-lactams, glycopeptides, and fosfomycin. They are also intrinsically resistant to polymyxins, sulphonamides, trimethoprim, rifampicin, and quinolones of first generation (nalidixic acid).

Reduced susceptibility levels or resistances to several families of antimicrobials have been reported in field isolates of pathogenic *Mycoplasma* species of major veterinary interest: *M. gallisepticum* and *M. synoviae* in poultry; *M. hyopneumoniae*, *M. hyorhinis*, and *M. hyosynoviae* in swine; *M. bovis* in cattle; and *M. agalactiae* in small ruminants (Gautier-Bouchardon 2018). The main genetic pathway described so far for the emergence of AMR in these organisms is the occurrence, selection, and fixation of chromosomal mutations in target genes (Gautier-Bouchardon 2018; Waites et al. 2014). For instance—mutations conferring quinolone resistance have been reported in pathogenic *Mycoplasma* species. It has been also proven that passages through sub-inhibitory concentrations can select for resistance in macrolides (e.g. in *M. hyopneumoniae* and tylosin with five to seven passages, high MICs levels occurred), fluoroquinolones (rapid selection, five to seven passages proven to select in *M. bovis*), tetracyclines or pleuromutilins (slower selection, e.g. for valnemulin/tiamulin in 10 passages), depending also on exact species of Mycoplasma (Hannan et al. 1997). *Mycoplasma* does not harbour plasmids. Mutations, as well as integrative and conjugative elements (ICEs), were described to participate in resistance or decreased susceptibility.

Mobile genetic elements—especially integrative and conjugative elements (ICEs) are being identified in a growing number of *Mycoplasma* species. Besides similarities with other bacterial ICEs (chromosomal integration and the use of a type IV secretion system to mediate horizontal dissemination) Mycoplasma ICEs (MICEs) revealed unique features. Chromosomal integration of MICEs is totally random and driven by a specific (DDE) recombinase related to the Mutator-like superfamily. *Mycoplasma* conjugation also involves the transfer of large chromosomal fragments that generate progenies with mosaic genomes. Therefore, nearly every position of the chromosome is mobile. As currently described, mycoplasmas can via above-described ways access to a variety of genetic resources distributed among a huge number of bacterial species (Citti et al. 2018). The high prevalence of ICEs or vICEs in some *Mycoplasma* species isolated both from human and animals are largely due to their vertical chromosomal inheritance. These characters were described also in food-producing animals isolates, e.g. *M. hyopneumoniae*, a swine pathogen, or poultry *M. synoviae* (Minion et al. 2004; Vasconcelos et al. 2005; Liu et al. 2011), and in ruminant species, e.g. *M. bovis* (Wise et al. 2011) and *M. agalactiae* (Marenda et al. 2006; Nouvel et al. 2010).

New research (Faucher et al. 2019) chose as a model isogenic lineages of the ruminant pathogen *Mycoplasma agalactiae*. Experiments confirmed that under antibiotic selective pressure, the time scale of the mutational pathway leading to high-level of enrofloxacin resistance can exist as a single conjugative step, in which several $Enro^R$ alleles were transferred from resistant to susceptible mycoplasma cells. *Mycoplasma* chromosomal transfer was proven to create a mosaic of resistant sub-populations with unpredicted and unrelated features, which in addition can promote antimicrobial resistance dissemination.

Resistance to tetracyclines was determined mainly via mutations of the genes encoding 16SrRNA (*rrs3* and *rrs4* alleles). Cross-resistance detected in in vitro study (tetracyclines and spectinomycin) was also reported by Sulyok et al. (2017).

In macrolides, most of the point mutations in the 23S r RNA genes, domain V (interacting mainly with A2058 nucleotide) and domain II (interacting mainly with G748 nucleotide) and with the surface L4 and L22 protein, lead to decreased susceptibility or resistance to macrolides and or lincosamides (i.e. cross-resistance in macrolides/lincosamide proven, that can be also selected via passages with subinhibitory concentrations—*M. hyorhinis*). In *M. synoviae*, *M. hyorhinis*, *M. hyopneumoniae* intrinsic resistance to 14-membered macrolides was described; *M. hyopneumonioae* acquired point mutation cause resistance to tylosin (16-membered macrolide) and lincosamides. In *M. bovis*, decreased susceptibility to tylosin and tilmicosin due to point mutation/s (domain V or II), combination of two mutations need to achieve higher MICs (Gautier-Bouchardon 2018). Substitutions in L22 protein were proven to influence susceptibility to tylosin, tilmicosin, and lincosamides in *M. agalactiae*. Resistance can also result of methylation of key nucleotides in domains II and/or V.

Pleuromutilins are antimicrobials that are also used in the prevention and treatment of mycoplasma caused infection in poultry (van Duijkeren et al. 2014). Similarly as in macrolides point mutations in the 23S rRNA (V domain) are associated with decreased susceptibility in *M. gallisepticum*. Those strains with such mutation/s were resistant not only to pleuromutilins (tiamulin, valnemulin), but also lincomycin, chloramphenicol, and florfenicol. Other mutants (A2058G/A2059G) show, except resistance to pleuromutilins also resistant to erythromycin, tylosin, and tilmicosin). Another, L3 protein targeted change, can also lead to pleuromutilin resistance in some bacterial species, but has not been proven in *Mycoplasma* yet (Gautier-Bouchardon 2018).

Resistance to fluoroquinolones in different *Mycoplasma* species is due to the alterations in the quinolone resistance-determining regions (QRDR) of the *gyr*A and *gyr*B genes encoding DNA gyrase as well as the *par*C and *par*E, encoding topoisomerase IV. Mainly target mutations to those above-mentioned regions are responsible for fluoroquinolone resistance in *M. bovis* and *M. agalactiae*—cattle, sheep, goats origin (Lysnyansky and Ayling 2016; Tatay-Dualde et al. 2017); *M. synoviae* and *M. gallisepticum*—turkey and chicken isolates (Beylefeld et al. 2018), *M. hyopneumoniae*—swine origin (Felde et al. 2018), but also active efflux has been proven to be a possible mechanism of resistance to fluoroquinolones in Mycoplasma.

It should be noted that for the clinical practice, mainly resistance to macrolides and tetracyclines is of concern, as for fluoroquinolones despite in vitro resistance proven, they still remain very potent antimicrobial for the treatment, with high probability due to their mycoplasmacidal effect, good pharmacokinetic, and also possibility to cover multifactorial infections.

Despite new methods coming into susceptibility testing of *Mycoplasma*, and also we are aware of different resistance mechanisms, there should be noted that further work on the setting of appropriate clinical breakpoints attempting to make better links among in vitro results and in vivo outcomes key for the treatment is of high importance.

1.7 *Pasteurella multocida*

Pasteurella multocida is Gram-negative, non-motile, penicillin-sensitive coccobacillus belonging to the Pasteurellaceae family. Classified into five serogroups (A, B, D, E, F) based on capsular composition and 16 somatic serotypes (1–16). More recently, Harper et al. (2015) simplified the typing of LPS antigens (L1–L8) and developed a multiplex PCR targeting the genes encoding the LPS structures (LPS-mPCR). Gene technologies such as 16S rRNA, restriction endonuclease analysis (REA), ribotyping, random amplification of polymorphic DNA (RAPD)-PCR, pulsed-field gel electrophoresis (PFGE) and multilocus sequence type analysis (MLST), have pushed forward studies of molecular epidemiology and genetic diversity of *P. multocida*.

P. multocida is the cause of a range of diseases in mammals and birds, including fowl cholera in poultry, atrophic rhinitis in swine, haemorrhagic septicaemia in bovine and buffaloes, snuffles in rabbits and zoonotic infection (wounds after bites or scratches of domestic pets, which can harbour *Pasteurella* as a part of their normal respiratory microbiota) (Li et al. 2018b). Whole-genome sequence (WGS) data of isolates (China) found that a capsular: lipopolysaccharide (LPS): multilocus sequence typing (MLST) genotype A: L1: ST129 (43.75%) was predominant in avian *P. multocida*; while genotypes B: L2: ST122 (60.00%) and A: L3: ST79 (30.00%) were predominant in bovine *P. multocida*; genotype D: L6: ST50 (37.50%) in porcine *P. multocida*; and genotype A: L3: ST9 (76.47%) in rabbit *P. multocida*. Comparative genomic analysis of *P. multocida* from different host species found that there are no genes in the *P. multocida* genome that are specific to any type of host (Peng et al. 2018). Despite of that, it can be promising from the therapeutic point of view knowledge of these antigens, as specific lytic bacteriophages can be in the future used for therapy (Chen et al. 2019).

Although *P. multocida* subspecies *multocida* is the most common cause of fowl cholera, the *P. multocida* subspecies *septica* and *gallicida* can also cause fowl cholera disease (Peng et al. 2017).

The pathogenicity of *P. multocida* is associated with different virulence factors. A number of virulence factors have been identified to date and include fimbriae, adherence, and colonisation factors (*ptf*A, *hsf*-1, *fim*A, *pfh*A, and *tad*D), iron-regulated and acquisition proteins (*ton*B, *hgb*A, *hgb*B, *tbp*A, and *fur*), extracellular enzymes such as neuraminidase (*nan*B and *nan*H), superoxide dismutase (*sod*A and *sod*C), hyaluronidase (*pm*HAS), dermonecrotoxin (*tox*A) and a variety of outermembrane proteins (OMPs) such as protectins (*omp*A, *omp*H, and *plp*B) (Shirzad Aski and Tabatabaei 2016; Tang et al. 2009).

Phenotypic susceptibility testing of Pasteurella can be performed according to CLSI (2018), either by disc diffusion and by agar dilution with using of Mueller–Hinton agar supplemented with 5% defibrinated sheep blood, or as broth dilution: CAMHB (strains of *P. multocida* that fail to grow in CAMHB may be retested using the reference method for *Streptococcus* spp. (which incorporates 2.5% to 5% LHB). Inoculum should be adjusted to 0.5 McFarland, using colonies from an overnight (18 to 24 h) sheep blood agar plate incubated in ambient air or 5% CO_2.

Resistance to tetracyclines is mostly via two mechanisms: tetracycline exporters encoded by mainly by *tet* genes coding for membrane-associated proteins of the major facilitator family and ribosome protective proteins encoded by the genes *tet*(M) and *tet*(O). The gene *tet*(M) is associated with conjugative transposon TN*916* family, and is considered as one of the most spread *tet* gene among bacteria (Michael et al. 2018; Rice 1998). In tetracycline exporters, where genes *tet* (B, G, H, and L) were detected to be responsible for in *Pasteurella* they are located on plasmids or transposons; plasmids carrying genes *sul*2, *tet* R-*tet*(H), and *str*1 *str*2 are responsible for resistance to sulphonamides, tetracyclines, and streptomycin. Also, ICEs have been detected in *Pasteurella* isolates.

Resistance to penicillins, caused by beta-lactamases, was identified in *Pasteurella* and linked to the genes (bla_{CMY-2}, bla_{OXA-2}, bla_{PSE-1}, bla_{ROB-1}, and bla_{TEM-1}) and is associated mainly with small (4.1–5.7 kb) plasmids as summarised by Michael et al. (2018). Interestingly gene bla_{OXA-2} (as a part of ICE*Pmu1* was found to be non-functional in *P. multocida*, but functional in *E. coli*; in the case of ROB-1 and TEM enzymes, they are considered responsible for resistance to penicillins and first-generation cephalosporins, but sensitive to inhibitors of beta-lactamases as clavulanic acid; the role of CMY-2 towards ceftiofur remains questionable.

For aminoglycoside—streptomycin genes *str*A and *str*B (phosphotransferase, and genes conferring resistance to streptomycin as well as spectinomycin (adenyltransferase) *aad*A, *aad*A14 and *aad*A25, but for spectinomycin also resistance via mutation in 16S rRNA or in *rps*E encoding for ribosomal protein *S5* can be responsible for resistance in *Pasteurella*. In Gentamicin *aad*B, encoding for adenyltransferase is of importance (Klima et al. 2014b; Michael et al. 2018). Genes for streptomycin resistance are usually linked with small non-conjugative plasmids (up to 15 kb in *Pasteurella multocida*). Despite many small-sized plasmids, also plasmids harbouring, except aminoglycoside (mainly *str*), also determinants for resistance to sulphonamides, kanamycin/neomycin, chloramphenicol, and ampicillin were detected.

In sulphonamides (gene sul2, sulphonamide-resistant dihydropteroate synthase) and in trimethoprim *dfr*A1 and *dfr*A14 are considered to be responsible for resistance, to sole substances or combination of them (Klima et al. 2014a, b; Wu et al. 2003). Of importance is published data by Kehrenberg and Schwarz (2005) who described firstly trimethoprim resistance gene *dfr*A20 on the 11-kb plasmid pCCK154 from bovine *P. multocida*. This plasmid was transferable into *E. coli*, where it replicated and expressed high-level resistance to sulphonamides and trimethoprim. Sequence analysis identified the gene *sul*2 for sulphonamide resistance and above-mentioned trimethoprim resistance encoding gene, designated *dfr*A20. On this gene can be shown, how cross-linked can be resistance determinants among different bacterial genera, species or even big groups (G+/G−). The *dfr*A20 gene codes for a trimethoprim resistant dihydrofolate reductase of 169 amino acids, which is only distantly related to the dihydrofolate reductases of Gram-negative bacteria (recently in calves isolate *E. coli* described gene *dfr*A35 encoding for protein with high homology (Wüthrich et al. 2019), but upon cluster analysis appears to be related to those found in the Gram-positive genera *Staphylococcus*, *Bacillus*, and *Listeria*.

To date, different studies (Kadlec et al. 2001; Michael et al. 2012a; Klima et al. 2014a, b; Noyes et al. 2015) have described several mechanisms of resistance to macrolides including genes encoding macrolide efflux proteins as well as phosphotransferases (*mrs*(E)—*mph*(E) and rRNA monomethylase—gene *erm*(42). Those strains with *erm*(42) exhibit high levels of resistance to multiple macrolides. If *erm*(42) is found in isolates harbouring also *mrs*(E)—*mph*(E) genes in one operon, the highest resistance to tulathromycin, tilmicosin, gamithromycin, and also clindamycin was reported. Such a combination was detected as a part of ICE*Pmu1*, which might explain their dissemination across strain, species, and genus boundaries (Michael et al. 2018). Also, mutations in A2058G and A2059G (23S rRNA) as published by Dayao et al. (2016) and Olsen et al. (2015) are of importance due to the link to very high MICs (>64 mg/L) to multiple macrolides including erythromycin, tilmicosin, tildipirosin, tulathromycin, and gamithromycin.

For amphenicols (chloramphenicol—genes *cat*A1, *cat*A3, and *cat*B2 for acetyltransferases A and B classes (Vassort-Bruneau et al. 1996; Peng et al. 2019a); for florfenicol gene *flo*R for efflux protein (Klima et al. 2014b) was reported as responsible for resistance. As mentioned above, some of those determinants are located together with other antimicrobial-resistant genes on small (up to 15 kb) plasmids.

The study indicated that the fluoroquinolone resistance of *P. multocida* is mainly due to multiple target gene mutations in *gyr*A and *par*C and the overexpression of efflux pump genes was published by Kong et al. (2014). Authors also mentioned that clinical isolates of *P. multocida*, with MICs lower or equal to 0.25 mg/L has no quinolone resistance-determining region (QRDR) mutations, but strains with MICs of 0.5 mg/L for both enrofloxacin and ciprofloxacin, were found to have Asp87Asn or Ala84Pro mutations in *gyr*A. Also should be mentioned that in isolates further selected for resistance by passage on sub-inhibitory concentrations of fluoroquinolones, multiple mutations in *gyr*A, *gyr*B, and *par*C, but not *par*E were found to be associated with high-level fluoroquinolone resistance (MICs >4 mg/L).

Think About

Do we have an example of the genetic information transfer from bacteria to bacteria (different bacterial species or even genera)? And how can be the genes moved around the world? And why some antimicrobials are not used, but genes and resistance are still there?

One of the very illustrative examples is a "success story" of plasmid pCCK381 harbouring gene *flo*R. It was detected from one calf in the United Kingdom both in *P. multocida* and in *Salmonella enterica* subsp. *enterica* serovar Dublin, as well from other sources and also *E. coli* isolates. Hypothesis of possibile exchange of plasmid carrying resistance to florfenicol among bacteria from different genera has been confirmed by thorough molecular analysis (Michael et al. 2018). Moreover exists segments that show extended similarity to plasmids isolated also from other pathogens of cattle: pDN1 from *Dichelobacter nodosus* (Whittle et al. 2000) and pMBSF1 from *E. coli* (Blickwede and Schwarz 2004), and of fish isolates: pRVS1 from *Vibrio salmonicida* (Sørum et al. 1992). Genes *flo*R from *P. multocida* plasmid pCCK381 was also detected in multi-resistance plasmids being parts of ICEs in *M. haemolytica* and *H. somni*. Plasmid pHPSGC with the region containing *flo*R and *lys*R that shared 99% similarity with the corresponding region of pCCK381 occurred in pig isolates of *Haemophilus parasuis* (Zhang et al. 2018a). Moreover, another mobilisable plasmid (pM3446F) containing the *flo*R gene, isolated from a florfenicol resistant pig isolate of *Actinobacillus pleuropneumoniae*, showed similarity to plasmids (pCCK381 and pMH1405) found in Pasteurellaceae (Bossé et al. 2015). All these facts confirm that the question of transferability among different bacterial species founded in different animal species is of high relevance. Persistence of resistance in the bacteria is specially promoted once there is same plasmid location of different genes encoding resistance to different antimicrobials and predominantly in the cases of cluster organisation of such resistance genes.

These above results can be linked also with knowledge of florfenicol use fish for coldwater vibriosis (*V. salmonicida*; e.g. Korea and Japan) or infectious pododermatitis of cattle (*D. nodosus*) veterinary medicinal products approved in several non-EU countries (as commented by Michael et al. 2018). But how the genes from fish pathogens can be transferred to pathogens of pigs or poultry? Suggestion that the transfer of *Vibrio* genes via feeding containing non-EU origin fish products to pigs and poultry seems to be a reality.

The answer to the last question, related to co-selection of resistance by one antimicrobial to another one from the different pharmacological group, can be demonstrated on multi-resistance gene cluster consisting of genes *sul*2, *cat*A3, and *str*A organised as a transcriptional unit. Several different plasmids, as well as

(continued)

chromosomal DNA were proven to contain this unit. The probability that individual genes from such a cluster will be lost is very low. As described by scientists from Germany (Kehrenberg and Schwarz 2001) who detected the isolates of *Pasteurella* and *Mannheimia* with dominant chloramphenicol resistance gene *cat*A3 located in a *sul*2-*cat*A3-*str*A cluster, the persistence of resistance gene encoding for chloramphenicol resistance remains, despite its ban in food-producing animals in EU (1994). Sulphonamides used broadly in the EU as well aminoglycosides (streptomycin (and dihydrostreptomycin) included) caused selective pressure that ensures the maintenance of the entire cluster. This brought the evidence that even without direct selective pressure of exact individual substance, the genes within the cluster are kept and could be also spread among bacteria via co-selection by other substances (Michael et al. 2018). Moreover the co-selectors are not antimicrobials only. Generally speaking co-selecting factor can be any factor giving to bacteria the ability to survive/to have advantage among the other bacteria in ecological niche. Once the proper genes of resistance encoding co-selector are in the same genetic cluster "owned" by bacteria (genes of resistance to antibiotic/s, heavy metal/s, disinfectant/s or gene for virulence factor/s or gene for enzyme allowing utilisation of some substance promoting growth), especially once these genes are located together with genes for resistance on mobile genetic elements, despite exact antimicrobial having gene located in such cluster is not used for a long time, the resistance gene is still there and still spread.

1.8 *Mannheimia haemolytica*

The Gram-negative bacterium *Mannheimia haemolytica* is the primary bacterial species associated with bovine respiratory disease (BRD) and is responsible for significant economic losses in feedlot cattle (Amat 2019). It belongs by taxonomy, in the same family Pasteurellaceae (previously named *Pasteurella haemolytica*) many of mechanisms of resistance are the same. Therefore, only a brief summary of genetic determinants of resistance is given. Recently was proven (Snyder et al. 2019) the existence of three different ICEs with resistance gene modules in *M. haemolytica*. The resistance genes *aph*A1, *str*A, *str*B, *sul*2, *flo*R, *erm* (42), tetH/R, *aad*B, *aad*A25, $bla_{OXA\text{-}2}$, *msr*E, and *mph*E were all located within an ICE. The gene $bla_{ROB\text{-}1}$ was also present in the isolates, but was not located within an ICE.

To date, 12 different capsular serotypes have been identified within *M. haemolytica* (Amat 2019; Rice et al. 2007). Among these serotypes, serotype 1 (S1), serotype 2 (S2), and serotype 6 (S6) are most frequently isolated from feedlot cattle, with the S1 and S6 being the most prevalent in bovine infection (Klima et al. 2014a, b; Klima et al. 2016). *M. haemolytica* residing in the upper respiratory tract of healthy cattle maintains a commensal relationship until the host immunity gets disrupted by stress and viral infections.

In pathogenesis of the diseases is involved *M. haemolytica* virulence factors including outer membrane proteins (adhesins and fimbriae), leukotoxin (Lkt) that attracts neutrophils and macrophages to the site of infection at low concentrations but at high concentrations induce cell death of leukocytes and phagocytes. That allows *M. haemolytica* to evade the detection and destruction by the host immune system, neuraminidases, lipopolysaccharide (LPS), and lipoproteins that also play a role in haemorrhage, oedema, hypoxemia, and acute inflammation. Seventy-two genes encoding virulence factors have been confirmed yet (Rice et al. 2007; Griffin et al. 2010; Klima et al. 2016).

Cozens et al. (2019), very precisely described the model, of healthy cattle colonisation, including bringing the evidence that cattle are frequently colonised by commensal serotype A2 strains, but the disease is usually caused by pathogenic strains of serotype A1. For not fully elucidated reasons, but known to be associated with crowding, stress, and/or viral infection, a sudden explosive proliferation occurs in the number of serotype A1 bacteria present in the upper respiratory tract of susceptible animals (Singh et al. 2011). The colonisation of the mucosal surfaces leads to inhalation of bacterium-containing aerosol droplets into the lungs and predisposes the animals to the onset of pneumonic disease (Cozens et al. 2019). The bacterium comprises 12 capsular serotypes (Angen et al. 1999).

Resistance to tetracyclines was proven to be via tetracycline exporters encoded by mainly by *tet* genes coding for membrane-associated proteins (*tet* (B, G, H, L) were detected located on plasmids or transposons.

Resistance to penicillins, caused by beta-lactamases, was identified in *Mannheimia* to be linked with genes ($bla_{\text{OXA-2}}$ and $bla_{\text{ROB -1}}$) and is associated mainly with small (4.1–5.7 kb) plasmids as summarised by Michael et al. (2018).

For aminoglycoside—streptomycin genes *str*A and *str*B (phosphotransferase, and genes conferring resistance to streptomycin as well as spectinomycin (adenyltransferase) *aad*A25. In gentamicin *aad*B, encoding for adenyltransferase is of importance (Klima et al. 2014a, b; Michael et al. 2018). Gene *aph*A1 is responsible for resistance to kanamycin/neomycin.

In sulphonamides (gene *sul*2, sulphonamide-resistant dihydropteroate synthase) exist (Michael et al. 2018).

In the case of macrolides, several mechanisms of resistance were described yet, including genes encoding macrolide efflux proteins as well as phosphotransferases (*mrs*(E)—*mph*(E) and rRNA monomethylase—gene *erm*(42) and also mutation in 23S rRNA was detected in *Mannheimia* (Noyes et al. 2015).

For amphenicols (chloramphenicol—genes *cat*A1 and *cat*A3 for acetyltransferases A (Vassort-Bruneau et al. 1996, Michael et al. 2012a); and for florfenicol gene *flo*R for efflux protein (Klima et al. 2014a, b) was reported as responsible for resistance and harboured by plasmids (catA3 also on chromosomes of bovine *M. haemolytica*). As mentioned above, some of those determinants are located together with other antimicrobials resistance genes on small (up to 15 kb) plasmids.

Resistance to fluoroquinolones was reported to be due to mutations of *gyr*A and *par*C. In *M. haemolytica*, resistance to nalidixic acid was associated with at least one amino acid substitution in one or both of *gyr*A and *par*C, whereas all of the strains with fluoroquinolone MICs ≥8.0 mg/L had two mutations in *gyr*A and one additional change in *par*C (Katsuda et al. 2009).

1.9 *Staphylococci*

Genus *Staphylococcus* is of importance for both human and veterinary medicine. As for taxonomy, *Staphylococcus* is Gram-positive, facultative anaerobic (except *S. aureus* subsp. *anaerobicus* and *S. saccharolyticus* (Mathema et al. 2009), coccal (round-shaped) bacterium, non-motile, non-spore forming, member of the *Firmicutes*, usual member of the microbiota of the body, frequently found in the upper respiratory tract and on the skin, but also pathogens seriously affecting the health of humans and animals. Despite the fact that currently there are identified 81 species and subspecies of the genus *Staphylococcus* (https://www.dsmz.de/services/online-tools/prokaryotic-nomenclature-up-to-date/prokaryotic-nomenclature-up-to-date/genus/516664, Accessed 25 August 2019), below it will be paid the attention mainly to *Staphylococcus aureus*, *Staphylococcus hyicus*, and very briefly will be mentioned group of coagulase-negative staphylococci (CoaNS). It should be also mentioned *Staphylococcus pseudintermedius*, which is of importance especially considering dogs and their skin, soft tissues and ear diseases, as well as possibility to carry the bacteria without significant clinical signs and the close contact with human bearing in mind this species can be considered as an important reservoir of the resistance genes. These above mentioned three species and one group of is considered as the most significant importance in the animal diseases pathogenesis, whilst other species of *Staphylococcus* spp. are considered to be predominantly associated with opportunistic infections (Coetzer and Tustin 2004; Schmidt et al. 2015).

As the group of the staphylococci is of great importance, but on the other hand this book aim is to give rather a summary and overview than very in-depth insight, for the staphylococci, should be referred to very comprehensive and very recent articles by Schwarz et al. (2018) and specifically for *Staphylococcus aureus* to the work by Haag et al. (2019), but also further specific articles of Feßler et al. (2018) or Kadlec et al. (2019).

Staphylococcus aureus is a mammalian commensal and opportunistic pathogen that colonises niches such as skin, nares, and diverse mucosal membranes. Interestingly while about 20–30% of the human population can be carriers of *Staphylococcus aureus* (Graveland et al. 2011) in animals, as mentioned by some bibliography, up to 90% of chickens, 42% of pigs, 29% of sheep and among 14–35% of cows and heifers are carriers (Haag et al. 2019). Originally 4 biotypes, considering also host specificity, have been currently broadened to 6 biotypes: human, beta-haemolytic human, bovine, caprine, avian-abattoir, and non-specific (Devriese 1984; Hennekinne et al. 2003; Piechowicz and Garbacz 2016; Haag et al. 2019). Of importance from the perspective of colonisation success is that staphylococci produce two general groups of virulence factors, namely surface-associated factors and degradative enzymes, including exotoxins that play a different role and have a different level of importance in the pathogenesis of diseases. The microbial surface components of *S. aureus* recognising the adhesive matrix molecular components (MSCRAMMs)—as, e.g. fibrinogen-, fibronectin-, collagen-binding proteins as well as iron-regulated surface determinant A (Zhou et al. 2018), consist surface proteins that promote colonisation and are important for initial stages of infection. During the progress of infection, the expression of tissue-binding proteins has started to be downregulated/switched to the synthesis of extracellular toxins and tissue-degrading enzymes, both of these groups help to aid the acquisition of nutrients and the dissemination of the bacteria (Schmidt et al. 2015). The survival and proliferation of *S. aureus* within cells via preventing the combination of phagosome and lysosome, subversion autophagy, and others should also be taken into account (Foster et al. 2014). The toxin factors of *S. aureus* play a pivotal role in the processes of penetration into cell membrane and intracellular survival. Both β-toxin (in first step hydrolyse sphingomyelin) and δ-toxin (as a second step permeabilise cytomembranes) are reported to relate to the penetration across cell membrane. It was also reported that α-toxin, a pore-forming toxin, can penetrate host cell membranes, and subsequently cause osmotic swelling, rupture, lysis, and cell death (Zhou et al. 2018).

It can be of "pragmatic" importance that staphylococci can grow in a wide pH range (4.8–9.4) and can survive temperatures of up to 60 °C for 30 min, many of them are also tolerant of high salt concentrations (7.5–10%) due to the production of osmoprotectants (Somerville and Proctor 2009).

As for *S. aureus* genome, whole-genome sequencing of a number of strains has revealed that approximately 75% of the bacterium's genome comprises a core component, common to all strains (Lindsay and Holden 2004), of which the majority are those associated with central metabolism and other "housekeeping" functions (Shittu and Lin 2007). The remaining 25% of the *S. aureus* genome are acquired genes encoding non-essential functions ranging from virulence, antimicrobial, heavy metal, and disinfectant resistance (Wendlandt et al. 2013), to substrate utilisation, and miscellaneous metabolisms. Many of them can be allocated to groups of mobile genetic elements (MGEs), such as chromosomal cassettes, pathogenicity islands, plasmids, prophages, and transposons (Lindsay 2010). It has been also proven that gain and loss of gene function are linked to *S. aureus* host adaptation (difference in isolates from cows vs. isolates from birds and pigs and also correlation of resistance to different antimicrobials with host specificity was investigated (Bacigalupe et al. 2019). Interestingly the authors (Richardson et al. 2018) concluded that they identify humans as a major reservoir for the spread of *S. aureus* to livestock—they commented that it reflects the role of humans in domestication of animals, and subsequent opportunities for cross-species switch/transmission events being in line with previous study/analysis using MLST (Weinert et al. 2012). Human population epidemic clones CC97 11 of *S. aureus* were identified to be bovine origin (Spoor et al. 2013). Discussion who was the original host for famous CC398, founded mostly in pigs or veal calves as well as in human is ongoing, as a report from the Netherlands identified 15% human cases of all LA-MRSA CC398 were isolated from people having no direct contact to pigs or veal calves (Lekkerkerk et al. 2015) and also another work (Price et al. 2012) bring quite interesting insight in this issue hypothesise that origin of LA-MRSA CC398 is from a human MSSA strain that acquired tetracycline and methicillin resistance (based on the collection of MRSA and MSSA CC398 isolates from animals and humans—19 different countries/4 continents/ using WGS. Most ancestral clade upon phylogenetic analysis was identified as human MSSA, while the LA-MRSA was composed of the most derived lineages, with three different SCCmec types, IV, V, and VII-like. LA-MRSA, moreover, largely missing the phages encoding human innate immune modulators detected human strains from the basal clades. Results were therefore considered by the authors as giving the evidence that origin of LA-MRSA CC398 was from humans as MSSA and that the jump from humans to livestock was associated with a loss of phage-carried human virulence genes. The evolution, epidemiology as well as molecular characterisation is also comprehensively summarised in the very recent work by Lakhundi and Zhang (2018).

Common mechanisms of resistance to antimicrobials in *S. aureus* as well in coagulase-negative staphylococci (CoNS) can be structured, according to the current level of knowledge, in five groups:

1. Production of enzymes that inactivate or destroy the antimicrobial (e.g. beta-lactamases affecting beta-lactams, acetyl-, adenyl, phosphor-transferases—affecting aminoglycosides)
2. Reduction of the bacterial cell wall permeability limiting the antimicrobial access into the cell (e.g. modification of the cell wall: d-alanylation of the cell wall teichoic acids—affecting efficacy of daptomycin) (Bayer et al. 2013).
3. Active elimination of the antimicrobial from the bacterial cell or the target site (e.g. efflux pumps—amphenicols)
4. Target site:
   (a) Protection (e.g. ribosome protection affecting macrolides efficacy)
   (b) Modification (e.g. rRNA methylases—amphenicols, lincosamides, pleuromutilins)
   (c) Replacement (e.g. alternatives to Penicillin-Binding Proteins (PBPs) affecting the efficacy of beta-lactams; mupirocin-insensitive isoleucyl-tRNA synthase)
5. Development of alternative metabolic pathways to those inhibited by the antimicrobial (e.g. affecting the efficacy of sulphonamide/trimethoprim (S/T) combination by several strategies as amino acid changes in the dihydropteroate acid synthase (DHPS) and/or dihydrofolate reductase (DHFR), acquisition of external genes encoding DHPS or DHFR that are less sensitive to inhibition by S/T (target bypass), a "clever" bypass strategy is the overproduction of DHFR or DHPS through mutations in the promoter region of the DNA encoding these enzymes (Munita and Arias 2016).

Genes located in different parts of genome of staphylococci (mainly on plasmids and transposons) mediate resistance to many classes of antimicrobial agents approved for use in animals, such as penicillins, cephalosporins, tetracyclines, macrolides, lincosamides, amphenicols, aminoglycosides, aminocyclitols, pleuromutilins, and diaminopyrimidines. In addition, numerous mutations have been identified that confer resistance to specific antimicrobial agents, such as ansamycins and fluoroquinolones. The gene products of some of these resistance genes confer resistance to only specific members of a class of antimicrobial agents, whereas others confer resistance to the entire class or even to members of different classes of antimicrobial agents (Schwarz et al. 2018). Moreover mobile genetic elements (MGE), many times containing several genes, proven to harbouring resistance, linked to different mechanisms, to antimicrobials, heavy metals, disinfectants—what is of importance from the co-selection of resistance as well as persistence of resistance that not necessarily need the use of certain (group) of antimicrobial. MGE are responsible for the exchange of resistance genes among members of the same and/or different staphylococcal species, but also between staphylococci and other Gram-positive bacteria (Schwarz et al. 2018).

Resistance to beta-lactams in staphylococci is mainly due to the enzymatic inactivation (*bla*Z- or bla_{ARL}- encoded beta-lactamases hydrolysing beta-lactam ring) and target site replacement (products encoded by the genes *mecA*, *mecB*, and *mecC*). Beta-lactamases can be encoded both genes located on plasmids and chromosomes, *bla*Z genes were detected in *S. aureus*, *S. hyicus* as well as CoaNS, of different animal origin, conferring resistance to penicillins (except isoxazolyl-penicillins). The gene *bla*Z was reported to be a part of a transposable element located on a large plasmid, which often carries additional antimicrobial resistance genes, e.g. those conferring resistance to erythromycin, fusidic acid, and gentamicin (Lowy 2003). The plasmid may also carry genes encoding resistance to disinfectants (quaternary ammonium compounds), dyes (acriflavine and ethidium bromide) or heavy metals (cadmium, lead, and mercury) as well as several virulence-related genes encoding, e.g. exfoliatins (*eta* and *etb*), leukotoxins (*luk*PV, *luk*ED, and *luk*M), hemolysins (*hla*, *hlb*, and *hld*) and others as indicated by Pantosti et al. (2007) and Jarraud et al. (2002). Operon bla_{ARL} is located on the chromosome. Those staphylococci carrying genes *mec*, that coding for alternative PBPs causing significantly reduced affinity to practically all beta-lactams (known exceptions, e.g. ceftobiprole, ceftaroline, approved for human use only) and therefore should be interpreted as having resistance to all beta-lactam antibiotics, including cephalosporins, cefamycins, and carbapenems. Genes *mec*A and *mec*C are located in "staphylococcal chromosome cassette (SCC*mec*)". Currently is known 13 major groups, which can be found by *SCCmecFinder* (tool able to identify all SCCmec element types, designated I to XIII, with subtyping of SCCmec types IV (2B) and V (5C2) (Kaya et al. 2018). Gene *mec*A was reported in *S. aureus* and *S. hyicus* (including food-producing species) as well as in *S. pseudintermedius* (mainly dogs and cats). The presence of gene should be accompanied with phenotypic detection of oxacillin resistance in *S. sciuri* and *S. vitulinus*, because in those species *mec*A alleles not conferring beta-lactam resistance are present). Gene *mec*C originally found in humans and cattle was, later on determined in many other domestic and wildlife animal species. Interestingly, Schwarz et al. (2018) also published the comment related to reduced oxacillin susceptibility in equine isolates of *S. aureus* not harbouring either *mec*A or *mec*C genes that can be either due to overproduction of beta-lactamase, mutation in PBPs (Ba et al. 2014), or decreased expression of *fem*A and *fem*B (factors essential for methicillin resistance) that can lead to heterogenous profile of oxacillin resistance, with subpopulation of staphylococci which are highly resistant to oxacillin (Lindsay 2008; Haag et al. 2019). It should be noted that of high importance and high threat from the perspective of successful treatment are strains of MRSA and MRSP harbouring genes for resistance to the broad portfolio of antimicrobials from different pharmacological groups.

Resistance to tetracyclines in animal staphylococci is most frequently associated with genes *tet*(K) and *tet*(L), often located on plasmids, encoding membrane-associated efflux proteins of the major facilitator superfamily and gene *tet*(M), commonly found on chromosome, but also located on conjugative transposon, encoding ribosome protective protein. Plasmids *tet*(K) rarely carry resistance to other antimicrobials, diversely from *tet*(L) carrying plasmids harbouring one or more additional resistance genes. Interestingly, *tet*(M) origin of conjugative transposons is of enterococci origin (*Tn 916* and *Tn1545*). Recently, a novel small *tet*(T)-*tet*(L)-*aad*D-carrying plasmid from MRSA and MSSA ST9 isolates of swine origin has been detected by Jiang et al. (2019), also *tet*(S) gene (encoding ribosome protective protein) was confirmed in *S. aureus* (carrying *mec*B, isolated from human) was proven (Schwarz et al. 2018).

Resistance to folate inhibitors, namely in trimethoprim, including genetic determinants was described in *S. aureus* (MMSA as well as

MRSA) *S. hyicus*, CNS, and *S. pseudintermedius* (MRSP). Among the genes *dfr*A (also reported as *dfr*S1), *dfr*D, *dfr*G, and *dfr*K are responsible for encoding dihydrofolate reductase enzyme the most frequently determined are dfrK gene (linked with *tet*(L) in MRSA ST398 from pigs; in MSSA CC398 located on transposon *Tn559*, except in pigs further found also in MRSA from cattle, horses, chickens, turkeys, but also in porcine *S. hyicus* isolates and CoNS from cattle (Argudin et al. 2017; Schwarz et al. 2018).

Resistance to macrolides, lincosamides, and streptogramins (M-L-S), where the *erm* genes are of the great importance (speaking about MLS_B resistance) can be in the genetic material of staphylococci present by one or more erm genes (A, B, C, F, T, and 33, 43, 44, 45, and 48), all these genes encoding for methylase (modification of 23S rRNA disability for any of M-L-S to bind on ribosome and block proteosynthesis). In MRSA and MRSP transposon, *Tn554* was frequently included in SCCmec type II elements. In *S. hyicus* (pig), and CoaNS in poultry and cattle genes *erm*(A) were found. LA-MRSA as well as in CoaNS (pigs, cattle, chicken, turkey or pigs, cattle, chicken, ducks, horses, respectively) own the gene *erm*(B). This gene was also detected in *S. hyicus* and *S. pseudintermedius*. Further genes *erm* (C) has been identified in a broad spectrum of different *Staphylococcus* species, *erm*(F)—mainly from CoaNS and *erm* (T) for the first time found on the large plasmid of MRSA CC398 (pigs), but later on found also in cattle, chickens, turkeys as well as in bovine CoaNS (Schwarz et al. 2018; Lindsay 2008; Feßler et al. 2018; Haag et al. 2019).

Furthermore should be commented on the existence of genes causing resistance to macrolides only: *mph*(C) encoding for macrolide phosphotransferase and *ere*(A) encoding for macrolide esterase. Resistance to lincosamides only is located on plasmids of small size carrying the genes *lnu*(A) and *lnu*(B) (lincosamide nucleotidyltransferase encoded). For specific staphylococci (*S. sciuri*, *S. simulans*, and *S. warneri*) also plasmid-borne gene *lsa*(B) responsible for decreased susceptibility to lincosamides due to the expression of ABC-F protein was described. Gene *lsa*(E) causes resistance to lincosamides, pleuromutilins, and streptogramins. Gene *msr*(A) was described to confer resistance to macrolides and streptogramin B (Golkar et al. 2018).

Resistance to aminoglycosides (in many *Staphylococcus* species) is conferred by widely spread acetyltransferase/phosphotransferase bifunctional gene located on transposon: *aac*A-*aph*D (resistance to gentamicin, kanamycin, tobramycin, and when overexpressed also to amikacin); furthermore *aad*D gene (adenyltrasferase encoding) cause resistance to several aminoglycosides—kanamycin, neomycin, and tobramycin; gene *aph*A3 (phosphotransferase encoding) is known as causing resistance to kanamycin, neomycin, and amikacin; gene *aad*E is responsible for resistance to streptomycin and is interestingly also a part of the multiresistance gene clusters of streptococcal origin identified in MRSA ST9 and CC398 in pigs and also porcine *S. hyicus*. Gene *aad* (6) also classified as *str* (adenyltransferase encoding and mediating streptomycin resistance is also widely spread among animal origin staphylococci (Wendlandt et al. 2015; Schwarz et al. 2018; Lindsay 2008; Feßler et al. 2018; Haag et al. 2019).

Resistance to (aminoglycosides related) aminocyclitols have been identified to be caused by genes encoding for different adenylyltransferases classified as *spc* (linked also with *erm(A)* gene on *Tn 554* transposon, in MRSA and also CoNS, *S. hyicus*), *spw* and *spd* (spectinomycin) and acetyltransferase *apm*A (apramycin resistance and decreased susceptibility to gentamicin). Plasmid carrying *apm*A carry multiresistance (*ica*-like cluster—AMR to antibiotics as well as copper and cadmium) (Wendlandt et al. 2015).

Resistance to fluoroquinolones is mainly caused by mutations topoisomerase encoding genes *gyr*A, *gyr*B, *grl*A, and *grl*B. Experimentally was also demonstrated that the plasmid-mediated multidrug efflux pump QacB variant QacBIII confers the capability for fluoroquinolone efflux in *S. aureus* (Foster 2017; Schwarz et al. 2018).

Resistance pleuromutilins, as well as lincosamides and streptogramins in staphylococci, is also conferred by a family of proteins called ABC-F that have two ATP binding cassette (ABC) domains: where lincosamide, streptogramin A (LSA) and lincosamide, streptogramin A, pleuromutilin (LS_AP) with determinants Vga, Lsa, and Sal were described (Foster 2017). Of importance is gene *lsa* (E) due to resistance conferring for pleuromutilins, lincosamides, and also streptogramin, but multidrug resistance is also an issue mainly due to the resistance linked to all antimicrobials binding to 50S subunit of ribosome ($PhLOPS_A$ = (phenicol, lincosamide, oxazolidinone, pleuromutilin, and streptogramin A) with the mechanism of action as Cfr protein (rRNA methyltransferase) influencing the binding activity of above-listed antimicrobials (Kehrenberg et al. 2005).

Resistance to ansamycins in *S. aureus* is conferred by a mutation in the *rpo*B gene (Aubry-Damon et al. 1998).

Resistance to amphenicols is due to enzymatic inactivation (acetyltransferases (R to non-fluorinated amphenicols): cat_{pC221} cat_{pC223} cat_{pC194}), active efflux (major facilitator superfamily; gene *fex*A located either on plasmid or in the chromosomal DNA), target site modification rRNA methylase Cfr or ribosome protection (ABC-F protein). Also, OptrA is functional in the case of phenicol transferable resistance, well described in enterococci and CoaNS. OptrA was likely selected due to extensive usage of florfenicol in intensive animal farming in China (Foster 2017).

1.10 Streptococci

Streptococcus belongs to the Gram-positive facultative anaerobic cocci. For cells of streptococci are typical chain arrangement. According to the in vitro culturing on blood-containing media, they can be distinguished to the beta-haemolytic (e.g. S. *agalactiae*, *S. equi*, *S. canis*, and *S. pyogenes*), or alpha-haemolytic (due to the lysis of erythrocytes, followed by oxidation of hemin with greenish creating effect on blood agar, with most known and occurring *S. dysgalactiae* and *S. pneumoniae*. Classification used for many years distinguish streptococci to the groups A to W, based on antigenic reaction (introduced by Rebecca Lancefield, 1930) and yet been used widely in practice. In the table below, please find some common streptococci classified according to this "Lancefield groups" with relevance in

Species	Lancefield	Hemolyse	Animal	Disease
S. pyogenes	A	β	Foal	Lymphadenitis
S. agalactiae	B	β (α; γ)	Cattle, goat, sheep	Mastitis
S. dysgalactiae subsp. *dysgalactiae*	C	α (β; γ)	Cattle Lamb	Mastitis Polyarthritis
S. dysgalactiae subsp. *equisimilis*	C (A,G, L)	β	Horse	Abscesses, endometritis, abortion Mastitis
			Pig, cattle	Various suppurative
S. equi subsp. *equi*	C	β	Horse	Strangle Genitourinary tract infections Purpura haemorhagic
S. equi subsp. zooepidemicus	C	β	Horse	Joint diseases, mastitis, abortion Secondary pneumonia
			Cattle	Metritis and mastitis
			Pig	Septicaemia, arthritis
			Poultry	Septicaemia, vegetative endocarditis
			Lamb	Pericarditis, pneumonia
Enterococcus	D	α (β; γ)		See further subchapter
S. suis	R	β or non-hem	Pig	Meningitis, arthritis, pneumonia, endocarditis, septicaemia
S. porcinus	E (P,U,V)	β	Pig	Abscesses, lymphadenitis
S. uberis	undefined	β	Cows Horses	Mastitis

veterinary/human medicine. Another classification can be done by phylogenetic relationships among streptococci species based on the analysis of 16S rRNA gene sequences (e.g. Facklam 2002).

Streptococci can be commensals, pathogens, and opportunistic pathogens for humans and animals. Some of them have zoonotic potential—as *S. suis*, causing specific diseases in humans (in-contact or food-transmitted), S. *agalactiae* is reported to be rarely zoonotic, but pathogenesis is known both in humans (mother to new-borne baby transmission) and in animals—but independently. Of specific importance are streptococci associated with mastitis (recently mostly *S. uberis*, *S. agalactiae*, *S. dysgalactiae,* and *S. parauberis*).

Transposons or integrative and conjugative elements (ICE) can disseminate resistance genes among streptococci. On ICEs gens as *tet*, *erm*, *ant6*, and *aphA*, known to confer multiple resistances to tetracycline, erythromycin, and aminoglycosides have been described as well as genes predicted to confer resistance to spectinomycin and lincomycin, *ant9* and *lnuB* (Campisi et al. 2016). Interspecies gene exchanges have been described, with a most recent hypothesis, supported also by exact data, that *S. suis* can be the source of genes of resistance for *Streptococcus agalactiae*, *Streptococcus pneumoniae*, *Streptococcus pyogenes*, and different serotypes of *S. suis*. Existence of different groups of MGEs, including *Tn*5252, ICE*Sp*1108, and *Tn*GBS2 groups ICEs, Φm46.1 group prophage, ICE_ICE, and ICE_prophage tandem MGEs have been analysed prior this hypothesise (Huang et al. 2016a). Antibiotic resistance in most of the streptococci is predominantly caused by resistance gene acquisition for majority of antimicrobials, with known exception as for fluoroquinolones, which instead arise mostly by mutations in the *gyr*A and *par*C genes and penicillins (Metcalf et al. 2017). Close genetic linkage has been also described between *tet*M and the AlpST-1 virulence gene (encoding adhesin), which give the clones containing colonisation advantage (Flores et al. 2015). More information is needed for the role of mobile genetic elements (MGEs), including ICEs, plasmids, transposons, phages, and prophages. Especially ICEs are within the scope of many recent studies in *S. agalactiae, S. suis, S. dysgalactiae*, and also other streptococci mostly of importance from human medicine perspective (Huang et al. 2016a; Zhang et al. 2018b; Du et al. 2019).

Other reasons underlying ineffective responses to antimicrobial treatment of diseases caused by streptococci, namely *S. suis* disease might include biofilm formation and the production of persistent cells (Seitz et al. 2016).

Different pharmacological groups of antimicrobials are used to treat animals with streptococcus aetiology of disease. Of importance is also pharmaceutical form (injectable and oral—with systemic concentrations, as well as intramammary/intrauterine with mostly topical effect).

Penicillins (including aminopenicillins, alone or in combination with aminoglycosides) can be considered as the drugs of first choice not only in human medicine, but mostly also in veterinary sector. Macrolides and lincosamides, as well as tetracyclines, are also possible choices. From the critically important in human medicine fluoroquinolones are also effective, but should but used only as last choice, once the previous mentioned are not effective.

As for resistance to penicillins, there can be noted rarity of decreased susceptibility to beta-lactams in group B streptococci, with resistance-conferring alleles arising through point mutation rather than inter-species transformation events (Metcalf et al. 2017). Resistance is due to the PBP2 alleles and due to the alterations of the penicillin-binding proteins. Resistance to beta-lactams gained via acquiring exogenous genes is unique in streptococci, but mutation of PBPs is possible, especially those from class 2, PBP2B, or PBP2X (described mostly in human isolates of *S. pneumoniae*, but also in Penicillin resistance due to modifications of PBPs were suggested in *S. suis* (Cain et al. 1995) and in *S. uberis* (Haenni et al. 2010). As for level of resistance, most of the strains were considered as intermediary resistant. The shift in antimicrobial susceptibility towards the decreased susceptibility, or even resistance is significantly less dynamic and speedy in streptococci comparing to Gram-negative bacteria, but further investigations are needed, especially for beta-lactams (Haenni et al. 2018). Latest studies with clinical isolates of *S. dysgalactiae* described 3 resistance genes (bla_{TEM}, bla_{IMP}, and bla_{SPM-1}) encoding for beta-lactams, but with different levels of genes expression. Hypothesis of existence of selection pressure caused by the use of cephalosporins (namely ceftiofur) was mentioned by the authors of this study (Zhang et al. 2018a, b). Despite the resistance mechanisms described for beta-lactams, including cephalosporins, penicillins (including aminopenicillins for selected cases) still remain the drug of choice for infections caused by streptococci both in human and in veterinary medicine.

Resistance to macrolides, lincosamides, and streptogramins is mostly due to the mechanisms as:

Target modification (ribosomal subunit modification, less common, probably due to the need of mutations in all the operon copies encoding ribosomal subunit).

Target protection (one of the most frequent and with broad clinical impact, where the *erm* genes (>40 variants) encoding different Erm methylases, constitutive or inducible; confer resistance to macrolides, lincosamides, streptogramins, MLS), despite the prevalence of this resistance differ across the European region, this is of importance for mastitis-causing pathogens (*S. dysgalactiae*, *S. uberis*, and *S. agalactiae*; *erm*(A) and *erm*(B) genes) as well as for pathogens of porcine origin (predominantly *erm*B genes) (Haenni et al. 2018).

Efflux (of the greatest importance are family Mef and family Msr. Mef family genes confer resistance to 14- and 15-membered, but not to 16-membered, macrolides, lincosamides, and streptogramins B, well-described are genes *mef* (B) and *mef*(E), having medium homology to *mef* (A), *mef*(I), and further *mef*(O) with high homology to *mef*(A) (Dinos 2017). Mef genes are frequently harboured on so-called MEGA (macrolide efflux genetic assembly), transposons, and composite mobile genetic elements (Haenni et al. 2018). The Msr family confers resistance to 14- and 15-membered macrolides and at low level also to ketolides (Canton et al. 2005). The understanding of the exact mechanism of resistance due the Msr is subject to scrutiny as for a long time, the Msr family was thought acting as efflux pumps (Fyfe et al. 2016). Now, it seems that chasing the bound macrolide from the ribosome is rather the exact mechanism of action (Sharkey et al. 2016; Su et al. 2018), who show that MsrE protein can bind to a stalled ribosome in which a peptidyl-tRNA is in the P-site. But the latest works speak also about M resistance phenotype to macrolides in streptococci due to an efflux transport system of the ATP-binding cassette (ABC) superfamily, where plays the role gene *mef*(A) encoding the transmembrane channel, and gene *msr*(D) encoding the two ATP-binding domains (Iannelli et al. 2018). It should be noted that among efflux-based mechanism of resistance belongs also those encoded by *lsa* family being of clinical importance due to possible cross resistance of lincosamides and pleuromutilins and streptogramin B.

Last mechanism of macrolide resistance to be mentioned is ***drug modification*** (Haenni et al. 2018; Golkar et al. 2018), where three classes of enzymes have been described to play major role: macrolide adenylases, phosphotransferases, and esterases. Firstly described *lin*B (encoding adenylase, inactivating lincosamides) newly renamed to *lnu*B. Further very interesting suggestion explaining multiresistance and resistance to spectinomycin in *Streptococcus suis* is based on investigation of the novel integrative and conjugative element *spw*_like-*aad*E-*lnu*(B)-*lsa* (E)cluster and a cadmium-resistance operon, where gene for resistance *lnu*B to macrolides is also involved (Huang et al. 2016a, b). Another, quite surprising result (considering that in most phenotypic tests clindamycin substrate is used to detect resistance to lincosamides) has been gained for *S. agalactiae* with gene *lnu*C (conferring resistance to licomycin, but remaining susceptible to clindamycin (Achard et al. 2005). Recently has been described also *lnu*B multidrug resistance gene cluster possibly acting as composite transposon flanked by IS1216, that can have an impact on the spread of resistance among *S. agalactiae* (Zhou et al. 2019).

Resistance to tetracyclines, mostly due to the harbouring *tet* genes (in streptococci *tet*(K), *tet* (L), *tet*(M), *tet*(O), *tet*(Q), and *tet* (T)). Interestingly *tet*(M) and *erm*(B) were described to be associated in multiresistant strains of streptococci. Mechanism of resistance caused by above-mentioned genes described in streptococci is either due to the efflux or due to ribosome protective proteins. In mastitis-causing streptococci *tet* genes are common (e.g. *S. dysgalactiae* mostly genes *tet(M), tet(L), and tet(O)*, has been described (Zhang et al. 2018a, b) and Haenni et al. (2018) mentioned combinations of *tet*(M)/*tet*(O), *tet*(M)/*tet*(K), or *tet*(O)/*tet*(K), but mostly phenotypic results are available for the European region for the recent period, and prevalence of resistance significantly differ (for more details refer to Haenni et al. 2018). As for the *S. suis* most commonly described are *tet*(M) and *tet*(O), *tet*(M) also described to be harboured by transposons (Haenni et al. 2018; Zhang et al. 2018).

Resistance to fluoroquinolones has rarely been reported and has broad clinical impact in isolates of streptococci coming from clinically diseased animals in Europe. Due to the recent works indicating strains of *S. dysgalactiae* harbouring resistance genes *gyr*(A) with impact on susceptibility (Zhang et al. 2018a, b) as well as genes *gyr* (A) and *par*(C) and importance of ICEs in resistance transfer in/from *S. suis* (Du et al. 2019) further work is of need within this area.

As for other resistance in streptococci, five resistance genes (*aphA-1*, *aphA-2*, *aphA-3*, and *aad-6* as well as *aadA1*/*aadA2* combination) encoding for aminoglycoside resistance was described in *S. dysgalactiae* recently, as well as 1 gene (*rrs*) encoding for streptomycin resistance and 2 genes (*sul1*, *sul2*) encoding for sulphonamide resistance (Zhang et al. 2018b).

1.11 *Trueperella pyogenes*

Gram-positive, pleomorphic, non-spore forming, non-motile, non-capsulated, facultative anaerobic rod, which is characterised by a fermentative metabolism and strong proteolytic activity. The species has been reclassified several times—*Arcanobacterium pyogenes* (formerly *Actinomyces pyogenes* (formerly *Corynebacterium pyogenes*).

Its growth requirements are not excessive, but media enriched with blood or serum need to be used for the culture, newly methods as loop-mediated isothermal amplification (LAMP) assay, matrix-assisted laser desorption/ionisation time-of-flight (MALDI-TOF) mass spectrometry, Fourier transform infrared (FT-IR) spectroscopy, or 16S rRNA gene sequencing may be used in the diagnostics of *T. pyogenes*. Improvement in methodology brings also more data on the diseases, where *T. pyogenes* can be either primarily causative pathogen or associated one (please refer to Table 3 according to Rzewuska et al. 2019).

Infections associated with *T. pyogenes* occur in both domestic and wild animals worldwide, but are rare in humans, infection can be primarily caused by *Trueperella*, but more frequently this species is involved in polymicrobial diseases, such as mastitis, uterine infections, interdigital phlegmon, or liver abscesses (please refer also to Table 2). This bacterium may be recovered from a mixed infection of various bacterial species, but especially frequently with Gram-negative anaerobes, such as *Fusobacterium necrophorum (FN)*, *Bacteroides* spp., or *Peptoniphilus* (formerly Peptostreptococcus) *indolicus*. In these cases, purulent and necrotic lesions are usually observed, leading to systemic signs and resulting in animal death (A particularly strong synergistic interaction occurs between *T. pyogenes* (creating environmental conditions suitable for FN) and *F. necrophorum*, which produce leukotoxin, protecting TP against phagocytosis, because of its ability to lysis leukocytes or to induce their apoptosis, depending on its concentration (Tadepalli et al. 2009). Other bacteria, especially *Escherichia coli*, are also often associated with *T. pyogenes* coinfections, mostly postpartum uterine infections.

Only a few virulence factors in *T. pyogenes* are recognised to date and significance in the pathogenicity/host specificity many of them have not been fully elucidated yet. They include pyolysin

Table 2 Example of diseases of cattle, swine, poultry, and horses caused by/associated with *T. pyogenes* (summarised from Rzewuska et al. (2019))

Species	Frequently detected	Sporadic/rare
Cattle	Infections of the reproductive tract (metritis, endometritis) and the mammary gland, as well as pneumonia and liver abscessation	Pneumonia, encephalitis, pyelonephritis and kidney abscesses, lymphadenitis, endocarditis, abscesses of various localisation, septicaemia, and abortion
Swine	Pneumonia, pleuritis, endocarditis, osteoarthritis, polyarthritis, mastitis, reproductive tract infections, and septicaemia Abscesses—superficial, muscular, or located in different organs—occur frequently, and may lead to the development of systemic purulent infection and inflammation of lungs, liver, kidneys, muscles, bones, joints, or other tissues	
Poultry		Very rare clinical lameness and osteomyelitis in turkeys
Horses		Sporadic/single cases of metritis, orchitis, mastitis, septicaemia, umbilical infection in foals, abscesses, and wound infection

(PLO), the only known toxin of this bacterium (belongs to the family of cholesterol-dependent cytolysins active through transmembrane pore formation); some adhesive factors, such as fimbriae neuraminidases and extracellular matrix-binding proteins; different exoenzymes, such as serine proteases with gelatinase and caseinase activity, or DNAses. Strains of *T. pyogenes* have also the ability to invade host cells and to create biofilm (Jost and Billington 2005).

As for antimicrobial susceptibility and resistance, it should be noted that previously very different methods [broth dilution, agar dilution, disc diffusion, Calgary Biofilm Device, E-test—as summarised in Feßler and Schwarz (2017)] were utilised for AST in *T. pyogenes*, what makes difficult any comparisons and trends evaluation during the time course. A specific method for AST in *T. pyogenes* has been developed and published by CLSI (2017) using broth dilution with CAMHB (Cation Adjusted Mueller Hinton Broth) + 2.5–5% (v/v) Lysed Horse Blood, using direct colony suspension, and incubation in 5% CO_2. Based on the MIC distributions, breakpoints only for category "susceptible" was proposed for penicillin, ampicillin, erythromycin, and sulfamethoxazole–trimethoprim.

Especially in metritis associated *T. pyogenes* information on resistance genes located on Class 1 integrons are available indicating that aminoglycoside resistance such as *aad*A1, *aad*A2, *aad*A5, *aad*A24, *aad*B, and *aaC*C genes, beta-lactamase $bla_{\text{PSE-1}}$ genes, chloramphenicol *cml*A6, and trimethoprim resistance genes such as *dfr*B2a exists (Liu et al. 2009; Zhao et al. 2011). As within the samples tested for resistance of trimethoprim in 28.6% of resistant strains only was detected *dfr*B2a gene, it can be expected that another gene/mechanism of resistance is responsible for resistance to trimethoprim (Feßler and Schwarz 2017). Isolates from metritis in cattle have been characterised also within the study performed by Ashrafi Tamai et al. (2018), with 65 isolates (coming from Iran), where in 48 either single or MDR was identified, with resistance mostly against trimethoprim–sulfamethoxazole, macrolides, and streptomycin. In 30 isolates *tet*(W) was confirmed and macrolide resistance genes *erm*(B) and *erm* (X) were detected 18 and 25 isolates, respectively. Virulence genes *fim*A and *plo* were identified in all tested isolates.

Inducible macrolide/lincosamide resistance was detected in *T. pyogenes*, but strains also exist with non-inducible type of resistance [e.g. tylosin and clindamycin (Jost et al. 2003)]. *T. pyogenes* can also harbour genes *erm*(X) or transposon located genes *erm*(B), which is of importance for the co-resistance to macrolides–

lincosamides and streptogramin B. As shown by Zastempowska and Lassa (2012), isolates non-susceptible to erythromycin simultaneously exhibited increased MIC of pirlimycin. Also, ribosomal mutations were proven to be responsible for macrolide resistance in *T. pyogenes* (Jost et al. 2004).

Tetracycline resistance is due to mobile *tet* (W) genes that code ribosome protective proteins (Billington and Jost 2006), Moreover *tet*(Z) gene was detected by Alešík (2006). And efflux protein associated mechanism is encoded by the gene *tet*A(33) harboured on the same plasmid as *erm* (X) (Jost et al. 2003, 2004).

Due to the lack of available interpretive criteria is hard to interpret intermediate susceptibility/resistance to fluoroquinolones, despite the fact that some data are available, especially for ciprofloxacin and enrofloxacin (Liu et al. 2009).

There can be used and also was tested in some studies rifampicin (Alkasir et al. 2016) and novobiocin (Watts et al. 1995). In the case of rifampicin, resistance was detected only in two from a total of 50 bovine mastitis isolates, in novobiocin low MICs (0.25 mg/L) were detected. Moreover, the combination of penicillin and novobiocin in 55 mastitis isolates reported by Zastempowska and Lassa (2012) shown MICs within the range 0.5–1 mg/L. Surprisingly resistance to zinc-bacitracin was reported in isolates from metritis and endometritis in bovine isolates with MICs $\geq$32 mg/L (Liu et al. 2009).

2 Specific Consideration for Pathogens and Antimicrobials with Importance in Pigs

2.1 *Acinetobacter* spp.

Pigs may harbour *A. baumannii*. Healthy pigs sampled at slaughterhouses in Scotland were found to contain genetically related strains as determined by PFGE and *bla*$_{OXA-51}$-like gene sequencing (Hamouda et al. 2011). The pig isolates had different PFGE patterns from those in humans and were grouped in three different clusters (A, B, and C) with genetic similarity ranging between 82% and 90%. One *A. baumannii* strain isolated in China from a lung sample of a pig with pneumonia and sepsis was found to harbour the carbapenemase gene *bla*$_{NDM-1}$ on a plasmid (Beceiro et al. 2011). In Lebanon, a *bla*$_{OXA-23}$-producing *A. baumannii* ST491 was recovered from the faeces of a healthy pig (Moffatt et al. 2010).

2.2 *Actinobacillus pleuropneumoniae*

Actinobacillus pleuropneumoniae is a small, Gram-negative, encapsulated rod with typical coccobacillary morphology. To date, 18 serovars have been recognised, 15 of them mainly on the basis of the antigenic properties of capsular polysaccharides and the O-polysaccharide (O-PS)3, 5, 24, 25, and another one, serovar 16, was proposed based on serology alone, serovars 17 (previously NT) and 18 (previously "K2:O7") was proposed based on serological and genomic results (Sárközi et al. 2015; Bossé et al. 2018). It is useful to classify isolates of *A. pleuropneumoniae* not only for epidemiological purposes, but also to inform vaccine development (including autogenous vaccines). Cytoplasmic glycoengineering of Apx toxin fragments in the development of *Actinobacillus pleuropneumoniae* glycoconjugate vaccines seems to be a new promising approach currently investigated (Passmore et al. 2019). Please refer to the overview of Apx toxins.

A variety of virulence factors have been described for *A. pleuropneumoniae*, and can be allocated to the categories of adhesion, acquisition of nutrients, induction of lung lesions, evasion of immune system, and persistence (Bossé et al. 2002; Chiers et al. 2010). Iron metabolism is of high importance for the pathogen to survive and multiply in the host, more than 50 genes are involved in iron uptake and metabolism (Xu et al. 2008), some of which are differentially expressed during infection (Deslandes et al. 2010; Klitgaard et al. 2012). Recent work described catecholamine binding to facilitates iron uptake, although iron availability is highly decreased during acute

Table 3 Overview of Apx toxins (modified according to Sassu et al. 2018)

Group	Haemolytic	Cytotoxic	Produced by serovars	Note
Apx I	Strongly	Strongly	1, 5a, 5b, 9, 10, 11, 14, 16	
Apx II	Weakly	Moderately	All serovars except for 10 and 14	
Apx III	Non-haemolytic	Strongly	Serovars 2, 3, 4, 6, 8, and 15	
ApX IV			All serovars in vivo	Diagnostic use[a] Some non-producers reported[b]

[a]Dreyfus et al. (2004)
[b]Tegetmeyer et al. (2008)

infection as a physiological acute reaction during inflammation (Humann-Ziehank et al. 2014; Li et al. 2015).

Of major importance, in regard to virulence, are the Apx toxins, with different degrees of cytotoxicity, haemolytic activity, and distribution among serovars (see Table 3).

Recent study results confirm that *A. pleuropneumoniae* is capable of integrating into biofilms formed by environmental bacteria (including *E. coli*), indicative of a possible survival strategy in the environment and a mechanism for disease dispersion (Ramírez-Castillo et al. 2018).

A wide range of antimicrobials is effective against the pathogen, although an increase of resistance to non-critical antimicrobials such as tetracyclines, penicillins, and trimethoprim/sulphonamides have been observed (Vanni et al. 2012; Bossé et al. 2017). In a Spanish retrospective study from 1994 to 2009, a high or an increasing trend for resistance against beta-lactam antibiotics, tetracyclines, and tilmicosin was recorded, while most isolates were susceptible to phenicols, fluoroquinolones, and ceftiofur (Vanni et al. 2012). In a recent study, only 33% of UK App isolates were negative for resistance genes, while 57% of the isolates were resistant (as adjudged by MICs) to tetracycline (with confirmed genes *tet*(B) or *tet*(H), 48% to sulfisoxazole (gene *sul2*), 20% to ampicillin (gene $bla_{\text{ROB-1}}$), 17% to trimethoprim (*dfr*A14), and 6% to enrofloxacin (with GyrAS83F mutation). In addition to presence on plasmid(s), the *tet*(B) gene was also found chromosomally either as part of a 56-kb integrative conjugative element (ICEApl1) in 21, or as part of a Tn7 insertion in 15 isolates (Bossé et al. 2017). Recently, plasmids conferring resistance to florfenicol and chloramphenicol were isolated from clinical isolates from Greece and Brazil (Bossé et al. 2015; da Silva et al. 2017), and enrofloxacin resistant strains have been reported (Bossé et al. 2017). It has been demonstrated that whole-genome sequencing can be used as a predictor for *Actinobacillus pleuropneumoniae* resistance to antimicrobial substances and genotype-based machine learning model can it even improve (Bossé et al. 2017; Liu et al. 2020). It is of importance that variation in levels of antimicrobial resistance of isolates even within the same herd can be high (Dayao et al. 2016), and there is not always an association between in vitro test results and success after treatment of disease.

Very recent data published by Holmer et al. (2019) discuss, except the data for *A. pleuropneumoniae* isolates from Danish pigs (where except erythromycin isolates showed full susceptibility or low levels of resistance to antimicrobial compounds tested (tetracycline, florfenicol, ampicillin, penicillin, ceftiofur, sulf/trim, tulathromycin, tilmicosin, ciprofloxacin, tiamulin, and spectinomycin). Similar observations were obtained for isolates from Poland, The Netherlands, France, and England (incl. Wales), but with notable differences, e.g. isolates from England: more resistance to tetracycline (22–37%) and trimethoprim–sulphonamide (13–46%), England and Poland isolates: higher resistance to ampicillin (2–7 and 8%, respectively) (Hendriksen et al. 2008). Despite the Czech Republic reported a high prevalence of resistance in 2011 (23%), the resistance to tetracyclines has significantly dropped down during the time course to 6.3% in 2018, where also resistance to other antimicrobials (MICs tested) was very low, i.e. no resistance of isolates to ceftiofur, florfenicol, enrofloxacin, very low to macrolides generally (Kucerova et al. 2011; CZ NMTP 2018), what could be also the case for countries reported in the past high prevalence of resistance to tetracyclines as Spain and Italy (Gutiérrez-Martín et al. 2006; Vanni et al. 2012), especially considering that many countries significantly reduce the use of tetracyclines in the past decade (EMA 2011; EMA 2019).

2.3 *Bordetella bronchiseptica*

Bordetella bronchiseptica is a Gram-negative bacterium closely related to *Bordetella pertussis* and *Bordetella parapertussis* with a broad host range that naturally infects a wide variety of wild, domestic, and companion animals. In swine, *B. bronchiseptica* is widespread and is an important contributor to respiratory disease. In young pigs, it is a primary cause of bronchopneumonia, and in older pigs, it contributes to secondary pneumonia *B. bronchiseptica* is currently a well-known pathogen in swine and is associated with a disease designated as atrophic rhinitis. In the pathogenesis of atrophic rhinitis, infection with *B. bronchiseptica* predisposes the animals to infections with toxigenic strains of *Pasteurella multocida*. This may lead to a severe form of the disease (Horiguchi 2012).

Bordetella species produce an exopolysaccharide, known as the *Bordetella* polysaccharide (Bps), which is encoded by the *bps*ABCD operon (39). Previous studies (Conover et al. 2010) have demonstrated that Bps is required for Bordetella biofilm formation, recent work confirmed that *bps*ABCD locus was found to enhance survival in the lower respiratory tract of swine (Nicholson et al. 2017).

A comparison of results from susceptibility testing after 20 h (upper limit of the CLSI recommended incubation times for non-fastidious bacteria) and 24 h incubation time as previously proposed (Prüller et al. 2015a) for *B. bronchiseptica* was performed by Prüller et al. (2015b). Out of 24 antimicrobial agents tested, the MIC_{50} values of porcine isolates showed slightly (1 to 2 dilution steps) higher values after 24 h incubation for ampicillin, cefquinome, cefoperazone, cefotaxime, enrofloxacin, tiamulin, and tetracycline, while MIC_{90} values of ceftiofur, cefquinome, cefotaxime, chloramphenicol, ciprofloxacin, and trimethoprim/sulfamethoxazole increased by one dilution step after 24 h of incubation. Reading the MIC values of porcine *B. bronchiseptica* after 24 h of incubation has recently been shown to be advantageous over reading the values after 16–20 h incubation due to a higher reproducibility of broth microdilution susceptibility testing results (Prüller et al. 2015a, b).

Recent bibliography brings new information on the resistance genes as well as resistance mechanisms in *B. bronchiseptica*. A gene cassette harbouring the beta-lactamase gene $bla_{\text{OXA-2}}$ was identified in this bacterial species, but despite it, beta-lactam resistance in *B. bronchiseptica* seems to be more likely due to reduced influx combined with the species-specific beta-lactamase (chromosomally located $bla_{\text{BOR-1}}$ gene. Detection of this beta-lactamase convenes with the presence of MICs indicating resistance to ampicillin (Prüller et al. 2015a, b). Intrinsic resistance or higher resistance to beta-lactams rates are both present in penicillins and cephalosporins. One of the mechanisms of resistance described in *B. bronchiseptica* is excretion of antimicrobial using efflux pumps. These efflux pumps belong to major facilitator superfamily encoded in the case of resistance to tetracyclines by genes *tet* (A), *tet*(C), and *tet*(31), to chloramphenicol by genes *s cml*B1 or to chloramphenicol and florfenicol by genes *flo*R. According to very recent publication (Borselli et al. 2019), another florfenicol resistance mechanism and gene different from *flo*R mechanism probably exists. Moreover two class B chloramphenicol acetyltransferase genes (*cat*B1 and *cat*B3) conferring resistance to non-fluorinated phenicols by enzymatic inactivation, with both genes located on gene cassettes and found in class 1 integrons and also harbouring the sulphonamide resistance gene *sul*1 were confirmed by genetic analysis. With respect to sulphonamide resistance, gene *sul*2 has also been detected—for both sulphonamide resistance genes is valid that encoding sulphonamide-insensitive dihydropteroate synthases. Genes *dfr*A1 and *dfr*B1, which code for trimethoprim-insensitive dihydrofolate reductase create a genetic background of resistance to trimethoprim. Genes of resistance to streptomycin *str*1 and *str*2 were also detected. The resistance genes were mostly located on conjugative plasmids (Prüller et al. 2015a, b; Kadlec and Schwarz 2018).

For the treatment of respiratory tract infections of swine, antimicrobial agents such as tetracyclines, doxycycline, tiamulin, amoxicillin, and the combination trimethoprim/sulfamethoxazole were frequently used, to some of those antimicrobials trends of elevated MICs were reported (Prüller et al. 2015a, b). Authors hypothesised, that since almost all *strA*-carrying *B. bronchiseptica* isolates were also positive for *sul2*, co-selection imposed by the use of sulphonamides might explain the frequent occurrence of *strA* resistance genes in our strain collection. Despite the fact that streptomycin is not used in swine respiratory disease, dihydrostreptomycin in combination with penicillin is used in some indications in sows, so the question related to the factors contributing to the occurrence of streptomycin resistance genes remains opened.

Very recent bibliography (Holmer et al. 2019) summarised results of B. *bronchiseptica* isolates from Danish pig production [No. of strains tested 90 (2004–7), 60 (2008–11), 116 (2012–15)], which were in 100% resistant to ampicillin (intrinsic resistance due to production of beta-lactamase (Prüller et al. 2015a, b), but only one isolate was detected to be resistant to florfenicol. Susceptibility trends to other antimicrobials tested remain similar. The question for the future remains, how to interpret the susceptibility/resistance for clinical purposes in the case of lack any internationally harmonised breakpoints. Also, panels of antimicrobials to be tested for MICs might need revision, e.g. consideration to involve also tulathromycine and tylosine (Holmer et al. 2019).

2.4 *Brachyspira*

Brachyspira hyodysenteriae is Gram-negative, microaerophilic anaerobic *Spirochaeta*. It is considered as the most relevant pathogen causing swine dysentery, diarrhoea with muco-hemorrhagic signs in the European continent. In pigs, also *B. pilosicoli* can cause production affecting pathology—spirochetal colitis (non-haemorrhagic, loose stools, milder disease course). From diagnostic perspective seems relevant to appropriately distinguish *B. hyodysenteriae* (e.g. by PCR) from other two strongly haemolytic species—*B. suanatina* and *B. hampsonii.* Multilocus sequence typing (MLST) is recently used to identify different major clonal groups among *B. hyodysenteriae* isolates—interestingly some clonal complexes are associated with the tiamulin susceptible strains (Archambault and Rubin 2018).

As for virulence factors, of significant importance are genes encoding: hemolysins—with particular importance for lesion production (e.g. *hlyA*—considered to be responsible for the pathogenesis of SD by causing disruptions to the colonic epithelium; *tlyA*, *tlyB*, *tlyC; BHWA1_RS02885*, *BHWA1_RS09085*, B*HWA1_RS04705*, and *BHWA1_RS02195*); inner (*clp*X); and outer membrane proteins (*bhlp16*, *bhlp17.6*, *bhlp29.7*, *bhmp39f*, and *bhmp39h*) potentially involved in adhesion and interactions with host cells as well as iron acquisition factors (*ftnA* and *bitC*); aerotolerance (*nox*); flagellin; and NADH-oxidase. Also attributes such as motility and chemotaxis are vital for the *Brachyspira* to allow them to colonise the large intestine (Joerling et al. 2018).

Study (Joerling et al. 2018) working with 116 isolates from Germany (100 farms, 1990–2016) investigated possible associations among the clonal origin, pleuromutilin susceptibility, and virulence gene profile of those *B. hyodysenteriae* isolates. The study confirmed predominance of three STs in Germany, namely ST52 (41.4%; detected in DE, but also AT, BE, ES, IT), ST8 (12.1%; detected also in BE, ES, PL, SE, UK), and the newly assigned ST112 (25.9%; known from DE and BE isolates) and detected 12 other STs to be present in the investigated samples ST113–118, ST120–123, ST131, and ST193. Due to the fact that some clones were described also in other European countries, it seems that spread of the clonal complexes is promoted by intracommunity trade with pigs. On the other hand as in some countries with significant pig production, e.g. NL, FR, data on *B. hyodysenteriae* are scarce or missing, it hardly to make unbiased overview. Based on the results obtained (e.g. 19 month persistence of the same ST on farm) authors of the study also hypothesised a long-term persistence of certain *B. hyodysenteriae* genotypes in pig either on farm or via transmission by other animals (as mices).

Mostly two pleuromutilins, tiamulin and valnemulin, as well as tylosin and lincomycin (i.e. macrolide/lincosamides group) and more limited also doxycyclin have been used across Europe for the treatment of swine dysentery caused by *B. hyodysenteriae*. Therefore selection pressure on resistance development, especially for the pleuromutilin and macrolide/lincosamides group was high, therefore in many European countries (CZ, ES, DE, SE, IT) the significant resistance was reported, especially from *B. hyodysenteriae* (Lobova et al. 2004; Hidalgo et al. 2009; Price et al. 2012; Mahu et al. 2017; De Luca et al. 2018). Single nucleotide polymorphism in rRNA gene sequences was proven to be linked to MICs increase for pleuromutilins, macrolides, lincosamides as well as doxycycline. Substitutions of adenine (A), being replaced by Guanin (G) in specific positions of the 23S rRNA gene was determined as the reason for the decreased susceptibility. Decreased susceptibility of two pleuromutilins, tiamulin, and valnemulin target is not only due to the peptidyl transferase centre (PTC) affection, including parts of the 23S rRNA, but also is associated with the ribosomal protein L2, L3, L4, and L22 (e.g. studies demonstrated a significant association between pleuromutilin susceptibility and a single nucleotide change in the ribosomal protein L3 gene at position 443—amino acid change Asn148Ser). Recently (Card et al. 2018) also new pleuromutilin resistance gene, *tva*(A)—tiamulin valnemulin antibiotic resistance, encoding a predicted ABC-F transporter was discovered by WGS techniques. Presence of *tva*(A) confers reduced pleuromutilin susceptibility not leading directly to clinical resistance but facilitating the development of higher-level resistance through mutations in genes encoding ribosome-associated functions. It is also under scrutiny, if the *tva* (A) can be mobilisable. Interestingly in *B. pilosicoli* variant of the above-mentioned gene, called *tva*(B), has been also recently described (Card et al. 2018). Based on the investigation of MSW and MPC, Card et al. also hypothesised that *tva*(A) presence can be critical for the development of clinical resistance, especially in cases where lower concentrations of tiamulin are used for metaphylaxis. The existence and potential mobilisation of *tva*(A) can have also broader impact, thinking within "One health perspective" as for the treatment of human bacterial infections by pleuromutilins as members of this group, retapamulin was approved for topical use (anti-MRSA, 2007 FDA approval) and lefamulin was recently approved for systemic use in human (anti-MDR *S. pneumoniae* and *S. aureus*, 2019 FDA approval) (FDA 2007, 2019). Last study (García-Martín et al. 2019) tried to investigate the effect of the expression of the cloned *tva* (A) gene in strains *Escherichia coli* AG100A and *Staphylococcus aureus* RN4220, while in *E. coli* was conferred decreased susceptibility to pleuromutilin and streptogramin A, data gained shown a minor effect on *S. aureus*.

With regard to lincomycin, recently (De Luca et al. 2018) was described resistance gene *lnu* (C) located on the small 1724-bp transposon MTnSag, associated with resistance to lincosamides in *B. hyodysenteriae* (ST 83 strain, which also contains an A to T substitution at position 2058 (A2058T) in the 23S rRNA gene which is known to be associated with macrolide and lincosamide resistance). Existence and properties of the above transposon harbouring *lnu*(C) brought the evidence that *B. hyodysenteriae* is able to acquire resistance to antibiotics via mobile genetic elements.

Decreased susceptibility to doxycycline is described to be associated with polymorphism (G1058T) in the 16S rRNA gene of some *B. hyodysenteriae* isolates.

Broad resistance of *B. hyodysenteriae* leads to necessity to use depopulation and elimination of infection through cleaning and disinfection, and then restocking as the only effective course of action (Hampson 2012; Strugnell et al. 2013)—unfortunately with enormous costs.

2.5 *Clostridium*

Kiu and Hall (2018) described isolates from pigs—*C. perfringens* type A strains (also less frequently, CPE-harbouring type F strains) are widely considered as the invasive agent of non-haemorrhagic enterocolitis in piglets with yet not well-described pathogenesis. Similar to other *C. perfringens*, 1-week-old piglets suffered from intestinal infections, suspected to gain the bacterium from the microbiota of mother sows via vertical transmission during the birth (Songer and Uzal 2005). Severe diarrhoea (non-haemorrhagic, accompanied by necrotic mucosa and atrophy of intestinal villi are associated with these infections. Frequently used diagnostic marker: β2-toxin was initially believed to drive the development of this disease (recently re-assessed), especially the type C strains linked to haemorrhagic enterocolitis of up to 4 days old piglets. Type C-infection is characterised by necrotic enteritis with haemorrhagy, which is proposed to be driven by the presence of type C strains and low trypsin secretion (trypsin can inactivate β-toxin) in the immature host gut, what makes it different from type A infection (or to a lesser degree, type F-infection)—Kiu and Hall (2018).

Macrolide and lincosamide resistance (mainly erythromycin and lincomycin) appears widespread (Slavić et al. 2011), and therefore is considered ineffective in treating *C. perfringens* infections. This is supported by a recent multidrug-resistance study of 260 strains of *C. perfringens* isolated from diarrhoeal neonatal piglets in Thailand, where higher resistance was observed for erythromycin, lincomycin, and tylosin (Ngamwongsatit et al. 2016).

2.6 *Escherichia coli*

To make information more comprehensive, considering together above data for *E. coli* animal isolates and genes for resistance here can be noted, specifically for pigs that very recent data from pathogenic *E. coli* isolates from Danish pigs show certain trends (Holmer et al. 2019). As those data speak not only about AMR but giving the AMR also into link with the use of antimicrobial, thanks to both intensive pig farming as well as precise evidence of AMR and use, those investigations are a good example of the interlink s among AMR and AMU. High occurrence of resistance was present, especially in streptomycin and tetracycline (around 70%) and further in ampicillin, spectinomycin, and sulphonamides slightly lower, but still high resistance was reported in general in the investigated time period. Results for aminoglycosides (neomycin) with decreasing trends from 31.3% (2004–2007 isolates) to 9.6% (2012–2015) and then again increasing to 13.9% (2017) as well as data for florfenicol indicating steadily increase 2.1% from (2004–2007 isolates) to 18.1% in 2017 were noted. Authors provide the data that shows clear interlink with a decrease/increase of AMU and decrease/increase of AMR, might be with certain delay as for the timing, but confirming strong correlation of both parameters. Luckily most of the isolates tested in Danish study show with exception of few isolates full susceptibility to colistin and fluoroquinolones, but interestingly, genes of resistance *gyr*A and *par*C occurred and phenotypic resistance to nalidixic acid was reported. Interesting is also difference among serovars shown O149 and O138 with similar resistance patterns, but differing from O139. Above mentioned results of Holmer and co-authors correlate with ARBAO-II study published by Hendriksen et al. (2008).

2.7 *Glaeserrella parasuis*

Glaeserrella parasuis, previously *Haemophilus parasuis*, Gram-negative, non-mobile, small pleomorphic bacterium belonging to Pasteurellaceae family, detected as epiphytic bacteria of the upper respiratory tract of pigs. Fifteen serovars of *G. parasuis* have been described at present, but pathogenic potential has not been exactly described in the full extent yet also due to the fact that individual serovars differ in virulence and virulent strains can particularly participate as secondary causative agents in already existing pneumonia, but also can cause septicaemia without polyserositis or Glässer's disease characterised by polyserositis, pericarditis, arthritis, and meningitis. Clinical symptoms of this disease are highly variable; the brain, joints, and polyserositis samples are of importance for diagnosis. The disease caused by *G. parasuis* can be treated with antibiotics; however, oral or parenteral administration of very high doses of antibiotics is necessary (Nedbalcova et al. 2006).

As for the resistance to antimicrobials, horizontal gene transfer can be expected to play a significant role based on the recent results of resistance patterns and diversity of strains determined in addition to existing clonal dissemination (Zhao et al. 2018).

Antimicrobial resistance genes detected yet include those genes conferring resistance to beta-lactams (e.g. $bla_{\text{TEM-1}}$, $bla_{\text{ROB-1}}$), macrolides, lincosamides, and streptogramins (e.g. *erm*(B), *erm*(A)), amphenicols (where novel small plasmid harbouring *flo*R was described, as well as gene *cat*l), genes encoding tetracyclines efflux pumps as *tet*(B), *tet*(C)), different genes causing via different mechanisms resistance to aminoglycosides (*rmt*(B), *rmt*(D), *aad*(A1), *aac*(3′)-llc and genes encoding resistance to sulphonamides *sul*1, and *sul*2. Resistance to fluoroquinolones is due to the *par*C mutations that can be accompanied with *gyr*A mutation. MDR strains were also described (Zhao et al. 2018).

2.8 *Lawsonia intracellularis*

Lawsonia intracellularis, anaerobic obligate intracellular bacterium, causative agent of small intestine disease ileitis/proliferative hemorrhagic/necrotic enteropathy of pigs mostly under 4 months of age with significant health and economic impact. Lawsonia has been also described in hamsters and horses. Once established and validated in vitro culture system for *L. intracellularis*, screening for antimicrobial susceptibility using the rat enterocyte-based culture system (IEC-18; ATCC® CRL-1589™) allow to test antimicrobials, from which penicillin, erythromycin, difloxacin, virginiamycin, and chlortetracycline had the highest potential to be bacteriostatic followed by tiamulin and tilmicosin. In the past, many antimicrobials, such as tiamulin, tylosin, tetracycline, lincomycin, and some quinoxalines were used in preventive programmes, even in subtherapeutic, long-lasting dosing schedules and mostly prior the onset of clinical disease outbreaks. For clinically diseased animals tylosin, enrofloxacin, tetracyclines, tiamulin, and tilmicosin are commonly used at higher doses with clinical effect (Karuppannan and Opriessnig 2018). Very recent study of MICs performed via technique of counting the number of heavily infected cells (HICs; means >30 bacteria per cell) using an immunoperoxidase monolayer assay, determined as in vitro most effective fluoroquinolones (enrofloxacin and marbofloxacin), followed by colistin and tylvalosin (Seo et al. 2019). As ineffective against clinical forms of disease outbreaks are considered penicillin, bacitracin, and aminoglycosides (probably also due the pharmacokinetic properties and disability to penetrate intracellularly), virginiamycin, and from other antiinfectives ionophores and zinc and copper compounds (Karuppannan and Opriessnig 2018). In another recent study in vitro tests shown weak susceptibility for amoxicillin, penicillin G, chlortetracycline, oxytetracycline, tiamulin, tilmicosin, and tylosin (Seo et al. 2019). A gene encoding ribosomal protection protein associated with tetracycline resistance was found in the prophage-associated island of *Lawsonia intracellularis*, but with expression of genes in extracellular phase, what causes that chlortetracycline can be still effective to those strains in clinical infections (Vannucci et al. 2013), therefore for the discussion also remains, especially within this pathogen with specific pathogenesis and ecology in vivo, how the results of in vitro susceptibility testing, as well as detection of resistance genes, correspond to clinically efficacy. Moreover, another very promising tool is new vaccines against Lawsonia that have been released and started to be used in pig production (Roerink et al. 2018).

2.9 *Staphylococci*

Pathogens of importance also in pigs, as for resistance patterns and transfer mainly described above in general part. This part is more closely targeted on *S. hyicus*, a coagulase-negative pathogen, where recently (Holmer et al. 2019) released data on trends in susceptibility in isolates from Danish pigs that show susceptibility to amphenicols and ciprofloxacin in all isolates, but high resistance to penicillin, tetracyclines, and tiamulin. Based on the fact that no isolates were found resistant to cefoxitin is assumed, that none of *S. hyicus* from the isolates tested, was methicillin resistant. Penicillin resistance was detected as the most prevalent (82.2%) with MIC values of $\leq$0.06 to $>$ 16 μg/ml determined. High resistance levels were detected also in tetracycline and tiamulin (period 2004–2015). In the period 2008–2011 for erythromycin ($p < 0.0014$), streptomycin ($p < 0.01$), and spectinomycin ($p < 0.00022$) was analysed statistically important resistance increase, compared to 2004–2007. Increase of resistance to trimethoprim was noted in the period 2004 to 2015. As further programmes monitor resistance in target pathogens, results for *S. hyicus* are also available for BE, CZ, FR, NL, and SE, mostly in national reports (Schrijver et al. 2018).

Above mentioned phenotypic results are supported also by molecular analysis of the both chromosome and MGE harboured genes. From genetic determinants reported in *S. hyicus* of porcine origin was determined *tet*(K) and *tet* (L) encoding for membrane-associated efflux proteins of the major facilitator superfamily and *tet*(M) causative for target site protection (Schwarz and Noble 1994; Aarestrup and Jensen 2002; Wendlandt et al. 2015). As for resistance to beta-lactams, gene *bla*Z, encoding beta-lactamase (Aarestrup and Jensen 2002; Wendlandt et al. 2015) was detected in *S. hyicus* isolated from diseased pigs, but also gene *mec*A encoding for alternative PBS was described in strains of *S. hyicus* isolated from pig carcasses (Hassler et al. 2008). As for phenicol resistance, gene Cat_{p221} encoding for acetyltransferase inactivating chloramphenicol was described by Schwarz et al. (1990) and gene *fex*A by Kehrenberg and Schwarz (2006). Aminoglycoside efficacy can be affected, once the strains of *S. hyicus* harbour and express gene *aad*E (kanamycin and neomycin are targeted) and gene *str* (streptomycin) (Schwarz and Noble 1994; Aarestrup and Jensen 2002; Wendlandt et al. 2015). Genes *spc*, *spw*, *spd* conferring resistance towards aminocyclitols (spectinomycin) (Wendlandt et al. 2015). Genes *erm*(A), *erm*(B), and *erm*(C) mostly localised on transposons or plasmids are responsible for rRNA methylase and cause resistance to macrolides, lincosamides and streptogramin B (MLS_B)—all of them were also described in porcine isolates of *S. hyicus* (Aarestrup and Jensen 2002; Lüthje and Schwarz 2007), Wendlandt et al. 2015). Enzymatic inactivation of lincosamides is caused once genes *lnu* (A) and *lnu*(B) are expressed (Lüthje and Schwarz 2007; Wendlandt et al. 2015), but more broader effect have genes *vga*(A) and *vga*(C)—affecting the efficacy of lincosamides, pleuromutilins, and streptogramin A due to ribosome protective ABC-F protein (Wendlandt et al. 2015). Even more broad-spectrum impact is once gen *cfr* (Kehrenberg and Schwarz 2006) is translated to rRNA methylase affecting phenicols, lincosamides, oxazolidinones, pleuromutilins, and streptogramin A. *S. hyicus* of porcine origin harbour also *dfr*K gene, as was confirmed by Kadlec et al. (2012).

2.10 *Streptococcus suis*

In very recent bibliographical reference (Holmer et al. 2019), data has been released following the long-term trends in resistance for pathogenic *S. suis* isolated from pigs produced in Denmark.

Bimodal distribution of MICs was detected for tetracycline, sulphonamides, trimethoprim, erythromycin, streptomycin, spectinomycin, and tiamulin, bimodal MIC distributions occurred. Tetracycline was indicated as an antimicrobial with highest detected resistance, around 75% throughout the whole period 2004–2017. Increasing resistance trend was determined for erythromycin (macrolide), trimethoprim, and tiamulin, in which number of strains with highest MICs increased during the time. This correlates also with the mostly used groups of antimicrobials: "Top 3" was identified: tetracyclines, macrolides, and tiamulin. The resistance level increased considerably for erythromycin (from 26.1% in 2004–2007 to 48.0% in 2017) and for trimethoprim (from 1.8% in 2004–2007 to 23.0% in 2017; with MIC_{90} increase from ≤ 1 to 8 μg/ml). Both MIC_{50} and MIC_{90} for penicillin remain low but a few isolates had MIC values above the clinical breakpoint. Comparing the results with other EU countries, high occurrence of tetracycline resistance (48 to 92%) was found in France, England, The Netherlands, Poland, and Portugal. Despite excellent susceptibility to penicillin in many countries, 8.1% and 13% of the isolates in Poland and Portugal, respectively, was reported as resistant to penicillin, trend data from the Czech Republic shows min 7.7%–max 11.5% penicillin-resistant isolates from diseased pigs in the period 2015–2018 (Hendriksen et al. 2008; CZ NMTP 2016, 2017, 2018).

Resistance to macrolides in *S. suis* is mainly associated with *erm* genes encoding ribosomal methylase or *mef* genes encoding macrolide efflux protein (14- and 15-membered macrolides). Generally, gene *erm*(B) is one of the most common genes found in macrolide-resistant *S. suis* and recently published data from Thailand list also *erm*(T) and *erm*(A) as genes gained from porcine isolates of S. suis resistant to azithromycin and erythromycin (Yongkiettrakul et al. 2019). Further research is needed as the current Thai study of *S. suis* strains indicate phenotypes of macrolide resistance, where new, unknown resistant genes were found (Yongkiettrakul et al. 2019).

Tetracycline resistance mechanism associated with *tet* genes encoding either tetracycline-resistant ribosomal protection protein *tet*(B), *tet* (L), *tet*(O), or tetracycline efflux protein—*tet* (M) and *tet*(W), specific *tet*(40)—as an efflux gene detected in *C. saccharolyticum* in tandem with a mosaic tet gene *tet*(O/32/O), and mosaic *tet* (O/W/32/O) have been identified (Palmieri et al. 2011). Interestingly *tet*(W)-carrying elements can carry also *erm*(B) and some further determinants conferring resistance to macrolide, aminoglycoside, and streptothricin and heavy metal (cadmium). Unstable, highly transferable, genetic element could be found inside an integrative and conjugative elements (ICE) containing *tet*(O/W/32/O) can also carry macrolide *erm* (B) and aminoglycoside (*aad*E, *aph*A) resistance genes (Palmieri et al. 2012).

One of the alarming information published (Huang et al. 2016a, b) is that *S. suis* mobilome (i.e. plasmids, transposons, ICEs, integrons, genomic islands, and prophages) can be the source of resistance for other species of streptococci. This is supported also by the fact that, higher prevalence and broader diversity of MGEs have been reported for *S. suis*, compared to other pathogenic *Streptococcus* species yet.

Another recent study has just reported a novel membrane transporter module SstFEG. With function as efflux pump for bacitracin resistance as well as a virulence-related protein involved in *S. suis* pathogenicity (Ma et al. 2019).

Considering that *S. suis* is more and more discussed as re-emerging zoonotic agent causing severe diseases, mostly meningitis, in pigs, and in humans having occupational contact with pigs and pork, such as farmers, slaughterhouse workers, and butchers (Dutkiewicz et al. 2017), the gaps in knowledge of virulence factors and resistance genes and their transfer among other species of streptococci should be filled by further research.

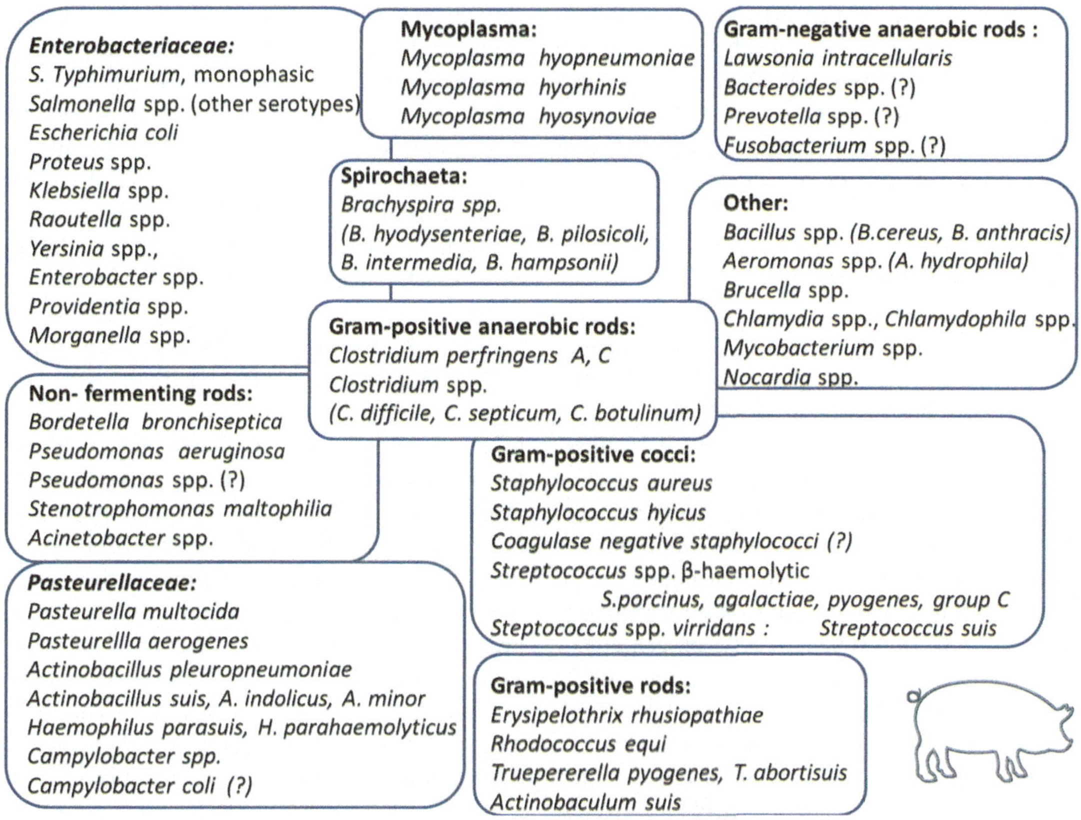

Fig. 2 Bacteria being considered as having pathogenicity in pigs and associated with pig diseases

3 Specific Consideration for Pathogens and Antimicrobials with Importance in Poultry

3.1 *Avibacterium paragallinarum*

Avibacterium paragallinarum (previously *H. paragallinarum*) is a capsulated, rod-shaped, Gram-negative facultative anaerobe. It is a member of Pasteurellaceae family, the causative agent of infectious coryza, an acute disease of the upper respiratory tract of chickens.

There is a lack of studies mapping genes of resistance in *A. paragallinarum.* Study by Hsu et al. (2007) reported results of 18 *Avibacterium paragallinarum* isolates collected in Taiwan (1990–2003), in which serotype and resistance (phenotypes/genotypes) were determined. About two-thirds of isolates contained plasmids—plasmid pYMH5-encoded functional streptomycin, sulphonamide, kanamycin, and neomycin resistance genes, plasmid pA14 encoded a putative MglA protein and RNase II, both possibly associated with virulence. Haemocin activity was proven in 7 isolates. Plasmid pYMH5 is the first multidrug-resistant plasmid reported in *A. paragallinarum.* Study made by Byarugaba et al. (2011) analysed four isolates from poultry in Tanzania, determined genes encoding for resistance to streptomycin (*str*(A), ampicillin (bla_{TEM}), tetracycline (*tet*(C) and *tet*(A), and sulfamethoxazole (*sul*2).)

Recent study (Requena et al. 2013) providing a draft genome sequence as well as insight to the genome of *A. paragallinarum* proves the existence of the Tn10 transposon, which was also found in plasmids from several chicken pathogens, including *Escherichia coli*, and *Salmonella enterica* serovar Typhimurium. Tn10 transposon comprising, among the others, also four genes associated with tetracycline resistance (*tet*(R), *tet*(A), *tet*(C), and *tet*(D) with potential to cause resistance to tetracyclines.

3.2 *Bordetella avium*

Bordetella avium is a small, coccoid-shaped, Gram-negative, motile by peritrichum flagella, strictly aerobic bacteria form the family Alcaligenaceae. It is a pathogen of poultry, mainly mentioned as causative agent of turkey respiratory disease—coryza, but has been also described in human patients with cystic fibrosis (Kadlec and Schwarz 2018; Harrington et al. 2009). Recently released study results (Sebaihia et al. 2006) show differences between genome of *B. avium* (investigated pathogenic strain N197) and *B. bronchiseptica* differing in more than 1100 genes and give insight at least to adaptation to host through specific factors. Surface or secreted proteins (agglutinins/adhesins, LPSs, capsules, and extracellular polymers, fimbriae and pili, autotransporters, large secreted proteins, secretion systems, and toxins) seems to be encoded by these different genes, that probably, among the other factors, cause-specific adaptation for survival and pathogenesis in the avian rather than the mammalian respiratory ciliated tissues and trachea. Synthesis of a polysaccharide capsule as well as hemagglutinins is also based on the code of these genes. Unique genes for both lipopolysaccharide and fimbrial biogenesis were also identified. Interestingly 3 prophages, which have some similarities, e.g. with bacteriophages of *B. bronchiseptica* were detected. The BvgAS virulence regulatory system appears to have polymorphisms at a poly(C) tract that is involved in phase variation in other *Bordetella*.

Information on this genus, especially those related to the mechanisms of resistance is scarce, but some non-EU studies recently appear. The only publicly available results from the European up to now come from (Szabó et al. 2015) coming with phenotypic results, where *B. avium* strains were reported to be resistant to ceftiofur and lincomycin and susceptible to doxycycline, gentamicin, polymyxin B, spectinomycin, and sulphonamides. MIC values for amoxicillin were $\leq$0.03 μg/ml to 1 μg/ml, for doxycycline $\leq$0.03 μg/ml to 0.12 μg/ml and for erythromycin 8 μg/ml to 16 μg/ml. Another recent publication describing isolates from turkey coryza in Egypt identified *B. avium* strains from clinical cases. According to genetic analysis, the authors declared identity with German *B. avium* ATCC 35086 strain as well as with the American strain 197N. Phenotypic antimicrobial susceptibility testing of the isolates showed fluoroquinolones (norfloxacin, ciprofloxacin), cefotaxime, florfenicol, and (gentamicin—for only one strain) in vitro effectiveness and resistance was detected in ampicillin, erythromycin, oxytetracycline, sulphamethoxazole/trimethoprim, and lincomycin. Investigation of the genes of resistance confirms the presence of bla_{TEM} (resistance to beta-lactams), *tet*(A)(resistance to tetracyclines), *aadA*1 (encoding aryloxyalkanoate dioxygenase for streptomycin resistance), *sul*1 (resistance to sulphonamides), and *dfr*A (resistance to trimethoprim) (Erfan et al. 2018).

3.3 *Clostridium perfringens*

C. perfringens is identified as the key aetiological organism of necrotic enteritis (NE) in broiler chickens Global financial impact of NE has been assessed and indicated to be of importance, with an estimated economic loss of 6 billion US$ in 2015. Gaseous lesions and mucosa necrosis in the gas-filled distended small intestine are pathologies accompanied by the disease as well affection of kidney and liver, once becoming systemic. Factors that play a role in pathologies are mainly hydrolytic enzymes (e.g. collagenase), toxin production (firstly described α-toxin, and more recently NetB and TpeL, both pore-forming toxins). Individual genes responsible for toxin production as well as their combinations were described as, e.g. *cpa*-*net*B-*cpb*2-*lux*S-*col*A-*vir*S (Forti et al. 2020) *C. perfringens* has been reclassified into seven toxinotypes (A–G) depending on the toxin they produce: alpha (CPA), beta (CPB), enterotoxin (CPE), or the necrotic enteritis beta-like toxin (NetB), epsilon (ETX), iota (ITX). Large plasmids encode most of these toxins, with the exception of CPA and perfringolysin O (PFO), which are harboured on the clostridial chromosome. Either on the chromosome or on plasmids are located CPE. Most of these plasmids are conjugative and have a tcp locus possibly functional in the spread of toxin genes and resistance determinants (Freedman et al. 2016). From groups of antimicrobials used in poultry in different parts of the world, resistance was described to aminoglycosides (streptomycin, gentamicin), erythromycin, lincomycin (*lnu)*, penicillin, ampicillin, amoxicillin/clavulanic acid, chloramphenicol, tetracycline (*tet*A(P), *tet*B(P), and *tet*(M), and bacitracin (*bcr*ABDR on chromosome) not only by phenotypes, by in some cases also with known gene background (Mwangi et al. 2019; Lee 2016; Charlebois et al. 2012). Moreover, it has been described that mobile genetic elements exist in *C. perfringens* harbouring different genes of resistance to antimicrobials: chloramphenicol (transposon Tn4451), bacitracin (resistance integrative conjugative element ICECp1), and lincomycin (resistance transferable insertion sequence tISCpe8) (Adams et al. 2018).

3.4 *Enterococcus cecorum* and *Enterococcus faecalis*

Enterococci are Gram-positive facultative anaerobic bacteria (previously ranked as streptococci of the group D according to Lacefield). They are ubiquitous (present in humans, animals, plants, and food, environment—water and soil). Some species can cause pathology in poultry as septicaemia, endocarditis, spondylitis, osteomyelitis, arthritis, and other diseases (Gilmore et al. 2002; Mazur-Gonkowska et al. 2006). According to ECDC (2011) mostly (up to 80% of isolates from pathologies) are *E. faecalis* and *E. faecium*, but other species of enterococci has been recognised to be important primary or secondary pathogens, among them, e.g. *E. cecorum*, *E. hirae*, *E. durans*, *E. gallinarum*, *E. casseliflavus*, and *E. avium*, even that from the taxonomy perspective much more species have been described yet. Enterococci are listed as the third and fourth most prevalent human pathogen worldwide (ECDC 2011) and considered as the third most frequent in causing bacteremia in Europe and North America, responsible for approximately 11–13% of all bacteraemia cases (Ammerlaan et al. 2013; de Kraker et al. 2013). Enterococci are also considered to be a good indicator of the emerging antimicrobial resistance in different surveillance systems, among the other factors, due to their zoonotic potential and possibility to transfer genes intra- and inter-bacterial species and spread antimicrobial resistance among different sources and environments. Recent bibliography reports in isolates from broilers (Canada) MDR strains of *E. faecium* with accumulation of resistance genes *bcr*R-*erm*A/B-*msr*C-*mef*A-*aac*-*aph*A-*tet*L-*tet*M (Rehman et al. 2018); in Polish isolates of enterococci from turkeys by phenotype testing strains containing combinations of resistance to ampicillin–amoxicillin/clavulanic acid–vancomycin–tetracycline and either ciprofloxacin or erythromycin were documented (Woźniak-Biel et al. 2019). These studies, as well as below mentioned detailed data on transposons, plasmids, and other mobile genetic elements can serve as evidence, that enterococci are a significant source of the genes of resistance.

Transposons such as Tn916/Tn1545, Tn917/Tn551, and Tn5397 that have been reported to disseminate resistant genes, including *tet*(M), *erm* (B), and *aph*A-III, by enterococci belong to the enterococcal mobilome (Hegstad et al. 2010). Furthermore, in the absence of antimicrobials, specific pheromone production was reported to induce a high-frequency plasmid transfer in *E. faecalis* (Hirt et al. 2018).

It should be also mentioned that a lot of specific strains of enterococci are used as probiotic "feed additives".

As for poultry, the most significant are *E. cecorum* (be found as dominant part of gastrointestinal microbiota of mature chicken (Gong et al. 2002). But also *E. faecalis* is frequently founded in pathologies.

The most common and well-described virulence determinants in enterococci are aggregation proteins involved in adherence to host cells (*agg*, *asa*1), genes associated with activation, transportation, and modifications of the cytolysin (*cyl*), extracellular surface protein (*esp*), adhesion to collagen (*ace*, *acm*), and adhesion-like endocarditis antigens ($efaA_{fs}$, $efaA_{fm}$) and endocarditis and biofilm-associated pili gene locus *ebp*ABC (Hanchi et al. 2018; Ben Braïek and Smaoui 2019; Rehman et al. 2018), chromosomal gelatinase (gelE) in *E. faecalis* were also described gelE-bearing isolates with the locus *fsr*B, encoding a processing enzyme that liberates GBAP (Gelatinase Biosynthesis Activating Pheromone) peptide important for virulence and present in strains causing pathology and diseases (Hancock and Perego 2004). *E. faecalis* isolates from Canadian broilers (Rehman et al. 2018) carried moreover a chromosomal *hyl*A/B gene (encoding enzyme hyaluronidases) and *elr* gene (encoding leucine-rich protein A facilitating escape of enterococci from host immune defenses and gene tpx the thiol peroxidase encoding with protectivity against oxidative stress.

Antimicrobial resistance patterns vary greatly geographically and in time. But there can be highlighted exact genes that are frequently detected and encoding resistance to different groups of antimicrobials. Also should be noted, that plasmid conferring antimicrobial resistance in enterococci has been detected towards glycopeptides (*van*), macrolides, and lincosamides (different genes), tetracyclines, aminoglycosides, zinc bacitracin, and also heavy metals (copper, cadmium). Presence of strains with such plasmids is promoted by the use of the antimicrobials belonging to above pharmacological groups, either as direct selection or as co-selection. Two species *E. faecalis* (namely ST 82 and ST16 lineages) and *E. cecorum* seem to be of importance for the transfer of resistance among poultry, food of poultry origin, and human (Torres et al. 2018).

Aminoglycosides: Intrinsic resistance to aminoglycosides in achievable concentrations due to low permeability of their cell wall, some species *E. faecium*, *E. durans*, and *E. hirae* has been described to intrinsically express acetyltransferase (encoded by *aac*) harboured on chromosome, resistant therefore to amikacin, tobramycin, kanamycin (resistance to these two antimicrobials can be also due to chromosomally encoded (rRNA) methyltransferase known as EfmM methyltransferase (encoded by *efm*M) described in *E. faecium*). In practice, it means that for cases of non HLAR strains, mostly gentamicin and streptomycin in combination with beta-lactam (e.g. combination gentamicin/ampicillin or amoxicillin and streptomycin/penicillin can be effectively used in practice, providing that dose will be carefully monitored not only from efficacy, but also toxicity perspective. Except this above-mentioned resistance, acquired resistance, so-called HLAR (High-Level Aminoglycoside Resistance) is known in human and later on detected also in isolates from animals—bifunctional aminoglycoside-modifying enzyme gene encoding HLAR is *aac(6′)-Ie-aph(2″)-Ia*, which is located on the Tn5281 transposon (Lebreton et al. 2013; Shete et al. 2017). Other genes conferring resistance to aminoglycosides are, e.g. phosphotransferases (2″-O phosphotransferase = *APH(2′)*; 3′-O phosphotransferase = APH(3′), adenyltransferases (3′-O adenyltransferase = ANT (3′), 4′-O adenyltransferase = ANT(4′) and 6′-O adenyltransferase = (ANT(6′), and N acetyltransferases (AAC).

Beta-lactams: There should be mentioned intrinsic resistance to cephalosporins and reduced susceptibility to penicillins (due to low affinity of Penicillin Binding proteins (PBPs), according to Miller et al. (2014). Despite this fact in human medicine is used effectively combination ampicillin/ceftriaxone for serious infections caused by enterococci. Usually, *E. faecium* MICs >MICs of *E. faecalis* and 6 PBP genes were described: *pon*A, *pbp*F, *pbp*Z (class A), and *pbp*5, *pbp*A, *pbp*B (class B). Acquired high-level resistance to ampicillin in *E. faecium* is linked to *pbp*5, but the level of expression and modification of PBPs differ and together with this fact also MICs differ.

Of importance is that horizontal transfer of *pbp*5 was described (Novais et al. 2016). In *E. faecalis* acquired ampicillin resistance is rare, but mutation of *pbp*4 is usual.

Glycopeptides: Group contains three major representatives (vancomycin, teicoplanin: used in human medicine and avoparcin: previously extensively used as an antimicrobial growth promoter (AGP) in animals (banned since 1997 in EU). Vancomycin resistance in enterococci is mediated by *van* operons, 9 operons have been described yet *van* A, –B (B1–B3), –C, –D (D1–D5), –E, –G(G1–G2), –L, –M, and N (*van*C is intrinsic, subgroups C1–C3). Operons generally consist of genes encoding two-component signal transduction systems, which activate the genes responsible for the synthesis of modified peptidoglycan precursors and destruct "normal" (D-alanine ending) precursors (Miller et al. 2014). *Paenibacillus popilliae* has been suggested to be a source of vancomycin resistance in enterococci, but other sources (Patel 2003; Ogawara 2019) speaking also about *Amylolactopsis orientalis*). Some of these operons are harboured by transposons (e.g. *van*A by Tn*1546*), *van*B (Tn*1547*, Tn*1549*, Tn*5382*). Some of them, located on chromosomes, some located on plasmids. As was mentioned, avoparcin was banned as AGP in EU, and after that was reported decrease of the VRE carriage in poultry and healthy humans (Klare et al. 1999), despite that, probably due to being harboured by same transferable plasmids as other frequently used antimicrobials, tetracyclines and/or macrolides—namely tylosin (gene *erm*(B), *van* genes still remain in the enterococci isolated from animals (Aarestrup et al. 2000). Recently published work investigated isolates both from human and turkeys in Poland (Woźniak-Biel et al. 2019) indicate 25% of positive human isolates with detected resistance to vancomycin (*van*A, *van*B, and *van*C-1 detected) and 15.69% of positive turkey's isolates (*van*A and *van*C-1 detected) in enterococci (*E. faecalis*, *E. faecium*, *and E. gallinarum*). Interestingly multidrug resistance of isolates phenotypically resistant to vancomycin was confirmed in following combinations (AMP + AMC + VAN + TET + ERY; AMP + AMC + VAN + CIP + TET; AMP + VAN + CIP + TET; VAN + CIP + TET; VAN + TET + ERY, where AMC = amoxicillin/clavulanic acid) in both human and turkey isolates. It will be of importance to gain knowledge, by which (mobile) genetic element are those genes harboured.

Macrolides (and lincosamides and streptogramins): *E. faecium* has been described to be intrinsically resistant to macrolides by *msr* (A) and lincosamides (clindamycin and lincomycin(*lin*(B). Moreover, transposon Tn*917* harbouring gene *erm*(B) broadly occurs both in isolates from human, animals, and food. Further genes conferring resistance to macrolides are efflux genes *mef*(A); resistance to virginiamycin is based on gene *vgb*(A) or vgb(B) that is linked to enzymatic cleavage of the ring structure of streptogramin B by the lactonases VgbA and VgbB and resistance to streptogramins is due to expression of the genes *vat*(D) or *vat*(E).

Oxazolidinones: Linezolid: Despite the fact that the any of the oxazolidinones have never been authorised and probably also not used in food-producing animals in Europe, functional genes *cfr* and *optr*A were detected both in enterococci and *S. aureus* of human and food as well as animal origin. The risk of this resistance persistence and transfer is enforced by the fact that in some cases it is located together with other genes encoding resistance to other antimicrobials used in animals [e.g. florfenicol (gene *fex*A)], novel *erm*(A)-like gene as well that those genes conferring resistance to oxazolidinones can be harboured on transposons or plasmids. Gene *optr*A was detected mainly in *E. faecalis* of pig and chicken origin as well as from food in European, South American as well as Asian countries. It should be noted that oxazolidinones belong among the last-resort antimicrobials reserved for use in human medicine against VRE (Vancomycin-Resistant Enterococci) and MRSA (Methicillin-Resistant *Staphylococcus aureus*) and therefore spread of resistance to these antimicrobials can narrow the spectrum of effective treatment options for the human life-threatening infections (Torres et al. 2018), despite the fact that new N-(1,3,4-oxadiazol-2-yl) benzamide analogs, bacteriostatic agents against methicillin- and vancomycin-resistant bacteria are in the research pipeline (Opoku-Temeng et al. 2018).

Quinolone resistance is a consequence of either mutations in *gyr*A and *par*C and *par*E genes affecting so-called "quinolone resistance-determining regions," which presumably alter the binding affinity of the quinolone molecule. Another resistance mechanism is due to acquisition of *qnr* genes encoding for protein, which is likely to protect DNA gyrase by decreasing DNA binding of the quinolone and the subsequent formation of the quinolone–gyrase complex. Also, efflux pumps were described in certain species: *E. faecalis*: EmeA, *efr*AB; *E. faecium* NorA-like (Miller et al. 2014; Torres et al. 2018; Shiadeh et al. 2019).

Tetracyclines: Resistance is mediated by multiple genes, involving two general principles: efflux of the antibiotic and ribosomal protection. Efflux pumps encoded by *tet*(K) and *tet*(L) are plasmid-borne determinants. Ribosomal target is relevant for the genes *tet*(M), *tet*(O), and *tet* (S) being a chromosomal resistance determinant encoding for a protein with significant homology to bacterial elongation factors (EFs), which confer resistance to tetracycline, doxycycline, and minocycline and can be located and transferred by Tn*916* transposon or in the case of the *tet* (M) gene also by Tn*1545 transposon together with erm*(B) (Miller et al. 2014; Torres et al. 2018).

Trimethoprim/sulphonamides: Susceptibility appeared when combination SXT/TRI is tested in vitro; however, these compounds are ineffective in vivo due to the ability of enterococci to utilise exogenous sources of folate which synthesis in different steps is the target of individual substances from this combination (Miller et al. 2014). In *E. faecalis* isolates from broilers in Canada, *dfr*E gene homolog was recently described (Rehman et al. 2018).

Zinc bacitracin: Resistance has been proven in enterococci isolates in China (*E. faecalis*) as well in isolates of *E. cecorum* from chicken, in both with a high level of resistance to zinc bacitracin (MIC >256 μg/ml) linked to *bcr* ABDR cluster, which can be located either on transferable plasmids or chromosomes. This structure can be also transferred by recombination intra- and interspecies. Specific pheromone responsive plasmid containing *bcr*ABDR in some cases linked to other genes of resistance are disseminated in Chinese farms mainly on clone *E. faecalis* ST16 (Chen et al. 2016). More recent study demonstrate even more spread of *bcr*ABDR—on pheromone-responsive conjugative multiresistance plasmid pE211 carrying the novel *optr*A locus from *E. faecalis* harbouring a mobile *bcr*ABDR locus. It should be highlighted that acquiring *optr*A gene encoding for ribosomal protection protein of the ABC-F family and causing cross-resistance to linezolid and florfenicol pose a real and serious threat to both human and veterinary medicine (Shang et al. 2019). This can be seen also by the optic, that Zn bacitracin is still allowed in many countries worldwide as a growth promoter.

3.5 *Gallibacterium anatis*

Gallibacterium anatis is Gram-negative, rod-shaped or pleomorphic, non-motile, microaerophilic bacterium of the Pasteurellaceae family. Upper respiratory and lower reproductive tract of chickens colonising, but also causing salpingitis, oophoritis, peritonitis, septicaemia, pericarditis, hepatitis, and upper respiratory tract lesions (Christensen et al. 2003). Isolated not only from chickens, but also from turkeys, ducks, geese, pheasants, and partridges altogether with other *Gallibacterium* species and subspecies (Bisgaard et al. 2009). Abnormalities in egg shell, decreased laying performance and increased mortality in pullets were also reported (Paudel et al. 2015) as associated with *Gallibacterium* infections especially with other co-factors participating (impaired host immunity, co-infections, and bad ventilation, overcrowdings, and climatic changes enabling more easily the infection pressure) (Roberts et al. 2011).

Among the virulence factors belong several specific proteins, haemagglutinins, adhesins, capsular extracellular polysaccharides, and outer membrane vesicles, which are more or less described and need to be further investigated both as for exact mechanisms and functions and for genetic background. The GtxA (*Gallibacterium* toxin A) is a protein expressed by *G. anatis* with haemolytic activity against erythrocytes from a wide variety of hosts, and leukotoxic activity against the chicken macrophage cell line HD11 (Kristensen et al. 2010; Persson and Bojesen 2015). As also adhesins are of importance for attachment to the mucosal/epithelial surfaces of the host, *G. anatis* strain genomes were analysed and in 3 strains F-17-like fimbriae clusters were identified (Johnson et al. 2013). It was determined that some of them belonging to a group of fimbriae that bind N-acetyl-D-glucosamine (Glc-NAc)-containing receptors on the surface of host cells (avian pathogenic *E. coli*), but also that there are the F17-like fimbria, FlfA protein (Bager et al. 2013). There were also described so-called Outer Membrane Vesicles (OMV) spherical, bilayer membrane structures, with different, yet not fully understood functions, among them, e.g. being transportation vehicles for the delivery of lipids, membrane proteins, insoluble compounds, or compounds that are easily degraded, including toxins and DNA. Further research is needed, but recent work (Pors et al. 2016) signalise possible use in the serotype independent vaccines due to immunisation with GtxA-N and FlfA, but later work (Persson et al. 2018) indicate that more effectively OMV in combination with FlfA for cross-protective immune response. Some strains of *G. anatis* were also proven to produce metalloproteases degrading immunoglobulins, and hemagglutinins, which may promote biofilm formation (Persson and Bojesen 2015).

Phenotype, microdilution, MIC determination study (Jones et al. 2013) analysed susceptibility patterns of *Gallibacterium anatis* (US, 2006–2011, 84 isolates) and demonstrated almost complete resistance to novobiocin, tylosin, lincosamide, and tetracycline with moderate to high susceptibility to sulphonamides, fluoroquinolones, and florfenicol; intermediate susceptibility was recorded towards spectinomycin and erythromycin; variable levels of resistance were described for beta-lactams and aminoglycosides. Recent study of *G. anatis* isolates coming from Germany—15 isolates tested for susceptibility, MICs testing (El-Adawy et al. 2018) show following results as fully susceptible were assessed apramycin and neomycin, 26.7% resistance was reported to gentamicin, full susceptibility to florfenicol seems in line with previous results for chloramphenicol published by Danish team (Bojesen et al. 2011). Chlortetracycline, as well as oxytetracycline, showed 73–80% resistance, what again corresponding to the above-cited Danish study (Bojesen et al. 2011). As for beta-lactams, relatively low resistance was reported for ampicillin (13.3%) and for ceftiofur (20%), but all tested strains were resistant to penicillin. Macrolides show high resistance (erythromycin (66.70%) and tylosin (86.7%), what correspond to other previous, but also recent studies (tylosin resistance 94.8% in 213 Austrian isolates from laying hens), where macrolides were reported as fully resistant (Lin et al. 2001; Hess et al. 2019) as well as clindamycin, where all isolates were resistant. Enrofloxacin resistance was reported in 33%, tiamulin in 26.7%), but very recently in Austria even 58.2% ($N = 213$, laying hens; Hess et al. 2019). Results of sulphonamides differ—a high percentage of resistance was reported in German isolates (sulphathiazole 100%, sulphamethoxim 93.3%) as well in Danish samples for sulhamethoxazole 97% and 77.4% in Austrian isolates (El-Adawy et al. 2018; Bojesen et al. 2011; Hess et al. 2019), but in German isolates only 20% resistance was determined, when tested fixed combination of sulphamethoxazole/trimethoprim (El-Adawy et al. 2018). What is also of importance, that Hess et al. (2019) described that resistance against antimicrobial substances increased significantly in isolates from older birds and high variability was described even in isolates from the same bird.

Very recent study made by Peng et al. (2019a, b) demonstrated in the investigated *Gallibacterium anatis* the presence of class 1 integron harbouring genes $bla_{\text{OXA-10}}$ and PSE-1. According to the authors, this is of importance due to possibility of acceleration of the spread of ESBLs among different Gram-negative bacteria. It should be noted that $bla_{\text{OXA-10}}$ and PSE-1-containing integrons have been already described in *Pseudomonas aeruginosa* but also *Salmonella* spp., both having zoonotic potential—i.e. risky from human health perspective. Considering above information it seems of high importance to make more detailed genetic analysis and investigate the location of genes as well as mechanisms of resistance and mechanisms of transfer of resistance for other antimicrobials, which were identified to be of concern form the antimicrobial resistance perspective in abovementioned studies.

3.6 *Ornithobacterium rhinotracheale*

*Ornithobacterium rhinotrache*ale is a Gram-negative, rod-shaped, causative agent of respiratory diseases, but able to disseminate and result in osteitis, meningitis, and joint infections) in turkeys, chickens, geese, ducks, and other avian species. In France in 2017, most samples analysed by RESAPATH programme came from turkeys (ANSES 2019). *O. rhinotracheale* can be classified into serotypes (A through R), with A serovar being the most prevalent among chicken and turkey isolates HU and DE (Szabó et al. 2017; Gashe 2017). The presence of different serotypes (A, B, C, D, and E) with variable adherence profiles suggest that these serotypes have different virulence factors. Using multilocus enzyme electrophoresis *O. rhinotracheale* isolates from different parts of the world were distinguished into six electrophoretic types (ET) that were later confirmed by 16S rRNA gene sequencing and rep-PCR analysis (Amonsin et al. 1997; Montes et al. 2018).

The most recent publicly available European data on susceptibility come from the study of Szabó et al. (2015), in which 36 strains of *O. rhinotracheale* (Hungary, 2009–2013, mostly from turkeys 28, 4 and 2 isolates from chickens and pigeons, respectively) were tested by the Kirby-Bauer disc diffusion method, and MICs of amoxicillin, doxycycline, and erythromycin were also determined. Strains were resistant to nalidixic acid, sulphamethoxazole–trimethoprim and gentamicin, and were susceptible to ampicillin, chloramphenicol, spectinomycin, and tilmicosin; MICs reported were for amoxicillin and erythromycin within the range of 0.12 μg/ml to 32 μg/ml, and for doxycycline 0.6 μg/ml to 32 μg/ml. Comparison with the results of other studies performed in different parts of the world is difficult, because a lack of standardisation in the methodology of testing as well as interpretative criteria for this microorganism, therefore more effort should be paid to the methodology/interpretation standardisation. The resistance patterns of Szabó et al. and older studies relevant for the European region are summarised in Table 4, which show high variability in susceptibility/resistance.

According to the authors best knowledge till 2019, results published, that bring knowledge on genes encoding for resistance via different mechanisms are scarce, despite the fact that, especially according MICs results seems that for certain substances resistance can be a concern (but no internationally standardised interpretive criteria has been published yet). Recently published data (Smith et al. 2020) brings the evidence that chromosomally encoded proteins can be associated with different mechanims of resistance: macrolide export protein (MacA) and macrolide export ATP153 biding/permease protein (MacB), Penicillin-binding proteins (PBP1a, PBP4) as well as multidrug resistance proteins (MdtA and MdtN, NorM, YheL).

Table 4 Phenotypes/MICs described in *Ornithobacterium rhinotracheale* from different regions in Europe according to individual publications (extracted/modified from Gashe 2017)

Antimicrobial	Results of susceptibility testing in %			Animals	Comments	References
	Susceptible	Resistant	Intermed			
Penicillins						
Penicillin	30		46.7	B	HU	Szabó et al. (2015)
	$MIC_{50} = 0.75$			B	NL	van Veen et al. (2001)
	$MIC_{90} = 3$					
	$MIC_{50} = 0.5$			T,L,PH	DE	Popp (2003)
	$MIC_{90} = 2$					
	$MIC_{50} = 4$	$MIC_{90} = 64$		T	DE	Waldow (2009)
Amoxicillin	40	36.7	23.3	B	HU	Szabó et al. (2015)
			63.2			
	$MIC_{50} = 0.5$			T,L,PH	DE	Popp (2003)
	$MIC_{90} = 4$					
Ampicillin	$MIC_{50} = 4$	$MIC_{90} = 64$		T	DE	Waldow (2009)
	97.6	2.4		T	FR	Dudouyt et al. (1995)
	40	36.7	23.3	B	HU	Szabó et al. (2015)
		100			BE	Devriese et al. (2001)
Aminoglycosides						
Gentamicin		100			HU	Szabó et al. (2015)
Neomycin	96.66	3.33		T,B,L	Turkey	Erganis et al. (2012)
Sulphamethoxazol/ trimethoprim		25	33		HU	Szabó et al. (2015)
		89.3			NL	van Veen et al. (2001)
Tetracyclines						
Tetracycline		60.9			NL	van Veen et al. (2001)
		$MIC_{50} = 16$		T,L,PH	DE	Popp (2003)
		$MIC_{90} = 16$				
		$MIC_{50} \geq 16$		T	DE	Waldow (2009)
		$MIC_{90} \geq 16$				
Doxycycline		30	16.5		HU	Szabó et al. (2015)
		80			BE	Devriese et al. (2001)
Oxytetracycline	96.66	3.33		T,B,L	Turkey	Erganis et al. (2012)
Macrolides						
Erythromycin		66	3.3		HU	Szabó et al. (2015)
	90	10		T,B,L	Turkey	Erganis et al. (2012)
		$MIC_{50} \geq 64$		B	NL	van Veen et al. (2001)
		$MIC_{90} \geq 64$				
		$MIC_{50} \geq 32$		T	DE	Waldow (2009)
		$MIC_{90} \geq 32$				

(continued)

Table 4 (continued)

Antimicrobial	Results of susceptibility testing in %			Animals	Comments	References
	Susceptible	Resistant	Intermed			
Tylosin	$MIC_{50} = 4$ $MIC_{90} = 8$			B	NL	van Veen et al. (2001)
		$MIC_{50} \geq 64$ $MIC_{90} \geq 64$		B	BE	Devriese et al. (2001)
Tilmicosin		$MIC_{50} = 32$ $MIC_{90} = 32$		T,L,PH	DE	Popp (2003)
		$MIC_{50} \geq 64$ $MIC_{90} \geq 64$		B	NL	van Veen et al. (2001)
		$MIC_{50} \geq 64$ $MIC_{90} \geq 64$		B	BE	Devriese et al. (2001)
Spiramycin		$MIC_{50} \geq 64$ $MIC_{90} \geq 64$		B	BE	Devriese et al. (2001)
	13.3	86.3		T,B,L	Turkey	Erganis et al. (2012)
(Fluoro) quinolones						
Enrofloxacin		16.7	63.3		HU	Szabó et al. (2015)
	98	2		T	FR	Dudouyt et al. 1995
	96.6	3.33		T,B,L	Turkey	Erganis et al. (2012)
	$MIC_{50} = 4$ $MIC_{90} = 8$			T,L,PH	DE	Popp (2003)
	$MIC_{50} = 2$ $MIC_{90} \geq 8$			T	DE	Waldow (2009)
		95.6			BE	Devriese et al. (2001)
		91.6			NL	van Veen et al. (2001)
Ciprofloxacin		70			HU	Szabó et al. (2015)
Nalidixic acid		100			HU	Szabó et al. (2015)
Tiamulin	$MIC_{50} = 0{,}5$	$MIC_{90} = 32$		T,L,PH		Popp (2003)
	$MIC_{50} \leq 0.5$ $MIC_{90} = 2$			T		Waldow (2009)

3.7 *Pasteurella multocida*

Pasteurella multocida is a Gram-negative, non--motile, facultative anaerobic, causative agent of fowl cholera mostly acute fatal septicaemia of adult birds, but can be also an asymptomatic or mild chronic sinusitis and conjunctivitis or pneumonia-like pasteurellosis. Mostly capsular types A are associated with fowl cholera, but types F and D were also reported (Peng et al. 2019a). Serotype B:3 was reported to be linked with avian sinusitis.

As for virulence factors genes associated to outer membrane proteins (*omp*H, *oma*87, *psl*), *plp*B, and *plp*E encoding for protective surface antigens, adhesion (*ptf*A, *pfh*A, *tad*D, *hsf*-1), iron metabolism (*exb*D-*ton*B, *fur*, *hgb*A), sialidases (*nan*B) and dismutases (*sod*A, *sod*C), capsule biosynthesis (*hya*D-*hya*C), and hyaluronic acid synthetase (*pmH*AS), were reported in poultry isolates (Wilson and Ho 2013; Furian et al. 2016).

In article written by Nhung et al. (2017) summarisation of phenotypic resistance in 617 isolates from poultry is given, commenting median and interquartile range (IQR) of resistance for ampicillin (median 2.3%; IQR 0.6–13.5%), gentamicin (4.3%; IQR 1.8–11.1%), erythromycin (18.0%; IQR 2.7–64.1%), florfenicol (0.6%; IQR 0–1.6%), tetracycline (13.8%; IQR 7.6–40.0%), co-trimoxazole (10.8%; IQR 0–20.0%), and enrofloxacin (4.7%; IQR 1.0–22.0%). From the above-listed results seems that most frequent is resistance to erythromycin and tetracyclines in *P. multocida* isolates from poultry.

Authors (Wu et al. 2003) of the study targeted on avian isolates of *Pasteurella multocida* from outbreaks of fowl cholera in Taiwan described sequences of the two plasmids, designated as pJR1 and pJR2. Mobilisable plasmid pJR1 contained among the other genes, those encoding for resistance to sulphonamides (*sul*II), tetracyclines (*tet*(G), and chloramphenicol (*cat*B2). The plasmid pJR2 involved genes encoding an aminoglycoside adenylyltransferase that confers resistance to streptomycin and spectinomycin (*aad*A1), beta-lactamase that confers resistance to ampicillin and carbenicillin (*bla*P1) Sequence comparisons showed that the high degree of homology was proven for antibiotic resistance genes found in both plasmids to the corresponding genes found in a great variety of Gram-negative bacteria.

3.8 *Riemerella anatipestifer*

Riemerella anatipestifer is a Gram-negative, non-motile, non-spore-forming, and rod shaped. Causative agent of disease ducks, geese, turkeys, and other avian species. This pathogen is well described especially in ducks, where is etiological agent is known to cause serositis, air-saculitis, meningitis, salpingitis, or septicaemia with high mortality rates (Zhong et al. 2009; Sun et al. 2012a, b; Li et al. 2016; Chen et al. 2018). Most studies come from Asia (China, India, Taiwan), where ducks are traditionally kept. Publicly available are results of study reporting results of susceptibility testing via Kirby-Bauer disc diffusion test (from 224 isolates of *R. anatipestifer* from China, 1998 and 2005, where 50% of the isolates were resistant against ceftazidime, aztreonam, cefazolin, cefepime, cefuroxime, oxacillin, penicillin G, rifampicin, and sulphonamide/trimethoprim (Zhong et al. 2009). It should be noted that from the perspective of orally administered approved antimicrobials in the EU region, phenoxypenicillin and sulphonamide/trimethoprim combination only can be considered to be used for treatment in practice. Another study (Sun et al. 2012a, b) from China (103 *R. anatipestifer* isolates from ducks, 2008 and 2010, agar dilution method) brought results of MIC_{50} and MIC_{90}: high levels (32 to ≥128 µg/ml) reported for aminoglycosides (streptomycin, kanamycin, gentamicin, apramycin, amikacin, and neomycin), nalidixic acid, and sulphadimidine. MIC_{90} 8 µg/ml was detected for ampicillin and florfenicol. Genes encoding for resistance as well as integrons was determined using PCR. The genes bla_{TEM-1}, several genes conferring resistance to aminoglycosides: *aph* (3′)-VII, *aad*A1, *aad*A2, *aac*(3′)-IV, *aac*(3′)-IIc, *aac*(6′)-Ib, phenicols: *cat*2, *cml*A, *flo*R, sulphonamides: *sul*1, and *sul*2 (newly also *sul*3) and tetracyclines: *tet*(A), *tet*(B), *tet*(C) were described by Sun et al. (2012a, b). Also should be noted that further study performed some years later shown *tet*(A), *tet*(M), *tet*(Q), *tet*(O), *tet*(B), and *tet*(O/W/32/O) genes in the *R. anatipestifer* Chinese duck isolates (Zhu et al. 2018). Strains in which nalidixic acid MICs ≥32 µg/ml (No 43) showed mutations in *gyr*A leading to the amino acid exchanges Ser83-Ile (in 86% tested isolates). In five isolates with a ciprofloxacin MIC of

>16 μg/ml was proven point mutations in *par*C (Arg120-Glu) (Sun et al. 2012b). As for resistance to fluoroquinolones study by Chen et al. (2018) brought the information on PMQR resistance genes *qnr*S and *qnr*D. Macrolide resistance, rRNA modification mediated by the *erm*F methyltransferase is proven to be the predominant mechanism of resistance to erythromycin in *R. anatipestifer* Chinese isolates (Luo et al. 2015). Study by Li et al. (2016) and Chen et al. (2018) demonstrated the role of efflux pumps on resistance in *R. anatipestifer*, as well as show the possible efflux inhibitors role.

3.9 *Staphylococcus aureus*

Staphylococcus aureus causes a wide range of chicken diseases, including septic arthritis, subdermal abscesses, and gangrenous dermatitis (Bystroń et al. 2010). Clusters of related isolates, grouped into clonal complexes (CCs) that share five or more alleles at seven MLST loci were described in *S. aureus* and clonal complexes pose variability, which can be linked to the sources of isolates—e.g. complex CC385 were isolated in birds including poultry, but not among human and mammalian species (Lowder et al. 2009), other lineages, such as CC398, CC5, and CC9 were isolated from chickens, turkeys, humans, and other hosts (Monecke et al. 2013; Sharma et al. 2019; Anjum et al. 2019), especially CC5 is of great importance as a chicken pathogen (Lowder et al. 2009; Bystroń et al. 2010), interestingly acquisition of this CC5 lineage is deemed to be by a single human to poultry "host jump"decades ago, after which the genome of this lineage has been changed to be more adapted to avian species as well as gaining novel mobile genetic elements (Lowder et al. 2009). Therefore, additional 47 specific genes were identified exclusively in poultry and not in human isolates, from these all 47 genes were also detected in poultry CC385; 41 genes in poultry CC398; 38 genes were also present in poultry CC1 in the various extent of strains investigated. Very interesting finding was also proved as for adaptation of poultry isolates, in which enhanced growth and erythrocyte lysis in avian body temperature were proven. Poultry accessory genome contains also genes encoding, e.g. *fnbB* fibronectin-binding protein facilitating colonisation and attachment, thiol protease *scpA* contributing virulence and hemolysis as well as *S. aureus* pathogenicity islands (Murray et al. 2017). Study by Argudín et al. (2013) worked with 34 isolates *of S. aureus* isolated from turkeys in France can be an example of the profile of *S. aureus* resistance and virulence genes. Isolates were classified into clonal complexes CC398, CC5, CC101, and CC121, one isolate was MRSA. All methicillin-sensitive isolates (MSSA) carried specific φAvβ prophage avian-niche-specific genes, what is in line also with another study by Price et al. (2012) that investigated broiler chicks *S. aureus* isolates prophage. All strains were resistant to penicillin (*bla*Z, *bla*I, and *bla*R) and tetracycline (mostly gene *tet* M and 3 isolates *tet*K), gene *erm*C was detected in strains resistant to erythromycin). In MRSA isolate streptogramin resistance gene vgaA, the quarter ammonium compounds resistance gene *qac*C encoding for efflux pump, and some SCCmec genes were detected. All MSSA strains harboured an intact beta-haemolysin gene (*hlb*), while the MRSA isolate *hlb* gene truncated after the probable insertion of the immune-evasion phage-borne genes *sak* (staphylokinase), *chp* (chemotaxis inhibitory protein), and *scn* (staphylococcal complement inhibitor), what was in line with the study from Germany (Monecke et al. 2013). Moreover, the MRSA strain also carried the genes encoding enterotoxin G (*ent*G) adhesion (bbp and sdrD), immune evasion proteins (*mpr*F), and the site-specific deoxyribonuclease subunit 2 (*hsd*S2). In individual isolated genes for resistance to copper (*cop*B) and apramycin (*amp*A) were proven.

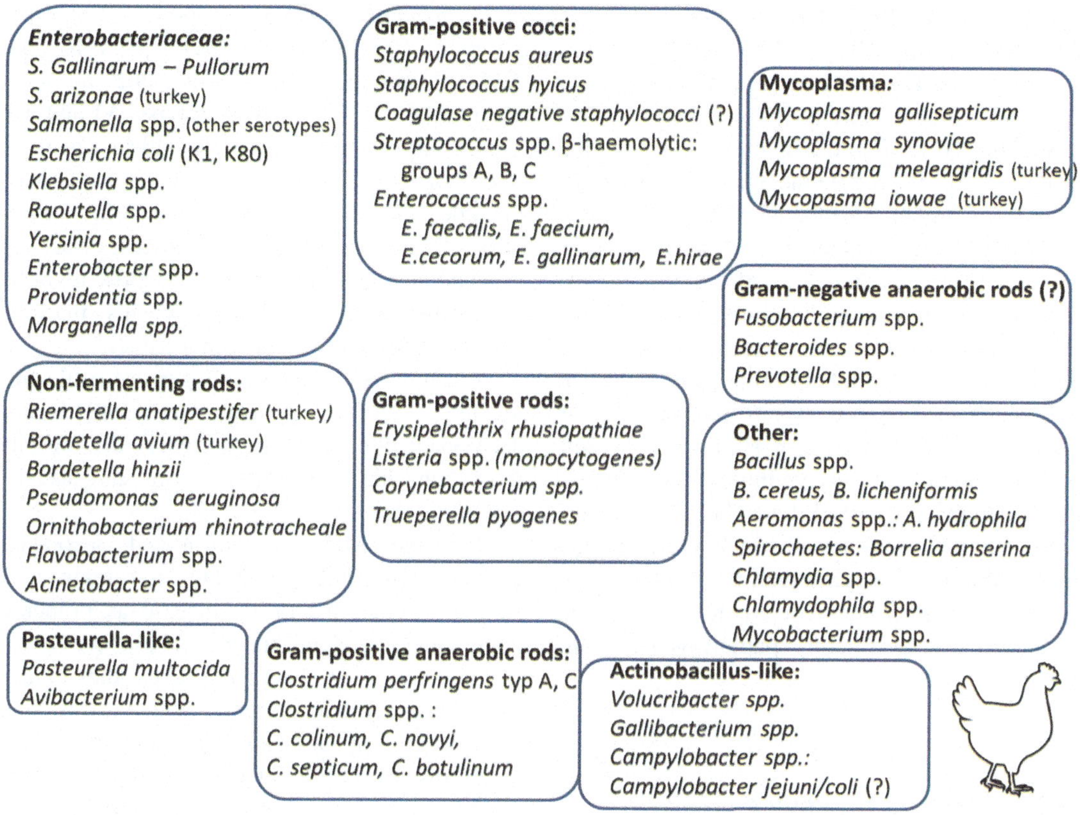

Fig. 3 Bacteria being considered as having pathogenicity in poultry and associated with poultry diseases

4 Specific Consideration for Pathogens and Antimicrobials with Importance in Cattle

4.1 *Acinetobacter* spp.

From the European isolates, of 50 faecal samples from a French dairy herd, 9 revealed *A. variabilis* (formerly 15 TU)—Bentz et al. 2002) possessing the *bla*$_{OXA-23}$ gene on Tn2008 (Poirel et al. 2012), and 2 of 45 nasal and rectal samples from cattle in Germany revealed *Acinetobacter indicus*-like isolates harbouring *bla*$_{OXA-23}$ localised on an interrupted Tn2008 transposon (Klotz et al. 2017), suggesting that these *Acinetobacter* spp. may play a role in the dissemination of *bla*$_{OXA-23}$ to *A. baumannii*. Further studies investigating Acinetobacters and their role not only as for pathogenesis in animals but also in transfer of resistance genes among human–animals–environment should be performed to fill the knowledge gap.

4.2 *Moraxella* spp.

Moraxella bovis is a Gram-negative, aerobic, oxidase-positive diplococcus, belonging to the order Pseudomonadaceae. It is the causative agent of infectious bovine keratoconjunctivitis, an eye disease of cattle. Limited data of phenotype—MICs testing of resistance to antimicrobials are available, while up to date no genetic profiles of genes encoding for resistance are publicly available. Data for 106 isolates of *M. bovis* show MIC testing results against 18 antimicrobials with a range of concentrations The MIC_{90} values were as follows: beta-lactams: penicillin ≤0.12 μg/ml; ampicillin ≤0.25 μg/ml; ceftiofur ≤0.25 μg/ml; tetracyclines: chlortetracycline = 1 μg/ml; Oxytetracycline = 1 μg/ml; aminoglycosides: gentamicin = 1 μg/ml; neomycin = 4 μg/ml; macrolides: tulathromycin = 2 μg/ml; and tylosin = 8 μg/ml; clindamycin = 2 μg/ml; danofloxacin and enrofloxacin ≤0.12 μg/ml; florfenicol = 0.5 μg/ml; spectinomycin = 16 μg/ml; sulphadimethoxine ≤256 μg/ml; tiamulin ≤0.5 μg/ml; trimethoprim–sulphamethoxazole ≤2 μg/ml. Lowest susceptibility was indicated for clindamycin (most of the isolates fell to category intermediate), for all other antimicrobials susceptibility was higher than 91%, the authors used CLSI interpretative criteria for bovine Gram-negative respiratory pathogens (Loy and Brodersen 2014).

4.3 *Mycoplasma bovis*

Based on phenotype tests (MIC levels) antimicrobial resistance by *M. bovis* to aminoglycosides, fluoroquinolones, lincosamides, macrolides, phenicols, pleuromutilins, and tetracyclines has been reported and appears to be increasing. The mechanisms of *M. bovis* antimicrobial resistance are considered to be largely based on genetic point mutations; only few studies have examined efflux mechanisms and no plasmids have so far been detected in *M. bovis* (Cai et al. 2019).

4.4 Pasteurellaceae

4.4.1 *Pasteurella multocida, Mannheimia haemolytica,* and *Histophilus somni*: Bovine Isolates

Pasteurella multocida (PM)—Capsular antigens A (cattle), B and E (causative agents of cattle haemorrhagic septicaemia), capsular antigen F may be involved also in fatal peritonitis of calves.

M. haemolytica (MH) comprises 12 capsular serotypes (A1, A2, A5–9, A12–14, A16, and A17). Respiratory diseases in cattle are most commonly associated with Serotypes A1 and A6.

H. somni (HS) is the etiological agent of thromboembolic meningoencephalitis (TEME) in cattle. It has also been associated with various other diseases in sheep, and diseases such as bronchopneumonia, necrotic laryngitis, myocarditis, arthritis, conjunctivitis, myositis, mastitis, abortion, and lightweight feeder calves.

All above-mentioned pathogens of cattle frequently occur as peracute or acute forms, accompanied by a high mortality rate, despite those subacute and chronic forms can also be caused by them.

PM and MH are considered secondary pathogens associated with the final progress of serious bovine respiratory disease as bronchopneumonia and pleuropneumonia or claves enzootic pneumonia. As being secondary, they invade the respiratory tract together with viruses/other bacteria (e.g. *Mycoplasma* spp.). The seriousness of the disease is promoted by factors like stress conditions (transport, mixing of animals from different herds, overstocking, and poor ventilation).

As for the genetic background of antimicrobial resistance, the importance of ICE*Pmu1* harbouring 12 antibiotic resistance genes, which confer resistance to streptomycin-spectinomycin (*aadA25*), streptomycin (*str*A and *str*B), gentamicin (*aad*B), kanamycin-neomycin (*aph*A1), tetracycline [*tet*R-*tet*(H)], chloramphenicol-florfenicol (*flo*R), sulfonamides (*sul*2), tilmicosin-clindamycin [*erm*(42)], and tilmicosin-tulathromycin [*msr*(E)-*mph*(E) should be highlighted (Michael et al. 2012a) and for groups of antimicrobials used in cattle following information can be summarised:

Aminocyclitols: Ribosomal mutations in 16S rRNA, spectinomycin binding site, conferring spectinomycin resistance, as well as mutations in *rps*E encoding for ribosomal protein S5, was described (Michael et al. 2018).

Aminoglycosides: Enzymatic inactivation (adenylation, acetylation, or phosphorylation) is a common mechanism of resistance as well as chromosomal mutations. Resistance to the oldest used aminoglycoside streptomycin is harboured either on plasmids (small non-conjugative as well as conjugative plasmid harbouring resistance to multiple antimicrobials). A novel streptomycin–spectinomycin resistance gene, designated *aadA14*, was identified on a small 5.2-kb plasmid from a bovine fatal peritonitis *P. multocida* capsular type F isolate from Belgium (Kehrenberg et al. 2005). Two ICEs from Pasteurellaceae, ICE*Pmu1* and ICE*Mh1*, have been shown to contain genes associated with resistance to streptomycin and other aminoglycosides and aminocyclitols. ICE*Pmu1* carries genes encoding resistance to streptomycin (*str*A and *str*B) and also adenyltransferase (*aad*A25) and *aad*B and phosphotransferases (*aph*A1), what causes that the strains with such elements are resistant to streptomycin, spectinomycin, gentamicin, kanamycin, and neomycin. A novel *aadA31* gene encodes a spectinomycin/streptomycin adenylyltransferase and was located in a variant of the integrative and conjugative element ICE*Mh1*, a mobile genetic element transmissible among members of the family Pasteurellaceae (Cameron et al. 2018).

Beta-lactams: This resistance among Pasteurellaceae is often associated with small plasmids. Recently Kadlec et al. (2019) have been reported plasmid harboured extended-spectrum beta-lactamase gene $bla_{\text{ROB-2}}$ in *M. haemolytica* with high MICs to cephalosporins.

Fluoroquinolones: Resistance is commonly due to mutational alterations in the genes coding for the enzymes gyrase and topoisomerase IV, but also due to active efflux or protection of the enzymes. Quinolone resistance determining region (QRDR) of the proteins encoded by the genes *gyrA*, *gyr*B, and *parC* in *P. multocida* and in *M. haemolytica* genes *gyrA* and *parC* (Katsuda et al. 2009; Kong et al. 2014).

Macrolides: *P. multocida* and *M. haemolytica* strains that possessed one or more of the macrolide resistance genes *erm*(42)—encoding rRNA methylase, *msr*(E) encoding macrolide efflux and *mph*(E) encoding phosphotransferase were reported by Beker et al. (2018). Different combinations of these genes conferred distinct resistance phenotypes to the 15-membered ring macrolides (gamithromycin and tulathromycin) and the 16-membered tilmicosine and tildipirosine) (Rose et al. 2012). Full battery of all three genes was required to attain high-level resistance to all these drugs, whereas just *mrs*(E)—*mph*(E) genes are present in isolates not resistant to clindamycin, with lower MICs for tilmicosin and higher MICs for gamithromycin and tulathromycin, compared to strains harbouring *erm*(42) gene.

High-level macrolide resistance can arise from 23S rRNA mutations in *P. multocida and M. haemolytica* (Olsen et al. 2015).

Phenicols: chloramphenicol resistance is most frequently due to enzymatic inactivation of the drug by chloramphenicol acetyltransferases *cat*A and *cat*B, and florfenicol resistance is based on the efflux pump system and associated with the presence of mobile genetic elements (plasmids, transpozones, gene cassettes) located gene *flo*R (Kehrenberg et al. 2008; Katsuda et al. 2012).

Sulphonamide/Trimethoprim resistance: Gene *sul2*, is frequently found on small plasmids in PM, MH, and HS, but also has been described on conjugative or non-conjugative plasmids and on ICE*Pmu1* (Michael et al. 2012b) together with genes encoding resistance to further antimicrobials. Also clusters of genes *sul2–catA3–strA* was described in *Mannheimia*.

Tetracycline resistance: Mostly due to *tet* genes (*tet*(A), *tet*(B), *tet*(C), *tet*(G), *tet*(H), *tet*(L), and *tet*(K) (Peng et al. 2019a). Transposon, Tn*5706*, was identified in 1998 on plasmid pPMT1 from a bovine *P. multocida* isolate. Interestingly some elements of this transposon were also detected as part of ICEs in isolates of *P. multocida*, *H. somni*, and *M. haemolytica*

(Michael et al. 2012a; Eidam et al. 2015; Klima et al. 2014a). *M. haemolytica* and *P. multocida* strains originating from cattle (Belgium isolates) harbour also gene *tet*(L).

Genes tet(M) encoding for ribosome protective proteins, have been identified in *P. multocida*, is located on conjugative transposon Tn916 and interestingly is considered as one of the most common *tet* gene among bacteria.

An integrative conjugative element; ICE*Hs1*, was identified containing 83 genes, including tetracycline resistance gene *tet*(H), a multidrug efflux pump gene *ebr*B (which was detected in neomycin-resistant isolates), and metal tolerant genes *mco* encoding for multicopper oxidase, *czc*D associated with Cu and Zn tolerance and *acr*3 encoding an arsenical efflux protein that pumps outside of bacterial cell arsenite. The ICE*Hs1* is an active element capable of intra- and inter-genus transfer as demonstrated by successful transfer to *H. somni* and *P. multocida* recipients and high homology (90% identity) with previously described ICEs of *H. somni*, *P. multocida*, and *M. haemolytica* was proven (Bhatt et al. 2018).

Within the study by Beker et al. (2018), the combination of *aphA1-strB-strA-sul2* genes was observed in several different ICEs, and these clusters with *flo*R and *erm*(42) in individual strains of *P. multocida* (Pmu3358) and *M. haemolytica* Mh6055. Moreover, same strains of *P. multocida* (and one more PmXX) and *M. haemolytica* harbour the combination of *aadB-aadA-bla*$_{OXA-2}$*–msr(*E*)-mph(*E*)-tetH* genes.

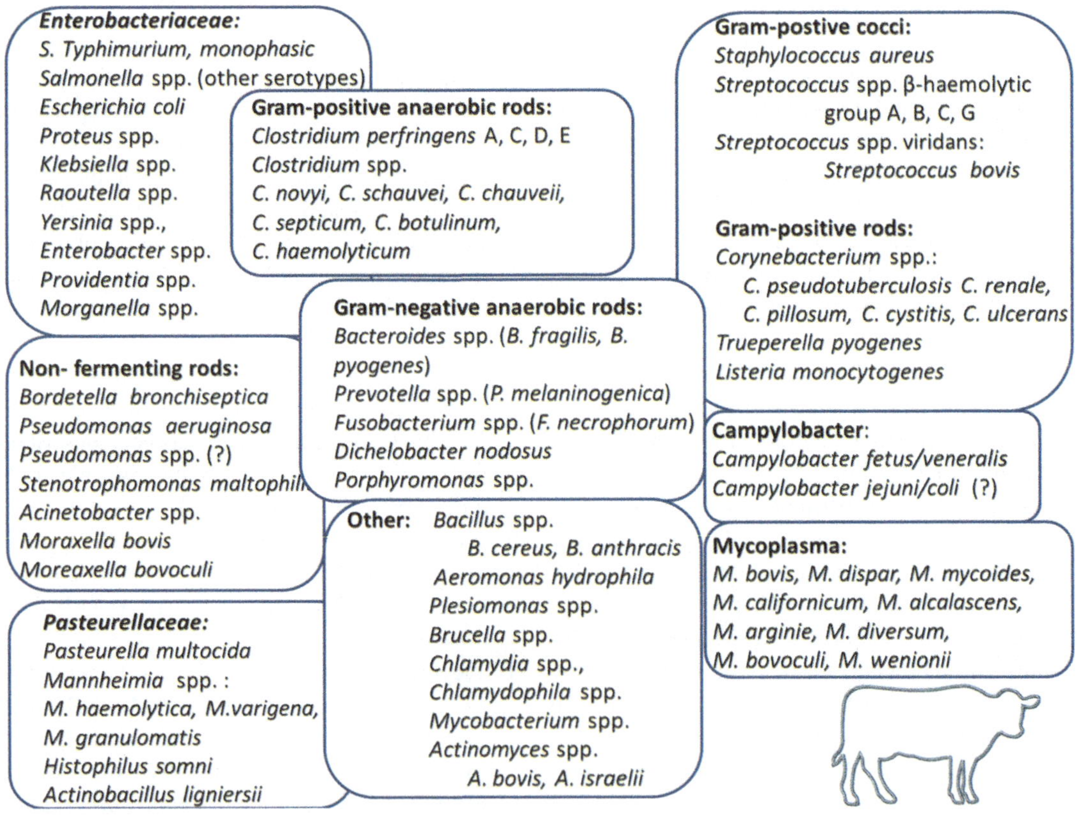

Fig. 4 Bacteria being considered as having pathogenicity in cattle and associated with cattle diseases

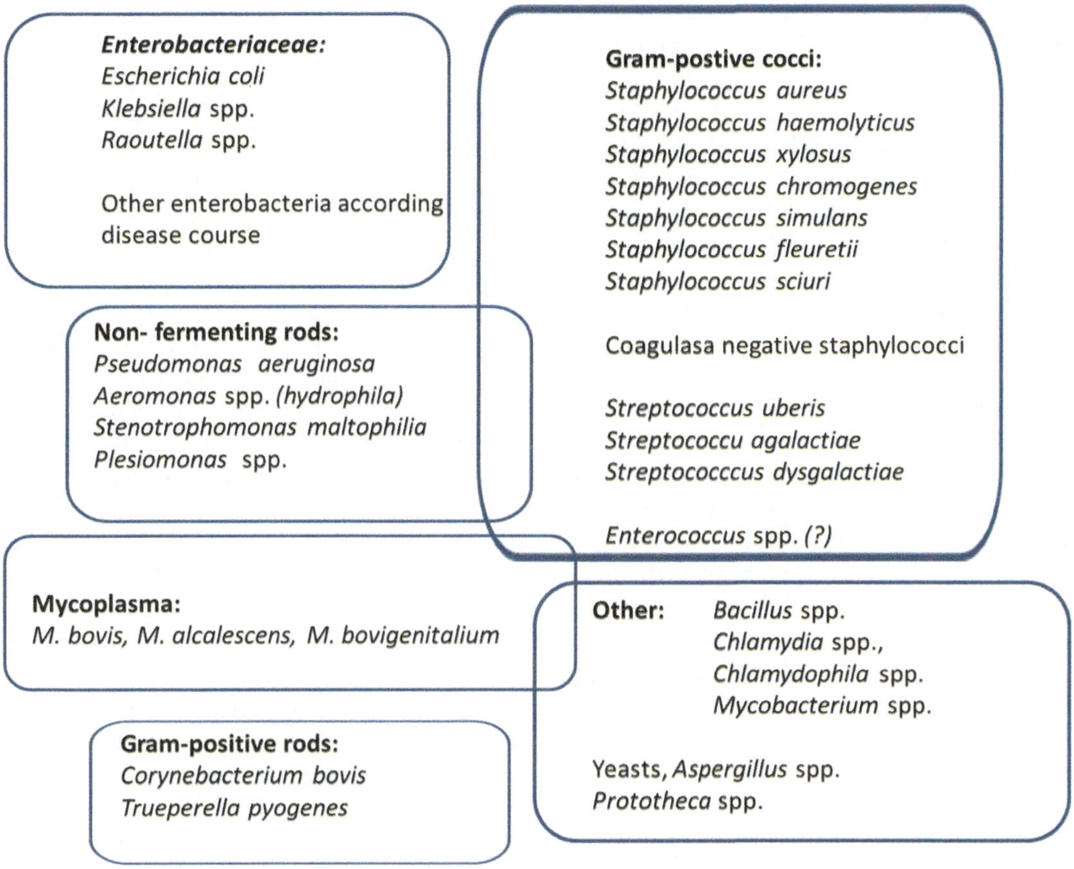

Fig. 5 Bacteria being considered as associated with cattle mastitis

4.5 Cattle mastitis pathogens

4.5.1 *Streptococcus uberis*

Streptococcus uberis belongs among Gram-positive, catalase-negative cocci, it is facultatively anaerobic and during last years is in many European studies reported as the most frequent Gram-positive causative agent of bovine mastitis (Botrel et al. 2010; Vakkamäki et al. 2017; Käppeli et al. 2019), e.g. in the French surveillance system RESAPATH is reported that *S. uberis* created 80% of isolates in 2018 and in the Czech Republic it was 53.8% of all isolates from milk in clinical mastitis.

All *S. uberis* strains produce free hyaluronidase that enhances the distribution of the pathogen within tissues and further important virulence factors/genes have also been detected in *S. uberis* together with their function: *pur*B (adenylosuccinate lyase responsible for nucleotide transport and metabolism), *lep*A (GTP-binding protein; associated with cell envelope biogenesis, outer membrane located), *gid*A (tRNA (uracil-5-)-methyltransferase, probably involved in translation, ribosomal structure, and biogenesis), *fba* (fructose-bisphosphate aldolase, transport, and metabolism of carbohydrates, associated with bacterial adhesion), and *ccp*A (catabolite control protein, transcriptional regulators) (Reyes et al. 2019).

Resistance to different groups of antimicrobials is based on the following genes in the respective pharmacological classes of antimicrobials. Phenotypic resistance to aminoglycosides is reported to be low for a long term in *S. uberis*, what can be seen from the

Dutch data (MARAN, isolates 2002–2008, rare resistance), in French isolates, according to RESAPATH dataset 2018, streptomycin resistance was 14%, gentamicin and kanamycin 2% and 7%, respectively. *APH*(3′) and *ANT*(6) genes encoding for enzymes inactivating kanamycin and neomycin are located in transposons in Gram-positive bacteria, ANT(6) was exclusively detected in *S. uberis*. In the case of beta-lactams, phenotypic resistance reported was low for a long term in *S. uberis*, what is supported by the data from the Netherlands (MARAN, isolates 2002–2008, rare resistance), in France—RESAPATH (ANSES 2019), where oxacillin susceptibility account for 88% and in the Czech Republic, where susceptibility of isolates of *S. uberis* ($N = 301$) to penicillin was 93.7% in 2016 (oxacillin 99.7%), (CZ NMTP 2016). Resistance to this group of antimicrobials is usually encoded by genes translated to beta-lactamases (*bl2b*, *TEM*—family). Recent study (Vélez et al. 2017) investigating isolates of *S. uberis* from clinical mastitis in Canada identified *bl2b* as well as series of TEM family genes: *TEM-1*, *TEM-127*, *TEM-136*, *TEM-157*, *TEM-163*, *TEM-47*, *TEM-71*, *TEM-89*, and *TEM-95*). In fluoroquinolones: Gene *gyr*A was identified as encoding for fluoroquinolone resistance (Vélez et al. 2017), interestingly phenotypic data from RESAPATH dataset 2018 indicate significant difference among enrofloxacin (37% resistance) and marbofloxacin (8% resistance). In the case of lincosamides, phenotypic resistance seems to be moderate according to the data from the Netherlands (MARAN, isolates 2002–2008, around 40%), French recent data from RESAPATH (ANSES 2019) indicate in lincomycin 18% resistance, and the data from the Czech Republic shows 33.7% isolates ($N = 362$) reported as resistant to clindamycin (CZ NMTP 2018). Genes *linB (newly lnuB) and linD* (adenylation of clindamycin) create a genetic background for resistance to lincosamides in *S. uberis* (Haenni et al. 2011; Vélez et al. 2017). Some authors mentioned also mph gene family phosphotransferases as a possibility, once resistance to lincosamides is appeared in streptococci (Haenni et al. 2018). In the case of pirlimycin, recent work studied occurrence of resistance genes in urine and slurry from cows treated with pirlimycin and detected genes *mefA*, *tet(W), and cfxA* (Li et al. 2019b). Phenotypic resistance to macrolides as well as associated resistances to related substances seems to be low to moderate according to the data from NL (MARAN, isolates 2002–2008, around 20%), RESAPATH isolates are reported as resistant in 15% for erythromycin, in 21% for tylosine, and in 10% for tulathromycin, while CZ data 4.7% of isolates ($N = 362$) resistant to erythromycin (CZ NMTP 2018). The ribosomal methylase, encoded by the *erm*(B) and *erm*(C) genes has been identified as the main determinant of macrolide–lincosamide–streptogramin (MLS) resistance in *S. uberis* and can be horizontally transferred as plasmids between bacteria (Vélez et al. 2017). The *erm* (B) was reported to be located on mobile genetic elements together with *tet* determinants (Haenni et al. 2018). Second mechanism of resistance to macrolides in pathogenic streptococci is associated with Mef efflux pumps and has been explicitly commented, especially in *S. pyogenes* and *S. pneumoniae* of human origin, but a study by Entorf et al. (2016) determined in isolates of *S. uberis* of bovine origin genes *mef*A and *mef*E. The combination of *mph*(B) (encoding for phosphotransferase) and *rdm*C-like genes resulted in a resistance to spiramycin in *S. uberis* in the case described by a study by Achard et al. (2008), but surprisingly not lead to resistance to tylosin and erythromycin (and also azithromycin and josamycin, which are, used in human medicine). Phenotypic resistance towards tetracyclines seems to be moderate to high according to the data from NL (MARAN, isolates 2002–2008, around 40%), FR (RESAPATH, increasing trend, isolates 2006, 14% and 2018, 20% (ANSES 2019), and CZ isolates 2018, $N = 362$, 55.8% (CZ NMTP 2018). Gene *tet* (M), as well as *tet*(S), was present in *S. uberis* (Vélez et al. 2017) isolated from mastitis milk in Canada; data on genetic background of tetracycline resistance specifically in *S. uberis* recent European isolates are not publicly available, according to the authors' best knowledge. Study of *S. uberis* from Canada describes also the most frequent combination of genes encoding for resistance to antimicrobials that were detected in 62 isolates from clinical and subclinical mastitis: *bl2b—Tem157–linB–lnuB–TetM* (Reyes et al. 2019). Rifamycins: Genetic background for resistance to ansamycins (of importance as antituberculotic drugs used in human medicine) *rpo*B gene has been detected in isolates of *S. uberis* by Vélez et al. (2017) and recent data from RESAPATH dataset 2018 show 50% resistance, while in the Czech Republic 44.5% of isolates was classified as susceptible, 52.5 as Intermediate and 3% as resistant according to the MIC profile identified in $N = 362$ isolates (CZ NMTP 2018).

4.5.2 *Streptococcus dysgalactiae*

Streptococcus dysgalactiae can be recently considered as the second most prevalent causative agent for bovine mastitis (Bigs (2018), CZ data (Slosarkova et al. 2019), FR data—RESAPATH (ANSES 2019) shows 13% prevalence of *S. dysgalactiae* ($n = 207/1572$). For aminoglycosides—low phenotypic resistance (up to 3% for streptomycin and even less for gentamicin (1%) and kanamycin (2%) was detected in RESAPATH (ANSES 2019) reporting French isolates. According to Vélez et al. (2017) ***APH(3′)*** phosphotransferase is responsible for resistance to aminoglycosides. In the case of beta-lactams: Phenotypic resistance detected among isolates from NL, 2002–2008 was 25% according to MARAN (2011) with data 2009, but in the last RESAPATH report (2018) 98% of isolates were considered susceptible. Genes *TEM-1*, *TEM-47*, *TEM-71*, *TEM-136*, *TEM-157*, *and bl2b* encoding for beta-lactamases were detected in Canadian strains *(*Vélez et al. 2017*). For fl*uoroquinolones—phenotypic resistance detected among isolates from FR was around 50% (RESAPATH 2018) Gene *gyr*A encoding for DNA gyrase subunit A and *par*E encoding for topoisomerase IV subunit B was declared as responsible for resistance in the strains of *S. dysgalactiae* (Vélez et al. 2017). Phenotypic resistance to macrolides reported in RESAPATH (ANSES 2019) was 16% and trend of decrease comparing to 2015 (22%) is proven. Vélez et al. (2017) did not detect any macrolide-resistant genes in their collection of 25 isolates of *S. dysgalactiae*, but other authors detected *erm*(B) genes encoding for resistance to macrolides, lincosamides, and streptogramins (Entorf et al. 2016*;* Zhang et al. 2018a, b). Phenotypic resistance to lincosamides reported in RESAPATH (ANSES 2019) was 12%. Interestingly *pho*P gene encoding for phosphate regulon transcriptional regulatory protein PhoP causing resistance to polymyxins was identified in *S. dysgalactiae* (Vélez et al. 2017) *as well as rpo*B encoding for DNA directed RNA polymerase beta subunit important for resistance to rifampicin and derivatives (Vélez et al. 2017). As for phenotypic resistance to tetracyclines, the group with the highest rate of resistance in streptococci of bovine origin, data from NL indicate up to 70% resistance among the isolates 2002–2008 and in the last RESAPATH report indicate 81% ($N = 183$, 2018), hypothesise that this is mainly due to transposone Tn*916* harbouring *tet* gene (ANSES 2019). Resistance gene *tet*(M) has been identified in *S. dysgalactiae (*Vélez et al. 2017*)*.

4.5.3 *Streptococcus agalactiae*

Streptococcus agalactiae a Gram-positive cocci, considered among the major streptococci pathogen, associated mainly with subclinical and mild-to-moderate clinical mastitis, but the incidence of the bovine mastitis caused by *S. agalactiae* dramatically decrease during last two decades predominantly due to the improved zoohygienic practices in dairy cow farming (Haenni et al. 2018). Current report, e.g. from the European countries show prevalence far below 10%, e.g. French surveillance report RESAPATH dataset 2018 shows 2% only, but studies coming from South America or China reported over 50%, or even 90% prevalence, respectively. On the other hand, *S. agalactiae* can be a very serious pathogen, considering infections of newborn babies, where transmission comes from mothers to babies (e.g. in Latin American countries up to 26% of positivity in pregnant women. Therefore, despite recent European data on genetic background of resistance are mostly targeted on *S. uberis* and *S. dysgalactiae*, as for

S. agalactiae some data still exist including the comments of possible transfer/homology of bovine and human isolates). Isolates CC103 were reported mostly for bovine/mastitis and not for humans, but from some regions (Asia, Colombia) isolates CC61/67 were reported in both cattle and human (Li et al. 2018a; Cobo-Angel et al. 2019). This is quite surprising as, originally this clonal complex was considered as mostly adapted to bovine, udder conditions by bovine-specific virulence factors, e.g. the lactose operon (Richards et al. 2011) as well as loss of human-specific virulence factors, e.g. capsule (Almeida et al. 2016). As for determinants of resistance, mostly *erm*(A) and *erm*(B) were reported as for resistance to macrolides (conferring resistance also to lincosamides and streptogramins) according to summary of the data from different parts of the world (Haenni et al. 2018). A study by Cobo-Angel et al. (2019) tetracycline resistance was encoded by genes *tet*(M) and rarely by genes *tet*(K).

Streptococcus, especially *S. agalactiae*, possesses a variety of virulence factors that contribute to pathogenicity. Several surface proteins and polysaccharide capsules were identified within this species. The *scp*B gene (only found in group B Streptococci) encodes for surface enzyme ScpB—C5a peptidase impairing of neutrophil recruitment and bind fibronectin to promote bacterial invasion of epithelial cells (Beckmann et al. 2002). The *bca* gene encodes for α-protein, a surface protein important for entering to the host cells, the *lmb* gene encodes for (laminin-binding protein) determined in *S. agalactiae* playing an important role in the adherence to host cells.

Further genes as *cyl* (encoding for β-haemolysin, that plays a role in tissue damage, spread to body tissues, including to the memings), *gln*A (glutamine metabolism, important for virulence) *cfb* (Christie–Atkins–Munch–Peterson (CAMP) factor impairing host immune defence response), *hyl*B (hyaluronate lyase contributing to invasion to the host), and *sca*A (aggregation factor) (Dmitriev et al. 2002; Ding et al. 2016).

4.5.4 *Klebsiella* spp.

Gram-negative rod-shaped, encapsulated, facultative anaerobe, belonging to the Enterobacteriaceae, among them seral species of *Klebsiella* genus can cause clinical mastitis. *Klebsiella pneumoniae* represents among these pathogens, mostly isolated and frequent pathogen, that can be characterised as an opportunistic environmental pathogen, with important transmission routes via contaminated faeces and bedding materials. *K. pneumoniae* is also clinically important in human medicine as a causative pathogen of serious nosocomial infections, such as septicaemia, pneumonia, urinary tract infections, surgical site infections, and soft tissue infections. Other species represented and reported in association with mastitis are *Klebsiella oxytoca*, *Klebsiella terrigena,* and rarely in the specific area also *Klebsiella variicola* (Podder et al. 2014).

Virulence genes and factors were identified by Yang et al. (2019). The enterobactin loci, adhesion-related gene clusters, secretion system-related gene clusters, and fimbria gene clusters were found in both human and veterinary (mastitis) isolates. A total of 135 genes were described in mastitis isolates (among them 26 were exclusive to the subclinical mastitis cases and 6 were exclusive to the clinical mastitis cases), in human isolates bigger portfolio of virulence genes was detected. Interestingly 132 genes were present in both human and bovine isolates. Genes expressing proteins related to metals' metabolism (iron, zinc, and calcium ions) were significantly more prevalent among investigated bovine isolates, moreover, in isolates from clinical mastitis ferric uptake operon *kfuABC was confirmed (*Yang et al. 2019*)*, the importance of this gene in *K. pneumoniae* was confirmed by authors of study (Gao et al. 2019) dealing with Chinese isolates, where *kfu* was detected in 31% of isolates ($N = 124$). This Chinese study identified in *K. pneumoniae* also further genes encoding for factors playing role in virulence as *ent*B (78%), *fim*H1 (55%), and *mrk*D (24%).

Recently, study results (Yang et al. 2019) have been released mapping resistance genes in

Klebsiella pneumoniae show that 40% (57/143) of isolates were resistant to one or more antimicrobial agents. Streptomycin (resistance to which was encoded by genes *str*A *and str*B) was identified as antimicrobial with the highest prevalence of resistance (29.4%), followed by tetracycline (5.6%), and gentamicin (4.2%). Comparing these data to isolates from the Czech Republic, gentamicin resistance was not reported in *K. pneumoniae* in 2018 ($N = 37$), and 2.4% of resistant isolates ($N = 42$) were detected in 2016, where also 2.4% isolates were reported as resistant to cefotaxime (CZ NMTP 2016, 2018). Within the isolates selected for genotyping 17 ARG with various prevalence of resistance genes was determined in bovine mastitis isolates described by Yang et al. (2019)—among the common resistance identified: two encoding for beta-lactam resistance bla_{AmpH} *and* bla_{SHV} was detected. Gene *oqx*AB quinolone resistance encoding gene was detected both by Yang et al. (2019) in mastitis isolates (US), but recently also within *K. pneumoniae* ST101 isolate from human in Italy genes *oqx*A and *oqx*B were described as well as fosfomycin resistance gene *fosA* described in both studies from mastitis and human isolates (Yang et al. 2019; Roe et al. 2019). Human isolates of ST101 lineage contains also genes for resistance to phenicols (*cat*B4, *cat*A2, *cml*A1, and *flo*R), macrolide, lincosamide, and streptogramin B resistance (*mph*E, *msr*E, *ere*A, *ere*B, and *mph*A), rifampicin (*arr*-2 and *arr*-3), sulphonamide (*sul*1, *sul*2, and *sul*3), tetracycline (*tet*(D), *tet*(X), and *tet*(A), and trimethoprim (*dfr*A1, *dfr*A5, *dfr*A14, *dfr*A16, and *dfr*A27). Moreover, significant resistance was detected towards beta-lactams (genes $bla_{\text{CTX-M-15}}$ and $bla_{\text{CTX-M-14}}$; $bla_{\text{SHV-1}}$ and specifically also towards carbapenems ($bla_{\text{KPC-2}}$ and $bla_{\text{OXA-48}}$ genes) in some strains belonging to lineage ST101 (Roe et al. 2019). In a study by He et al. (2016), *K. pneumoniae* harbouring $bla_{\text{NDM-5}}$ gene was isolated from milk and faecal samples of dairy cows with mastitis in China. In a study by Yang et al. (2019) plasmid *IncHI1B*-type harboured genes for streptomycin resistance in 27 of the 96 strains from clinical mastitis were detected and precisely described. Genes *str*A *and str*B were settled on transposon Tn*5393*, the most common mobile element playing most commonly a significant role in interspecies *str* resistance gene transmission. Moreover, in strains resistant to ceftiofur and macrolides, plasmid co-harboured the $bla_{\text{CTX-M-1}}$ and *mph*(A) genes were identified (Yang et al. 2019). Inc-N-like plasmid with a high level of homology was also described in *E. coli* isolates from lamb and human (Wang et al. 2014; Dolejska et al. 2013).

4.5.5 *Escherichia coli*

Escherichia coli is a Gram-negative, rod-shaped, facultative anaerobe, together with *Klebsiella* spp. one of the most notable cause of mastitis among Gram negatives. *E. coli* usually infected the mammary gland of cows at parturition and early lactation period which could lead to local and acute, in some cases systemic, severe mastitis.

Despite the fact that in some bibliography remain opinion that severity of mastitis caused by *E. coli* is mostly influenced by the host (cow) predisposition, also role of *E. coli* strains and their factors enabling and promoting the invasion are of high importance. Mastitis *E. coli* strains are, according to the recent data characterised by the lack of virulence genes, rather than by the presence of a combination of virulence genes (Blum et al. 2015), however, *E. coli* strains isolated from persistent mastitis cases showed increased adherence and persistence in a mammary epithelial cell line (Dogan et al. 2006). Of great importance is also ability of the *E. coli* strains to grow in milk (but standing alone, is not able to initiate the process of pathology). Polymorphonuclears (PMNs) are considered as key immunodefense factor for the resolution of mastitis. Within the study performed by Roussel et al. (2017) strain B1171 was proven to be fully resistant to phagocytosis by bovine PMNs. That can be partially explained by the presence of capsula. Another factor of high importance is that different *E. coli* strains stimulate mammary epithelial cells (MECs), which are key for the initiation of the innate immune response. It means that *E. coli* strains with low stimulation power and with high production of proinflamatory cytokines (TNF-α, IL-6, and IL-1β) and chemokines, producing capsular factors can cause more severe and quickly progressing infections. Moreover, those strains with PMNs resistance, low stimulation of MECs, and presence of gene *est*A1 encoding a heat-stable enterotoxin are finally responsible for mammary tissue lesions (Roussel et al. 2017). Very recent study working with 82 *E. coli* strains—isolates from mastitis (82 cows/66 acute mastitis/2 subclinical/1 chronic/13 unspecified/total 49 herds/2017/Zurich canton).

If tested, set of virulence genes can include, e.g. *afa (e.g. D-8 and E-8)*, *bfp*A (bundle-forming pilus); *cnf*1 and *cnf*2, *eae*A (intimin), *fyu*A (ferric yersiniabactin uptake protein), *hly*A (haemolysin), *iut*A (aerobactin siderophore receptor), *iuc*D, KpsMII (group 2 polysaccharide capsule), *pap*AH (pyelonephritis-associated major pilin protein), *pap*C (outer membrane usher protein), EF(fimbrial protein subunit), *sfa*S, *sub*AB (subtilase cytotoxin), PAI and *tra*T lipoprotein involved in serum resistance, *vat* and *yfcV* f17A (Marashifard et al. 2018; Nüesch-Inderbinen et al. 2019).

Genes of resistance to beta-lactamase $bla_{\text{CTX-M-1}}$ (isolates from FR, DE), $bla_{\text{CTX-M-14}}$ (FR), $bla_{\text{CTX-M-15}}$(DE), encoding for so-called ESBLs (extended-spectrum beta-lactamases) were identified in dairy cattle and are frequently located on plasmids, where most prevalent replicons are IncF, IncI1, IncN, IncH11, and IncH12. Phenotypic resistance in ampicillin was 17% (Germvet, data 2012, $N = 323$), and 16.9% (Czech NMTP, 2018, $N = 65$), but 37.7% (Switzerland data, 2018, $N = 53$); amoxicillin/clavulanic acid 2.5% (Germvet), 1.5% (Czech NMTP), and 1.9% (Switzerland); ceftiofur 8.7% (Germvet), 1.5% (Czech NMTP) and 0% (Switzerland). Also, AmpC Cephalosporinases (CMY-2) has been already confirmed in bovine mastitis milk (Endimiani et al. 2012). In recent work from China bla_{TEM}, $bla_{\text{CTX-M}}$, bla_{CMY}, and bla_{SHV} were found in 46.7% of PMQR genes were detected in isolates resistant to quinolones; *aac-(6′)-Ib-cr* was one of the determinant detected, exact genes harboured by isolates were *oqx*A/B, *qep*A4, *qnr*S, and *qnr*B2, respectively. Some of *E. coli* isolates were found to harbour both genes of resistance to quinolones as well as extended-spectrum β-lactamase. Resistance to aminoglycosides through target modification was described also in *E. coli* isolates from mastitis milk. 16S RNA methylase encoded by *Rmt*B was detected in isolates from mastitis in China (Yang et al. 2018). Also, genes encoding enzymes inactivating aminoglycosides were described in *E. coli* from mastitis in the Czech Republic: *aph (6)-Ia* gene known as *str*A (or *aph*D), *aph(6)-Id* gene also known as *str*B or *orf*I (Pyatov et al. 2017). Phenotypic resistance to gentamicin was reported as 1.5% (Germvet, data 2012, $N = 323$), 0% (CZ NMTP 2018, $N = 65$), and 11.3% (Switzerland data, $N = 53$, Nüesch-Inderbinen et al. 2019). Sulphonamide resistance genes that are located on plasmids: *sul*1, *sul*2 (in further bibliography, e.g. Poirel et al. 2018) is mentioned also the presence of this sul genes on Class 1 integrons, multiresistant plasmids as well as frequent association with *str*A and *str*B, what was also the case in the study by Pyatov et al. (2017). Phenotypic resistance to trimethoprim/sulfamethoxazole was 7.4% (Germvet), 6.2% (CZ NMTP 2018) and 28.3% in Switzerland. Efflux pump encoded by the genes *tet*A and *tet* B that were confirmed in Czech isolates of *E. coli* from mastitis (Pyatov et al. 2017). Phenotypic resistance to tetracycline was 10.5% (Germvet), 20% (CZ NMTP 2018), and 30.2% in Switzerland. Since 2015, when first reports on transferable colistin resistance encoding genes *mcr* were released a lot of studies reported different mcr genes (until now *mcr-1* to *mcr-10* have been recognised and reported, see Chap. 8). In *E. coli* isolates from cattle mostly it was detected in faeces of calves, but recent reports documented findings in mastitic milk both gene *mcr-1* encoding for phosphoethanolamine transferase enzymes and extended-spectrum β-lactamases (ESBLs) $bla_{\text{CTX-M-15}}$ gene. Both genes were harboured by conjugative plasmids IncP and IncF, respectively (Liu et al. 2019).

4.5.6 *Pseudomonas aeruginosa*

Pseudomonas aeruginosa is a Gram-negative, non-fermenting, aerobic, motile rod, an opportunistic pathogen often found in water and soil that is pathogenic to human beings, farm and companion animals as well as plants. It is frequently reported as life-threatening nosocomial infections causative pathogen, especially in immunocompromised patients or those with cystic fibrosis. In dairy cattle can cause serious, hardly treatable mastitis, in dogs it cause pyoderma, otitis, and urinary tract infections and has been described also in horses causing endometritis. *P. aeruginosa* has a big arsenal of resistance mechanism including overexpression of drug efflux pumps as well as cell wall with low permeability, porins, and inactivating enzymes. *P. aeruginosa* is known to be adaptive resistant. Also intrinsic resistance to a wide range of antimicrobials including penicillin (i.e. benzylpenicillin), aminopenicillins, amoxicillin/clavulanic acid, cephalosporins of the first generation (e.g. cephalothin and cefazolin), cephalosporins of the second generation (e.g. cefuroxime), cephamycins (cefoxitin and cefotetan), clindamycin, fusidic acid, glycopeptides (e.g. vancomycin), macrolides (erythromycin, azithromycin, and clarithromycin), chloramphenicol, sulphonamides and sulphonamide/trimethoprim combination, tigecycline, ertapenem, and rifampicin (CLSI 2018) is an issue. Moreover, this bacterium is able to form biofilms, is motile, and is able to acquire diverse resistance mechanisms either via horizontal gene transfer or by mutation. Among the main virulence factors reported from human medicine Lipopolysaccharide, Flagellum, Type IV Pili, Type III Secretion System, Exotoxin A, Proteases, Alginate, Quorum Sensing, Biofilm Formation, Type VI Secretion Systems, Iron acquisition, Elastase, Pyocyanin, and swarming factors (Rocha et al. 2019), but role in exact udder conditions need to be considered in each individual factor and further investigated. As for the data bringing the proof of evidence of exact genes encoding resistance to antimicrobials, a lot of work was done studying the isolates coming from human medicine either nosocomial infections or cystic fibrosis patients. Recent data from the veterinary side with open access, mastitis specific, reporting *P. aeruginosa* phenotypes and genes encoding for resistance are missing. Data from older studies, e.g. Ohnishi et al. (2011), reported phenotype results for 171 isolates of *P. aeruginosa* from Japan. High susceptibilities of $\geq$95.7% to ciprofloxacin and gentamicin were detected in a similar proportion also for imipenem, meropenem, piperacillin, ceftazidime, cefepime, cefoperazone/sulbactam, amikacin, tobramycin, which are not authorised as veterinary medicinal products in EU. Resistances to ceftriaxone, enrofloxacin, cefotaxime, and moxalactam were reported. In reality, only a few (3) antimicrobials tested with relevant susceptibility are authorised in veterinary medicinal products for dairy cattle across Europe and can be used in clinical practice.

4.5.7 *Staphylococcus aureus*

Staphylococcus aureus is a Gram-positive, facultative anaerobic, catalase-positive, and coccus-shaped zoonotic pathogen. Mammary gland of infected cows becomes the major reservoir and source of the pathogen, which can be spread by milking equipment, hands of the staff and is mostly present on the skin of the udder and orifices of cows, as well as in the environment—litter, feedstuff, and equipment, but also other animals on the farm (Constable et al. 2017).

The study by Monistero et al. (2018) investigated 120 isolates from eight different countries for 26 different virulence factors via RS-PCR. Genes related to host adhesion and invasion (*clf*A, *cna*, and *fmt*B), host defence mechanisms interfering genes (*tsst*, *scn*, *chp*, *sak*, enterotoxins from *sea*, *sei*, *she*, *sel*, and leukotoxins) were confirmed.

In bovine isolates from mastitis milk following genes of resistance were identified in *S. aureus* isolated from Czech farms ($N = 52$) *erm*B, *erm*C encoding for macrolide resistance, *msr*A encoding for macrolide and lincosamide resistance and *tet*(M), *tet*(K) encoding for tetracycline resistance (Pyatov et al. 2017; Yang et al. 2016) confirmed in isolates from Northwest China ($N = 44$, 2014) genes encoding resistance to rifampicin (*rpo*B), penicillin (*bla*Z), tetracycline (*tet*K, *tet*M, alone or in combination), confirmed in erythromycin (*erm*B or *erm*C), gentamicin/tobramycin/kanamycin (*aac*A-*aph*D), methicillin (*mec*A), and vancomycin (*van*A). More information on genes of resistance is placed above in general chapter related *S. aureus*. Phenotypic results show in German isolates of *S. aureus* ($N = 205$, 2013) resistance to ampicillin and penicillin in 16.1% of isolates, for oxacillin 5.9% resistance (also finally MRSA positivity) was detected. In the Czech Republic, 106 isolates of *S. aureus* from mastitis were tested and resistance to penicillin detected was 26.4% and to tetracycline 18.7%, to gentamicin 5.5%. As there were detected also isolates resistant to cefoxitin ($N = 5/106$), further confirmation was done, with positive results and detection of *mec*A gene. In Switzerland data 2018, 16% resistance ($N = 9/56$) to penicillin and ampicillin, and only 1 isolate was considered to be resistant to ciprofloxacin and 2 isolates to tetracycline (Anonymus 2018).

5 Conclusion

A variety of bacterial species are involved as the causative agents of the most important (from the perspective of prevalence, seriousness, and economic loss) in major food-producing species: pigs, cattle, and poultry. Despite the fact that some of them are etiological agents in different animal species and even having zoonotic potential, current genetic methods bring the evidence on the differences (host and virulence specificity) among the isolates from different animal species as well as from different tissues/organ systems. Increasing numbers of resistant or multiresistant strains were described among those bacterial isolates that impact the efficacy of antimicrobial agents approved for use in animals. With high probability, it can be anticipated that in the near future, there will be no classes of antimicrobial agents, with completely new mechanisms of action, approved for use in veterinary medicine. Therefore, a lot of work should be done to investigate how to protect the animals against diseases caused by resistant bacteria as for the nearest future, veterinarians will have to rely on those antimicrobial agents already available.

Many of the resistance genes known to be present in above-mentioned bacteria are associated with mobile genetic elements, mainly plasmids or transposons, but the exact mechanisms of the gene expression, as well as the link with host-specific factors and virulence factors, should be further investigated. Might be, some results of these investigations can lead to the discovery of the newly targeted antimicrobials blocking, e.g. expression of genes of virulence/factors enabling the successful colonisation of animal host. Also, deeper knowledge, especially of the arrangement of the resistance genes on mobile genetic elements, as well as the circumstances for their transfer, co-selection, and persistence, will be valuable information for veterinarians and could help them to select more rationally the most efficacious antimicrobial agents. There should be also effort paid to the building of some "alert systems of early detection" of changes in the susceptibility status and signalisation of the new, emerging resistance/multiresistance profiles and research capacities should be targeted also on analysis of the selected strains with those newly acquired/developed resistance genes and resistance-mediating mutations. There should be highlighted that this is the space not only for "pure scientists" from the area of microbiology, genetics, epidemiology, veterinary medicine, and bioinformatics, but for the cooperation with veterinarians in practice. This is because each scientific analysis should start in proper design of experiment, precise sampling, and test performance ending with analysis and interpretation of results considering all circumstances known from practice and finally, if performed in the right manner, should bring a missing piece to mosaic of knowledge and impact somehow the practice.

References

Aarestrup FM, Kruse H, Tast E, Hammerum AM, Jensen LB (2000) Associations between the use of antimicrobial agents for growth promotion and the occurrence of resistance among *Enterococcus faecium* from broilers and pigs in Denmark, Finland, and Norway. Microb Drug Resist 6(1):63–70

Aarestrup FM, Jensen LB (2002) Trends in antimicrobial susceptibility in relation to antimicrobial usage and presence of resistance genes in *Staphylococcus hyicus* isolated from exudative epidermitis in pigs. Vet Microbiol 89(1):83–94

Achard A, Villers C, Pichereau V, Leclercq R (2005) New lnu(C) gene conferring resistance to lincomycin by nucleotidylation in Streptococcus agalactiae UCN36. Antimicrob Agents Chemother 49(7):2716–2719

Achard A, Guérin-Faublée V, Pichereau V, Villers C, Leclercq R (2008) Emergence of macrolide resistance gene mph(B) in *Streptococcus uberis* and cooperative effects with rdmC-like gene. Antimicrob Agents Chemother 52(8):2767–2770

Adams V, Han X, Lyras D, Rood I (2018) Antibiotic resistance plasmids and mobile genetic elements of *Clostridium perfringens*. Plasmid 99:32–39

Al Bayssari C, Dabboussi F, Hamze M, Rolain JM (2015) Emergence of carbapene-mase-producing *Pseudomonas aeruginosa* and *Acinetobacter baumannii* in livestock animals in Lebanon. J Antimicrob Chemother 70:950–951

Alešík E (2006) Wirksamkeit antimikrobieller Wirkstoffe bei *Arcanobacterium pyogenes*: Etablierung und Anwendung der Empfindlichkeitsbestimmung mittels Bouillon mikrodilution sowie genotypische Charakterisierung tetracyclinresistenter Stämme, Disertation, Munchen, 189 p. https://edoc.ub.uni-muenchen.de/6300/1/Alesik_Eva.pdf. Accessed 31 May 2020

Alkasir R, Wang J, Gao J, Ali T, Zhang L, Szenci O, Bajcsy ÁC, Han B (2016) Properties and antimicrobial susceptibility of *Trueperella pyogenes* isolated from bovine mastitis in China. Acta Vet Hung 64(1):1–12

Almeida A, Alves-Barroco C, Sauvage E, Bexiga R, Albuquerque P, Tavares F, Santos-Sanches I, Glaser P (2016) Persistence of a dominant bovine lineage of group B Streptococcus reveals genomic signatures of host adaptation. Environ Microbiol 18(11):4216–4229

Alvarez-Perez S, Blanco JL, Bouza E, Alba P, Gibert X, Maldonado J, Garcia ME (2009) Prevalence of *Clostridium difficile* in diarrhoeic and non-diarrhoeic piglets. Vet Microbiol 137(3–4):302–305

Amat S (2019) Bovine respiratory disease in feedlot cattle: antimicrobial resistance in bovine respiratory bacterial pathogens and alternative antimicrobial approaches, Bacterial cattle diseases, Hussein Abdel hay El-Sayed Kaoud, IntechOpen. https://www.intechopen.com/books/bacterial-cattle-diseases/bovine-respiratory-disease-in-feedlot-cattle-antimicrobial-resistance-in-bovine-respiratory-bacteria. Accessed 6 Jan 2020

Ammerlaan HS, Harbarth S, Buiting AG, Crook DW, Fitzpatrick F, Hanberger H, Herwaldt LA, van Keulen PHJ, Kluytmans JAJW, Kola A, Kuchenbecker RS, Lingaas E, Meessen N, Morris-Downes MM, Pottinger JM, Rohner P, dos Santos RP, Seifert H, Wisplinghoff H, Ziesing S, Walker AS, Bonten MJM (2013) Secular trends in nosocomial bloodstream infections: antibiotic-resistant bacteria increase the total burden of infection. Clin Infect Dis 56 (6):798–805

Amonsin A, Wellehan JF, Li LL, Vandamme P, Lindeman C, Edman M, Robinson RA, Kapur V (1997) Molecular epidemiology of *Ornithobacterium rhinotracheale*. J Clin Microbiol 35(11):2894–2898

Angen O, Mutters R, Caugant DA, Olsen JE, Bisgaard M (1999) Taxonomic relationships of the Pasteurella haemolytica complex as evaluated by DNA-DNA hybridizations and 16S rRNA sequencing with proposal of *Mannheimia haemolytica* gen. nov., comb. nov., *Mannheimia granulomatis* comb. nov., *Mannheimia glucosida* sp. nov., *Mannheimia ruminalis* sp. nov. and *Mannheimia varigena* sp. nov. Int J Syst Bacteriol 49:67–86

Anjum MF, Marco-Jimenez F, Duncan D, Marín C, Smith RP, Evans SJ (2019) Livestock-associated methicillin-resistant *Staphylococcus aureus* from animals and animal products in the UK. Front Microbiol 10:2136

Anonymus (2018) Schlussbericht zum Pilotprojekt über die Überwachung von Antibiotika-resistenzen bei tierpathogenen Erregern, Eidgenössisches Departement des Innern EDI Bundesamt für Lebensmittelsicherheit und Veterinärwesen BLV Tiergesundheit (in German) August 2018, pp 1–24. https://www.blv.admin.ch/blv/de/home/tiere/tierarzneimittel/antibiotika/antibiotikaresistenzen.html. Accessed 26 Jan 2020

ANSES (2019) RESAPATH French surveillance network for antimicrobial resistance in bacteria from diseased animals 2018 Annual Report https://resapath.anses.fr/resapath_uploadfiles/files/Documents/Rapport%20annuel/2018_RESAPATH%20Rapport%20Annuel.pdfAccessed 28 May 2020

Archambault M, Rubin J (2018) Antimicrobial resistance in *Clostridium* and *Brachyspira* spp. and other anaerobes. In: Schwarz S, Cavaco L, Shen J (eds) Antimicrobial resistance in bacteria from livestock and companion animals. ASM Press, Washington, DC, pp 447–470

Argudín MA, Cariou N, Salandre O, Le Guennec J, Nemeghaire S, Butaye P (2013) Genotyping and antimicrobial resistance of *Staphylococcus aureus* isolates from diseased turkeys. Avian Pathol 42(6):572–580

Argudin MA, Deplano A, Meghraoui A, Dodemont M, Heinrichs A, Denis O et al (2017) Bacteria from animals as a pool of antimicrobial resistance genes. Antibiotics 6:E12. https://doi.org/10.3390/antibiotics6020012

Ashrafi Tamai I, Mohammadzadeh A, Zahraei Salehi T, Mahmoodi P (2018) Genomic characterisation, detection of genes encoding virulence factors and evaluation of antibiotic resistance of *Trueperella pyogenes* isolated from cattle with clinical metritis. Antonie Van Leeuwenhoek 111:2441–2453

Aubry-Damon H, Soussy CJ, Courvalin P (1998) Characterization of mutations in the rpoB gene that confer rifampin resistance in *Staphylococcus aureus*. Antimicrob Agents Chemother 42:2590–2594

Ba X, Harrison EM, Edwards GF, Holden MT, Larsen AR, Petersen A, Skov RL, Peacock SJ, Parkhill J, Paterson GK, Holmes MA (2014) Novel mutations in penicillin-binding protein genes in clinical *Staphylococcus aureus* isolates that are methicillin resistant on susceptibility testing, but lack the mec gene. J Antimicrob Chemother 69(3):594–597

Bacigalupe R, Tormo-Mas MÁ, Penadés JR, Fitzgerald JR (2019) A multihost bacterial pathogen overcomes continuous population bottlenecks to adapt to new host species. Sci Adv 5(11):eaax0063

Bager RJ, Nesta B, Pors SE, Soriani M, Serino L, Boyce JD, Adler B, Bojesen AM (2013) The fimbrial protein FlfA from *Gallibacterium anatis* is a virulence factor and vaccine candidate. Infect Immun 81(6):1964–1973

Bayer AS, Schneider T, Sahl HG (2013) Mechanisms of daptomycin resistance in *Staphylococcus aureus*: role of the cell membrane and cell wall. Ann N Y Acad Sci 1277:139–158

Beceiro A, Llobet E, Aranda J, Bengoechea JA, Doumith M, Hornsey M, Dhanji H, Chart H, Bou G, Livermore DM, Woodford N (2011) Phosphoethanolamine modification of lipid A in colistin-resistant variants of *Acinetobacter baumannii* mediated by the pmrAB two-component regulatory system. Antimicrob Agents Chemother 55:3370–3379

Beckmann C, Waggoner JD, Harris TO, Tamura GS, Rubens CE (2002) Identification of novel adhesins from group B Streptococci by use of phage, display reveals that c5a peptidase mediates fibronectin binding. Infect Immun 70:2869–2876

Beker M, Rose S, Lykkebo CA, Douthwaite S (2018) Integrative and conjugative elements (ICEs) in Pasteurellaceae species and their detection by multiplex PCR. Front Microbiol 9:1329

Ben Braïek O, Smaoui S (2019) Enterococci: between emerging pathogens and potential probiotics. Biomed Res Int 2019:5938210

Bentz AI, Wilkins PA, MacGillivray KC, Barr BS, Palmer JE (2002) Severe thrombocytopenia in 2 thoroughbred foals with sepsis and neonatal encephalopathy. J Vet Intern Med 16:494–497

Beylefeld A, Wambulawaye P, Bwala DG, Gouws JJ, Lukhele OM, Wandrag DBR, Abolnik C (2018) Evidence for multidrug resistance in nonpathogenic Mycoplasma species isolated from South African poultry. Appl Environ Microbiol 84(21):e01660–e01618

Bhatt K, Timsit E, Rawlyk N, Potter A, Liljebjelke K (2018) Integrative conjugative element ICEHs1 encodes for antimicrobial resistance and metal tolerance in *Histophilus somni*. Front Vet Sci. 5:153

Bigs A (2018) Streptococcus uberis: environmental, contagious or both? How best to manage it? Cattle Pract 26 (2):61–68

Billington SJ, Jost BH (2006) Multiple genetic elements carry the tetracycline resistance gene tet(W) in the animal pathogen *Arcanobacterium pyogenes*. Antimicrob Agents Chemother 50(11):3580–3587

Bisgaard M, Korczak BM, Busse HJ, Kuhnert P, Bojesen AM, Christensen H (2009) Classification of the taxon 2 and taxon 3 complex of Bisgaard within Gallibacterium and description of *Gallibacterium melopsittaci* sp. nov., *Gallibacterium trehalosifermentans* sp. nov. and *Gallibacterium salpingitidis* sp. nov. Int J Syst Evol Microbiol 59:735–744

Blesa A, Baquedano I, Quintáns NG, Mata CP, Castón JR, Berenguer J (2017) The transjugation machinery of *Thermus thermophilus*: identification of TdtA, an ATPase involved in DNA donation. PLoS Genet 13 (3):e1006669

Blickwede M, Schwarz S (2004) Molecular analysis of florfenicol-resistant *Escherichia coli* from pigs. J Antimicrob Chemother 53:58–64

Blum SE, Heller ED, Sela S, Elad D, Edery N, Leitner G (2015) Genomic and phenomic study of mammary pathogenic *Escherichia coli*. PLoS One 10(9): e0136387

Bojesen AM, Vazquez ME, Bager RJ, Ifrah D, Gonzalez C, Aarestrup FM (2011) Antimicrobial susceptibility and tetracycline resistance determinant genotyping of *Gallibacterium anatis*. Vet Microbiol 148(1):105–110

Bonnin RA, Nordmann P, Poirel L (2013) Screening and deciphering antibiotic resistance in *Acinetobacter baumannii*: a state of the art. Expert Rev Anti-Infect Ther 11:571–583

Boritsch EC, Khanna V, Pawlik A, Honoré N, Navas VH, Ma L, Bouchier C, Seemann T, Supply P, Stinear TP, Brosch R (2016) Key experimental evidence of chromosomal DNA transfer among selected tuberculosis-causing mycobacteria. Proc Natl Acad Sci U S A 113:9876–9881

Borselli D, Brunel JM, Gorgé O, Bolla JM (2019) Polyamino-isoprenyl derivatives as antibiotic adjuvants and motility inhibitors for Bordetella bronchiseptica porcine pulmonary infection treatment. Front Microbiol 10:1771

Bossé JT, Janson H, Sheehan BJ, Beddek A, Rycroft AN, Kroll JS, Langford PR (2002) *Actinobacillus pleuropneumoniae*: pathobiology and pathogenesis of infection. Microbes Infect 4(2):225–235

Bossé JT, Li Y, Atherton TG et al (2015) Characterisation of a mobilisable plasmid conferring florfenicol and chloramphenicol resistance in *Actinobacillus pleuropneumoniae*. Vet Microbiol 178:279–282

Bossé JT, Li Y, Rogers J, Fernandez Crespo R, Li Y, Chaudhuri RR et al (2017) Whole genome sequencing for surveillance of antimicrobial resistance in *Actinobacillus pleuropneumoniae*. Front Microbiol 8:311

Bossé JT, Li Y, Sárközi R, Fodor L, Lacouture S, Gottschalk M, Casas Amoribieta M, Angen Ø, Nedbalcova K, Holden MTG, Maskell DJ, Tucker AW, Wren BW, Rycroft AN, Langford PR, BRaDP1T consortium (2018) Proposal of serovars 17 and 18 of *Actinobacillus pleuropneumoniae* based on serological and genotypic analysis. Vet Microbiol 217:1–6

Botrel M-A, Haenni M, Morignat E, Sulpice P, Madec J-Y, Calavas D (2010) Distribution and antimicrobial resistance of clinical and subclinical mastitis pathogens in dairy cows in Rhône-Alpes, France. Foodborne Pathog Dis 7:479–487

Brown AWW, Wilson RB (2018) *Clostridium difficile* colitis and zoonotic origins – a narrative review. Gastroenterol Rep (Oxf) 6(3):157–166

Byarugaba DK, Minga UM, Gwakisa PS, Katunguka-Rwakishaya E, Bisgaard M, Christensen H, Olsen JE (2011) Demonstration of antibiotic resistance genes *str*A, *bla*TEM, *tet*A, *tet*C and *sul*2 in *Avibacterium paragallinarum*. Afr J Microbiol Res 5(22):3624–3627

Bystroń J, Podkowik M, Piasecki T, Wieliczko A, Molenda J, Bania J (2010) Genotypes and enterotoxin gene content of *S. aureus* isolates from poultry. Vet Microbiol 144(3–4):498–501

Cai HY, McDowall R, Parker L, Kaufman EI, Caswell JL (2019) Changes in antimicrobial susceptibility profiles of *Mycoplasma bovis* over time. Can J Vet Res 83 (1):34–41

Cain D, Malouin F, Dargis M, Harel J, Gottschalk M (1995) Alterations in penicillin-binding proteins in strains of *Streptococcus suis* possessing moderate and high levels of resistance to penicillin. FEMS Microbiol Lett 130:12–17. https://doi.org/10.1111/j.1574-6968.1995.tb07708.x

Cameron A, Klima CL, Ha R, Gruninger RJ, Zaheer R, McAllister TA (2018) A Novel *aad*A aminoglycoside resistance gene in bovine and porcine pathogens. mSphere 3(1)

Campisi E, Rosini R, Ji W, Guidotti S, Rojas-López M, Geng G, Deng Q, Zhong H, Wang W, Liu H, Nan C, Margarit I, Rinaudo CD (2016) Genomic analysis reveals multi-drug resistance clusters in group B Streptococcus CC17 hypervirulent isolates causing neonatal Iivasive disease in southern Mainland China. Front Microbiol 7:1265

Canton R, Mazzariol A, Morosini MI, Baquero F, Cornaglia G (2005) Telithromycin activity is reduced by efflux in *Streptococcus pyogenes*. J Antimicrob Chemother 55:489–495

Card RM, Stubberfield E, Rogers J, Nunez-Garcia J, Ellis RJ, AbuOun M, Strugnell B, Teale C, Williamson S, Anjum MF (2018) Identification of a new antimicrobial resistance gene provides fresh insights into pleuromutilin resistance in *Brachyspira hyodysenteriae*, aetiological agent of swine dysentery. Front Microbiol 9:1183

Carter GP, Cheung JK, Larcombe S, Lyras D (2014) Regulation of toxin production in the pathogenic clostridia. Mol Microbiol 91(2):221–231

Cerqueira GM, Peleg AY (2011) Insights into *Acinetobacter baumannii* pathogenicity. IUBMB Life 63:1055–1060

Charlebois A, Jacques M, Archambault M (2014) Biofilm formation of Clostridium perfringens and its exposure to low-dose antimicrobials. Front Microbiol 5:183

Charlebois A, Jalbert LA, Harel J, Masson L, Archambault M (2012) Characterization of genes encoding for acquired bacitracin resistance in *Clostridium perfringens*. PLoS One 7(9):e44449

Chen MY, Lira F, Liang HQ, Wu RT, Duan JH, Liao XP, Martínez JL, Liu YH, Sun J (2016) Multilevel selection of bcrABDR-mediated bacitracin resistance in *Enterococcus faecalis* from chicken farms. Sci Rep 6(1)

Chen Q, Gong X, Zheng F, Ji G, Li S, Stipkovits L, Szathmary S, Liu Y (2018) Interplay between the phenotype and genotype, and efflux pumps in drug-resistant strains of *Riemerella anatipestifer*. Front Microbiol 9:2136

Chen Y, Guo G, Sun E, Song J, Yang L, Zhu L, Liang W, Hua L, Peng Z, Tang X, Chen H, Wu B (2019) Isolation of a T7-like lytic Pasteurella bacteriophage vB_PmuP_PHB01 and its potential use in therapy against *Pasteurella multocida* infections. Viruses 11 (1):86

Chiers K, De Waele T, Pasmans F, Ducatelle R, Haesebrouck F (2010) Virulence factors of *Actinobacillus pleuropneumoniae* involved in colonization, persistence and induction of lesions in its porcine host. Vet Res 41:65. https://doi.org/10.1051/vetres/2010037

Christensen H, Bisgaard M, Bojesen AM, Mutters R, Olsen JE (2003) Genetic relationships among avian isolates classified as *Pasteurella haemolytica*, *'Actinobacillus salpingitidis'* or *Pasteurella anatis* with proposal of *Gallibacterium anatis* gen. nov., comb. nov. and description of additional genomospecies within *Gallibacterium* gen. nov. Int J Syst Evol Microbiol 53:275–287

Citti C, Dordet-Frisoni E, Nouvel LX, Kuo CH, Baranowski E (2018) Horizontal gene transfers in mycoplasmas (Mollicutes). Curr Issues Mol Biol 29:3–22

Clark NM, Zhanel GG, Lynch JP III (2016) Emergence of antimicrobial resistance among Acinetobacter species: a global threat. Curr Opin Crit Care 22:491–499

CLSI (2011) Methods for antimicrobial susceptibility testing for human mycoplasmas (1st edn), M43AE, ISBN Number: 1-56238-769-3

CLSI (2017) Methods for antimicrobial testing of infrequently isolated or fastidious bacteria isolated from animals, 1st edn. CLSI supplement VET06. Clinical and Laboratory Standards Institute, Wayne, PA

CLSI (2018) Performance standards for antimicrobial disk and dilution susceptibility tests for bacteria isolated from animals (4th edn). CLSI document Vet 08. Clinical and Laboratory Standards Institute, Wayne, p 170

Cobo-Angel CG, Jaramillo-Jaramillo AS, Palacio-Aguilera M, Jurado-Vargas L, Calvo-Villegas EA, Ospina-Loaiza DA, Rodriguez-Lecompte JC, Sanchez J, Zadoks R, Ceballos-Marquez A (2019) Potential group B Streptococcus interspecies transmission between cattle and people in Colombian dairy farms. Sci Rep 9:14025

Coetzer JAW, Tustin RC (2004) Infectious diseases of livestock, 2nd edn. Oxford University Press, Cape Town, p 2159

Conover MS, Sloan GP, Love CF, Sukumar N, Deora R (2010) The Bps polysaccharide of *Bordetella pertussis* promotes colonization and biofilm formation in the nose by functioning as an adhesin. Mol Microbiol 77 (6):1439–1455

Constable PD, Hinchcliff KW, Done SH, Grunbergh W (2017) Chapter 3. Veterinary medicine: a textbook of the diseases of cattle, horses, sheep, pigs, and goats, 11th edn. Elseviers Publications, Amsterdam

Corver J, Bakker D, Brouwer MS, Harmanus C, Hensgens MP, Roberts AP, Lipman LJ, Kuijper EJ, van Leeuwen HC (2012) Analysis of a *Clostridium difficile* PCR ribotype 078 100 kilobase island reveals the presence of a novel transposon, Tn6164. BMC Microbiol 12:130

Coyne S, Courvalin P, Perichon B (2011) Efflux-mediated antibiotic resistance in *Acinetobacter* spp. Antimicrob Agents Chemother 55:947–953

Cozens D, Sutherland E, Lauder M, Taylor G, Berry CC, Davies RL (2019) Pathogenic *Mannheimia haemolytica* invades differentiated bovine airway epithelial cells. Infect Immun 87(6)

CZ NMTP (2016) National programme of monitoring antimicrobial resistance of pathogens with veterinary importance, data 2015 and 2016 (in Czech). https://www.svscr.cz/narodni-program-sledovani-rezistenci-k-antimikrobikum-u-veterinarne-vyznamnych-patogenu/. Accessed 31 Jan 2020

CZ NMTP (2017) National programme of monitoring antimicrobial resistance of pathogens with veterinary importance, data 2017 (in Czech). https://www.svscr.cz/wp-content/files/dokumenty-a-publikace/Narodni-program-sledovani-rezistenci-k-antimikrobikum-u-veterinarne-vyznamnych-patogenu-2017-cast-I.pdf. Accessed 31 Jan 2020

CZ NMTP (2018) National programme of monitoring antimicrobial resistance of pathogens with veterinary importance, data 2018 (in Czech). https://www.svscr.cz/wp-content/files/dokumenty-a-publikace/Zprava-RL-antibioticke-centrum-pro-veterinarni-klinickou-praxi-za-rok-2018.pdf. Accessed 31 Jan 2020

Damier-Piolle L, Magnet S (2008) AdeIJK, a resistance-nodulation-cell division pump effluxing multiple antibiotics in *Acinetobacter baumannii*. Antimicrob Agents Chemother 52(2):557–562

da Silva GC, Rossi CC, Santana MF, Langford PR, Bossé JT, Bazzolli DMS (2017) p518, a small *flo*R plasmid from a south American isolate of *Actinobacillus pleuropneumoniae*. Vet Microbiol 204:129–132

Daubin V, Szöllősi GJ (2016) Horizontal Gene Transfer and the History of Life. Cold Spring Harb Perspect Biol 8(4):a018036

Dayao DAE, Seddon JM, Gibson JS, Blackall PJ, Turni C (2016) Whole genome sequence analysis of pig respiratory bacterial pathogens with elevated minimum inhibitory concentrations for macrolides. Microb Drug Resist 22:531–537

De Luca S, Nicholson P, Magistrali CF, García-Martín AB, Rychener L, Zeeh F, Frey J, Perreten V (2018) Transposon-associated lincosamide resistance lnu (C) gene identified in *Brachyspira hyodysenteriae* ST83. Vet Microbiol 214:51–55

de Kraker ME, Jarlier V, Monen JC, Heuer OE, van de Sande N, Grundmann H (2013) The changing epidemiology of bacteraemias in Europe: trends from the European antimicrobial resistance surveillance system. Clin Microbiol Infect 19(9):860–868

Deslandes V, Denicourt M, Girard C, Harel J, Nash JH, Jacques M (2010) Transcriptional profiling of *Actinobacillus pleuropneumoniae* during the acute phase of a natural infection in pigs. BMC Genomics 11(1):98

Devriese LA (1984) A simplified system for biotyping *Staphylococcus aureus* strains isolated from different animals species. J Appl Bacteriol 56:215–220

Devriese LA, Herdt PD, Haesebrouck F (2001) Antibiotic sensitivity and resistance in *Ornithobacterium rhinotracheale* strains from Belgian broiler chickens. Avian Pathol 30:197–200

Diab SS, Songer G, Uzal FA (2013) *Clostridium difficile* infection in horses: a review. Vet Microbiol 167:42–49

Ding Y, Zhao J, He X, Li M, Guan H, Zhang Z, Li P (2016) Antimicrobial resistance and virulence-related genes of Streptococcus obtained from dairy cows with mastitis in Inner Mongolia. Pharm Biol 54(1):162–167

Dinos GP (2017) The macrolide antibiotic renaissance. Br J Pharmacol 174:2967–2983

Dmitriev A, Shakleina E, Tkáciková L, Mikula I, Totolian A (2002) Genetic heterogeneity of the pathogenic potentials of human and bovine group B Streptococci. Folia Microbiol (Praha) 47(3):291–295

Dogan B, Klaessig S, Rishniw M, Almeida RA, Oliver SP, Simpson K, Schukken YH (2006) Adherent and invasive *Escherichia coli* are associated with persistent bovine mastitis. Vet Microbiol 116(4):270–282

Doi Y, Murray GL, Peleg AY (2015) *Acinetobacter baumannii*: evolution of antimicrobial resistence- treatment options. Semin Respir Crit Care Med 36:85–98

Dolejska M, Villa L, Hasman H, Hansen L, Carattoli A (2013) Characterization of IncN plasmids carrying bla CTX-M-1 and qnr genes in *Escherichia coli* and Salmonella from animals, the environment and humans. J Antimicrob Chemother 68(2):333–339

Dordet-Frisoni E, Sagné E, Baranowski E, Breton M, Nouvel LX, Blanchard A, Marenda MS, Tardy F, Sirand-Pugnet P, Citti C (2014) Chromosomal transfers in Mycoplasmas: When minimal genomes go mobile. mBio 5(6):1958

Dordet-Frisoni E, Faucher M, Sagné E, Baranowski E, Tardy F, Nouvel LX, Citti C (2019) Mycoplasma chromosomal transfer: a distributive, conjugative process creating an infinite variety of mosaic genomes. Front Microbiol 10. https://wwwfrontiersinorg/article/103389/fmicb201902441. Acessed 6 Jan 2020

Dreyfus A, Schaller A, Nivollet S, Segers RPAM, Kobisch M, Mieli L, Soerensen V, Hüssy D, Miserez R, Zimmermann W, Inderbitzin F, Frey J (2004) Use of recombinant ApxIV in serodiagnosis of *Actinobacillus pleuropneumoniae* infections, development and prevalidation of the ApxIV ELISA. Vet Microbiol 99(3–4):227–238

Du F, Lv X, Duan D, Wang L, Huang J (2019) Characterization of a Linezolid- and Vancomycin-Resistant Streptococcus suis Isolate That Harbors optrA and vanG Operons. Front Microbiol 10:2026

Dudouyt J, Léorat J, van Empel P, Gardin Y, Céline D (1995) Isolement d'un nouvel pathogenechez la dinde: *Ornithobacterium rhinotracheale*; Conduite a tenir. In Proceedings of the Journéesde la Recherche Avicole, Angers, pp 240–243 (in French)

Dutkiewicz J, Sroka J, Zając V, Wasiński B, Cisak E, Sawczyn A, Kloc A, Wójcik-Fatla A (2017) *Streptococcus suis*: a re-emerging pathogen associated with occupational exposure to pigs or pork products. Part I – Epidemiology. Ann Agric Environ Med 24 (4):683–695

Eckburg PB, Bik EM, Bernstein CN, Purdom E, Dethlefsen L, Sargent M, Gill SR, Nelson KE, Relman DA (2005) Diversity of the human intestinal microbial flora. Science 308(5728):1635–1638

Eidam C, Poehlein A, Leimbach A, Michael GB, Kadlec K, Liesegang H, et al. (2015) Analysis and comparative genomics of ICE Mh1, a novel integrative and conjugative element (ICE) of Mannheimia haemolytica. J Antimicrob Chemother. 70: 93–97. pmid:25239467

El-Adawy H, Bocklisch H, Neubauer H, Hafez HM, Hotzel H (2018) Identification, differentiation and antibiotic susceptibility of Gallibacterium isolates from diseased poultry. Ir Vet J 71:5

Endimiani A, Bertschy I, Perreten V (2012) *Escherichia coli* producing CMY-2 β-lactamase in bovine mastitis milk. J Food Prot 75(1):137–138

Entorf M, Feßler AT, Kaspar H, Kadlec K, Peters T, Schwarz S (2016) Comparative erythromycin and tylosin susceptibility testing of streptococci from bovine mastitis. Vet Microbiol 194:36–42

EMA (2011) European Medicines Agency, 2011. 'Trends in the sales of veterinary antimicrobial agents in nine European countries (2005–2009)' (EMA/238630/ 2011)

EMA (2019) European Medicines Agency, European Surveillance of Veterinary Antimicrobial Consumption, 2019. 'Sales of veterinary antimicrobial agents in 31 European countries in 2017' (EMA/294674/2019)

Erfan A, Badr JM, Abd-elhalim M (2018) First record of Bordetella avium in Egyptian turkey flocks. Biosci Res 15(3):2583–2590

Erganis O, Hadimli HH, Kav K, Sayn Z (2012) The antimicrobial susceptibility of *Ornithobacterium rhinotracheale* isolates. Eurasian J Vet Sci 28:27–30

EUCAST (2020) The European Committee on Antimicrobial Susceptibility Testing. Breakpoint tables for interpretation of MICs and zone diameters Version 100, valid from 2020-01-01. http://wwweucastorg/fileadmin/src/media/PDFs/EUCAST_files/Breakpoint_tables/v_100_Breakpoint_Tablespdf. Last accessed 18 Jan 2019

EU-JAMRAI (2019) The desing of EARS-vet in action; https://eu-jamrai.eu/help-build-ears-vet/. Last accessed 6 Jan 2019

European Food Safety Authority (EFSA), Aerts M, Battisti A, Hendriksen R, Kempf I, Teale C, Tenhagen BA, Veldman K, Wasyl D, Guerra B, Liebana E, Thomas-Lopez D, Beloeil PA (2019) Scientific report on the technical specifications on harmonised monitoring of antimicrobial resistance in zoonotic and indicator bacteria from food-producing animals and food. EFSA J 17(6):5709, 122 pp

Ewers C, Klotz P, Scheufen S, Leidner U, Göttig S, Semmler T (2016) Genome sequence of OXA-23 producing *Acinetobacter baumannii* IHIT7853, a carbapenem-resistant strain from a cat belonging to international clone IC1. Gut Pathog 8:37

Ewers C, Klotz P, Leidner U, Stamm I, Prenger-Berninghoff E, Göttig S, Semmler T, Scheufen S (2017) OXA-23 and IS Aba1–OXA-66 class D β-lactamases in *Acinetobacter baumannii* isolates from companion animals. Int J Antimicrob Agents 49 (1):37–44

Facklam R (2002) What happened to the streptococci: overview of taxonomic and nomenclature changes. Clin Microbiol Rev 15:613–630

Faucher M, Nouvel L-X, Dordet-Frisoni E, Sagne E, Baranowski E, Hygonenq MC, Marenda MS, Tardy F, Citti C (2019) Mycoplasmas under experimental antimicrobial selection: the unpredicted contribution of horizontalchromosomal transfer. PLoS Genet 15(1)

FDA, U.S. Food and Drug Administration (2007) Drug approval package. https://www.accessdata.fda.gov/drugsatfda_docs/nda/2007/022055s000TOC.cfm. Accessed 6 Jan 2020

FDA, U.S. Food and Drug Administration (2019) FDA approves new antibiotic to treat community-acquired bacterial pneumonia. https://www.fda.gov/news-events/press-announcements/fda-approves-new-antibiotic-treat-community-acquired-bacterial-pneumonia. Accessed 6 Jan 2020

Felde O, Kreizinger Z, Sulyok KM, Hrivnák V, Kiss K, Jerzsele Á et al (2018) Antibiotic susceptibility testing of *Mycoplasma hyopneumoniae* field isolates from Central Europe for fifteen antibiotics by microbroth dilution method. PLoS One 13(12):e0209030

Feßler A, Schwarz S (2017) Antimicrobial resistance in *Corynebacterium* spp., *Arcanobacterium* spp., and *Trueperella pyogenes*. Microbiol Spectr 5(6)

Feßler A, Kadlec K, Wang Y, Zhang WJ, Wu C, Shen J, Schwarz S (2018) Small antimicrobial resistance plasmids in livestock-associated methicillin-resistant *Staphylococcus aureus* CC398. Front Microbiol 9:2063

Flores AR, Galloway-Peña J, Sahasrabhojane P, Saldaña M, Yao H, Su X, Ajami NJ, Holder ME, Petrosino JF, Thompson E, Margarit Y, Ros I, Rosini R, Grandi G, Horstmann N, Teatero S, McGeer A, Fittipaldi N, Rappuoli R, Baker CJ,

Shelburne SA (2015) Sequence type 1 group B Streptococcus, an emerging cause of invasive disease in adults, evolves by small genetic changes. Proc Natl Acad Sci U S A 112(20):6431–6436

Forti K, Ferroni L, Pellegrini M, Cruciani D, De Giuseppe A, Crotti S, Papa P, Maresca C, Severi G, Marenzoni ML, Cagiola M. (2020) Molecular Characterization of Clostridium perfringens Strains Isolated in Italy. Toxins 2020, 12, 650.

Foster TJ (2017) Antibiotic resistance in *Staphylococcus aureus*. Current status and future prospects. FEMS Microbiol Rev 41(3):430–449

Foster TJ, Geoghegan JA, Ganesh VK, Höök M (2014) Adhesion, invasion and evasion: the many functions of the surface proteins of *Staphylococcus aureus*. Nat Rev Microbiol 12(1):49–62

Freedman JC, Shrestha A, McClane BA (2016) *Clostridium perfringens* enterotoxin: action, genetics, and translational applications. Toxins (Basel) 8(3)

Furian TQ, Borges KA, Laviniki V, Rocha SL, de Almeida CN, do Nascimento VP, Salle CT, Moraes HL (2016) Virulence genes and antimicrobial resistance of *Pasteurella multocida* isolated from poultry and swine. Braz J Microbiol 47(1):210–216

Fyfe C, Grossman TH, Kerstein K, Sutcliffe J (2016) Resistance to macrolide antibiotics in public health pathogens. Cold Spring Harb Perspect Med 6(10)

Gao J, Li S, Zhang J, Zhou Y, Xu S, Barkema HW, Nobrega DB, Zhu C, Han B (2019) Prevalence of potential virulence genes in *Klebsiella* spp. isolated from cows with clinical mastitis on large Chinese dairy farms. Foodborne Pathog Dis 16(12):856–863

García-Martín AB, Schwendener S, Perreten V (2019) The tva(A) gene from *Brachyspira hyodysenteriae* confers decreased susceptibility to pleuromutilins and streptogramin A in *Escherichia coli*. Antimicrob Agents Chemother 63(9)

Gashe S (2017) Untersuchungen zum Resistenzverhalten von *Riemerella anatipestifer* und *Ornithobacterium rhinotracheale* des Wirtschaftsgeflügels (Doctoral dissertation), Freien Universität in Berlin, Germany, Journal Nr.: 3967 (in German). https://refubium.fu-berlin.de/bitstream/handle/fub188/1601/Gasche_online.pdf?sequence=1&isAllowed=y. Accessed 6 Jan 2020

Gautier-Bouchardon AV (2018) Antimicrobial resistance in *Mycoplasma* spp. Microbiol Spectr 6(4)

Gilmore MS, Coburn PS, Nallapareddy SR, Murray BE (2002) Enterococcal virulence. In: Gilmore M, Clewell D, Courvalin P, Dunny G, Murray B, Rice L (eds) The enterococci. ASM Press, Washington, DC, pp 301–354

Golkar T, Zieliński M, Berghuis AM (2018) Look and outlook on enzyme-mediated macrolide resistance. Front Microbiol 9:1942

Gong J, Forster RJ, Yu H, Chambers JR, Wheatcroft R, Sabour PM et al (2002) Molecular analysis of bacterial populations in the ileum of broiler chickens and comparison with bacteria in the cecum. FEMS Microbiol Ecol 41(3):171–179

Graveland H, Duim B, van Duijkeren E, Heederik D, Wagenaar J (2011) Livestock-associated methicillin-resistant *Staphylococcus aureus* in animals and humans. Int J Med Microbiol 301:630–634

Gray TA, Krywy JA, Harold J, Palumbo MJ, Derbyshire KM (2013) Distributive conjugal transfer in mycobacteria generates progeny with meiotic-like genome-wide mosaicism, allowing mapping of a mating identity locus. PLoS Biol 11(7)

Griffin D, Chengappa M, Kuszak J, McVey DS (2010) Bacterial pathogens of the bovine respiratory disease complex. Vet Clin North Am Food Anim Pract 26:381–394

Guardabassi L, Dijkshoorn L, Collard JM, Olsen JE, Dalsgaard A (2000) Distribution and in-vitro transfer of tetracycline resistance determinants in clinical and aquatic Acinetobacter strains. J Med Microbiol 49:929–936

Gutiérrez-Martín CB, del Blanco NG, Blanco M, Navas J, Rodríguez-Ferri EF (2006) Changes in antimicrobial susceptibility of *Actinobacillus pleuropneumoniae* isolated from pigs in Spain during the last decade. Vet Microbiol 115(1–3):218–222

Haag A, Fitzgerald J, Penadés J (2019) *Staphylococcus aureus* in animals. Microbiol Spectr 7(3):GPP3-0060-2019

Haenni M, Galofaro L, Ythier M, Giddey M, Majcherczyk P, Moreillon P, Madec JY (2010) Penicillin-binding protein gene alterations in *Streptococcus uberis* isolates presenting decreased susceptibility to penicillin. Antimicrob Agents Chemother 54 (3):1140–1145

Haenni M, Saras E, Chaussière S, Treilles M, Madec JY (2011) *erm*B-mediated erythromycin resistance in *Streptococcus uberis* from bovine mastitis. Vet J 189:356–358

Haenni M, Lupo A, Madec J (2018) Antimicrobial resistance in *Streptococcus* spp. Microbiol Spectr 6:1–25

Hamouda A, Findlay J, Al Hassan L, Amyes SG (2011) Epidemiology of *Acinetobacter baumannii* of animal origin. Int J Antimicrob Agents 38(4):314–318

Hampson DJ (2012) *Brachyspiral colitis*. In: Zimmerman JJ, Karriker LA, Ramirez A, Schwartz KJ, Stevenson GW (eds) Diseases of Swine, 10th edn. Wiley Blackwell, Ames, IA, pp 680–696

Hanchi H, Mottawea W, Sebei K, Hammami R (2018) The genus Enterococcus: between probiotic potential and safety concerns-an update. Front Microbiol 9:1791

Hancock LE, Perego M (2004) The *Enterococcus faecalis fsr* two-component system controls biofilm development through production of gelatinase. J Bacteriol 186:5629–5639

Hannan PCT, Windsor HM, Ripley PH (1997) In vitro susceptibilities of recent field isolates of *Mycoplasma hyopneumoniae* and *Mycoplasma hyosynoviae* to valnemulin (Econor), tiamulin and enrofloxacin and the *in vitro* development of resistance to certain antimicrobial agents in *Mycoplasma hyopneumoniae*. Res Vet Sci 63:157–160

Harper M, John M, Turni C, Edmunds M, St Michael F, Adler B, Blackall PJ, Cox AD, Boyce JD (2015) Development of a rapid multiplex PCR assay to genotype *Pasteurella multocida* strains by use of the lipopolysaccharide outer core biosynthesis locus. J Clin Microbiol 53:477–485

Harrington AT, Castellanos JA, Ziedalski TM, Clarridge JE III, Cookson BT (2009) Isolation of *Bordetella avium* and novel Bordetella strain from patients with respiratory disease. Emerg Infect Dis 15(1):72–74

Hassler C, Nitzsche S, Iversen C, Zweifel C, Stephan R (2008) Characteristics of *Staphylococcus hyicus* strains isolated from pig carcasses in two different slaughterhouses. Meat Sci 80(2):505–510

Hata E, Harada T, Itoh M (2019) Relationship between Antimicrobial susceptibility and multilocus sequence type of *Mycoplasma bovis* isolates and development of a method for rapid detection of point mutations involved in decreased susceptibility to macrolides, lincosamides, tetracyclines, and spectinomycin. Appl Environ Microbiol 85(13):e00575-19

He T, Wang Y, Sun L, Pang M, Zhang L, Wang R (2016) Occurrence and characterization of $bla_{\text{NDM-5}}$-positive *Klebsiella pneumoniae* isolates from dairy cows in Jiangsu, China. J Antimicrob Chemother 72(1):90–94

Hegstad K, Mikalsen T, Coque TM, Werner G, Sundsfjord A (2010) Mobile genetic elements and their contribution to the emergence of antimicrobial resistant *Enterococcus faecalis* and *Enterococus faecium*. Clin Microbiol Infect 16(6):451–554

Hendriksen RS, Mevius DJ, Schroeter A, Teale C, Jouy E, Butaye P, Franco A, Utinane A, Amado A, Moreno M, Greko C, Stärk KD, Berghold C, Myllyniemi AL, Hoszowski A, Sunde M, Aarestrup FM (2008) Occurrence of antimicrobial resistance among bacterial pathogens and indicator bacteria in pigs in different European countries from year 2002-2004: the ARBAO-II study. Acta Vet Scand 50:19

Hennekinne J, Kerouanton A, Brisabois A, de Buyser ML (2003) Discrimination of *Staphylococcus aureus* biotypes by pulsed-field gel electrophoresis of DNA macro-restriction fragments. J Appl Microbiol 94:321–329

Hess C, Beatrice G, Bagheri S, Kaesbohrer A, Zloch A, Hess M (2019) Number of multiresistant strains and substantial variability even between isolates from the same organ. Microbial Drug Resistance 26(2):169–177

Hidalgo A, Carvajal A, García-Feliz C, Osorio J, Rubio P (2009) Antimicrobial susceptibility testing of Spanish field isolates of *Brachyspira hyodysenteriae*. Res Vet Sci 87:7–12. https://doi.org/10.1016/j.rvsc.2008.10.01

Hirt H, Greenwood-Quaintance KE, Karau MJ, Till LM, Kashyap PC, Patel R, Dunny GM (2018) *Enterococcus faecalis* sex pheromone cCF10 enhances conjugative plasmid transfer *in vivo*. MBio 9(1)

Holmer I, Salomonsen CM, Jorsal SE, Astrup LB, Jensen VF, Høg BB, Pedersen K (2019) Antibiotic resistance in porcine pathogenic bacteria and relation to antibiotic usage. BMC Vet Res 15(1):449

Horiguchi Y (2012) Swine atrophic rhinitis caused by *Pasteurella multocida* toxin and Bordetella dermonecrotic toxin. Curr Top Microbiol Immunol 361:113–129

Hsu YM, Shieh HK, Chen WH, Sun TY, Shiang JH (2007) Antimicrobial susceptibility, plasmid profiles and haemocin activities of *Avibacterium paragallinarum* strains. Vet Microbiol 124(3–4):209–218

Huang J, Ma J, Shang K, Hu X, Liang Y, Li D, Wu Z, Dai L, Chen L, Wang L (2016a) Evolution and diversity of the antimicrobial resistance associated mobilome in *Streptococcus suis*: a probable mobile genetic elements reservoir for other streptococci. Front Cell Infect Microbiol 6:118

Huang K, Zhang Q, Song Y, Zhang Z, Zhang A, Xiao J, Jin M (2016b) Characterization of Spectinomycin resistance in *Streptococcus suis* leads to two novel insights into drug resistance formation and dissemination mechanism. Antimicrob Agents Chemother 60 (10):6390–6392

Hujer KM, Hamza NS, Hujer AM, Perez F, Helfand MS, Bethel CR, Thomson JM, Anderson VE, Barlow M, Rice LB, Tenover FC, Bonomo RA (2005) Identification of a new allelic variant of the *Acinetobacter baumannii* cephalosporinase, ADC-7 β-lactamase: defining a unique family of class C enzymes. Antimicrob Agents Chemother 49:2941–2948

Humann-Ziehank E, Menzel A, Roehrig P, Schwert B, Ganter M, Hennig-Pauka I (2014) Acute and subacute response of iron, zinc, copper and selenium in pigs experimentally infected with *Actinobacillus pleuropneumoniae*. Metallomics 6:1869–1879

Husain F, Tang K, Veeranagouda Y, Boente R, Patrick S, Blakely G, Wexler HM (2017) Novel large-scale chromosomal transfer in *Bacteroides fragilis* contributes to its pan-genome and rapid environmental adaptation. Microb Genom 3(11)

Iannelli F, Santoro F, Santagati M, Docquier JD, Lazzeri E, Pastore G, Cassone M, Oggioni MR, Rossolini GM, Stefani S, Pozzi G (2018) Type M resistance to macrolides is due to a two-gene efflux transport system of the ATP-binding cassette (ABC) superfamily. Front Microbiol 9:1670

Jarraud S, Mougel C, Thioulouse J, Lina G, Meugnier H, Forey F, Nesme X, Etienne J, Vandenesch F (2002) Relationships between *Staphylococcus aureus* genetic background, virulence factors, agr groups (alleles), and human disease. Infect Immun 70(2):631–641

Jiang N, Li J, Feßler AT, Wang Y, Schwarz S, Wu C (2019) A novel small *tet*(T)–*tet*(L)–*aad*D-carrying plasmid from MRSA and MSSA ST9 isolates of swine origin. J Antimicrob Chemother 74 (8):2462–2464

Joerling J, Barth SA, Schlez K, Willems H, Herbst W, Ewers C (2018) Phylogenetic diversity, antimicrobial susceptibility and virulence gene profiles of *Brachyspira hyodysenteriae* isolates from pigs in Germany. PLoS One 13(1)

Johnson TJ, Danzeisen JL, Trampel D, Nolan LK, Seemann T, Bager RJ, Bojesen AM (2013) Genome analysis and phylogenetic relatedness of *Gallibacterium anatis* strains from poultry. PLoS One 8(1):e54844

Jones KH, Thornton JK, Zhang Y, Mauel MJ (2013) A 5-year retrospective report of *Gallibacterium anatis* and *Pasteurella multocida* isolates from chickens in Mississippi. Poult Sci 92:3166–3171

Jost BH, Billington SJ (2005) *Arcanobacterium pyogenes*: molecular pathogenesis of an animal opportunist. Antonie Van Leeuwenhoek 88(2):87–102

Jost BH, Field AC, Trinh HT, Songer JG, Billington SJ (2003) Tylosin resistance in *Arcanobacterium pyogenes* is encoded by an erm X determinant. Antimicrob Agents Chemother 47(11):3519–3524

Jost BH, Trinh HT, Songer JG, Billington SJ (2004) Ribosomal mutations in *Arcanobacterium pyogenes* confer a unique spectrum of macrolide resistance. Antimicrob Agents Chemother 48(3):1021–1023

Kaas RS, Friis C, Ussery DW, Aarestrup FM (2012) Estimating variation within the genes and inferring the phylogeny of 186 sequenced diverse *Escherichia coli* genomes. BMC Genomics 13:577

Kadlec K, Schwarz S (2018) Antimicrobial resistance in *Bordetella bronchiseptica*. Microbiol Spectr 6(4)

Kadlec K, Brenner Michael G, Sweeney MT, Brzuszkiewicz E, Liesegang H, Daniel R, Watts JL, Schwarz S (2001) Molecular basis of macrolide, triamilide, and lincosamide resistance in *Pasteurella multocida* from bovine respiratory disease. Antimicrob Agents Chemother 55:2475–2477

Kadlec K, FeßlerAT CN, Pomba CF, Schwarz S (2012) Unusual small plasmids carrying the novel resistence genes dfrK or apmA isolated from methicillin-resistant or -susceptible staphylococci. J Antimicrob Chemother 67:2342–2345

Kadlec K, Entorf M, Peters T (2019) Occurrence and characteristics of livestock-associated methicillin-resistant *Staphylococcus aureus* in quarter milk samples from dairy cows in Germany. Front Microbiol 10:1295

Kaper JB, Nataro JP, Mobley HL (2004) Pathogenic *Escherichia coli*. Nat Rev Microbiol 2:123–140. https://doi.org/10.1038/nrmicro818

Käppeli N, Morach M, Zurfluh K, Corti S, Nüesch-Inderbinen M, Stephan R (2019) Sequence types and antimicrobial resistance profiles of *Streptococcus uberis* isolated from bovine astitis. Front Vet Sci 6:234

Karuppannan AK, Opriessnig T (2018) *Lawsonia intracellularis*: revisiting the disease ecology and control of this fastidious pathogen in pigs. Front Vet Sci 5:181

Katsuda K, Kohmoto M, Mikami O, Uchida I (2009) Antimicrobial resistance and genetic characterization of fluoroquinolone-resistant *Mannheimia haemolytica* isolates from cattle with bovine pneumonia. Vet Microbiol 139:74–79

Katsuda K, Kohmoto M, Mikami O, Tamamura Y, Uchida I (2012) Plasmid-mediated florfenicol resistance in *Mannheimia haemolytica* isolated from cattle. Vet Microbiol 155:444–447

Kaya H, Hasman H, Larsen J, Stegger M, Johannesen TB, Allesøe RL, Lemvigh CK, Aarestrup FM, Lund O, Larsen AR (2018) SCCmecFinder, a web-based tool for typing of Staphylococcal cassette chromosome mec in *Staphylococcus aureus* using whole-genome sequence data. mSphere 3(1):612–617

Kehrenberg C, Schwarz S (2001) Occurrence and linkage of genes coding for resistance to sulfonamides, streptomycin and chloramphenicol in bacteria of the genera *Pasteurella* and *Mannheimia*. FEMS Microbiol Lett 205(2):283–290

Kehrenberg C, Schwarz S (2005) Identification of *dfr*A20, a novel trimethoprim resistance gene from *Pasteurella multocida*. Antimicrob Agents Chemother 49:414–417

Kehrenberg C, Schwarz S (2006) Distribution of florfenicol resistance genes *fex*A and cfr among chloramphenicol-resistant Staphylococcus isolates. Antimicrob Agents Chemother 50(4):1156–1163

Kehrenberg C, Catry B, Haesebrouck F, de Kruif A, Schwarz S (2005) A novel spectinomycin/streptomycin resistance gene, aadA14, from *Pasteurella multocida*. Antimicrob Agents Chemother 49:3046–3049

Kehrenberg C, Wallmann J, Schwarz S (2008) Molecular analysis of florfenicol-resistant *Pasteurella multocida* isolates in Germany. J Antimicrob Chemother 62:951–955

Keyburn AL, Yan XX, Bannam TL, Van Immerseel F, Rood JI, Moore RJ (2010) Association between avian necrotic enteritis and *Clostridium perfringens* strains expressing netB toxin. Vet Res 41:21

Kiu R, Hall LJ (2018) An update on the human and animal enteric pathogen *Clostridium perfringens*. Emerg Microbes Infect 7(1):141

Kiu R, Caim S, Alexander S, Pachori P, Hall LJ (2017) Probing genomic aspects of the multi-host pathogen *Clostridium perfringens* reveals significant pangenome diversity, and a diverse array of virulence factors. Front Microbiol 8:2485

Klare I, Badstübner D, Konstabel C, Böhme G, Claus H, Witte W (1999) Decreased incidence of Van A-type vancomycin-resistant enterococci isolated from poultry meat and from fecal samples of humans in the community after discontinuation of avoparcin usage in animal husbandry. Microb Drug Resist 5(1):45–52

Klima CL, Alexander TW, Hendrick S, McAllister TA (2014a) Characterization of *Mannheimia haemolytica* isolated from feedlot cattle that were healthy or treated for bovine respirátory disease. Can J Vet Res 78 (1):38–45

Klima CL, Zaheer R, Cook SR, Booker CW, Hendrick S, Alexander TW, McAllister TA (2014b) Pathogens of bovine respiratory disease in North American feedlots conferring multidrug resistance via integrative conjugative elements. J Clin Microbiol 52:438–448

Klima CL, Cook SR, Zaheer R, Laing C, Gannon VP, Xu Y, Rasmussen J, Potter A, Hendrick S, Alexander TW, McAllister TA (2016) Comparative genomic analysis of *Mannheimia haemolytica* from bovine sources. PLoS One 11(2):e0149520

Klitgaard K, Friis C, Jensen TK, Angen O, Boye M (2012) Transcriptional portrait of *Actinobacillus pleuropneumoniae* during acute disease–potential strategies for survival and persistence in the host. PLoS One 7:e35549

Klotz P, Göttig S, Leidner U, Semmler T, Scheufen S, Ewers C (2017) Carbapenem resistance and pathogenicity of bovine *Acinetobacter indicus*-like isolates. PLoS One 12(2):e0171986

Knetsch CW, Connor TR, Mutreja A, van Dorp SM, Sanders IM, Browne HP, Harris D, Lipman L, Keessen EC, Corver J, Kuijper EJ, Lawley TD (2014) Whole genome sequencing reveals potential spread of *Clostridium difficile* between humans and farm animals in the Netherlands, 2002 to 2011. Euro Surveill 19 (45):20954

Knetsch CW, Kumar N, Forster SC, Connor TR, Browne HP, Harmanus C, Sanders IM, Harris SR, Turner L, Morris T, Perry M, Miyajima F, Roberts P, Pirmohamed M, Songer JG, Weese JS, Indra A, Corver J, Rupnik M, Wren BW, Riley TV, Kuijper EJ, Lawley TD (2018) Zoonotic transfer of *Clostridium difficile* harboring antimicrobial resistance between farm animals and humans. J Clin Microbiol 56(3)

Knight DR, Riley TV (2019) Genomic delineation of zoonotic origins of *Clostridium difficile*. Front Public Health 7:164. https://doi.org/10.3389/fpubh.2019.00164

Knight DR, Elliott B, Chang BJ, Perkins TT, Riley TV (2015) Diversity and evolution in the genome of *Clostridium difficile*. Clin Microbiol Rev 28(3):721–741

Knight DR, Squire MM, Collins DA, Riley TV (2016) Genome analysis of *Clostridium difficile* PCR ribotype 014 lineage in australian pigs and humans reveals a diverse genetic repertoire and signatures of long-range interspecies transmission. Front Microbiol 7:2138

Knight DR, Kullin B, Androga GO, Barbut F, Eckert C, Johnson S, Spigaglia P, Tateda K, Tsai PJ, Riley TV (2019) Evolutionary and genomic insights into Clostridioides difficile sequence type 11: a diverse, zoonotic and antimicrobial resistant lineage of global One Health importance. MBio 10(2)

Kong LC, Gao D, Gao YH, Liu SM, Ma HX (2014) Fluoroquinolone resistance mechanism of clinical isolates and selected mutants of *Pasteurella multocida* from bovine respiratory disease in China. J Vet Med Sci 76:1655–1657

Kristensen BM, Frees D, Bojesen AM (2010) GtxA from *Gallibacterium anatis*, a cytolytic RTX-toxin with a novel domain organisation. Vet Res 41(3):25

Krutova M, Zouharova M, Matejkova J, Tkadlec J, Krejčí J, Faldyna M, Nyc O, Bernardy J (2018) The emergence of *Clostridium difficile* PCR ribotype 078 in piglets in the Czech Republic clusters with *Clostridium difficile* PCR ribotype 078 isolates from Germany, Japan and Taiwan. Int J Med Microbiol 308 (7):770–775

Kucerova Z, Hradecka H, Nechvatalova K, Nedbalcova K (2011) Antimicrobial susceptibility of *Actinobacillus pleuropneumoniae* isolates from clinical outbreaks of porcine respiratory diseases. Vet Microbiol 150 (1–2):203–206

Lakhundi S, Zhang K (2018) Methicillin-resistant *Staphylococcus aureus*: molecular characterization, evolution, and epidemiology. Clin Microbiol Rev 31(4): e00020-18

Lebreton F, van Schaik W, McGuire AM, Godfrey P, Griggs A, Mazumdar V, Corander J, Cheng L, Saif S, Young S, Zeng Q, Wortman J, Birren B, Willems RJ, Earl AM, Gilmore MS (2013) Emergence of epidemic multidrug-resistant *Enterococcus faecium* from animal and commensal strains. mBio 4(4)

Leclercq R, Cantón R, Brown DF, Giske CG, Heisig P, MacGowan AP, Mouton JW, Nordmann P, Rodloff AC, Rossolini GM, Soussy CJ, Steinbakk M, Winstanley TG, Kahlmeter G (2013) EUCAST expert rules in antimicrobial susceptibility testing. Clin Microbiol Infect 19:141–160

Lee Y (2016) Antimicrobial resistance and molecular characterization of *Clostridium perfringens* isolated from chicken. J Prev Vet Med 40(2):71–79

Lee CR, Lee JH, Park M, Park KS, Bae IK, Kim YB, Cha C, Jeong BC, Lee SH (2017) Biology of *Acinetobacter baumannii*: pathogenesis, antibiotic resistance mechanisms, and prospective treatment options. Front Cell Infect Microbiol 7:55

Lekkerkerk WS, van Wamel WJ, Snijders SV, Willems RJ, van Duijkeren E, Broens EM, Wagenaar JA, Lindsay JA, Vos MC (2015) What is the origin of livestock-associated methicillin-resistant *Staphylococcus aureus* clonal complex 398 isolates from humans without livestock contact? An epidemiological and genetic analysis. J Clin Microbiol 53(6):1836–1841

Li L, Chen Z, Bei W, Su Z, Huang Q, Zhang L, Chen H, Zhou R (2015) Catecholamines promote *Actinobacillus pleuropneumoniae* growth by regulating iron metabolism. PLoS One 10(4): e0121887

Li Y, Jiang H, Xiang R, Sun N, Zhang Y, Zhao L, Gu P, Wang L, Zeng Z (2016) Effects of two efflux pump inhibitors on the drug susceptibility of *Riemerella anatipestifer* isolates from China. J Integr Agric 15 (4):929–933

Li C, Yan X, Lillehoj HS (2017) Complete genome sequence of *Clostridium perfringens* LLY_N11, a necrotic enteritis-inducing strain isolated from a healthy chicken intestine. Genome Announc 5(44)

Li L, Wang R, Huang Y, Huang T, Luo F, Huang W, Yang X, Lei A, Chen M, Gan X (2018a) High incidence of pathogenic *Streptococcus agalactiae* ST485 strain in pregnant/puerperal women and isolation of hyper-virulent human CC67 strain. Front Microbiol 9:50

Li Z, Cheng F, Lan S, Guo J, Liu W, Li X, Luo Z, Zhang M, Wu J, Shi Y (2018b) Investigation of genetic diversity and epidemiological characteristics of *Pasteurella multocida* isolates from poultry in southwest China by population structure, multi-locus sequence typing and virulence-associated gene profile analysis. J Vet Med Sci 80(6):921–929

Li J, Hao H, Dai M, Zhang H, Ning J, Cheng G, Shabbir MAB, Sajid A, Yuan Z (2019a) Resistance and virulence mechanisms of *Escherichia coli* selected by enrofloxacin in chicken. Antimicrob Agents Chemother 63:e01824-18

Li MM, Ray P, Knowlton KF, Pruden A, Xia K, Teets C, Du P (2019b) Fate of pirlimycin and antibiotic resistance genes in dairy manure slurries in response to temperature and pH adjustment. Sci Total Environ 710:136310

Lin MF, Lan CY (2014) Antimicrobial resistance in *Acinetobacter baumannii*: from bench to bedside. World J Clin Cases 2:787–814

Lin MY, Lin KJ, Lan YC, Liaw MF, Tung MC (2001) Pathogenicity and drug susceptibility of the *Pasteurella anatis* isolated in chickens in Taiwan. Avian Dis 45(3):655–658

Lindsay JA (2008) Staphylococcus – molecular genetics. Caister Academic Press, Norfolk

Lindsay JA (2010) Genomic variation and evolution of *Staphylococcus aureus*. Int J Med Microbiol 300:98–103

Lindsay JA, Holden M (2004) *Staphylococcus aureus*: superbug, super genome? Trends Microbiol 18 (8):378–385

Liou GF, Yoshizawa S, Courvalin P, Galimand M (2006) Aminoglycoside resistance by ArmA-mediated ribosomal 16S methylation in human bacterial pathogens. J Mol Biol 359:358–364

Liu MC, Wu CM, Liu YC, Zhao JC, Yang YL, Shen JZ (2009) Identification, susceptibility, and detection of integron-gene cassettes of *Arcanobacterium pyogenes* in bovine endometritis. J Dairy Sci 92(8):36

Liu W, Feng Z, Fang L et al (2011) Complete genome sequence of *Mycoplasma hyopneumoniae* strain 168. J Bacteriol 193:1016–1017

Liu G, Ali T, Gao J, Ur Rahman S, Yu D, Barkema HW, Huo W, Xu S, Shi Y, Kastelic JP, Han B (2019) Co-occurrence of plasmid-mediated colistin resistance (mcr-1) and extended-spectrum β-lactamase encoding genes in *Escherichia coli* from bovine mastitic milk in China. Microb Drug Resist. doi:https://doi.org/10.1089/mdr.2019.0333. Accessed 20 Jan 2020

Liu Z, Deng D, Lu H, Sun J, Lv L, Li S, Peng G, Ma X, Li J, Li Z, Rong T and Wang G (2020) Evaluation of Machine Learning Models for Predicting Antimicrobial Resistance of Actinobacillus pleuropneumoniae From Whole Genome Sequences. Front. Microbiol. 11:48.

Lobova D, Smola J, Cizek A (2004) Decreased susceptibility to tiamulin and valnemulin among Czech isolates of *Brachyspira hyodysenteriae*. J Med Microbiol 53 (Pt 4):287–291

Longo F, Vuotto C, Donelli G (2014) Biofilm formation in *Acinetobacter baumannii*. New Microbiol 37:119–127

Lowder BV, Guinane CM, Ben Zakour NL, Weinert LA, Conway-Morris A, Cartwright RA, Simpson AJ, Rambaut A, Nübel U, Fitzgerald JR (2009) Recent human-to-poultry host jump, adaptation, and pandemic spread of *Staphylococcus aureus*. Proc Natl Acad Sci U S A 106(46):19545–19550

Lowy FD (2003) Antimicrobial resistance: the example of *Staphylococcus aureus*. J Clin Invest 111 (9):1265–1273

Loy JD, Brodersen BW (2014) *Moraxella* spp. isolated from field outbreaks of infectious bovine keratoconjunctivitis: A retrospective study of case submissions from 2010 to 2013. J Vet Diagn Investig 26 (6):761–768

Luo H, Liu M, Wang L, Zhou W, Wang M, Cheng A, Jia R, Chen S, Sun K, Yang Q, Chen X, Zhu D (2015) Identification of ribosomal RNA methyltransferase gene ermF in *Riemerella anatipestifer*. Avian Pathol 44:162–168

Lupo A, Haenni M, Madec J-Y (2018) Antimicrobial resistence in *Acinetobacter* spp and *Pseudomonas* spp. Microbiol Spectr 6(3)

Lüthje P, Schwarz S (2007) Molecular basis of resistance to macrolides and lincosamides among staphylococci and streptococci from various animal sources collected in the resistance monitoring program BfT-GermVet. Int J Antimicrob Agents 29(5):528–535

Lysnyansky I, Ayling RD (2016) *Mycoplasma bovis*: mechanisms of resistance and trends in antimicrobial susceptibility. Front Microbiol 7:595. https://doi.org/10.3389/fmicb.2016.00595

Ma J, Liu J, Zhang Y, Wang D, Liu R, Liu G, Yao H, Pan Z (2019) Bacitracin resistance and enhanced virulence of *Streptococcus suis* via a novel efflux pump. BMC Vet Res 15(1):377

Magnet S, Courvalin P, Lambert T (2001) Resistance-nodulation cell division-type efflux pump involved in aminoglycoside resistance in *Acinetobacter baumannii* strain BM4454. Antimicrob Agents Chemother 45:3375–3380

Mahu M, Pasmans F, Vranckx K, De Pauw N, Vande Maele L, Vyt P, Vandersmissen T, Martel A, Haesebrouck F, Boyen F (2017) Presence and mechanisms of acquired antimicrobial resistance in Belgian *Brachyspira hyodysenteriae* isolates belonging to different clonal complexes. Vet Microbiol 207:125–132. https://doi.org/10.1016/j.vetmic.2017.05.022

MARAN (2011) Monitoring of antimicrobial resistance and antibiotic usage in animals in the Netherlands in 2009. https://www.wur.nl/upload_mm/d/7/a/cd29a66b-aaeb-4ba2-9284-ce21ff6acc6d_MARAN2009.pdf. Accessed 28 May 2020

Marashifard M, Karimi Aliabad Z, Malek Hosseini SAA, Darban-Sarokhalil D, Mirzaii M, Khoramrooz SS (2018) Determination of antibiotic resistance pattern and virulence genes in *Escherichia coli* isolated from bovine with subclinical mastitis in southwest of Iran. Trop Anim Health Prod 51(3):575–580

Marenda M, Barbe V, Gourgues G et al (2006) A new integrative conjugative element occurs in *Mycoplasma agalactiae* as chromosomal and free circular forms. J Bacteriol 188:4137–4141. https://doi.org/10.1128/JB.00114-06

Mathema B, Mediavilla J, Chen L, Kreiswirth B (2009) Evolution and taxonomy of Staphylococci. In: Crossley D, Jefferson K, Archer G, Fowler V (eds)

Staphylococci in human disease, 2nd edn. Singapore, Wiley-Blackwell, pp 31–64

Mazur-Gonkowska B, Krasnodębska-Depta A, Koncicki A (2006) Enterococci in the pathology of poultry. Med Weter 62(10):1108–1112

Metcalf BJ, Chochua S, Gertz RE Jr, Hawkins PA, Ricaldi J, Li Z, Walker H, Tran T, Rivers J, Mathis S, Jackson D, Glennen A, Lynfield R, McGee L, Beall B, Active Bacterial Core Surveillance Team (2017) Short-read whole genome sequencing for determination of antimicrobial resistance mechanisms and capsular serotypes of current invasive *Streptococcus agalactiae* recovered in the USA. Clin Microbiol Infect 23(8):574

Michael GB, Eidam C, Kadlec K, Meyer K, Sweeney MT, Murray RW, Watts JL, Schwarz S (2012a) Increased MICs of gamithromycin and tildipirosin in the presence of the genes *erm*(42) and *msr*(E)-mph(E) for bovine *Pasteurella multocida* and *Mannheimia haemolytica*. J Antimicrob Chemother 67:1555–1557

Michael GB, Kadlec K, Sweeney MT, Brzuszkiewicz E, Liesegang H, Daniel R, Murray RW, Watts JL, Schwarz S. (2012b). ICE *Pmu1*, an integrative conjugative element (ICE) of Pasteurella multocida: analysis of the regions that comprise 12 antimicrobial resistance genes. J Antimicrob Chemother 67:84–90. doi:10.1093/jac/dkr406.

Michael GB, Bossé JT, Schwarz S (2018) Antimicrobial resistance in *Pasteurellaceae* of veterinary origin. Microbiol Spectr 6(3)

Michalopoulos A, Falagas ME (2010) Treatment of Acinetobacter infections. Expert Opin Pharmacother 11:779–788

Miller WR, Munita JM, Arias CA (2014) Mechanisms of antibiotic resistance in enterococci. Expert Rev Anti-Infect Ther 12(10):1221–1236

Minion FC, Lefkowitz EJ, Madsen ML et al (2004) The genome sequence of *Mycoplasma hyopneumoniae* strain 232, the agent of swine mycoplasmosis. J Bacteriol 186:7123–7133. https://doi.org/10.1128/JB.186.21.7123-7133.2004

Moffatt JH, Harper M, Harrison P, Hale JD, Vinogradov E, Seemann T, Henry R, Crane B, St Michael F, Cox AD, Adler B, Nation RL, Li J, Boyce JD (2010) Colistin resistance in *Acinetobacter baumannii* is mediated by complete loss of lipopolysaccharide production. Antimicrob Agents Chemother 54:4971–4977

Monecke S, Ruppelt A, Wendlandt S, Schwarz S, Slickers P, Ehricht R, Jäckel SC (2013) Genotyping of *Staphylococcus aureus* isolates from diseased poultry. Vet Microbiol 162(2–4):806–812

Monistero V, Graber HU, Pollera C, Cremonesi P, Castiglioni B, Bottini E, Ceballos-Marquez A, Lasso-Rojas L, Kroemker V, Wente N, Petzer IM, Santisteban C, Runyan J, Veiga Dos Santos M, Alves BG, Piccinini R, Bronzo V, Abbassi MS, Said MB, Moroni P (2018) *Staphylococcus aureus* isolates from bovine mastitis in eight countries: genotypes, detection of genes encoding different toxins and other virulence genes. Toxins (Basel) 10(6):247

Montes DE, Oca-Jimenez R, Vega-Sanchez V, Morales-Erasto V, Salgado-Miranda C, Blackall PJ, Soriano-Vargas E (2018) Phylogenetic relationship of *Ornithobacterium rhinotracheale* strains. J Vet Med Sci 80(6):869–873

Munita JM, Arias CA (2016) Mechanisms of antibiotic resistance. Microbiol Spectr 4(2)

Murray S, Pascoe B, Méric G, Mageiros L, Yahara K, Hitchings MD, Friedmann Y, Wilkinson TS, Gormley FJ, Mack D, Bray JE, Lamble S, Bowden R, Jolley KA, Maiden MCJ, Wendlandt S, Schwarz S, Corander J, Fitzgerald JR, Sheppard SK (2017) Recombination-mediated host adaptation by avian *Staphylococcus aureus*. Genome Biol Evol 9 (4):830–842

Mwangi S, Timmons J, Fitz-Coy S, Parveen S (2019) Characterization of *Clostridium perfringens* recovered from broiler chicken affected by necrotic enteritis. Poult Sci 98(1):128–135

Naito M, Pawlowska TE (2016) The role of mobile genetic elements in evolutionary longevity of heritable endobacteria. Mob Genet Elem 6(1)

Nedbalcova K, Satran P, Jaglic Z, Ondriasova R, Kucerova Z (2006) *Haemophilus parasuis* and Glässer's disease in pigs: a review. Veterinarni Medicina 51(5):168–179

Ngamwongsatit B, Tanomsridachchai W, Suthienkul O, Urairong S, Navasakuljinda W, Janvilisri T (2016) Multidrug resistance in *Clostridium perfringens* isolated from diarrheal neonatal piglets in Thailand. Anaerobe 38:88–93

Nhung NT, Chansiripornchai N, Carrique-Mas JJ (2017) Antimicrobial resistance in bacterial poultry pathogens: a review. Front Vet Sci 4

Nicholson TL, Brockmeier SL, Sukumar N, Paharik AE, Lister JL, Horswill AR, Kehrli ME Jr, Loving CL, Shore SM, Deora R (2017) The Bordetella bps polysaccharide is required for biofilm formation and enhances survival in the lower respiratory tract of swine. Infect Immun 85(8)

Nouvel LX, Sirand-Pugnet P, Marenda MS et al (2010) Comparative genomic and proteomic analyses of two *Mycoplasma agalactiae* strains: clues to the macro- and micro-events that are shaping mycoplasma diversity. BMC Genomics 11:86. https://doi.org/10.1186/1471-2164-11-86

Novais C, Tedim AP, Lanza VF, Freitas AR, Silveira E, Escada R, Roberts AP, Al-Haroni M, Baquero F, Peixe L, Coque TM (2016) Co-diversification of *Enterococcus faecium* core genomes and PBP5: evidences of pbp5 horizontal transfer. Front Microbiol 7:1581

Noyes NR, Benedict KM, Gow SP, Booker CW, Hannon SJ, McAllister TA, Morley PS (2015) *Mannheimia haemolytica* in feedlot cattle: prevalence of recovery and associations with antimicrobial use, resistance, and health outcomes. J Vet Intern Med 29:705–713

Nüesch-Inderbinen M, Käppeli N, Morach M, Eicher C, Corti S, Stephan R (2019) Molecular types, virulence profiles and antimicrobial resistance of *Escherichia coli* causing bovine mastitis. Veterinary Record Open 6:e000369

Ogawara H (2019) Comparison of antibiotic resistance mechanisms in antibiotic-producing and pathogenic bacteria. Molecules 24(19):3430

Ohnishi M, Sawada T, Hirose K, Sato R, Hayashimoto M, Hata E, Yonezawa C, Kato H (2011) Antimicrobial susceptibilities and bacteriological characteristics of bovine *Pseudomonas aeruginosa* and *Serratia marcescens* isolates from mastitis. Vet Microbiol 154 (1–2):202–207

Olsen AS, Warrass R, Douthwaite S (2015) Macrolide resistance conferred by rRNA mutations in field isolates of *Mannheimia haemolytica* and *Pasteurella multocida*. J Antimicrob Chemother 70:420–423

Opoku-Temeng C, Naclerio G, Mohammad H, Dayal N, Abutaleb N, Seleem MN, Sintim H (2018) N-(1, 3, 4-oxadiazol-2-yl) benzamide analogs, bacteriostatic agents against methicillin-and vancomycin-resistant bacteria. Eur J Med Chem 155:797–805

Palmieri C, Varaldo PE, Facinelli B (2011) *Streptococcus suis*, an emerging drug-resistant animal and human pathogen. Front Microbiol 2:235

Palmieri C, Magi G, Mingoia M, Bagnarelli P, Ripa S, Varaldo PE, Facinelli B (2012) Characterization of a *Streptococcus suis tet*(O/W/32/O)-carrying element transferable to major streptococcal pathogens. Antimicrob Agents Chemother 56(9):4697–4702

Pantosti A, Sanchini A, Monaco M (2007) Mechanisms of antibiotic resistance in *Staphylococcus aureus*. Future Microbiol 2(3):323–334

Park M, Rafii F (2014) Global phenotypic characterization of effects of fluoroquinolone resistance selection on the metabolic activities and drug susceptibilities of *Clostridium perfringens* strains. Int J Microbiol 2014:456979

Passmore IJ, Andrejeva A, Wren BW, Cuccui J (2019) Cytoplasmic glycoengineering of Apx toxin fragments in the development of *Actinobacillus pleuropneumoniae* glycoconjugate vaccines. BMC Vet Res 15(1):6

Patel R (2003) Clinical impact of vancomycin-resistant enterococci. J Antimicrob Chemother 51 (90003):13iii-21

Paudel S, Hess C, Wernsdorf P, Kaser T, Meitz S, Jensen-Jarolim E, Hess M, Liebhart D (2015) The systemic multiplication of *Gallibacterium anatis* in experimentally infected chickens is promoted by immunosuppressive drugs which have a less specific effect on the depletion of leukocytes. Vet Immunol Immunopathol 166:22–32

Peleg AY, de Breij A, Adams MD, Cerqueira GM, Mocali S, Galardini M, Nibbering PH, Earl AM, Ward DV, Paterson DL, Seifert H, Dijkshoorn L (2016) The success of Acinetobacter species; genetic, metabolic and virulence attributes. PLoS One 7(10): e46984

Peleg AY, Seifert H, Paterson DL (2008) *Acinetobacter baumannii*: emergence of a successful pathogen. Clin Microbiol Rev 21(3):538–582

Peng Z, Liang W, Liu W, Chen H, Wu B (2017) Genome characterization of *Pasteurella multocida* subspecies septica and comparison with *Pasteurella multocida* subspecies *multocida* and *gallicida*. Arch Microbiol 199(4):635–640

Peng Z, Liang W, Wang F, Xu Z, Xie Z, Lian Z, Hua L, Zhou R, Chen H, Wu B (2018) Genetic and phylogenetic characteristics of *Pasteurella multocida* isolates from different host species. Front Microbiol 9:1408

Peng Z, Wang X, Zhou R, Chen H, Wilson BA, Wu B (2019a) *Pasteurella multocida*: genotypes and genomics. Microbiol Mol Biol Rev 83(4):e00014-19

Peng Z, Gao D, Yang X, Liu HY, Huangfu HP, Wang CQ (2019b) $Bla_{\text{OXA-10}}$ and PSE-1 genes located on class 1 integrons in *Gallibacterium anatis*. Curr Microbiol 76:959

Périchon B, Courvalin P, Galimand M (2007) Transferable resistance to aminoglycosides by methylation of G1405 in 16S rRNA and to hydrophilic fluoroquinolones by QepA-mediated efflux in *Escherichia coli*. Antimicrob Agents Chemother 51:2464–2469

Persson G, Bojesen AM (2015) Bacterial determinants of importance in the virulence of *Gallibacterium anatis* in poultry. Vet Res 46(1):57

Persson G, Pors SE, Thøfner ICN, Bojesen AM (2018) Vaccination with outer membrane vesicles and the fimbrial protein FlfA offers improved protection against lesions following challenge with *Gallibacterium anatis*. Vet Microbiol 217:104–111

Piechowicz L, Garbacz K (2016) Poultry-like pA+ biotype of *Staphylococcus aureus* CC346/084 clone in human population. Curr Microbiol 73(1):124–131

Podder MP, Rogers L, Daley PK, Keefe GP, Whitney HG, Tahlan K (2014) Klebsiella species associated with bovine mastitis in Newfoundland. PLoS One 9(9): e106518

Poirel L, Nordmann P (2006) Carbapenem resistance in *Acinetobacter baumannii* mechanisms and epidemiology. Clin Microbiol Infect 12(9):826–836

Poirel L, Berçot B, Millemann Y, Bonnin RA, Pannaux G, Nordmann P (2012) Carbapenemase-producing *Acinetobacter* spp. in cattle, France. Emerg Infect 18:523–525

Poirel L, Madec JY, Lupo A, Schink AK, Kieffer N, Nordmann P, Schwarz S (2018) Antimicrobial resistance in *Escherichia coli*. Microbiol Spectr 6(4)

Popp C (2003) *Ornithobacterium rhinotracheale*: Typisierung, Pathogenität, Resistenzverhalten und Bekämpfung (Ph.D Thesis), p. 245, Freie Universität Berlin, Germany (in German)

Pormohammad A, Nasiri MJ, Azimi T (2019) Prevalence of antibiotic resistance in *Escherichia coli* strains simultaneously isolated from humans, animals, food, and the environment: a systematic review and meta-analysis. Infect Drug Resist 12:1181–1197

Pors SE, Skjerning RB, Flachs EM, Bojesen AM (2016) Recombinant proteins from *Gallibacterium anatis* induces partial protection against heterologous challenge in egg-laying hens. Vet Res 47:36

Potron A, Bour M, Triponney P, Muller J, Koebel C, A Bonnin R, Plésiat P (2019) Sequential emergence of colistin and rifampicin resistance in an OXA-72- producing outbreak strain of *Acinetobacter baumannii*. Int J Antimicrob Agents 53(5):669–673

Price LB, Stegger M, Hasman H, Aziz M, Larsen J, Andersen PS, Pearson T, Waters AE, Foster JT, Schupp J, Gillece J, Driebe E, Liu CM, Springer B, Zdovc I, Battisti A, Franco A, Zmudzki J, Schwarz S, Butaye P, Jouy E, Pomba C, Porrero MC, Ruimy R, Smith TC, Robinson DA, Weese JS, Arriola CS, Yu F, Pringle M, Landen A, Unnerstad HE, Molander B, Bengtsson B (2012) Antimicrobial susceptibility of porcine *Brachyspira hyodysenteriae* and *Brachyspira pilosicoli* isolated in Sweden between 1990 and 2010. Acta Vet Scand 54:54. https://doi.org/10.1186/1751-0147-54-54

Prüller S, Frömke C, Kaspar H, Klein G, Kreienbrock L, Kehrenberg C (2015a) Method of broth microdilution susceptibility testing for porcine *Bordetella bronchiseptica*. PLoS One 10(4):e0123883

Prüller S, Rensch U, Meemken D, Kaspar H, Kopp PA, Klein G, Kehrenberg C (2015b) Antimicrobial susceptibility of *Bordetella bronchiseptica* isolates from swine and companion animals and detection of resistance genes. PLoS One 10(8)

Pyatov V, Vrtkova I, Knoll A (2017) Detection of selected antibiotic resistance genes using multiplex PCR assay in mastitis pathogens in the Czech Republic. Acta Vet Brno 86:167–174

Raimondi S, Righini L, Candeliere F, Musmeci E, Bonvicini F, Gentilomi G, Starčič Erjavec M, Amaretti A, Rossi M (2019) Antibiotic resistance, virulence factors, phenotyping, and genotyping of *E. coli* isolated from the feces of healthy subjects. Microorganisms 7(8):251

Rajamohan G, Srinivasan VB, Gebreyes WA (2010) Molecular and functional characterization of a novel efflux pump, AmvA, mediating antimicrobial and disinfectant resistance in *Acinetobacter baumannii*. J Antimicrob Chemother 65:1919–1925

Ramírez-Castillo FY, Loera-Muro A, Vargas-Padilla ND, Moreno-Flores AC, Avelar-González FJ, Harel J, Jacques M, Oropeza R, Barajas-García CC, Guerrero-Barrera AL (2018) Incorporation of *Actinobacillus pleuropneumoniae* in preformed biofilms by *Escherichia coli* isolated from drinking water of swine farms. Front Vet Sci 5:184

Rehman MA, Yin X, Zaheer R, Goji N, Amoako KK, McAllister T, Pritchard J, Topp E, Diarra MS (2018) Genotypes and phenotypes of Enterococci isolated from broiler chickens. Front Sustain Food Syst 2:83

Requena D, Chumbe A, Torres M, Alzamora O, Ramirez M, Valdivia-Olarte H, Gutierrez AH, Izquierdo-Lara R, Saravia LE, Zavaleta M, Tataje-Lavanda L, Best I, Fernández-Sánchez M, Icochea E, Zimic M, Fernández-Díaz M (2013) Genome sequence and comparative analysis of *Avibacterium paragallinarum*. Bioinformation 9(10):528–536

Revitt-Mills SA, Rood JI, Adams V (2015) *Clostridium perfringens* extracellular toxins and enzymes: 20 and counting. Microbiol Aust 36:114–117

Reyes J, Rodriguez-Lecompte JC, Blanchard A, McClure JT, Sánchez J (2019) Molecular variability of *Streptococcus uberis* isolates from intramammary infections in Canadian dairy farms from the Maritime region. Can J Vet Res 83(3):168–176

Reygaert WC (2017) Antimicrobial mechanisms of *Escherichia coli* in *Escherichia coli* – Recent Adv. Physiol. Pathog. Biotechnol. Appl. Intech open 2:64. https://www.intechopen.com/books/-i-escherichia-coli-irecent-advances-on-physiology-pathogenesis-and-biotechnological-applications/antimicrobial-mechanisms-of-i-escherichia-coli-i. Accessed 28 May 2020

Rice LB (1998) Tn916 family conjugative transposons and dissemination of antimicrobial resistance determinants. Antimicrob Agents Chemother 42(8):1871–1877

Rice JA, Carrasco-Medina L, Hodgins DC, Shewen PE (2007) *Mannheimia haemolytica* and bovine respiratory disease. Anim Health Res Rev 8:117–128

Richards VP, Lang P, Bitar PD, Lefébure T, Schukken YH, Zadoks RN, Stanhope MJ (2011) Comparative genomics and the role of lateral gene transfer in the evolution of bovine adapted *Streptococcus agalactiae*. Infect Genet Evol 11(6):1263–1275

Richardson EJ, Bacigalupe R, Harrison EM, Weinert LA, Lycett S, Vrieling M, Robb K, Hoskisson PA, Holden MTG, Feil EJ (2018) Gene exchange drives the ecological success of a multi-host bacterial pathogen. Nat Ecol Evol 2(9):1468–1478

Roberts JR, Souillard R, Bertin J (2011) Avian diseases which affect egg production and quality. In: Nys Y, Bain M, Van Immerseel F (eds) Improving the safety and quality of eggs and egg products. Elsevier. ISBN 9780857093912

Rocha AJ, Barsottini MR, Rocha RR, Laurindo MV, Moraes FLL, Rocha SL (2019) *Pseudomonas aeruginosa*: virulence factors and antibiotic resistance genes. Braz Arch Biol Technol, 62, e19180503. doi: https://doi.org/10.1590/1678-4324-2019180503. Accessed 20 Jan 2020

Rodriguez C, Hakimi DE, Vanleyssem R, Taminiau B, Van Broeck J, Delmée M, Korsak N, Daube G (2017) *Clostridium difficile* in beef cattle farms, farmers and their environment: assessing the spread of the bacterium. Vet Microbiol 210:183–187

Roe CC, Vazquez AJ, Esposito EP, Zarrilli R, Sahl JW (2019) Diversity, virulence, and antimicrobial resistance in isolates from the newly emerging *Klebsiella pneumoniae* ST101 lineage. Front Microbiol 10:542

Roerink F, Morgan CL, Knetter SM, Passat MH, Archibald AL, Ait-Ali T, Strait EL.(2018) A novel inactivated vaccine against Lawsonia intracellularis induces rapid induction of humoral immunity, reduction of bacterial shedding and provides robust gut barrier function. Vaccine.;36(11):1500-1508.

Rood JI, Adams V, Lacey J, Lyras D, McClane BA, Melville SB, Moore RJ, Popoff MR, Sarker MR, Songer JG, Uzal FA, Van Immerseel F (2018) Expansion of the *Clostridium perfringens* toxin-based typing scheme. Anaerobe 53:5–10. https://doi.org/10.1016/j.anaerobe.2018.04.011

Rose S, Desmolaize B, Jaju P, Wilhelm C, Warrass R, Douthwaite S (2012) Multiplex PCR to identify macrolide resistance determinants in *Mannheimia haemolytica* and *Pasteurella multocida*. Antimicrob Agents Chemother 56(7):3664–3669

Roussel P, Porcherie A, Répérant-Ferter M, Cunha P, Gitton C, Rainard P, Germon P (2017) *Escherichia coli* mastitis strains: in vitro phenotypes and severity of infection in vivo. PLoS One 12(7):e0178285

Rozwandowicz M, Brouwer MSM, Fischer J, Wagenaar JA, Gonzalez-Zorn B, Guerra B, Mevius DJ, Hordijk J (2018) Plasmids carrying antimicrobial resistance genes in *Enterobacteriaceae*. J Antimicrob Chemother 73:1121–1137

Ruzin A, Keeney D, Bradford PA (2007) Ade ABC multidrug efflux pump is associated with decreased susceptibility to tigecycline in *Acinetobacter calcoaceticus-Acinetobacter baumannii* complex. J Antimicrob Chemother 59:1001–1004

Rzewuska M, Kwiecień E, Chrobak-Chmiel D, Kizerwetter-Świda M, Stefańska I, Gieryńska M (2019) Pathogenicity and virulence of *Trueperella pyogenes*: a review. Int J Mol Sci 20(11):2737

Sárközi R, Markai L, Fodor L (2015) Identification of a proposed new serovar of *Actinobacillus pleuropneumoniae*: serovar 16. Acta Vet Hung 63:444–450

Sassu EL, Bosse JT, Tobias TJ, Gottschalk M, Langford PR, Hennig-Pauka I (2018) Update on *Actinobacillus pleuropneumoniae*-knowledge, gaps and challenges. Transbound Emerg Dis 65(Suppl. 1):72–90

Schmidt T, Kock MM, Ehlers MM (2015) Diversity and antimicrobial susceptibility profiling of staphylococci isolated from bovine mastitis cases and close human contacts. J Dairy Sci 98:6256–6269

Schrijver R, Stijntjes M, Rodríguez-Baño J, Tacconelli E, Babu Rajendran N, Voss A (2018) Review of antimicrobial resistance surveillance programmes in livestock and meat in EU with focus on humans. Clin Microbiol Infect 24(6):577–590

Schwarz S, Noble WC (1994) Tetracycline resistance in Staphylococci from the skin of pigs. J Appl Bacteriol 78:320–326

Schwarz S, Cardoso M, Blobel H (1990) Plasmid-encoded antibiotic resistance in *Staphylococcus hyicus*. Dtsch Tierarztl Wochenschr 97(11):498, 501–503

Schwarz S, Feßler AT, Loncaric I, Wu C, Kadlec K, Wang Y, Shen J (2018) Antimicrobial resistance among Staphylococci of animal origin. Microbiol Spectr 6(4)

Sebaihia M, Preston A, Maskell DJ, Kuzmiak H, Connell TD, King ND, Orndorff PE, Miyamoto DM, Thomson NR, Harris D, Goble A, Lord A, Murphy L, Quail MA, Rutter S, Squares R, Squares S, Woodward J, Parkhill J, Temple LM (2006) Comparison of the genome sequence of the poultry pathogen *Bordetella avium* with those of *B. bronchiseptica*, *B. pertussis*, and *B. parapertussis* reveals extensive diversity in surface structures associated with host interaction. J Bacteriol 188(16):6002–6015

Seitz M, Valentin-Weigand P, Willenborg J (2016) Use of antibiotics and antimicrobial resistance in veterinary medicine as exemplified by the swine pathogen *Streptococcus suis*. In: Stadler M, Dersch P (eds) How to overcome the antibiotic crisis: facts, challenges, technologies and future perspectives. Springer International Publishing, Cham, pp 103–121

Seo B, Sung-Hyun M, So-Min L, Seung-Yoon L, Byeong-Yeal J, Won-Il K, Chang-Seop L, Yeonsu O, Ho-Seong C (2019) Recent antimicrobial susceptibility of *Lawsonia intracellularis* field isolates from pigs with proliferative hemorrhagic enteropathy in Korea. Thai J Vet Med 49(1):81–85

Shang Y, Li D, Shan X, Schwarz S, Zhang SM, Chen YX, Ouyang W, Du XD (2019) Analysis of two pheromone-responsive conjugative multiresistance plasmids carrying the novel mobile *optr*A locus from *Enterococcus faecalis*. Infect Drug Resist 12:2355–2362

Sharkey LK, Edwards TA, O'Neill AJ (2016) ABC-F proteins mediate antibiotic resistance through ribosomal protection. MBio 7(2):e01975

Sharma M, AbuOun M, Nunez-Garcia J, Rogers J, Welchman D, Teale C, Anjum MF, Kearns AM, Pichon B, Foster G, Robb A, McMillan M (2019) MRSA spa type t*899* from food animals in the UK. Vet Rec 182:697–698

Shete V, Grover N, Kumar M (2017) Analysis of aminoglycoside modifying enzyme genes responsible for high-level aminoglycoside resistance among Enterococcal isolates. J Pathog 2017:3256952

Shiadeh SMJ, Hashemi A, Fallah F, Lak P, Azimi L, Rashidan M (2019) First detection of *efr*AB, an ABC multidrug efflux pump in *Enterococcus faecalis* in Tehran, Iran. Acta Microbiol Immunol Hung 66 (1):57–68

Shirzad Aski H, Tabatabaei M (2016) Occurrence of virulence-associated genes in *Pasteurella multocida* isolates obtained from different hosts. Microb Pathog 96:52–57

Shittu A, Lin J (2007) Insights on virulence and antibiotic resistance: a review of the accessory genome of *Staphylococcus aureus*. Wounds 19(9):237–244

Singh K, Ritchey JW, Confer AW (2011) *Mannheimia haemolytica*: bacterial-host interactions in bovine pneumonia. Vet Pathol 48:338–348

Singh H, Thangaraj P, Chakrabarti A (2013) *Acinetobacter baumannii*: a brief account of mechanisms of multidrug resistance and current and future therapeutic management. J Clin Diagn Res 7:2602–2605

Slavić D, Boerlin P, Fabri M, Klotins KC, Zoethout JK, Weir PE, Bateman D (2011) Antimicrobial susceptibility of *Clostridium perfringens* isolates of bovine, chicken, porcine, and turkey origin from Ontario. Can J Vet Res 75(2):89–97

Slosarkova S, Nedbalcova K, Bzdil J, Fleischer P, Zouharova M, Staněk S, Kasna E, Pechova A (2019) Antimicrobial susceptibility of streptococci most frequently isolated from czech dairy cows with mastitis. Ann Anim Sci 19(3):679–694

Smith EA, Miller EA, Weber BP, Munoz Aguayo J, Flores Figueroa C, Huisinga J, Nezworski J, Kromm M, Wileman B, Johnson TJ, Björkroth J, (2020) Genomic Landscape of Ornithobacterium rhinotracheale in Commercial Turkey Production in the United States. Applied and Environmental Microbiology 86 (11)

Snyder ER, Alvarez-Narvaez S, Credille BC (2019) Genetic characterization of susceptible and multi-drug resistant *Mannheimia haemolytica* isolated from high-risk stocker calves prior to and after antimicrobial metaphylaxis. Vet Microbiol 235:110–117

Somerville G, Proctor R (2009) The biology of staphylococci. In: Crossley K, Jefferson K, Archer G, Fowler V (eds) Staphylococci in human disease, 2nd edn. Singapore, Wiley-Blackwell, pp 3–18

Songer JG, Uzal FA (2005) Clostridial enteric infections in pigs. J Vet Diagn Investig 17:528–536. https://doi.org/10.1177/104063870501700602

Sørum H, Roberts MC, Crosa JH (1992) Identification and cloning of a tetracycline resistance gene from the fish pathogen *Vibrio salmonicida*. Antimicrob Agents Chemother 36:611–615

Spoor LE, McAdam PR, Weinert LA, Rambaut A, Hasman H, Aarestrup FM, Kearns AM, Larsen AR, Skov RL, Fitzgerald JR (2013) Livestock origin for a human pandemic clone of community-associated methicillin-resistant *Staphylococcus aureus*. mBio 4 (4):e00356-13

Strugnell BW, Ellis RJ, Thomson JR, Steventon A, Teale CJ, Williamson SM, Clarke H, Goodyear KL, Wall L (2013) Preliminary findings on the use of multi-locus sequence typing (MLST) to investigate outbreaks of swine dysentery in northern England. Pig J 68:82–87

Su W, Kumar V, Ding Y, Ero R, Serra A, Lee BST, Wong ASW, Shi J, Sze SK, Yang L, Gao YG (2018) Ribosome protection by antibiotic resistance ATP-binding cassette protein. Proc Natl Acad Sci U S A 115:5157–5162

Sulyok KM, Kreizinger Z, Wehmann E, Lysnyansky I, Bányai K, Marton S, Jerzsele Á, Rónai Z, Turcsányi I, Makrai L, Jánosi S, Nagy SÁ, Gyuranecz M (2017) Mutations associated with decreased susceptibility to seven antimicrobial families in field and laboratory-derived *Mycoplasma bovis* strains. Antimicrob Agents Chemother 61(2)

Sun JR, Perng CL, Chan MC, Morita Y, Lin JC, Su CM (2012a) A truncated AdeS kinase protein generated by ISAba1 insertion correlates with tigecycline resistance in *Acinetobacter baumannii*. PLoS One 7(11):e49534

Sun N, Liu JH, Yang F, Lin DC, Li GH, Chen ZL, Zeng ZL (2012b) Molecular characterization of the antimicrobial resistance of *Riemerella anatipestifer* isolated from ducks. Vet Microbiol 158:376–383

Szabó R, Wehmann E, Magyar T (2015) Antimicrobial susceptibility of *Bordetella avium* and *Ornithobacterium rhinotracheale* strains from wild and domesticated birds in Hungary. Acta Vet Hung 63(4):413–424

Szabó R, Wehmann E, Makrai L, Nemes C, Gyuris É, Thuma Á, Magyar T (2017) Characterization of *Ornithobacterium rhinotracheale* field isolates from Hungary. Avian Pathol 46(5):506–514

Tadepalli S, Narayanan SK, Stewart GC, Chengappa MM, Nagaraja TG (2009) *Fusobacterium necrophorum*: a ruminal bacterium that invades liver to cause abscesses in cattle. Anaerobe 15(1–2):36–43

Tang X, Zhao Z, Hu J, Wu B, Cai X, He Q, Chen H (2009) Isolation, antimicrobial resistance, and virulence genes of *Pasteurella multocida* strains from swine in China. J Clin Microbiol 47(4):951–958

Tatay-Dualde J, Prats-van der Ham M, de la Fe C, Paterna A, Sánchez A, Corrales JC, Contreras A, Gómez-Martín Á (2017) Mutations in the quinolone resistance determining region conferring resistance to fluoroquinolones in *Mycoplasma agalactiae*. Vet Microbiol 207:63–68

Tegetmeyer HE, Jones SC, Langford PR, Baltes N (2008) ISApl1, a novel insertion element of *Actinobacillus pleuropneumoniae*, prevents ApxIV-based serological detection of serotype 7 strain AP76. Vet Microbiol 128:342–353

Torres C, Alonso CA, Ruiz-Ripa L, Leon-Sampedro R, del Csmpo R, Coque TM (2018) Antimicrobial resistence in *Enterococcus* spp. Microbiol Spectr 6(2)

Udhayavel S, Thippichettypalayam Ramasamy G, Gowthaman V, Malmarugan S, Senthilvel K (2017) Occurrence of *Clostridium perfringens* contamination in poultry feed ingredients: isolation, identification and its antibiotic sensitivity pattern. Anim Nutr 3 (3):309–312

Vakkamäki J, Taponen S, Heikkilä AM, Pyörälä S (2017) Bacteriological etiology and treatment of mastitis in Finnish dairy herds. Acta Vet Scand 59:1–9

van Asten AJ, van der Wiel CW, Nikolaou G, Houwers DJ, Grone A (2009) A multiplex PCR for toxin typing of *Clostridium perfringens* isolates. Vet Microbiol 136:411–412

van der Kolk JH, Endimiani A, Graubner V, Gerber V, Perreten V (2018) Acinetobacter in veterinary medicine, with an emphasis on *Acinetobacter baumannii*. J Glob Antimicrob Resist 16:59–71

van Duijkeren E, Greko C, Pringle M, Baptiste KE, Catry B, Jukes H, Moreno MA, Pomba MC, Pyörälä S, Rantala M, Ružauskas M, Sanders P, Teale C, Threlfall EJ, Torren-Edo J, Törneke K (2014) Pleuromutilins: use in food-producing animals in the European Union, development of resistance and impact on human and animal health. J Antimicrob Chemother 69:2022–2031 van Veen L, Hartman E, Fabri T (2001) In vitro antibiotic sensitivity of strains of *Ornithobacterium rhinotracheale* isolated in the Netherlands between 1996 and 1999. Vet Rec 149:611–613

Vanni M, Merenda M, Barigazzi G, Garbarino C, Luppi A, Tognetti R, Intorre L (2012) Antimicrobial resistance of *Actinobacillus pleuropneumoniae* isolated from swine. Vet Microbiol 156(1–2):172–177

Vannucci FA, Kelley MR, Gebhart CJ (2013) Comparative genome sequencing identifies a prophage-associated genomic island linked to host adaptation of *Lawsonia intracellularis* infections. Vet Res 44(1):49

Varga JJ, Therit B, Melville SB (2008) Type IV pili and the CcpA protein are needed for maximal biofilm formation by the gram-positive anaerobic pathogen *Clostridium perfringens*. Infect Immun 76:4944–4951

Vasconcelos ATR, Ferreira HB, Bizarro CV et al (2005) Swine and poultry pathogens: the complete genome sequences of two strains of *Mycoplasma hyopneumoniae* and a strain of *Mycoplasma synoviae*. J Bacteriol 187:5568–5577

Vassort-Bruneau C, Lesage-Decauses M-C, Martel J-L, Lafont J-P, Chaslus-Dancla E (1996) CAT III chloramphenicol resistance in *Pasteurella haemolytica* und *Pasteurella multocida* isolated from calves. J Antimicro Chemother 38:205–213

Vélez JR, Cameron M, Rodríguez-Lecompte JC, Xia F, Heider LC, Saab M, McClure JT, Sánchez J (2017) Whole-genome sequence analysis of antimicrobial resistance genes in *Streptococcus uberis* and *Streptococcus dysgalactiae* isolates from Canadian dairy herds. Front Vet Sci 4:63

Waites KB, Lysnyansky I, Bébéar CM (2014) Emerging antimicrobial resistance in mycoplasmas of humans and animals. In Mollicutes: molecular biology and pathogenesis. Caister Academic Press, Norfolk

Waldow K (2009) Untersuchungen zur Embryoletalitat, Genotypisierung und Resistenzlage aktueller *Ornithobacterium rhinotracheale*-Isolate (Doctoral Dissertation, in German language). https://refubium.fuberlin.de/bitstream/handle/fub188/1152/Dissertation_Waldow_2009.pdf?sequence=1&isAllowed=y. Accessed 6 Jan 2020

Walther B, Klein KS, Barton AK, Semmler T, Huber C, Wolf SA, Tedin K, Merle R, Mitrach F, Guenther S, Lübke-Becker A, Gehlen H (2018) Extended-spectrum beta-lactamase (ESBL)-producing *Escherichia coli* and *Acinetobacter baumannii* among horses entering a veterinary teaching hospital: the contemporary "Trojan horse". PLoS One 13:e0191873

Wang J, Stephan R, Power K, Yan Q, Hächler H, Fanning S (2014) Nucleotide sequences of 16 transmissible plasmids identified in nine multidrug-resistant *Escherichia coli* isolates expressing an ESBL phenotype isolated from food-producing animals and healthy humans. J Antimicrob Chemother 69(10):2658–2668

Wareth G, Neubauer H, Sprague LD (2019) *Acinetobacter baumannii* – a neglected pathogen in veterinary and environmental health in Germany. Vet Res Commun 43:1. https://doi.org/10.1007/s11259-018-9742-0

Watts JL, Salmon SA, Yancey RJ Jr, Nickerson SC, Weaver LJ, Holmberg C, Pankey JW, Fox LK (1995) Antimicrobial susceptibility of microorganisms isolated from the mammary glands of dairy heifers. J Dairy Sci 78(7):1637–1648

Weinert LA, Welch JJ, Suchard MA, Lemey P, Rambaut A, Fitzgerald JR (2012) Molecular dating of human-to-bovid host jumps by *Staphylococcus aureus* reveals an association with the spread of domestication. Biol Lett 8:829–832

Wendlandt S, Feßler AT, Monecke S, Ehricht R, Schwarz S, Kadlec K (2013) The diversity of antimicrobial resistance genes among staphylococci of animal origin. Int J Med Microbiol 303(6–7):338–349

Wendlandt S, Kadlec K, Schwarz S (2015) Four novel plasmids from *Staphylococcus hyicus* and CoNS that carry a variant of the spectinomycin resistance gene spd. J Antimicrob Chemother 70(3):948–949

Whittle G, Katz ME, Clayton EH, Cheetham BF (2000) Identification and characterization of a native *Dichelobacter nodosus* plasmid, pDN1. Plasmid 43:230–234

Wilharm G, Skiebe E, Higgins PG, Poppel MT, Blaschke U, Leser S, Heider C, Heindorf M, Brauner P, Jäckel U, Böhland K, Cuny C, Łopińska A, Kaminski P, Kasprzak M, Bochenski M, Ciebiera O, Tobółka M, Żołnierowicz KM, Siekiera J, Seifert H, Gagné S, Salcedo SP, Kaatz M, Layer F, Bender JK, Fuchs S, Semmler T, Pfeifer Y, Jerzak L (2017) Relatedness of wildlife and livestock avian isolates of the nosocomial pathogen to lineages spread in hospitals worldwide. Environ Microbiol 19 (10):4349–4364

Wilson BA, Ho M (2013) *Pasteurella multocida:* from zoonosis to cellular microbiology. Clin Microbiol Rev 26(3):631–655

Wise KS, Calcutt MJ, Foecking MF et al (2011) Complete genome sequence of *Mycoplasma bovis* type strain PG45 (ATCC 25523). Infect Immun 79:982–983. https://doi.org/10.1128/IAI.00726-10

Wollman EL, Jacob F, Hayes W (1956) Conjugation and genetic recombination in *Escherichia coli* K-12. Cold Spring Harb Symp Quant Biol 21:141–162. https://doi.org/10.1101/sqb.1956.021.01.012

Woźniak-Biel A, Bugla-Płoskońska G, Burdzy J, Korzekwa K, Ploch S, Wieliczko A (2019) Antimicrobial resistance and biofilm formation in *Enterococcus* spp. isolated from humans and turkeys in Poland. Microb Drug Resist 25(2):277–286

Wu JR, Shieh HK, Shien JH, Gong SR, Chang PC (2003) Molecular characterization of plasmids with antimicrobial resistant genes in avian isolates of *Pasteurella multocida*. Avian Dis 47:1384–1392

Wüthrich D, Brilhante M, Hausherr A, Becker J, Meylan M, Perreten V (2019) A novel trimethoprim resistance gene, dfrA36, characterized from *Escherichia coli* from calves. mSphere 4(3):e00255-19

Xu Z, Zhou Y, Li L, Zhou R, Xiao S, Wan Y, Zhang S, Wang K, Li W, Li L, Jin H, Kang M, Dalai B, Li T, Liu L, Cheng Y, Zhang L, Xu T, Zheng H, Pu S, Wang B, Gu W, Zhang XL, Zhu GF, Wang S, Zhao

GP, Chen H, Ahmed N (2008) Genome biology of *Actinobacillus pleuropneumoniae* JL03, an isolate of serotype 3 prevalent in China. PLoS One 3(1):e1450

Yang F, Wang Q, Wang X, Wang L, Li X, Luo J, Zhang S, Li H (2016) Genetic characterization of antimicrobial resistance in *Staphylococcus aureus* isolated from bovine mastitis cases in Northwest China. J Integr Agric 15(12):2842–2847

Yang F, Zhang S, Shang X, Wang L, Li H, Wang X (2018) Characteristics of quinolone-resistant *Escherichia coli isolated* from bovine mastitis in China. J Dairy Sci 101 (7):6244–6252

Yang Y, Higgins CH, Rehman I, Galvao KN, Brito IL, Bicalho ML, Song J, Wang H, Bicalho RC (2019) Genomic diversity, virulence, and antimicrobial resistance of *Klebsiella pneumoniae* strains from cows and humans. Appl Environ Microbiol 85(6)

Yongkiettrakul S, Maneerat K, Arechanajan B, Malila Y, Srimanote P, Gottschalk M, Visessanguan W (2019) Antimicrobial susceptibility of *Streptococcus suis* isolated from diseased pigs, asymptomatic pigs, and human patients in Thailand. BMC Vet Res 15(1):5

Zaragoza NE, Orellana CA, Moonen GA, Moutafis G, Marcellin E (2019) Vaccine production to protect animals against pathogenic clostridia. Toxins (Basel) 11(9):525

Zastempowska E, Lassa H (2012) Genotypic characterization and evaluation of an antibiotic resistance of *Trueperella pyogenes (Arcanobacterium pyogenes)* isolated from milk of dairy cows with clinical mastitis. Vet Microbiol 161(1–2):153–158

Zhang Q, Zhou M, Song D, Zhao J, Zhang A, Jin M (2013a) Molecular characterisation of resistance to fluoroquinolones in *Haemophilus parasuis* isolated from China. Int J Antimicrob Agents 42:87–89

Zhang WJ, Lu Z, Schwarz S, Zhang RM, Wang XM, Si W, Yu S, Chen L, Liu S (2013b) Complete sequence of the blaNDM-1-carrying plasmid pNDM-AB from *Acinetobacter baumannii* of food animal origin. J Antimicrob Chemother 68:1681–1682

Zhang C, Zhang Z, Song L et al (2015) Antimicrobial resistance profile and genotypic characteristics of *Streptococcus suis* capsular type 2 isolated from clinical carrier sows and diseased pigs in China. Biomed Res Int:284303

Zhang JS, Xia YT, Zheng RC, Liang ZY, Shen YJ, Li YF, Nie M, Gu C, Wang H (2018a) Characterisation of a novel plasmid containing a florfenicol resistance gene in *Haemophilus parasuis*. Vet J 234:24–26

Zhang S, Piepers S, Shan R, Cai L, Mao S, Zou J, Ali T, De Vliegher S, Han B (2018b) Phenotypic and genotypic characterization of antimicrobial resistance profiles in *Streptococcus dysgalactiae* isolated from bovine clinical mastitis in 5 provinces of China. J Dairy Sci 101(4):3344–3355

Zhao KL, Liu Y, Zhang XY, Palahati P, Wang HN, Yue BS (2011) Detection and characterization of antibiotic-resistance genes in *Arcanobacterium pyogenes* strains from abscesses of forest musk deer. J Med Microbiol 60(Pt 12):1820–1826

Zhao YD, Guo LL, Li J, Huang XH, Fang BH (2018) Characterization of antimicrobial resistance genes in *Haemophilus parasuis* isolated from pigs in China. PeerJ 6:e4613

Zhong CY, Cheng AC, Wang MS, Zhu DK, Luo QH, Zhong CD, Li L, Duan Z (2009) Antibiotic susceptibility of *Riemerella anatipestifer* field isolates. Avian Dis 53(4):601–607

Zhou K, Li C, Chen D, Pan Y, Tao Y, Qu W, Liu Z, Wang X, Xie S (2018) A review on nanosystems as an effective approach against infections of *Staphylococcus aureus*. Int J Nanomedicine 13:7333–7347

Zhou K, Zhu D, Tao Y, Xie L, Han L, Zhang Y, Sun J (2019) New genetic context of lnu(B) composed of two multi-resistance gene clusters in clinical Streptococcus agalactiae ST-19 strains. Antimicrob Resist Infect Control 8:117

Zhu DK, Luo HY, Liu MF, Zhao XX, Jia RY, Chen S, Sun KF, Yang Q, Wu Y, Chen XY, Cheng AC, Wang MS (2018) Various profiles of tet genes addition to tet (X) in *Riemerella anatipestifer* isolates from ducks in China. Front Microbiol 9:585

Index

L. Pokludová (ed.), *Antimicrobials in Livestock 1: Regulation, Science, Practice*,
https://doi.org/10.1007/978-3-030-46721-0

B

D

The manufacturer's authorised representative in the EU is Springer Nature Customer Service Centre GmbH, Europaplatz 3, 69115 Heidelberg, Germany. If you have any concerns regarding our products, please contact ProductSafety@springernature.com

Printed and bound by CPI Group (UK) Ltd, Croydon, CR0 4YY
30/09/2026
02225291-0020